Geographies of Global Change

Reviews of the previous edition:

"A wonderfully rich and invigorating mapping of late modern geographies; essential reading for anyone striving to understand the complexity and diversity of the contemporary world at the end of the twentieth century – *Geographies of Global Change* is clearly written, rigorously argued, and gripping reading. It redefines what we mean by a 'textbook' and sets new standards for teachers and students alike."
John Pickles, Professor of Geography, University of Kentucky.

"This book is a remarkably coherent collection and altogether a significant accomplishment. It is notable for the high standards achieved by the individual contributions and also for the contemporary relevance of the arguments marshaled. Accessible and informative, it should be indispensable reading for every geography major. Teachers will enjoy using it. Editors and authors alike are to be congratulated on an impressive achievement."
Kevin R. Cox, Professor of Geography, Ohio State University.

"There is no better text for helping to grasp the breadth of issues implied by global change, and for getting a sense of what needs to be done."
Neil Smith, Professor of Geography, Rutgers University.

Geographies of Global Change

Remapping the World

Second Edition

Edited by
R. J. Johnston, Peter J. Taylor,
and Michael J. Watts

Blackwell
Publishing

First published 1995 by Blackwell Publishers Ltd
Reprinted 1996 (three times), 1997, 1998, 1999
Second edition published 2002

Library of Congress Cataloging-in-Publication Data

Geographies of global change : remapping the world / edited by R.J. Johnston,
Peter J. Taylor, and Michael J. Watts. — 2nd ed.
       p.   cm.
    Includes bibliographical references (p.).
    ISBN 0–631–22285–5 (hb : alk. paper) — ISBN 0–631–22286–3 (pbk. : alk.
paper)
    1. Geography.   2. Change.   I. Johnston, R. J. (Ronald John)   II. Taylor,
Peter J. (Peter James), 1944–   III. Watts, Michael

    G128 .G474 2002
    910—dc21
                                                          2002022136

A catalogue record for this title is available from the British Library.

Set in 10.5 on 12.5 Sabon
by Ace Filmsetting Ltd, Frome, Somerset
Printed and bound in the United Kingdom
by TJ International, Padstow, Cornwall

For further information on
Blackwell Publishing, visit our website:
www.blackwellpublishing.com

Contents

List of Figures   viii

List of Tables   x

List of Contributors   xi

Preface   xvii

Acknowledgments   xviii

1   Geography/Globalization   1
    Peter J. Taylor, Michael J. Watts, and R. J. Johnston

Part I   Geoeconomic Change   19

Introduction to Part I: The Reconfiguration of Late
Twentieth-Century Capitalism   21

2   A Hyperactive World   29
    Nigel Thrift

3   Trading Worlds   43
    Peter Dicken

4   From Farming to Agribusiness: Global Agri-Food Networks   57
    Sarah Whatmore

5   Transnational Corporations and Global Divisions of Labor 68
    Richard Wright

6   Global Change in the World of Organized Labor   78
    Andrew Herod

7   Trajectories of Development Theory: Capitalism, Socialism,
    and Beyond   88
    David Slater

Part II   Geopolitical Change   **101**

Introduction to Part II: After the Cold War   103

8   Democracy and Human Rights After the Cold War   117
     John Agnew

9   The Renaissance of Nationalism   130
     Nuala C. Johnson

10   Global Regulation and Trans-State Organization   143
     Susan M. Roberts

11   The Rise of the Workfare State   158
     Joe Painter

12   Post-Cold War Geopolitics: Contrasting Superpowers
     in a World of Global Dangers   174
     Gearóid ÓTuathail

Part III   Geosocial Change   **191**

Introduction to Part III: People in Turmoil   193

13   Population Crises: From the Global to the Local   198
     Elspeth Graham and Paul Boyle

14   Global Change and Patterns of Death and Disease   216
     John Eyles

15   Changing Women's Status in a Global Economy   236
     Susan Christopherson

16   Stuck in Place: Children and the Globalization of
     Social Reproduction   248
     Cindi Katz

17   Race and Globalization   261
     Ruth Wilson Gilmore

Part IV   Geocultural Change   **275**

Introduction to Part IV: Modernity, Identity, and Machineries
of Meaning   277

18   Consumption in a Globalizing World   283
     Peter Jackson

19   Understanding Diversity: The Problem of/for "Theory"   296
     Linda McDowell

20   Resisting and Reshaping Destructive Development:
     Social Movements and Globalizing Networks   310
     Paul Routledge

21 World Cities and the Organization of Global Space 328
Paul L. Knox

22 The Emerging Geographies of Cyberspace 340
Rob Kitchin and Martin Dodge

Part V   Geoenvironmental Change **355**

Introduction to Part V: A Burden Too Far? 357

23 The Earth Transformed: Trends, Trajectories, and Patterns 364
William B. Meyer and B. L. Turner II

24 The Earth as Input: Resources 377
Jody Emel, Gavin Bridge, and Rob Krueger

25 The Earth as Output: Pollution 391
David K. C. Jones

26 Sustainable Development? 412
W. M. Adams

27 Environmental Governance 427
Simon Dalby

Part VI   Conclusion **441**

28 Remapping the World: What Sort of Map? What Sort
of World? 443
Peter J. Taylor, Michael J. Watts, and R. J. Johnston

Bibliography 453

Index 499

Figures

1.1   The rise of globalization as a topic in the social sciences in
the 1990s.   2
2.1   Foreign exchange trading.   35
2.2   Paris to Detroit and back.   38
3.1   The roller-coaster of world trade, 1980–99.   47
3.2   The network of world merchandise trade, 1995.   49
4.1   An outline of the contemporary agri-food system.   61
4.2   Industrialization of the potato commodity chain.   63
11.1   Per capita public expenditure on health ca. 1998.   161
12.1   Military expenditures, 1999.   182
13.1   Estimated global population.   202
13.2   Projected global population.   203
13.3   The logistic curve.   205
13.4   National vulnerability to water scarcity.   208
14.1   DALYs attributable to conditions in the unfinished agenda
in low- and middle-income countries, estimates for 1998.   218
14.2   Infant mortality rate related to income.   219
14.3   The big four pandemics.   224
14.4   Direct and indirect health impacts of climate change.   230
18.1   "Commodity circuitry": the production and marketing of
Nike sports shoes.   292
20.1   Location of some contemporary non-violent social
movements.   317
21.1   Regional world cities and their spheres of influence.   331
21.2   The GaWC inventory of world cities.   334
22.1   Thirty days in active worlds.   346
22.2   ET-map: the hierarchical category map.   350
22.3   Webspace landscape.   351
23.1   Trends in selected forms of human-induced transformation
of environmental components.   368
23.2   Recency and rate of change in human-induced
transformation of environmental components.   370

23.3   Risk and blame.   373
24.1   Annual worldwide production of selected metals from
       1700 to 1983.   379
24.2   Annual worldwide production of selected commodities:
       recent trends.   380
25.1   The temporal and spatial variability of pollution:
       Chernobyl cloud (1986).   397
25.2   The progress of acidification.   402
25.3   The progress of global warming.   404
25.4   Changing levels of atmospheric CO_2 (160,000 BP–AD 2100).   405
25.5   Variations in sea-level across the globe and with time
       (1880–2100).   406
25.6   Variations in total ozone with latitude and season in
       Dobson units (1964–80 and 1984–93) and mean size of
       Antarctic ozone "hole" (1990–9 and 1999–2000).   407

Tables

3.1 The world league table of merchandise trade, 1963 and 1999. 50

10.1 Trans-state organizations formally related to the UN. 148

11.1 Active members of medical benefits insurance schemes as a percentage of the labor force, selected European countries. 163

11.2 Unemployment insurance: members as a percentage of the labor force, selected European countries. 163

11.3 Infant mortality (deaths under one year of age per 1,000 live births), selected countries. 164

11.4 General government expenditure as a percentage of GDP, selected European countries. 164

11.5 The social distribution of expenditure on public services in the UK. 165

11.6 Government consumption and total expenditure, selected industrialized capitalist countries. 167

12.1 Three distinctive "geopolitical worlds." 178

12.2 World nuclear arsenals, 2000. 184

13.1 Global demographic indicators, 1995–2000. 201

14.1 Global burden of disease and injury attributable to selected risk factors, 1990. 221

14.2 Spread of HIV/AIDS, 1998. 225

14.3 Recent examples of emerging infections and probable factors in their emergence. 226

14.4 Factors in infectious disease emergence. 228

14.5 Health implications of climate change for Canada. 231

14.6 Major tropical vector-borne diseases and the likelihood of change with climate change. 232

15.1 Gender-sensitive Human Development Indices. 241

23.1 The Great Transformation: selected forms of human-induced transformation of environmental components – chronologies of change. 369

25.1 The shrinking terrestrial wilderness habitats. 400

25.2 Primary causes of forest decline in the 1980s, ranked in order of importance by region. 403

Contributors

W. M. Adams is a Lecturer in Geography at the University of Cambridge. His research concerns environment, conservation, and development in the UK and the Third World. His books include *Green Development: Environment and Sustainability in the Third World* (1990) and *Wasting the Rain: Rivers, People and Planning in Africa* (1992).

John Agnew is a Professor and Chair of the Department of Geography at UCLA, Los Angeles. His books include *Place and Politics* (1987), *Mastering Space* (co-author, 1995), *American Space/American Place* (co-editor, 2002), and *Place and Politics in Modern Italy* (2002).

Paul Boyle is Professor of Human Geography in the School of Geography and Geosciences at the University of St. Andrews. He is also Director of the Scottish Longitudinal Study. His interests lie mainly in population and health geography and he has published widely on migration and gender, housing, health, and labor market issues. Relevant co-authored books include *Exploring Contemporary Migration* (1998), *Migration into Rural Areas: Theories and Issues* (1998), and *Migration and Gender in the Developed World* (1999).

Gavin Bridge is a graduate student at Clark University. He is engaged on a project to investigate the social and environmental impacts of economic and political transition in the former Soviet Union.

Susan Christopherson is an Associate Professor in the Department of City and Regional Planning, Cornell University. Her primary research field is the organization and regulation of labor markets, where she has published widely. She is actively engaged in international policy research for the United Nations Conference on Trade and Development, and the Development Working Party on Women's Role in the Economy.

Simon Dalby is Professor of Geography and Political Economy at Carleton

University in Ottawa. His research interests include political ecology, critical geopolitics, sustainability, and environmental security. He is author of *Creating the Second Cold War* (1990) and *Environmental Security* (2002). He is also co-editor of *Rethinking Geopolitics* (1998) and *The Geopolitics Reader* (second edition, 2003).

Peter Dicken is Professor of Geography at the University of Manchester. He has held visiting appointments in the USA, Canada, Mexico, Australia, Hong Kong, and Singapore. In addition to a large number of academic papers, his various books include *Global Shift: Transforming the World Economy* (third edition, 1998). His primary research interests are in global economic change and transnational corporations.

Martin Dodge works as a computer technician and researcher in the Centre for Advanced Spatial Analysis (CASA) at University College London. He maintains the Cyber-Geography Research website at <http://www.cybergeography.org>, which includes the online version of the *Atlas of Cyberspace*. With co-author Rob Kitchin he has also written the books *Mapping Cyberspace* (2000) and *Atlas of Cyberspace* (2001).

Jody Emel is an Associate Professor in the Geography Department and is Director of the Center for Land, Water, and Society in the George Perkins Marsh Institute at Clark University. Most of her research is within two broad areas: the cultural and economic history of resource production regions and the social and legal theory of resource development and allocation institutions.

Ruth Wilson Gilmore is Assistant Professor of Geography at the University of California, Berkeley. Her research interests include race, gender, and power; uneven development; urban–rural linkages. The author of numerous articles, she recently completed *Golden Gulag*, a book examining the political and economic geographies of California's prison expansion, and opposition to it, during the past two decades.

Elspeth Graham is Senior Lecturer in the School of Geography and Geosciences at the University of St. Andrews. The main focus of her research is on population and health geographies and she has published papers on both global and local population issues. She also has research interests in methodology and the philosophy of the social sciences and is co-editor of *Postmodernism and the Social Sciences* (1994).

Andrew Herod is Associate Professor of Geography at the University of Georgia, Athens, USA. He has published widely on the topics of globalization and its impacts upon workers and labor unions. He is the author of *Labor Geographies: Workers and the Landscapes of Capitalism* (2001), editor of *Organizing the Landscape: Geographical Perspectives on Labor Unionism* (1998), and co-editor (along with Gearóid Ó Tuathail and Susan Roberts) of *An Unruly World? Globalization, Governance and Geography* (1998). His forthcoming book, co-edited with Melissa W. Wright, is *Power, Politics, and Geography: Placing Scale* (Blackwell, 2002).

Peter Jackson is Professor of Human Geography at the University of Sheffield. His research focuses on the geography of consumption and the politics of identity. Recent publications include *Shopping, Place and Identity* (1998), *Commercial Cultures* (2000), and *Making Sense of Men's Magazines* (Polity Press, 2001).

Nuala C. Johnson is a Lecturer at Queen's University, Belfast. Here research focuses on the historical geographies of nationalism, social memory, and the heritage industry. She has published widely on the role of public monuments in the making of national identities and on the geography of war and public memory in Ireland.

R. J. Johnston is Professor of Geography at the University of Bristol. From 1974 to 1992 he was Professor of Geography at the University of Sheffield. From 1992 to 1995 he was Vice-Chancellor of the University of Essex. His works include *Environmental Problems* (1989), *A Question of Place* (Blackwell, 1991), and, as co-editor, *A World in Crisis?* (Blackwell, 1991) and *The Dictionary of Human Geography* (Blackwell, 1994).

David K. C. Jones is Professor of Physical Geography at the London School of Economics and Political Science (LSE) and currently Head of the Department of Geography and Environment. A geomorphologist by training, his research has ranged from studies of alluvium and landsliding to long-term landform evolution, applied geomorphology, geohazard management and risk. His main recent publications include joint authorship of *Landsliding in Great Britain* (1994) and *Accident and Design: Contemporary Debates in Risk Management* (1996).

Cindi Katz is Professor of Geography in Environmental Psychology and Women's Studies at the Graduate Center of the City University of New York. Her work concerns social reproduction and the production of space, place, and nature; children and the environment; and the consequences of global economic restructuring for everyday life. She is the editor (with Janice Monk) of *Full Circles: Geographies of Gender Over the Life Course* (1993) and recently completed *Disintegrating Developments: Global Economic Restructuring and Children's Everyday Lives* (to be published in 2003). She is currently working on a project called Retheorizing Childhood and another on the social wage.

Rob Kitchin is a Senior Lecturer in Human Geography in the Department of Geography and Director of the National Institute of Regional and Spatial Analysis at the National University of Ireland, Maynooth. His research interests center on the geographies of disability, sexuality, and cyberspace. He has written six books and is the managing editor of the journal *Social and Cultural Geography*.

Paul L. Knox is Associate Dean of the College of Architecture and Urban Studies and Professor of Urban Affairs at Virginia Polytechnic Institute and State University. His research focuses on urbanization and global change. His publications include *The Restless Urban Landscape* (1993), *Urbanization* (1994), and, as co-author, *The Geography of the World Economy* (1994).

Linda McDowell is a Lecturer in Geography at the University of Cambridge and a Fellow of Newnham College. Her main interests are in the social and economic restructuring of contemporary Britain, especially the feminization of the labor force, and in feminist theory. She is co-author of *Landlords and Property* (1989) and co-editor of *Defining Women* (Polity Press, 1993).

William B. Meyer is Research Assistant Professor in the George Perkins Marsh Institute at Clark University. His interests include the human dimensions of global change and environmental history. He is co-editor of *Changes in Land Use and Land Cover* (1994).

Gearóid Ó Tuathail is Director of Virginia Tech's Public and International Affairs program in Northern Virginia, USA. He is the author of *Critical Geopolitics* (1996) and a co-editor of *A Companion to Political Geography* (Blackwell, 2002) and *The Geopolitics Reader* (second edition, 2003) among other works. He also serves as Associate Editor of the journal *Geopolitics*. His current research interests are in the critical geopolitics of world risk society, and US foreign policy towards the Balkans in the 1990s.

Joe Painter is Reader in Geography at the University of Durham. He is currently working on questions of citizenship, democracy, and identity in relation to Europe and its regions. He has previously written and published on the geographical aspects of the state, urban politics and governance, and on the use of theories of regulation and governance in the analysis of political change. He is the author of *Politics, Geography and "Political Geography"* (1995).

Susan M. Roberts is Associate Professor in the Department of Geography at the University of Kentucky. Her main field of research is contemporary global political economy, particularly the international finance system. This includes work on offshore financial centers in the Caribbean.

Acknowledgments

The editors and publishers gratefully acknowledge permission to reproduce copyright material as follows.

Table 14.1, Global burden of disease and injury attributable to selected risk factors, 1990, from *The Global Burden of Disease* (ed. C. J. L. Murray and A. D. Lopez) (Harvard University Press, 1996, © Harvard Center for Population and Development Studies, Cambridge, MA); table 14.3, Recent examples of emerging infections and probable factors in their emergence, from "The Man in the Mirror: David Harvey's Condition of Postmodernity," *Theory, Culture and Society* (ed. M. Morse) (Sage Publications, London, 1992, reprinted by permission of Sage Publications Limited); table 14.4, Factors in infectious disease emergence, from "The Man in the Mirror: David Harvey's Condition of Postmodernity," *Theory, Culture and Society* (ed. M. Morse) (Sage Publications, London, 1992, reprinted by permission of Sage Publications Limited); table 25.1, The shrinking terrestrial wilderness habitats, from "A Reconnaissance Level Inventory of the Amount of Wilderness Remaining in the World" (ed. J. M. McCloskey and H. Spalding) *Ambio* 18: 4, 1989; figure 4.2, Industrialization of the potato commodity chain, S. Whatmore (1994) from *Holding Down the Global* (ed. A. Amin and N. Thrift) (Oxford University Press, Oxford, 1994); figure 14.1, DALYs attributable to conditions in the unfinished agenda in low- and middle-income countries, estimates for 1998, *Report on Infectious Diseases* (World Health Organization, 1999); figure 18.1, "Commodity circuitry: the production and marketing of Nike sports shoes, from *Nike Culture* (Sage Publications, 1998); figure 23.1, Trends in selected forms of human-induced transformation of environmental components, from *The Earth as Transformed by Human Action: Global and Regional Changes in the Biosphere over the Past 200 Years* (ed. B. L. Turner II, William C. Clark, Robert W. Kates, John F. Richards, Jessica T. Mathews, William B. Meyer) (Cambridge University Press, Cambridge, 1990); figure 23.2, Recency and rate of change in human-induced transformation of environmental components, from *The Earth as Transformed by Human Action: Global and Regional*

Preface

One of the key arguments of *Geographies of Global Change* is that the contemporary world is changing at an incredibly fast pace. On this reckoning this second edition is way overdue. We were very satisfied with the first edition and therefore much of this new edition consists of revised chapters. However, we have also taken the opportunity to introduce new topics, sometimes to cover obvious omissions from the first edition, other times to cover new themes that have come to the fore since the mid-1990s. Thus new chapters have been prepared for the new edition to cover trade, labor, geopolitics, children, race, consumption, cyberspace, and environmental governance. This has inevitably led to us dropping some original chapters to make room. In addition we have provided a completely new chapter 1 to introduce geography students to globalization.

In the introductory chapter we argue that whatever position one takes – as scholar or as citizen – with respect to globalization, the myriad processes that are creating an increasingly globalized world cannot be ignored. This book is designed to help young geographers formulate their positions on globalization. Our changes have made the volume even more comprehensive in the range of geographical topics covered and this is entirely consistent with the all-embracing nature of globalization, both as a contested concept and as multiple practices.

Ron Johnston, Peter Taylor, Michael Watts

currently engaged in an Economic and Social Research Council research fellowship in their Global Environmental Change Programme. She has published *Farming Women: Gender, Work and Family Enterprise* (1991).

Richard Wright received his Ph.D. in Geography from Indiana University in 1985 and has worked at Dartmouth College since then. His scholarship, with support from the NSF, SSRC, and the John Simon Guggenheim Foundation, concerns immigration: specifically questions about how immigrants "fit" into US society. Recent papers analyze labor market operations, "assimilation" broadly conceived, nativism, the racialization of immigrants (and the native born), transnationalism, and state sovereignty. Future research plans include analysis of the relationship between residential and labor market segregation, geographies of interracial partners, and finishing a book on nativism and immigrant settlement patterns in the United States.

Paul Routledge is Leverhulme Fellow in the Geography Department, University of Bristol. His research interests lie in development, culture, and resistance. He is author of *Terrains of Resistance: Nonviolent Social Movements and the Contestation of Place in India* (1993).

David Slater is Professor of Social and Political Geography at Loughborough University. He is author of *Territory and State Power in Latin America* (1989) and co-editor of *The American Century* (Blackwell, 1999). He is also an editor of *Political Geography*.

Peter J. Taylor is Professor of Geography at Loughborough University. His research interests focus on the changing nature of state and comparative hegemonies. His books include *The Way the Modern World Works* (1996), *Modernities: A Geohistorical Perspective* (1999), *Political Geography: World-Economy, Nation-State and Locality* (2000), and he has co-edited *World Cities in a World-System* (1995) and *American Century* (Blackwell, 2000).

Nigel Thrift is Professor of Geography at the University of Bristol. His research interests include the social and cultural determinants of the international financial system, geographies of financial exclusion, social and cultural theory, and countries of the Pacific Basin. He has co-edited *The Socialist Third World* (Blackwell, 1987), *Class and Space: The Making of Urban Society* (1987), *New Models in Geography* (1989), *Money, Power and Space* (Blackwell, 1994).

B. L. Turner II is Professor of Geography and Director of the George Perkins Marsh Institute at Clark University. He has interests in nature–science relationships ranging from ancient Maya agriculture and environment to contemporary agricultural change in the tropics and global land-use change. He has written *The Earth Transformed by Human Action* (1990) and is co-editor of *Changes in Land Use and Land Cover* (1994). He is co-chair of the joint International Geosphere–Biosphere Programme and Human Dimensions Programme planning a core research project on global land-use/cover changes.

Michael J. Watts is Director of the Institute of International Studies and Professor of Geography at the University of California, Berkeley. His research interests are in African agrarian change and the role of local, national, and international power and politics in Third World development. He is author of *Silent Violence: Food, Famine and Peasantry in Northern Nigeria* (1983) and co-author of *Reworking Modernity* (1992).

Sarah Whatmore is Reader in Human Geography at the University of Bristol. Her research interests include agriculture, property and nature. She is

Changes in the Biosphere over the Past 200 Years (ed. B. L. Turner II, William C. Clark, Robert W. Kates, John F. Richards, Jessica T. Mathews, William B. Meyer) (Cambridge University Press, Cambridge, 1990); figure 23.3, Risk and blame, from *Risk and Blame: Essays in Cultural Theory* (ed. Mary Douglas) (Routledge, London, 1992).

The publishers apologize for any errors or omissions in the above list and would be grateful to be notified of any corrections that should be incorporated in the next edition or reprint of this book.

1 Geography/Globalization

Peter J. Taylor, Michael J. Watts, and R. J. Johnston

The incessant talk about globalization – the word, the images associated with it, and arguments for and against "it" – both reflect and reinforce fascination in boundless connectivity. Yet scholars do not need to choose between a rhetoric of containers and a rhetoric of flows. Not least of the questions we should be asking concern the present: what is actually new? What are the limits and mechanisms of ongoing changes? And above all, can we develop a differentiated vocabulary that encourages thinking about connections and their limits?

Fred Cooper (2001)

Over a half century ago in his magnificent book *The Great Transformation*, Karl Polanyi (1944) pointed to the costs and consequences of unfettered free trade in a world market and to what he called the "tremendous hazards of planetary interdependence" (ibid: 53). Since the end of World War II, the connectivity and dependence within a world system that so concerned Polanyi has come to saturate, some might say dominate, virtually all aspects of contemporary life. Perceived for several decades in terms of "development," in "post-development" contemporary parlance Polanyi's (1944) planetary interdependence is expressed through "the incessant talk about globalization."

Globalization has emerged over the last decade as the lodestar of the new millennium. Dreamed up by business gurus and popularized by media pundits, policy wonks, and academics – in newspapers, magazines, current affairs books, radio and TV news and discussion programs, and internet discussion groups – the concept has seemingly acquired a life of its own. What is indisputable is that the vocabulary of globalization is a pervasive way of thinking and writing about modernity and contemporary society. As likely to appear in corporate advertising as in serious news media, globalization has inevitably attracted massive attention from those who are in the business of analyzing and interpreting social change. In an online search of the social science literature, for example, we have identified 4,239 publications on globalization through the 1990s (figure 1.1). Rising precipitately in the final years of the decade, by 1999 social science publications on globalization were appearing at the rate of 32 a week! Such a conceptual head of steam is unlikely to dissipate quickly; indeed globalization is set fair to become a critical social concept – what Raymond Williams (1976) called a "keyword" – in the new century. Globalization is a clear example of what Williams refers to as an indicative word – it indicates certain forms of thought.

The implications of such an enormous explosion of interest within and outwith social science are many. Most fundamental, globalization cannot be ignored. There are lively debates about the meaning of globalization ranging from its dismissal as "overblown hype" to its designation as a "unique global era" (see Held, McGrew, Goldblatt, and Perraton 1999, ch. 1 for a review). Whatever your own opinion may be, any intellectual engagement with social change in the twenty-first century has to address this concept seriously, and assess its capacity to explain the world we currently inhabit.

A second implication inevitably turns on definition and the parameters of the global. Any new and controversial concept will be subject to particular scrutiny in this regard, but in the case of globalization part of the debate referred to above revolves around different meanings that inevitably construct the world in radically different ways. Globalization is a very fluid and flexible concept; indeed it is a very slippery customer to deal with. Nevertheless, we can begin to grasp the concept by focusing on what we might call the "primitive core" of its meanings, what Raymond Williams (1976) referred to as a "cluster." First, and most obviously, globalization indicates a specific scale of social activity: it is worldwide or planetary in scope. But how is the global qualitatively different from the much older idea of "inter-

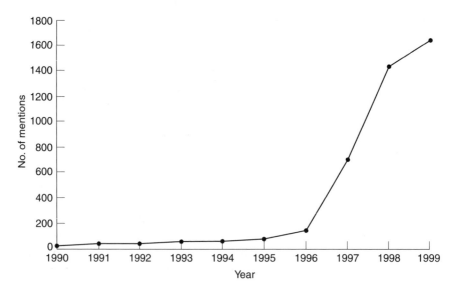

Figure 1.1   The rise of globalization as a topic in the social sciences in the 1990s.

national"? What makes the twenty-first century World Wide Web facilitating "global communication" substantively different from the nineteenth-century steamship facilitating "international trade"? Clearly simple reference to scale needs augmenting. Second, and more subtly, globalization transcends relations between states – the "international" or "cross-border" processes. Globalization is constituted by "trans-state" processes. These are activities and outcomes that occur beyond the confines of states as such: they do not merely cross borders, these processes operate as if borders were not there. Global financial markets and global warming are two very different but nevertheless classic examples of this trans-state dimension of contemporary world society.

Of particular note is the fact that these two primitive aspects of globalization are both geographical. We can agree, therefore, with Storper's (1997) claim that globalization is intrinsically geographical in nature. Using this as our starting point we identify four different ways geography relates to globalization.

First, under the heading "geography *and* globalization" we consider the fact that geography as a discipline has a global tradition that far precedes the rise of globalization as a concept. This allows geographers to approach globalization through a geohistorical perspective that identifies the specificities of contemporary globalization. An important corollary of this first view of globalization through the lenses of geography as a discipline is that we avoid the charge of "presentism," namely, assuming the distinctiveness of today without asking what is different about contemporary forms of globalization and what are its historical precedents.

Second, under the heading "geography *in* globalization" we introduce the idea that globalization is constituted by different types of spaces compared to earlier times. Here we discuss relations between "spaces of places" and "spaces of flows." An important corollary of this second view of globalization through the lenses of geographies of relations is that we avoid the notion of the global as a stage, an inert space on which events inevitably unfold.

Third, under the heading "geography *of* globalization" we describe the spatial distribution of things "global"; differing intensities of much that is considered under globalization reveals traditional power geographies not only maintained but enhanced by globalization. An important corollary of this third view of globalization through the lenses of geographical differences is that we can avoid the charge of "totalization," the idea that globalization is an all-encompassing narrative, a seamless field of connections which over-determines social change.

Fourth, under the heading "geography *for and against* globalization" we look at reactions to globalization, at place-based resistances to trans-state processes and cosmopolitan embraces of globalization. An important corollary of this fourth view of globalization through political geography lenses is that we avoid seeing globalization as heir to the cosmopolitan tradition, the Euro-Americo-centric progressive thinking that undervalues the essential variety of humanity.

We cannot know the future trajectories of globalization; they have yet to be made. But we can begin to understand contemporary geographies of global change, their past trajectories and current power relativities, so as to be in a position to understand future possibilities and contribute to the making of preferred options. One thing we can be sure of is that geographical knowledges will be at the heart of the debates about twenty-first century society.

Geography and Globalization

Geography literally means "Earth-description" and therefore the subject has normally been in some sense global in its intellectual concerns. For instance, for several centuries geography was closely associated with European exploration of the world – so much so that the terms "geography" and "exploration" came to appear to be almost synonymous (Baker 1931; Stoddart 1986). Cartography as one expression of the geographic vision has necessarily had global ambition, to chart the borders and limits of the known world. With the initial creation of geography as an academic discipline in German universities in the late nineteenth century, this global concern was maintained and reinforced. For instance, in the Netherlands the original university chair of geography was specifically designated "colonial geography."

As a modern discipline geography is a product of imperialism (Godlewska and Smith 1994). The expansionary activities of European states in the late nineteenth century "age of classical imperialism" created "global closure" – a politically integrated world spanning and linking continents. Global concerns stood at the heart of the development of a geography discipline in European universities in the years around 1900. For instance, A. J. Herbertson (1910) divided the world into large "natural regions" as an aid to imperial policy. He advocated that governments should set up "Statistical-Geographical Departments" staffed by geographers to map and evaluate the current and future economic value of these regions as tools for imperial planning. This aid to "statecraft" is as explicitly global in vision as any strategic plan to come out of the executive office of a contemporary multinational corporation. Herbertson's combination of imperial power with environmental knowledge is typical of much of the geography created in universities in the first half of the twentieth century (Taylor 1993a).

One of the most famous imperial power/environmental knowledge models devised in geography is Mackinder's (1904, 1919) "heartland thesis." This model identified the large central zone of Eurasia as a special "secure place" around which geopolitical policies for global strategies have to be devised. Initially a simple land power (heartland) versus sea power argument, the global vision of Mackinder survived the rise of air power (both planes and missiles) to underpin the geostrategy of the Cold War. With the USSR in the role of heartland, this has meant that Mackinder's model provided a dominant world spatial order for nearly all of the twentieth century. On the other side of the Atlantic, Isaiah Bowman, geographer, university president, and foreign policy strategist, played a similar role with respect to a model of world geopolitics appropriate to the growing hegemony of the United States in the first half of the century (Bowman 1921). However, during the second half of the century geopolitics was largely designed outside of the discipline of geography (Dalby 1990). Taken over by Cold War global strategists, Mackinder's heartland concept all but disappeared from mainstream geography. This was part of a process of human geography retreating from the global scale as it converted to social science norms that were predominantly concerned with social activities at the scale of the state and below (Taylor 1996a).

Geographers neglecting the global scale did not, of course, mean that actual "planetary interdependence" was put on a back burner. In the wake of World War II one can now see in hindsight how the foundation for a different global order, under US hegemony, was laid. President Truman's call for democratic fair dealing on a global scale initially led to a boom in "development" studies, not unlike globalization's current boom, consisting of country-by-country evaluation of economic growth potentials. This theoretical and practical emphasis on international comparison was buttressed by the United Nations and its related economic international institutions, the Bretton Woods regulatory organizations such as the IBRD (International Bank of Reconstruction and Development, usually referred to as the World Bank) and the IMF (International Monetary Fund). Decolonization provided the new opportunities for financial and productive investment as the old imperial system began to crack, creating new sorts of "planetary interdependence" premised on "development." By the 1960s media guru Marshall McLuhan (1962) was beginning to talk of a global village as innovations in the communications sector held the promise of worldwide integration. But it was in the late 1960s and 1970s that the global reemerged as a major scale of interest in the public consciousness for the first time since the age of imperialism. It occurred in three main guises. First, there was concern for environmental issues – a concern with the fragility of planet Earth – notably in relation to trans-state pollution and environmental harms (not the least of which was the nuclear threat). Second, there was the growing might and power of multinational corporations and their "global reach" creating a "new international division of labor." And third, there was concern for global inequalities, including a demand for a "new international economic order" precipitated by the devastating consequences of the oil price rises of 1973 and 1979 and by the collapse of commodity prices upon which many Third World economies depended. These issues were all initially approached through the auspices of various multilateral institutions, and therefore as matters of international relations, but in hindsight we can see now that they each represent a different aspect of globalization: global integration, global governance, and global inequity.

Social science in general and human geography in particular responded to these new concerns by developing global-scale studies alongside their more normal researches that focused on processes at or below the scale of the state. By the 1980s the new global scale of analysis was becoming commonplace in human geography, for instance by studies of multinational corporations (Taylor and Thrift 1982), in world-systems political geography (Taylor 1985), in reassessments of development geography (Corbridge 1986; Peet 1991), and through a new global economic geography (Dicken 1998). In an overview of this new "global geography" (Johnston and Taylor 1986), more topics were added, including cultural and social concerns. It soon became clear that the new global scale impinged upon the whole gamut of social science themes. This is what we were responding to with the first edition of this book with its "geoeconomic," "geopolitical," "geosocial," "geocultural," and "geoenvironmental" sections (Johnston, Taylor, and Watts 1995). Despite the interweaving of these themes in practice, we feel this organization still has pedagogic advantages and is used again in this volume. But why this global coincidence of all these topics at this particular time? To answer such questions we need to consider the geography *in* globalization.

Geography in Globalization

Contributions of geographers to the globalization debate should not be seen as simply "adding" a scale to conventional country-level analyses. Nor should they be seen as promoting the "global" above other scales of activity. In fact one of the most positive features to emerge from geography's latest engagement with the global has been to interrogate the meaning of scale within geographical study. We identify two key points in this new scrutinizing of scale.

The first point to make is that scales are not independent of one another. You can cross a boundary and leave a country but you can never relocate outside a given geographical scale. Geographers use phrases such as "glocalization" (Swyngedouw 1997) and "local-global" (Pred and Watts 1993) to reinforce the idea that geographical scales are relational in nature. Any analysis that pits local versus global is developing a very suspect research agenda. It is, of course, legitimate to ask about the variable salience of scale levels for particular activities and for different people. This is in fact the geography that is intrinsic to globalization. Globalization is about changing relationships between geographical scales: our social world is being "rescaled" through remapping the geographies of activities and functions. Sometimes this involves processes moving to institutions "above the state," other times to "below the state," and through all this the state itself is changing and adapting, not necessarily becoming less important.

The second point is that geographical scales in human activities do not just "emerge," they are constructed through human activities (Smith 1993). This is a very important geographical contribution to globalization debates because it means that there was nothing inevitable about an increasing scale of activities, nor of an increasingly global future. Globalization is no more "natural" than was imperialism: the latter was justified by a racial determinism, the former legitimated by market determinism, the need to roll back welfare states and to "get the prices right." The domination of pro-market rhetoric – TINA (there is no alternative) – by politicians of all political hues throughout the world is both a reflection of, and a contribution to, the development of contemporary globalization. Of course, globalization is much more than rhetoric: the rescaling is the latest round of "spatial fix" for a crisis-ridden capitalist world economy (Harvey 2000). Such fixes involve moving capital from low-return to high-return places in a restructuring of production and consumption in order to resolve profit crises.

These changing relationships between geographical scales can be seen for example in the rise of new and different forms of nationalism. In Nigeria, to take one example, the scale of the nation-state – itself a product of the colonial era and the declaration of independence in 1960 – has been undercut by the emergence of so-called "ethno- or sub-nationalisms." The impoverished oil-producing ethnic communities in the Niger delta are making claims in regard to their territory (a local or community scale) against the backdrop of global processes (the presence of oil companies and the role of the world market). At the same time, Nigeria is being drawn into the global space of the World Bank and the IMF through austerity and adjustment programs in a way that compromises the sovereignty and integrity of the nation-state.

At its most basic, therefore, the new global scale of activities has been a collective response of capital to the economic stagnation in the world economy that began in the 1970s. But how exactly did this rescaling come about? Most treatments of the "rise" of globalization begin with the concept of a "shrinking world," the idea being that over time different parts of the world have come closer together because of technological advances in transport and communications (Allen and Hamnett 1995). In transport, this "space-time compression" begins in earnest in the early nineteenth century, first with the railways and then steam ships, before moving on to cars, trucks, and jet planes in the twentieth century. However, for globalization it has been communications that have been particularly important because of the effective simultaneity produced in the linking of places (Hugill 1999). Beginning with the telegraph in the late nineteenth century, coded messages could travel long distances with little time lag in connecting sender and receiver. Development in both cable and radio transmission culminated in reliable trans-Atlantic telephone communication by the late 1950s. This had one very important effect: it allowed control functions to be centralized. For instance, in the early 1960s Ford dispensed with local executive management in its European operations and ran operations from its global headquarters in Detroit (Hugill 1999: 229). This presaged a massive growth in the activities of multinational corporations across the world: as noted earlier, by the 1970s they were deemed to have "global reach," with a select few economically larger than most countries (Barnet and Muller 1974). By 2000 the top 100 transnational corporations had assets many times larger than the collective GDP of the 100 poorest national economies.

Globalization, however, is not just the technological culmination of a shrinking world. New innovations in the 1970s led to a convergence between the communications industries and the computing industries (Cooke et al. 1992). Combining communications and computing created a powerful new enabling technology. Initially, it would aid in control functions, going far beyond telephone links to make information and analysis immediately available over long distances. But aiding centralization was not the only outcome of this technology: it enabled flexible production and distribution that could result in a decentralization of economic functions. Furthermore, these developments were not limited to economic processes: communication included the media industries, so that the new combined technologies had widespread cultural and social impacts, and began to realize what had earlier been called the "global village" (McLuhan 1962). And an important part of the media transmissions were advertising campaigns promoting more and more consumption with future environmental consequences. The World Wide Web with its e-marketing and e-commerce and with email and teleconferencing for "global managers" in their "smart offices" (Graham and Marvin 1996) – often referred to as the new "super information highway" existing in electronic "cyberspace" – represents the current culmination of this technological impulse.

Transfers of information have become a key element of global transmissions, so much so that some researchers have suggested that it represents a fundamental change in the nature of society. The notions of an "information age" and "knowledge society" have become commonplace. The most influential theorist in this area is Manuel Castells (1996) who argues that we live in an "informational age" (superseding the industrial age) which has produced a new "network society." His ideas are important to geographers because he identifies a new space at the heart of this new form of society. Until the 1970s, according to Castells, modern society was constituted as spaces of places, such as neighborhoods, regions, and states. Network society, on the other hand, is constituted as spaces of flows, a myriad of linkages, connections, and relations across space.

The space of flows consists of two prominent levels, the infrastructural and the organizational. The former is the "wired world," the hardware equipment linked to software which makes electronic transmissions around the world possible. The latter consists of the social patterns of linkages between the people and institutions who make the network society operate. Both of these networks have distinctive geographies that are just beginning to be understood. The electronic superhighway may define a cyberspace, but it is still grounded in real places where it operates, is maintained, and is continually being developed. The social networks come in many forms but the most conspicuous is the "world city network" where corporate service firms locate at the heaviest intersections of information which they can convert into "professional knowledge" and sell to their clients (Sassen 1991). Services such as devising new financial instruments, advising on inter-jurisdictional law, and creating global advertising campaigns are concentrated in these world cities. These myriad global spaces of flows are the critical geography in globalization.

Just like geographical scales, you cannot choose to be in one type of space but not the other. Spaces of places have not disappeared with the coming of network society. Cities, for instance, might be considered to be nodes in a network but they are also distinctive places. And, of course, there were numerous spaces of flows before the 1970s. What we are talking about here is a changing balance of importance between the two types of flows: as Fred Cooper (2001) says, "scholars do not need to choose between a rhetoric of containers and a rhetoric of flows." The enabling technology of communications and computing has created a situation in which flows have become relatively dominant at the expense of places. But human geography continues to be about unraveling the relations between places and flows. Contemporary globalization poses the tension between place and flow as a direct reflection of power relativities in a most unequal geography of globalization.

Geography of Globalization

At first glance, it might seem strange that globalization should have a geography, a variable distribution across space. Globalization is everywhere but it is not a homogeneous process: its outcomes vary markedly across the world. And the pattern is quite simple. The rise of globalization has been marked by increased material polarization between regions; for every global city in a network of global cities there is what Castells calls a "black hole" of marginalization and exclusion from the global network society. This has been called uneven globalization (Holm and Sorensen 1995).

Starting in the "era of development" we can note that the proportion of the world's population which enjoyed per capita income growth rates of over 5 percent tripled between 1965 and 1980. Some newly industrialized states in East Asia have experienced historically unprecedented rates of "industrial compression": Taiwan and South Korea were, after all, wartorn, impoverished, and archetypically postcolonial "underdeveloped" exporters of wigs and sugar and rice in the 1950s. But on balance the record is indeed uneven, and nowhere more so than with respect to the plight and privation – the structured inequality – of women. Of the 1.3 billion people in poverty, 70 percent are women. Between 1965 and 1988 the number of rural women living below the poverty line increased by 47 percent; the corresponding figure for men was less than 30 percent (UNDP 1996). Mass poverty has been stubbornly resistant to the changing fads and fashions of development policy. If the incidence of poverty declined as a proportion of the world's population in the postwar period (itself perhaps contestable), the total number falling below the absolute poverty line has unequivocally increased. In the period since 1980, economic growth in 15 countries has brought rapidly rising incomes to 1.5 billion people, yet 1 person in 3 still lives in poverty and basic social services are unavailable to more than 1 billion people.

Locating poverty on the larger canvas of postwar "development" allows us to see two important historical forces at work. First, some key constituencies did not participate in the growth and productivity achievements of the 1945–80 period (poor women and rural landless for example): that is to say growth was accompanied by *exclusion*. And second, the record of the poor in participating in the market successes of the post-1980 period was constricted in the absence of redistribution: market-driven growth was marked by *marginalization*. One hundred countries totaling 1.6 billion people have actually experienced economic decline; in almost half of them average incomes are lower now than in 1970. The gravity of these figures is only deepened by a recognition of the growing polarities within the global economy as a whole. According to the United Nations Development Program, between 1960 and 1991 the share of the richest 20 percent rose from 70 percent to 85 percent of global income – while that of the poorest fell from 2.3 percent to 1.4 percent. Between states, the ratio of the shares of the richest to the poorest increased from 30 : 1 to 61 : 1. The problem is polarization: the proportion of the globe experiencing low income growth rates per head has grown, and since the 1980s it has grown substantially (Pritchett 1997). And this same pattern of increased polarization is to be found at other scales. Within countries, the more affluent central regions have prospered relative to their outer, poorer regions. Urban regions have experienced increasing differences between rich and poor neighborhoods; world cities in particular are said to be marked by exceptional economic polarization among their residents (Sassen 1991). And at the individual level, Bill Gates of Microsoft was worth $100 billion at the recent stock market peak in technology shares (April 1, 1999), which is greater than the total sum of the GNP of all countries except the richest 18 (Cohen and Kennedy 2000: 111).

What are we to make of this very clear effect of globalization? First, at the most general level, we can say that capital's latest spatial fix has worked. For the century up to the 1970s, new structures were being constructed – democracy, welfare states, decolonization – to enable the world to become a more equal place. Probably for the only time in history, the lower economic strata were successfully using their collective power as workers and consumers, and as soldiers and voters, to gain a larger share of the growing global wealth. Late twentieth-century globalization marks a historical turn-around in this trend. Capital has regained its lost power. Lurking behind the political rhetoric of market opportunities, economic realism, and de-regulation, there is the threat, and sometimes the actuality, of firms moving to more capital-friendly places. Whereas states, by their very territorial na-ture, are stuck in the old space of places, capital can take full advantage of the new space of flows. Hence firms can play one country off against an-other in paying taxes, in investing in new plants, in pay negotiations, in levels of regulation, in getting grants and subsidies, and many other activi-ties that directly affect their overall global profits. Capital flight can be the harbinger of deindustrialization and what US presidential candidate Ross Perot called the "great sucking sound" of job loss. Industrialization of the Third World used to conjure up images of brand new iron and steel works (e.g., Mountjoy 1963); today it is about consumer brand sweatshops (Dicken 1998: 313–14).

The reassertion of the private over the public interest – what has been called the neoliberal counter-revolution – has become pervasive, greatly abetted by the demise of the communist "Second World" (including eco-nomic changes in the still formally communist China). Of course, the mo-bility of capital is nothing new – historically it was called the "runaway shop" – but the scale and magnitude of direct foreign investment (reaching $230 billion of "outflow" per annum in 1995) has conferred upon capital an unprecedented level of power. This enhanced leverage is to be found at all scales. Geographically it works through established uneven developments, such as core–periphery patterns at global and state scales, to accentuate past inequalities. Uneven globalization is more polarized than traditional geographical inequalities. Political policies to counter polarization are ma-jor victims of capital's power. Urban and regional development policies are likely to be converted to "private–public initiatives" in rich countries, and aid policies are likely to be converted into credits to buy arms in poor coun-tries. But the clearest example of policy turnaround has come in the form of structural adjustment programs (SAPs) and stabilization measures in the Third World.

Precipitated by the debt crises and economic recession in Latin America and Africa in the early 1980s, these programs were devised by the Interna-tional Monetary Fund (IMF) and World Bank to correct disarray in the finances of developing countries which were not developing. Involving mas-sive cutbacks in social programs to make the countries more attractive to capital investment, SAPs created mass impoverishment resulting in food riots in several countries. The 1980s – the high tide of austerity and adjust-ment – saw plummeting incomes and standards of living in Africa and Latin America; the so-called Lost Decade saw average African incomes plummet by 1990 to below their 1960 levels when many African states became

independent. Subsequently the establishment of the World Trade Organization (WTO) to promote a laissez-faire global economy has posed a further
threat in the 1990s to reversing Third World economic regression.

The SAPs in the Third World were part of a neoliberal agenda that originated in the First World in the late 1970s and 1980s. Here the cutting back
of the welfare state was central to the new politics appropriate for the new
global times. Led from the right in the 1980s by Reagan (Reaganomics) in
the USA, Kohl in Germany, and Thatcher (Thatcherism) in Britain, by the
1990s their politics based upon faith in the market had diffused across the
party spectrum to encompass the likes of the "New Democrats" under
Clinton and "New Labour" under Blair. Global neoliberalism's greatest
political success came with the demise of the USSR that led to a rapid adoption of market reforms. The resulting mafia-style capitalism – coupled with
a terrifying collapse of standards of living seen in the increase of poverty to
40 percent and sharp increases in child mortality and life expectancy – is
also the major political indictment of neoliberalism. Its truly human indictment is the millions of additional children who have died in the Second and
Third Worlds as a general result of impoverishment and a specific result of
reducing health services in the name of economic efficiency.

This uneven economic globalization has been masked to some degree by
cultural globalization. Whereas producers in sweatshops have often been
hidden from sight, consumers in designer shops have been highly visible in
globalization. Neoliberalism's promotion of the individual as an economic
agent in the reform of socialist or regulated market economies is simultaneously the bearer of the ideology of contemporary consumerism. It promotes
a vision of a homogeneous world of consumption that goes under various
well-known branded names: McWorld, Coca-colonization, Disnification,
and the Levi Generation, for example. The fact that these all refer to US
corporations is an artefact of mid-twentieth-century "Americanization":
today, consumerism is as likely to be fueled by European and Asian corporations. Although it has been shown that both initial Americanization and
subsequent cultural globalization are much more subtle than commonly
appreciated, producing many hybrid cultures ("the Maharaja Mac" in India!) rather than a single homogeneous culture, the image of "one world"
that consumerism presents has become as pervasive as neoliberalism itself.
Symbolically, the largest McDonald's in the world is on Red Square, Moscow.

There is a double irony in cultural globalization's one-world vision. It is a
vision shared both by past cosmopolitan internationalists, as expressed in
the name "United Nations," and by contemporary environmentalists in their
concept of the Earth as the home of humanity. Superseding the ordered
political world of the UN is one thing, but consumerism sharing its image
with environmentalism is quite another matter. It is the very consumption
promoted by one-world advertising that is threatening the home of humanity. Here we enter the heart of the contemporary politics of globalization.

Geography for and against Globalization

The political hegemony of neoliberalism has led many to equate this program of policies with globalization itself. This is understandable but mistaken. Globalization is the result of processes that are much broader than neoliberalism. In an earlier section we referred to the key technologies as "enabling" and we can see now that they have enabled capital to reassert its dominance of the world economy. This is hardly surprising given the way the technologies themselves have been developed by a select number of corporations to aid all corporations to enhance their profits. But this does not exhaust the enabling potential of these technologies. They can be used to resist globalization or applied to try and construct alternative globalizations.

Globalization has coincided with a rise in ethnic conflicts across the world. Although conventionally interpreted as long-standing local hatreds (for example, the purportedly deep and abiding hatreds among Hutu and Tutsi in Rwanda), this does not explain why the 1990s witnessed, not only the eruptions of civil wars, but also horrific ethnic cleansings (former Yugoslavia) and even a new genocide (Rwanda). Even Freud, a famous pessimist, might have been shocked by the violence which has attended the late twentieth century, the "age of extremes" as British historian Eric Hobsbawm (1994) has recently dubbed it. In the wake of the Cold War, amid much talk of "peace dividends" and a new world order, the atrocities in the Balkans, the genocide in Rwanda, the civil war in Algeria, and the communal violence in India must represent something like a "return of the repressed." These forms of violence are instances of what Nancy Fraser (1997), in her book *Justice Interruptus*, refers to as "politics of recognition," and as such they reveal that the politics of community – a word which as Raymond Williams (1976) noted long ago is never used unfavorably – can turn very sour indeed. There have been several celebrated "peace processes" but these have turned out to be problematic in their different ways: falling apart (Israel/Palestine), fragile (Northern Ireland/Ireland/UK), and profoundly disappointing (South Africa). These events remind us that globalization is not the only important bundle of processes to emerge in recent years. The end of the Cold War in 1989 "liberated" ethnic feelings into the political realm, both within and beyond the old Second World. At the same time a broader new politics of identity was emerging where individuals were asserting their chosen collective identities, undermining traditional class and orthodox party politics. Many of these identity movements adapted to the space of flows to themselves become globalized, notably the gay movement and feminism. But ethnic identities have always been more place-based. When transmuted into national claims they become wedded to a mosaic world space of places wherein nationalists demand their own "homeland." Hence the irony that globalization has occurred just as the number of states in the world has dramatically increased.

For nationalists, globalization with its homogenizing tendencies can be seen as a very real threat. But we must not read this as a spatial conflict of places versus flows. All spatial experiences, nationalist or not, are constituted by both places and flows. In fact, defense of places will always depend upon a myriad of flows both material and informational. A classic case of this is the resistance of Commandante Marcos against the effects of the North American Trade Organization on indigenous farmers in southern Mexico in which he combines place and flow strategies. As well as local military resistance, he has mobilized world opinion using email. This technology has been similarly important for the anti-capitalism demonstrations at world conferences. At the state level France's resistance to English becoming the world language has a similar two-pronged media approach. First, there are measures to conserve the French language within France through minimum quotas on non-French language items on radio and TV. Second, there is promotion of French as a language on the internet so that the World Wide Web does not become a new English-language realm.

It is in the social and employment sector of state policy that the question of places and flows is most intertwined. The conundrum for those on the political left is how to maintain economic gains made within states over many years while not eschewing the potential new benefits offered by globalization. In the social democracies of Europe this has taken the form of defending key aspects of the welfare state while simultaneously promoting market processes under the label "entrepreneurship" (Lipietz 1996). The European Union has been brought into the battle through its "Social Charter" which provides for minimum social and employment benefits so as to prevent blatant attempts by member states to attract global capital by severe erosion of social and employment rights. In the USA the conundrum is being felt most acutely by the trade unions, which have long experience of the "runaway shop" within their country (Herod 1997). Hence more recent threats to jobs from Mexico and Asia are part of a more continuous process of "spatial fix" by American capital. But whatever the context, globalization denial is not an option: there is the realization that globalization is not going to disappear in the near future and therefore a strategy of simple resistance can only be a partial and temporary response. Whatever strategies are adopted it seems clear that geographical knowledge of places and flows will be indispensable (Herod 2000).

There is an alternative positive way of looking at the trend towards globalization. This relates it to cosmopolitan ideals that have always been suspicious of the "parochialism" of the territorial state. The French Revolutionary and Napoleonic Wars stimulated both nationalistic and cosmopolitan politics. In the latter case, a pamphlet written by Immanuel Kant in 1795 is pivotal in advocating "perpetual peace" (Gallie 1978). He argued that in a world where countries were becoming more and more connected, peace was in the interests of both individuals and states. In the mid-nineteenth century this argument was taken up by the "Manchester liberals" who contrasted the "party of peace" based upon commercial interests with the "party of war" based upon territorial interests (Taylor 1996b). Although clearly underestimating the power of nationalism unleashed in the following century, nevertheless we can see that economic globalization can claim to be in this "perpetual peace" tradition: no pair of countries selling "Big Macs" have ever been to war against each other according to the chief executive of McDonald's (until the USA bombed Serbia).

More seriously, new cosmopolitan ideas abound in our globalizing world. As we would expect from a cosmopolitan perspective, it is not that the global scale is seen as the problem, but rather its particular realization today. Put simply, contemporary globalization is critically unbalanced. The ideal balanced modern society consists of a successful economy to provide affluence, a firm government to provide security, and a vibrant civil society to provide identity. In this trilogy, government regulates the economy and is legitimated by society through democracy, while the economy provides the goods for society to rise above basic needs. Clearly globalization is far from meeting these ideals worldwide on all three counts. However it is most developed economically, although not providing affluence for the vast majority of humanity, there is some global governance, but the development of a global civil society is most elusive.

The implications of this imbalance are profound. Starting in the middle, global governance has developed beyond the initial liberal internationalism of the United Nations. The most important institution is the series of G7 summits. These are regular meetings of the seven leading states (USA, Japan, Germany, France, UK, Italy, Canada – with the presidents of the European Union and Russia now normally in attendance) where the global economic agenda for the rest of the world is set. And that agenda is a neoliberal one. Without a global democratic mandate, this governance does not so much regulate the world economy as simply promotes it. This is explicitly illustrated by the setting up of the World Trade Organization (WTO) in 1997 with trans-state rules negotiated by the rich countries. To stay in the global space of flows poor countries have no choice but to join this organization, but in doing so they leave themselves open to economic punishment if they employ the economic trade policies the G7 countries themselves employed before they were rich (i.e., forms of trade protectionism). This global democratic deficit pervades the putative formation of a global civic society. The only plausible trans-state class is in the upper economic strata (Sklair 1995), a global capitalist class who are inevitably cheerleaders for the neoliberal world economy. Opposition comes from transnational (and globally networked) social movements which have been influential in several world conferences held under UN auspices. The pitched battles against the G7/IMF/IBRD/WTO in Seattle, Washington DC, Prague, and Genoa have unquestionably projected the globalization debate into the popular political consciousness in important ways.

These movements have in some respects been limited partly because the movements themselves have a severe democratic deficit: representing humanity ultimately requires legitimation through some sort of people's mandate. On the other hand they have helped keep the debates alive over reforming the global regulatory institutions and have the capacity to link up with national movements, for instance, around GMOs (genetically modified organisms) in Brazil, or dam construction in India, with surprising effect.

In fact the vast expansion of non-governmental organizations (NGOs) – one expression of voluntary self-government if not participatory democracy – is indisputably part of the post-1970 globalization process. The Johns Hopkins Center for Civil Society Studies estimates for a sample of only twenty-two countries, that NGOs generated $1.1 trillion in revenue, employed 19 million workers, and recruited 10 million volunteers. In contrast to Putnam's (2000) claim that social capital has collapsed over the last three decades, a deepening of associational and civic life is one of the hallmarks of the post-1968 generation. In 1960 each country had, on average, citizens participating in 122 NGOs; by 1990 the number had leapt to over 500. Significantly, two-thirds of the NGOs in Western Europe have been founded since 1970. There are now in excess of 2 million non-governmental organizations in the US, three-quarters of which have been established since 1968. In Eastern Europe 100,000 non-profit organizations appeared between 1989 and 1995; Kenya authorizes almost 250 new NGOs each year. Among international NGOs the growth and proliferation is no less explosive. In 1909 there were 176; currently there are over 29,000, virtually 90 percent of which have been established since the 1960s (*The Economist* January 29, 2000: 25–6). Putnam (2000) argues that many of these are far from democratic organizations since their "members" are mainly "subscribers" who support the organization's activities (indeed, in some cases the organizations tailor their campaigns and activities to those most likely to raise subscriber income), but who play little part in the NGOs' daily work. However, for someone like Melucci the emergence of such transnationally networked organizations, a sort of global civil society, marks a rupture, a shift from the new social movements of the 1970s to "an overarching system of closely interdependent transnational relations" (Melucci 1996: 224) and new forms of governance and "partial government." As if to drive home the point, the Rand Arroyo Center, in a recent study sponsored by the US Deputy Chief of Staff for Intelligence, published a report entitled "The Zapatista Social Netwar in Mexico," documenting the grave dangers of "electronic horizontal networks" of social mobilization which "confound fundamental beliefs" in virtue of their "epistemological" approach to politics!

The very idea of a "global civil society" is certainly limited at present, and there are good theoretical reasons for doubting its efficacy. Civil societies at the state level have operated best where there is one dominant national project. This is because democracy as a decision-making instrument works most efficiently when there is only one "demos"; where there is more than one "people" democracy is usually excessively divisive. It is not clear how a global civil society could ever cope with the vast variety that is humanity. Assuming it is anti-territorial in nature, we can envisage a transstate civil society organized in the space of flows through a world city network. But this immediately introduces hierarchy: are these cities to become new centers of "command and control" as some see their role in the contemporary world economy? This could promote the latent authoritarianism to be found in many democratically deficient transnational movements. Even if such cosmopolitanism injected more humane and environmentally sensitive policies on to global agendas, it would still be another "globalization from above."

The key problem is that we do not seem to have the political equipment for developing a "globalization from below" (Falk 2000). Castells (1999) talks of "grassrooting the space of flows" and certainly the technology is enabling of such a non-territorial political project. An immensely complex undertaking, any such "globalization from below" project would have to respect and promote cultural diversities. However, it is not clear how we could prevent such a fragmented "global civil society" becoming a victim of a more unified economic and political order. The ideal of harnessing globalization's potential for a better world by respecting diversity (spaces of places) while promoting cosmopolitan ideals of equity and freedom (spaces of flows) certainly seems a long way away. But we know the starting point: understand globalization today, to make a better tomorrow, for all humanity.

Part I
Geoeconomic Change

Introduction to Part I:
The Reconfiguration of Late Twentieth-Century Capitalism

When the history of the late twentieth century is written, there seems little doubt that mobility – of capital, labor, and meaning – will be one of its touchstones. The ceaseless search for profitability within the interstices of a world market has propelled a radical restructuring of national economies around the world. The global corporation dominates the economic iconography of the post-1945 period. Its leitmotifs are speed, innovation, integration, and what geographers have come to call space-time compression (Harvey 1990). Joseph Schumpeter's (1952) invocation of creative destruction as the "essential fact" of capitalism provides us with a powerful image to understand the geoeconomic restructuring wrought by capitalist impulses amid a worldwide rhetoric of monetarism, laissez-faire, deregulation, and market triumphalism.

Transnational Capital and Foreign Investment

By the late 1990s there were 60,000 transnational corporations (TNCs) in the world with almost 500,000 foreign affiliates (UN 1999) and accounting for 25 percent of global output. The total stock of foreign direct investment (FDI) accounted for by this universe of corporations was in excess of $2 trillion in 1998, and totalled over $11 trillion in worldwide sales. Transnational capital is, of course, highly concentrated geographically, sectorally, and in terms of the share of foreign assets controlled by the largest firms; 90 percent of TNCs are headquartered in the advanced capitalist states (five major home countries account for over half of the developed country total), and the triad of Japan, North America, and Western Europe produced 72 percent of global foreign investment inflows and 92 percent of global outflows in 1997 (between 1988 and 1998 the total outward stock of the triad increased from $1 trillion to $2.4 trillion). Roughly 1 percent of parent TNCs own half of the total FDI stock and the largest 100 TNCs account for $1.8 trillion in foreign assets, 6 million employees in foreign affiliates, and 14 percent of the total world stock of outward investment

(UN 1999). Since 1990 the average transnationality index – the measure of internationalization employed by the United Nations – increased from 51 percent to 55 percent.

Since the mid-1970s, flows of foreign investment have followed an upward growth trend averaging 13 percent per annum. Commonly understood in terms of the seemingly unstoppable globalization or "transnationalization" of the world economy, there have been in fact two striking foreign investment surges, one from 1978 to 1981 (growth averaged 15 percent per year) and another between 1986 and 1990 (averaging an astonishing 28 percent per annum). Between 1986 and 1990 the annual growth rate of FDI outward stock was 21.3 percent; even during the recession of the 1991–5 period it increased by 10 percent and is now running at 20 percent (UN 1999). For the developing world this has meant that the vast majority of total net resource flows is now overwhelmingly dominated by FDI inflows which have increased from $50 billion in 1991 to $200 billion in 1998. The unprecedented increase after 1985 was unquestionably facilitated by a period of economic growth in the wake of the recession of the early 1980s and by a period of intense mergers and acquisitions, but there were also long-term structural forces at work. In particular the accumulation of FDI stock since the first oil boom, the proliferation of integrated international production ("the world car"), and the radical changes in the macro-ideological environment which, under the auspices of monetarism in the core and the growing hegemony of the International Monetary Fund and the World Bank at the periphery, precipitated extensive trade liberalization, exchange rate reform, and privatization (particularly in Western Europe, Latin America, and increasingly in the former COMECON countries). The virtual disappearance of nationalization since 1974 and the explosion of privatization – in essence the selling off of state properties and enterprises – has been especially striking. Notwithstanding the 1997 GATT (General Agreement on Tariffs and Trade) agreement, leading to the creation of the World Trade Organization (WTO), 79 new legislative measures adopted during 1992 in 43 countries liberalized the rules on FDI (UN 1999). In 1998, the year for which the most recent data are available, the trend toward liberalization of regulatory regimes for FDI continued unabated, often complemented with proactive promotional measures. Out of 145 regulatory changes relating to FDI in that year by 60 countries, 94 percent created favorable conditions for investment. The number of bilateral investment agreements also increased markedly in 1998, 40 percent of which were between developing countries.

The rapid growth of FDI has been accompanied by important shifts in its sectoral composition. During the 1950s foreign investment was concentrated in primary and resource-based manufacturing. By the 1990s services accounted for close to 50 percent of all FDI and they absorbed almost two-thirds of annual flows. While integration and capital mobility are proceeding at different rates across different industries and functions, financial services is probably the most global of corporation activities, stimulated by electronic transfers and 24-hour trading. The rise of an integrated global agrifood system, propelled by recent regulatory changes imposed by NAFTA (North American Free Trade Agreement), GATT, and WTO, confirms, however, that all sectors have felt the press of global capital flows. Indeed, transnational capital in the food preparation and processing sector has emerged as one of the most dynamic fronts along which the market is opening up the former Soviet sphere. Nonetheless, hypermobility of capital and frictionless space is hardly the norm, and a truly global research and development and manufacturing system remains restricted to a relatively small number of firms in limited sectors and branches.

Worldwide FDI flows, however, slowed in the first part of the 1990s – for the first time in fact since 1982 – largely as a result of contractions by Japanese and Western European business operations. Developing countries received 25 percent of all inflows in 1991 (equal to their share in the first half of the 1980s!); the figure grew substantially in 1992. Inflows to East and Southeast Asia, and increasingly in Latin America, have been especially dramatic. Investment flows into Asia and the Pacific rose by almost 10 percent in 1992 and investment in Eastern and Central Europe, while spatially uneven, has been singularly impressive. In the latter part of the 1990s, however, in spite of the effects of the 1997 Asia crisis, the total outward flow of FDI per annum and the corresponding inward flows have grown steadily. In 1998 world FDI outflows reached a record level of $649 billion and inflows reached $644 billion. In 1998 inflows and outflows grew respectively by 39 percent and 37 percent! Further estimates for 1999 suggest further increases: cross-border mergers and acquisitions (M&As) in the first half of 1999 were $574 billion, equal to the total value of cross-border M&As in 1998. The paradox of FDI growth under adverse global circumstances in the latter part of the 1990s is explained, however, by sharp contrasts in regional trends. Virtually all of the increase in FDI was concentrated in the developed countries and inflows into the developed countries decreased by 4 percent (indeed 48 developing countries attracted less than $3 billion, 0.4 percent of world FDI flows). Nevertheless, on balance the transnationalization of capital has been a propulsive force in the rapid industrialization of some newly industrializing states (most recently in Thailand, China, and Malaysia, which represent a sort of second wave of newly industrialized countries (NICs) following on the heels of Taiwan, Singapore, South Korea, and Hong Kong). During their industrial revolutions of the nineteenth century, Britain and the US took a half century to double their incomes; the Asian Tigers and their new companions are achieving this within a decade. Whether this growing heterogeneity within the less developed world, and the rapidity of industrialization in particular, warrants the sort of optimism expressed by *The Economist* – "rapid changes in the balance of the world economy" and the NICs having "consigned to the rubbish bin the old notion that the rich world . . . towers over the whole world economy" (January 8, 1994: 16) – is, however, another matter entirely.

Uneven Globalization

Within the broad parameters of geoeconomic restructuring, capital flows and their effects are nonetheless highly uneven by region and by sector or firm. Some parts of the South – sub-Saharan Africa in particular – almost slid off the economic map during the 1980s. Indeed, the impact of structural adjustment and liberalization during the 1980s produced horrifying contraction and austerity in large parts of the Third World, with little evidence of subsequent economic recovery and growth. In sub-Saharan Africa economic output and exports tumbled by 30 percent between 1980 and 1986; private net transfers declined from a positive inflow of $2.5 billion in 1980–2 to a net outflow of $7 billion in 1985–7. The standard of living of the "average African" is, according to the World Bank, lower now than at independence, and several formerly prospering economies have rapidly declined (as in Zimbabwe). Finally recognized as a global policy issue, the 2001 G8 summit meeting at Genoa included a meeting with African leaders and the acceptance of a "plan for Africa" drawn up at a previous meeting of all African leaders (stimulated by President Gaddafi of Libya) which started moves towards an African economic union. Latin America did not fare much better (Watts 1991). So-called "shock therapy" in Russia and Eastern Europe has also taken its toll, reflected in the terrifying rates of unemployment, plant closures, and economic insecurity. Life expectancy in Russia dropped from 62 to 59 years over twelve months, the largest single-year drop ever recorded in a developed country, and the death rate soared by 20 percent in 1992. Plagued by inflation and economic chaos, Russia looks increasingly like an archetypical Third World state. More than anything, the demise of the former socialisms reveals that markets cannot simply be wished into existence but have to be constructed. Often they are not, and what emerges from the ashes of socialism is not a free market wonderland but what has been referred to as market Stalinism. Market liberalization as often as not produces illiberal politics, and not infrequently wildly distorted markets.

As if to drive home the point, 1997 witnessed an astonishing economic crisis which began in Southeast Asia with a series of fiscal problems in Thailand and Indonesia associated with the weak banking sector and the liberalization of the capital account (that is to say the ease with which financial flows can shape a national economy). The insolvency of the banking sector, coupled with short-term speculative activity, triggered capital flight and a stockmarket collapse in other parts of the world (South Korea, Brazil, Russia). Currency devaluation against the dollar in Southeast Asia was of the order of 35–40 percent and stock markets collapsed by 89 percent in Indonesia, by 75 percent in Korea, and 73 percent in Asia, that is to say in those countries that had provided a source of enormous vitality in the world economy over the last twenty years. The IMF rescue missions had the effect of facilitating vast transfers of assets to Western and Japanese corporations and destabilizing the so-called "Asian model" of industrialization (Wade and Veneroso 1998; Brenner 1998). The Asian economies were squeezed – millions were thrown into poverty in the region – and are now trying to export their way back to growth in the face of stiff competition from China. What was remarkable about the Asia crisis was the resiliency of capital flows into the region precisely because of the availability of cheap assets due to currency devaluation and FDI liberalization associated with IMF policies. While talk of the recovery of the region abounds – exports increase, capital flows rebound – the 2000–1 recession in the US economy must throw such speculations into question.

In short, amid the euphoria of liberalization and global economic restructuring driven by new trade agreements and international capital mobility, the end of the Cold War looks as though it has been replaced by new forms of North–South dependency. According to the United Nations Development Program, global polarization of income and wealth has increased substantially; between 1960 and 1990 country differentials between the wealthiest and the poorest 20 percent increased from 30 to more than 60, while real disparities between people, as opposed to country averages, became even more stark (UNDP 1993). Since the Third World was in one sense the creation of the Cold War, the new world order implies, as some have suggested, the end of the Third World. However, it has been replaced, after a decade of severe discipline meted out by the global regulatory agencies, by a map of North–South dependency which conveys an extremely bleak prospect for large parts of the global economy (Arnold 1993; Bello 1994). The costs of the economic restructuring during the 1980s are no less in evidence among the advanced capitalist states, of course. Unemployment in the OECD (Organization for Economic Cooperation and Development) states stands at a historic high, while in different ways and at different velocities the postwar Fordist arrangements – as corporatist systems of production, consumption, and regulation – have been dismantled and reconfigured (Sayer and Walker 1992). High monetarism and neoliberal orthodoxy perpetrated by Kohl, Thatcher, and Reagan have eroded the political and economic landscapes of both the EU and North America. Whether this represents a shift to a distinctively new system of post-Fordist or flexible accumulation is a matter of intense debate (Scott and Storper 1993). What is less arguable is that the shape and character of national economies, and the scope and nature of public intervention, have been radically altered since the mid-1970s. What new sorts of social contracts and private–public networks will emerge in the twenty-first century is anyone's guess. What seems clear, however, in debates over economic modernization in the North and South alike, is that the uncritical adoption of simple market or state models no longer carries much cachet. And to this extent the role of civil society – the third pole of economic development – in reviving flagging economies is drawing considerable attention (Putnam 1993a). The hysteria over the new economy – of dot.com mania for example – quickly evaporated during 2001 as the US high technology sector took a terrific beating. Silicon Valley and northern California – the ground zero of the new informational economy – is in a severe recession, and massive layoffs by Intel, Oracle, and others is one indication of the "hype" and classic speculative activity that surrounded the e-commerce revolution. None of this should imply that new forms of regional trade integration or deregulation (NAFTA, EU, GATT, WTO) or integrated international production are of no consequence or will not continue. But it would also be foolhardy to assume that the current trend toward further deregulation, free trade, regional integration, and hypermobile capital are historical inevitabilities. *Fin de siècle* capitalism cannot be so readily captured and contained.

Regional Governance

If capitalism is to be captured and contained, then political action will be needed, since all of the evidence points to the failure of capital to regulate itself. The state, with its monopoly over certain powers (such as sovereignty over territory and the monopoly over legalized violence), is the only institution that has so far had much success in regulating capitalism. But there is a major geographical tension now, because whereas the forces of globalization are no respecters of boundaries and distance, territorial states are still bounded containers, whose governments spend more time and energy competing with each other for the benefits of globalization than cooperating in regulating it for the more general good – as in George W. Bush's frequent arguments in the first year of his presidency that he would sign-up to no international agreements which harmed the interests of American corporations and threatened American jobs.

There are two possible responses to that tension. One is for the restructuring of the system of territorial states into larger units, which are better able – because of their size and power – to regulate global capitalist forces. The other is for international agreement on regulation – agreement that can then be implemented. Both are full of difficulties. Restructuring of the system of states raises many problems of loss of sovereignty for those citizens and governments of states which agree to pool their sovereignty. Only one new structure – the European Union – has gone very far in this direction, during its nearly fifty years of existence (under various names and with a slowly expanding membership from the original group of six). Its advances have been greatest in the creation of a single market, which stimulates trade and other movements. But there are strong voices within the Union that simply creating a single market is insufficient. For example, in order to facilitate successful regulation of the whole, it is necessary, they say, for major economic regulatory functions (such as the control of money supply and interest rates) to be centralized (and depoliticized, handed over to independent central banks). This is opposed by many – especially in the UK and the Scandinavian member states – as involving a major transfer of sovereignty, not only to a higher authority, but also to one that is not democratically accountable. It removes leverage over major economic policies from the governments of member states, and so prevents them from taking steps which they see as in the interests of their own citizens – to whom they are democratically accountable – irrespective of their impacts on others. And yet without this, it is argued – and as the experience of the NAFTA shows – the spatial dynamics of a single market are likely to favor the richer areas. It is for this reason that there are strong arguments within the EU for socioeconomic policies aimed at reducing spatial differentials, creating a uniform economic plain within the 15 states. Again, this is opposed by those who see it as an invasion of state sovereignty, putting the interests of the whole above those of some of the parts. And enlargement of the EU – to as many as 27 states if all current expressions of interest are realized – could exacerbate this perceived problem. Most of the potential new members are relatively poor, Eastern European (former communist) countries: were they to enter the EU, then policies aimed at removing regional inequalities would see them as the major beneficiaries, to the detriment of those countries which are now net financial gainers from EU policies – such as Spain, Portugal, and Greece – while the freedom of labor movement could, it is feared, lead to massive immigrations to the EU's rich core. Thus, while expansion is supported rhetorically, and on strategic terms, its likely geographical consequences make it fraught with difficulties.

The EU came about, as did NAFTA and other attempts at restructuring the global territorial structure, through the actions of governments promoting what they thought was in their own interests, and when they identified aspects of the restructuring which were against those interests they opposed them strongly – as in the UK's dealings with the EU throughout the 1980s and 1990s, which led to some major changes in policy and practice where the British view prevailed, usually on issues over which, because of EU rules, it had veto power. This illustrates the problem of the second response to the geographical tension of contemporary capitalism: the problem of obtaining binding international agreements between sovereign states seeking to promote their own, rather than collective, interests. Of course, many such agreements are reached, either where governments collectively realize that the general good is also their individual good or where powerful individual governments are able to impose such agreements on weaker others – the 1944 Bretton Woods agreement on post-World War II restructuring of currencies illustrates that. But many attempts at international agreements fail to achieve more than rhetorical support for promoting the general good because most of them are, in the end, toothless, since there is an absence of binding international regulatory systems to ensure that member states enact and implement that to which they agree. This is illustrated by the United Nations organization itself, and by many of the bodies and agreements it supports and tried to sustain; and ultimately, in many aspects of international economic, social, and political relations, it is the strongest who prevail, and ensure that their notions of an international order are imposed – to their benefit. Thus in July 2001, while the heads of the world's largest economies were meeting to discuss an economic agenda at the G8 summit in Genoa, the leaders of the world's poorest 41 countries were meeting in Zanzibar to produce and promote an alternative vision. One doesn't have to be cynical, just realistic, to identify which view is likely to prevail, whatever the protestations of the former group that they will act to help the latter, since "what is good for Africa is good for us all"!

These geographical tensions between global capitalism, on the one hand, and the system of territorial states, on the other, will be a major feature of the twenty-first century. In the long run, it will be the latter that has to be changed to meet the demands of the former – but within those changes some parts of the world (current countries and regions) will ensure that the restructuring creates a new world political map that is to their economic advantage.

2 A Hyperactive World

Nigel Thrift

Introduction

That light overhead, the one gliding slowly through the night sky, is a tele-communications satellite. In miniature, it contains the three main themes of this chapter. Through it, millions of bits of information are being passed back and forth. Because of it, money capital seems to have become an el-emental force, blowing backwards and forwards across the globe. As a re-sult of innovations like it, the world is shrinking – many places seem closer together than they once did.

The satellite is itself a sign of a world whose economies, societies, and cultures are becoming ever more closely intertwined – a process which usu-ally goes under the name of globalization (Giddens 1991). But what sense can we make of this process? Again, the satellite provides some clues. Those mil-lions of messages signify a fundamental problem of *representation*. Simply put, the world is becoming so complexly interconnected that some have be-gun to doubt its very legibility. The swash of money capital registering in the circuits of the satellite comes to signify the "hypermobility" of a new space of flows. In this space of flows, money capital has become like a hyperactive child, unable to keep still even for a second. Finally, the shrinking world that innovations like the satellite have helped to bring about is signified by time-space compression. Places are moving closer together in electronic space and, because of transport innovations, in physical space too. These three simple themes of legibility, the space of flows, and time-space compression can there-fore be seen as "barometers of modernity" (Descombes 1993), big ideas about what makes our modern world "modern."

In the first part of this chapter, I will examine these three barometers carefully because they inform so much current discussion about modern global geographies. My purpose is simple – to show that they are partial accounts made into a whole. In the second part, I hope to show, through a double-take on the world of international money, just how partial these accounts are, even in that sphere of the world's economy which might be expected most closely to approximate to them. Then in the last part of the chapter I propose the beginnings of a more moderate account of the signs of the times.

Preamble

I shall begin by taking a reading of the barometers of modernity. I will do this by examining illegibility, hypermobility, and time-space compression through the writings of the three authors who have most often deployed them: Fredric Jameson, Manuel Castells, and David Harvey.

Each of these authors paints a picture of the world as having come under the sway of a new form of capitalism – whether it is called late or multinational or informational or global capitalism. This form of capitalism usually involves a combination of ingredients, and most especially: the accelerated internationalization of economic processes; a frenetic international financial system; the use of new information technologies; new kinds of production; different modes of state intervention; and the increasing involvement of culture as a factor in and of production. The three authors disagree on how swiftly the new form of capitalism has taken hold of the world. To Harvey, for example, it has been a comparatively rapid process. For Jameson, in contrast, the new form has crept up on us: "we have gone through a transformation of the life world which is somehow decisive but incomparable with the older convulsions of modernization and industrialization, less perceptive and dramatic, somehow, but more permanent precisely because more thoroughgoing and all-pervasive" (Jameson 1991: xxi). But each of the three authors agrees that at the heart of this change has been *space*. The new system of capitalism attacks and suppresses distance, *and* our notions of distance, producing a new global economic space in which global capitalism can play.

These three cartographers of global capitalism map its presence in different ways. They use different means of locating its presence. For Fredric Jameson, one of the defining characteristics of global capitalism is simply that it is hard to locate. Global capitalism's labyrinthine complexity makes everything less and less legible, less and less easily read. This is a world where electronically generated images prevail. As a result it is a world which is increasingly difficult to touch or tie down; a world of confused senses which can only sense confusion. This brave new world of global capitalism has nefarious consequences. First, the *effects* of power are all too obvious – poverty, famine, war, disease – but the exact *causes* of oppression are more and more difficult to discern. The Four Horsemen of the Apocalypse still stalk the world but all we can see are the hoof-prints. Secondly, the modern city becomes suspended in a global space "in which people are unable to map (in their minds) either their own position or the urban totality in which they find themselves" (ibid: 51). Thus the observer experiences a kind of vertigo, a sense of an unseen abyss over which humanity teeters. Jameson's answer is to call for "new cognitive maps" that will help us to retreat from the edge.

The barometer of hypermobility is best expressed in the work of Manuel Castells, who identifies a new type of economic space – a mobile space of flows – which is the precondition for the coming into existence of a worldwide informational economy: "the enhancement of telecommunications has created the material infrastructure needed for the formation of a global economy, in a movement similar to that which lay behind the construction of the railways and the formation of national markets during the nineteenth century" (Castells 1993: 20). This space of flows "dominates the histori-cally constructed space of places, as the logic of dominant organizations detaches itself from the social constraints of cultural identities and local societies through the powerful medium of information technologies" (Castells 1989: 6). Increasingly, in other words, electronic trade winds blow across the globe, creating a new economic atmosphere.

Finally, the barometer of time-space compression is nowadays usually associated with the work of David Harvey, who uses the idea in two main ways: first, to express a marked increase in the pace of life brought about by innovations like modern telecommunications and the effects that this has on the topology of human communication – a seeming collapse of space and time; and, secondly, to signal the subsequent upheaval in our experi-ence and representation of space and time that this speed-up brings about; "time-space compression always exacts its toll on our capacity to grapple with the realities unfolding around us" (Harvey 1989: 306). Just as in Jameson, there is a call for cognitive maps that can be used to navigate "through a period of excessive ephemerality in the political and private as well as the social realm" (ibid; see also Harvey 2000).

These three accounts share much in common. Each is concerned with drawing out the lineaments of fundamental transformations in economies, societies, and cultures – and in the nature of time and space. Each displays a considerable degree of apprehension about our ability to comprehend these transformations. Each places much of the blame for this state of af-fairs on the roadrunner pace of modern life, which blurs our understand-ing. But each of them also believes that it is possible to find a theoretical space from which one can look out and explain what is going on.

What ought we to make of these accounts? I want to make a critique in two stages. First of all, I want to suggest that these accounts are simply the latest manifestation of a tradition of thinking that goes back a long, long way into history. They need to be taken with "a large pinch of *déjà vu*" (Porter 1993: 16). Secondly, I want to suggest that these accounts are in danger of constructing global capitalism as a more abstract system than it actually is. I will show this by reference to the world of international money and finance.

Critique 1: Antique Barometers?

Each of the three accounts briefly outlined above has a long history. In-deed, they go back so far in time that they may even have reached their historical sell-by date. To illustrate this point I will take each account in turn.

The debate over the idea that the world, as it becomes increasingly globalized, has become increasingly illegible has exact resonances in nineteenth-century reactions to the expanding metropolis as a disconcerting mixture of multiplicity, movement, and decenteredness (Prendergast 1992). These reactions were threefold. There was nostalgia for the old days, a nostalgia which was not much more than a demand for a return to the more secure and hierarchical social taxonomies of the past where everyone could be located in their "proper place." There was the idea that the city could be controlled through new forms of visualization which systematically refused to see significant forms of difference, division, and conflict – as in many Impressionist paintings. Then, lastly, there was a simple flight from the city, back into a "rural" world of psychic peace and soothed subjectivities. In each case, these reactions still exist. Further, they still exert their charms – look only at the content of some modern television advertisements.

Again, the account of a space of flows is growing hoary with age. It dates from at least the eighteenth century, and ideas of "circulation" – of desires and letters circulating in the body of the nation-state. But it comes into its own in the nineteenth century with the spread of the railway and then the telegraph (Thrift 1990). In France, for example, writers used it as a convention to describe the new spaces of continuous movement and circulation that were springing up as a result of these innovations. Modern life is drawn in terms of speed and flow – everything moves too fast. In the twentieth century the innovations may change but the phenomenology of speed and flow remains much the same.

This same phenomenology can be found in the account of time-space compression. One side of Harvey's account, the annihilation of space by time, was a favorite meditation of the early Victorian writer: "it was the topos which the early nineteenth century used to describe the new situation into which the railroad placed natural space after depriving it of its hitherto absolute powers. Motion was no longer dependent on the conditions of natural space, but on a mechanical power that created its own new spatiality" (Schivelsbuch 1986: 10). An article published in the *Quarterly News* in 1839 exactly captures the sense of struggle that resulted:

> For instance, supposing that railroads, even at our present simmering rate of travelling, were to be suddenly established all over England, the whole population of the country would, speaking metaphorically, at once advance en masse, and place their chairs nearer to the fireside of their metropolis by two thirds of the time which now separates them from it: they would also sit nearer to one another by two thirds of the time which now respectively alienates them. If the rate were to be sufficiently accelerated, this process would be repeated: our harbours, our dockyards, our towns, the whole of our rural population, would again not only draw nearer to each other by two thirds, all would proportionally approach the national hearth. As distances were thus annihilated the surface of our country would, as it were, shrivel in size until it

became not much bigger than one immense city. (Cited in Schivelsbuch 1986: 34)

The idea of the annihilation of space by time was recycled by writers like Marx later in the nineteenth century, surfaced again in geography textbooks of the 1920s and 1930s, and was then resurrected once more in the geography of the 1960s and 1970s as the phenomenon of "time-space convergence." The other side of Harvey's account, the effects that the upheaval of our experience of space and time have on our powers of representation and, by implication, identity, has clear resonances with ideas that date from at least the eighteenth century that the increased pace of life would lead to a kind of general hysteria in society (Porter 1993): time-space compression leads to time-space depression. It is the kind of depiction of volatile, fragmented subjects for volatile, fragmented times which Virginia Woolf captured so brilliantly in the 1920s in her discussion of the "atomism of the city" which is staged not only as "a problem of perception" but also as one "which raised problems of identity":

> After twenty minutes the body and mind were like scraps of torn paper tumbling from a sack and, indeed, the process of motoring fast out of London so much resembles the chopping up small of identity which precedes unconsciousness and perhaps death itself that it is an open question in what sense Orlando can be said to have existed at the present moment. (Woolf, 1926, cited in Prendergast 1992: 193)

Of course, just because these three accounts are starting to show their age does not mean that they are without a certain kind of narrative power. They make for a wonderful modernist detective story, full of mystery (illegibility), spirit (the space of flows), and pace (time-space compression). But do these three barometers of modernity convincingly represent the modern world? This is not a question that we can answer directly. What we can say instead is that they produce a partial representation which does not recognize its own partiality. Clearly the world has become more difficult to understand as it has become more and more complex. Certainly, there is an electronic space of flows. Of course, the world has speeded up. But because the world is more difficult to understand doesn't mean that nothing is understandable. The space of flows doesn't reach everyone. Speed isn't everything. The story is as much in what is missing from these accounts as in what is there.

These points can be made better by considering a real example: the world of international money, which forms the second part of this chapter. This world is often regarded as the most telling example of a brave new world of flows: abstract, complex, instantaneous. But what do we find? Nothing of the kind.

Critique 2: Hooked on Speed

Take 1: Masters of the universe?

An image which has become a cliché. A foreign exchange dealing room of a major bank in London, New York, or Tokyo. The mainly young men and women who inhabit these rooms for ten or eleven hours at a time are under pressure. They are under pressure to make profits. They are under pressure from their fellow traders – they don't want to lose face by screwing up a deal. They are under the pressure of constant surveillance – from managers, from video cameras, from tape recorders capturing all their calls. Above all, they are under pressure of time. Dealing itself is largely a matter of timing and dealers are expected not only to make profits from their deals but to make them quickly. You are only as good as your last deal.

To cap it all, these dealers are at the sharp end of time-space compression. Their world is a world where telecommunications have become more and more sophisticated and, as a result, space has virtually been annihilated by time. A dealer's world consists of a few immediate colleagues, the electronic screens which are the termini of electronic networks that reach round the world to other colleagues in other cities, and the electronic texts that can be read off the battery of screens (figure 2.1). If there is a space of flows, then this is it.

Certainly, what these dealers can do to the world is, in its own way, quite extraordinary. Over $300 trillion (or $300,000,000,000,000) is traded on average each year in the world's currency markets (Davis, 1998). On any working day that amounts to $1,269 billion, "a ratio to world trade of nearly 70/1, equal to the entire world's official gold and foreign exchange resources" (Eatwell and Taylor 2000: 4). To provide some kind of perspective, "it took the United States over 200 years to amass a government debt amounting to $1 trillion" (ibid.).

And what is the purpose of these almost incomprehensible sums? In effect, they are a part of a large global casino – much of the currency speculation is between banks and other financial institutions for the purpose of speculation. The bulk of trades are made in hope of capital gain, to seek gains from arbitrage, or to hedge against potential capital loss. The volume of trading is such that national governments – whose national currencies are after all the commodities being traded – find it very difficult to intervene successfully. Indeed, such is the power of the markets that they often have to work hard to produce economic policies that will sustain "market credibility" in a system which is increasingly premised on the judgments made by the markets. As one dealer (cited in Kahn and Cooper 1993: 10) put it, "If a government steps out of line they get their currency whacked." Plenty of governments can attest to this statement, as the 1990s showed almost year by year. The European Exchange Rate system (ERM) was introduced as a means of transforming the European Union into a zone of monetary

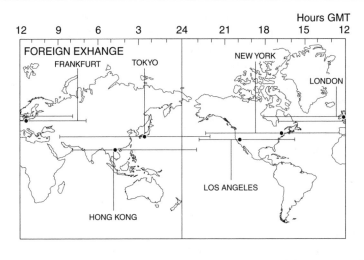

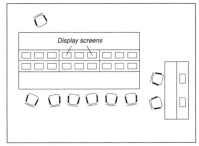

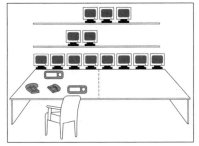

Part of a London foreign exchange dealing room: the "Number One" desk

Displays available to the "Number One" desk

Figure 2.1 Foreign exchange trading.

stability in which the exchange rate of the different European currencies could vary in an ordered and predictable way. But in 1992 and 1993 it came under a series of speculative assaults from currency traders. These assaults were aimed at forcing currency devaluations which would then guarantee windfall gains for the traders when the EU central banks intervened to prop up their currencies. The results of these speculative aspects were devastating. All but one ERM currency was devalued against the Deutschmark, the anchor currency of the system. Two currencies – sterling and the lira – were forced to leave the system altogether, and the rules of ERM membership were relaxed to such a degree that it became a pale shadow of its former self. This was just the first in a series of financial crises in which currency traders played an important part during the 1990s: the Mexican crisis of 1994–5, the Asian crisis of 1997–8, the Russian crisis of 1998, the Brazilian crises of 1999, all were in part the result of speculative assaults. Traders engaged in financial feeding frenzies "created by highly liquid capital being traded in huge volumes in an ever-expanding complex of markets for an evolving portfolio of instruments" (Eatwell and Taylor 2000: 5). Now it may be that the economic fundamentals of some of these countries were weak – in some cases very weak – but they hardly needed to be kicked when they were down.

Examples of this kind might be used to suggest that international money can flow where it will without let or hindrance – in other words, we see here a true space of flows. But like those apparently fraught young men and women in the dealing room (according to Kahn and Cooper 1993, they apparently actually suffer less stress than nurses dealing with the mentally handicapped), this is something of an exaggeration. First of all, the "phantom state" of international money is a nomad state (Thrift and Leyshon 1994) – it has no permanent spaces to call its own, only a series of transient sites in a few global cities. This constant mobility has its advantages. In particular, the world of international money is difficult to tie down. But it also has its disadvantages. The phantom state is always in danger of being trapped by nation-states which control territories, and are able to regulate what goes on within them. Thus this space of flows can be choked off by the rules nation-states impose – like capital adequacy ratios, which force banks to set aside a certain portion of their capital, or rules on how or what financial instruments to use. And it can also be restrained by developments like the establishment of the Euro zone on January 1, 2000, which in part is an attempt by ten (now eleven) EU countries to eliminate foreign exchange risk in Europe. The European Union, burnt by the speculative damage done to the ERM in 1992 and 1993, created a new currency, the Euro, which results from ten currencies having their parities fixed "forever." The new currency is managed by a European central bank that has total authority over monetary policy within "Euroland." But the difficulties which even a close-knit and committed grouping like the EU have faced in creating supranational economic institutions through which to manage the new supranational currency also serve as an example of the difficulties which still plague attempts to regulate the currency markets. Secondly, the phantom state has to be constantly in motion, chasing into all the nooks and crannies of the world economy that might produce a profit. Such a task requires an enormous investment of not only money but also communication. Nowadays money is essentially information, and getting that information, interpreting it, and using it at the right time requires constant human interaction. The result is that the hypermobile world of international money has become a rather strange amalgam of different forms of interaction. On one level, this is now a "post-social" world (Knorr-Cetina 2001), with new forms of togetherness based upon the minute, minute-by-minute interaction of human and information on a screen – in which software may well be becoming a more important part of the environment than humans; on another level, at one and the same time, this is also a "hypersocial" world, since to feed the voracious appetite for information also requires constant interchange between people, whether over electronic networks, or in face-to-face meetings, or at the end of often lengthy journeys. In this sense this world of flows is not abstract at all – it is the product of and it is produced by people communicating about what is going on.

So the barometers of illegibility, of a space of flows, and of time-space compression can clearly be seen to be partial, even when an example is chosen which should cast them in a favorable light. But one last point needs to be made, that repeats the historical lessons of the first section of this chapter: for the denizens of the world of international money these barometers do not represent some new condition. They are practiced in living with them. Since the international financial system has been in operation, its practitioners have had to live with uncertainty, using only limited information to assess the risks they run in investing money. Since international financial markets started to coalesce in the late nineteenth century, because of the telegraph and then the telephone, their practitioners have become well versed in living with time-space compression. It is all part of the game they play every day and it is a game they are good at.

Take 2: Networks and ghettos

The dealers in the international financial system live life in the fast lane. But what about those of us waiting for the bus or cursing the late train? For us, too, monetary transaction is speeding up. The installation of credit cards, automated teller machines (ATMs), and the like, means that life in the slow lane is moving faster (figure 2.2).

> I'm in Paris, it's late evening, and I need money quickly. The bank I go to is closed . . . but outside is an ATM . . . I insert my ATM card from my branch in Washington DC and punch in my identification number and the amount of 500 francs, roughly equivalent to $300. The French bank's computers detect that it is not their card, so my request goes to the Cirrus system's inter-European switching centre in Belgium which detects that it is not a European card. The electronic message is then transmitted to the global switching centre in Detroit, which recognizes that there's more than $300 in my account in Washington and deducts $300 plus a fee of $1.50. Then it's back to Detroit, to Belgium and to the Paris bank and its ATM and out comes $300 in French francs. Total elapsed time: 16 seconds. (*The Economist*, September 6, 1995: 100)

Increasingly, we are all dependent on the speed and processing power of telecommunications that examples like this illustrate. But the new space of telecommunications is not, in reality, a smooth global space over which messages can flow without friction. It is a skein of *networks* which are "neither local nor global but are more or less long and more or less connected" (Latour 1993: 122). To think otherwise is to mistake length or connection for differences in scale level, to believe that some things (like

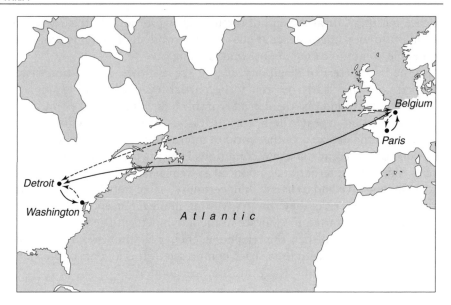

Figure 2.2   Paris to Detroit and back.

people or ideas or situations) are "local" while others (like organizations or laws or rules) are "global."

> The late twentieth century is covered by a lattice of networks. Public and private, civil and military, open and closed, the networks carry an unimaginable volume of messages, conversations, images and commands. By the early 1990s, the world's population of 600 million telephones and 600 million television sets will have been joined by over 100 million computer workstations, tens of millions of home computers, fax machines, cellular phones and pagers. (Mulgan 1991: 6)

But when we see this world of telecommunications as indeed a world of networks, we can also see other things. First of all, telecommunications networks still rely on many hundreds of thousands of people and machines all around the world who build, monitor, repair, and use them, just like the navvies and the telegraph engineers did of old. Second, modern telecommunications networks may be hybrid systems, in which machines and people are increasingly mixed together in queer combinations, but this does not make them more abstract or more abstracted. They break down. They stutter. They pause. They make errors. Whether it is a case of a vandalized ATM, a line fault, or an atmospheric disturbance, these networks are not self-sustaining. Third, these networks can be organized in different ways which are more or less effective (Mansell 1993). Fourth, and most importantly, not everyone is connected to these networks. The new telecommunications networks have produced "electronic ghettos" (Davis 1992) in which the only signs of globalization that can be found flicker across television screens, endlessly mocking their viewers by producing in front of them the lives and possessions that they will never be able to obtain. This is not even life in the slow lane. It is life on the hard shoulder.

In the electronic ghettos the space of flows comes to a full stop. Time-space compression means time to spare and the space to go nowhere. It is all horribly legible – as the example of the South Central area of Los Angeles shows only too well. There, access to normal monetary transactions and credit, represented by a network of facilities like ATMs, bank branches, and the like, is in decline as these facilities have been shut down by banks whose bottom line is under pressure, further fueling South Central's economic decline (Dymski and Veitch 1996; Pollard 1996). Its inhabitants are forced back on an informal system of check-cashing services, mortgage brokers, credit unions, and cash. Yet just a few miles from South Central are the recently constructed corporate towers of Los Angeles's financial district, a place with all the necessary connections with telecommunications networks that are long and interconnected all around the world. It is a district that is nearer in this network time and space to New York, or London, or Tokyo, than it is to South Central.

Conclusions

When we actually look at the space of flows, instead of taking it as a given, we find nothing very abstract or abstracted. What we find, as the example of the world of money shows only too well, is a new topology that makes it possible to go almost anywhere that networks reach (but without occupying anything but very narrow lines of force). But what we also find is a system where people are often in interaction with only four or five other people at a time – on a trading floor, or in a bank branch. What we find, in other words, are networks that are always both "global" and "local."

The problem is that writers like Jameson, Castells, and Harvey, with their ideas of a global capitalist order typified by barometers such as an illegible globalization, a space of flows, and time-space compression, are in constant danger of simply reproducing modernist views of an increasingly frantic commodified world filled with decontextualized rationalities like capitalism, various kinds of organizational bureaucracy, and markets – rationalities which are soulless and relentless and shiveringly impersonal. But things, as they say, ain't necessarily like that:

> An organization, a market, an institution are not supralunar objects made of a different matter from our poor local sublunar relations. The only difference stems from the fact that they are hybrid and have to mobilize a greater number of objects for their descriptions. The capitalism of Karl Marx or Fernand

> Braudel is not the total capitalism of (some) Marxists. It is a skein of longer networks that rather inadequately embrace a world on the basis of points that become centres of profit and calculation. In following it step by step, one never crosses the mysterious lines that should divide the local from the global. The organization of American big business described by Alfred Chandler is not the organization described by Kafka. It is a braid of networks materialized in order slips and flow charts, local procedures and special arrangements which permit it to spread to an entire continent so long as it does not cover that continent. One can follow the growth of an organization in its entirety without ever changing (scale) levels and without ever discovering "decontextualized" rationality . . . The markets . . . are indeed regulated and global, even though none of the causes of that regulation and that aggregation is itself either global or total. The aggregates are not made from some substance different from what they are aggregating. No visible or invisible hand suddenly descends to bring order to dispersed and chaotic individual atoms. (Latour 1993: 121–2)

In other words, what we need to produce geographies of global change is less exaggeration and more moderation.

But what would the more modest accounts look like? There are three closely related ways in which we need to ring the changes. First of all, we need to change the way that we do theory. In particular, that means recognizing that large-scale changes are always complex and contingent. They must be seen as

> a multiplicity of often minor processes, of different origin and scattered location, which overlap, repeat, or imitate one another, support one another, distinguish themselves from one another according to their domain of application, converge and gradually produce the blueprint of a general method. (Foucault 1977b: 138)

Secondly, we need to recognize that all networks of social relations, whether we are talking about capitalism, or firms, or markets, or any other institutions, are incomplete, tentative, and approximate. They are constantly in the process of ordering a somewhat intractable geography of different and often very diverse geographical contexts: "And ordering extends only so far into that geography. The very powerful learn this quickly" (Law 1994: 46). Accounts of these networks therefore need to recognize that it is a struggle to keep them at a particular size. There is no such thing as a scale. Rather, size is an uncertain effect generated by a network and its modes of interaction. A network of social relations has no natural tendency to be a particular size or operate at a particular scale: "some network configurations generate effects which, so long as everything is equal, last longer than others. So the tactics of ordering have to do, in general, with the construction of network arrangements that might last a little longer" (ibid: 103). Usually, this trick depends upon the invention and use of materials that can be easily carried about and retain their shape, "immutable mobiles" like writing, print, paper, money, a postal system, cartography, navigation, ocean-going vessels, cannons, gunpowder, xerographics, computing, and telephony. Electronic telecommunications can be interpreted as the latest of these mobiles, yet another means of allowing certain networks of social relations to retain their integrity by ordering distant events. Accounts of these networks also need to recognize that new forms of connection produce new forms of disconnection. We will never reach a totally connected world. As the example of the new electronic ghettos shows, new peripheries are constantly being created. Thirdly, and finally, we need to beware of confusing our theoretical ambitions with the reality that we are located *in* the world. Too often, the works of authors like Jameson, Castells, and Harvey seem to assume that there is a place, like a satellite, from which it is possible to get an overview of the whole world. Yet, as numerous feminist commentators have made clear, this assumption now looks increasingly like a classic masculine fantasy, a dream of being able to find a vantage point from which everything will become clear and a way of refusing to recognize that the world is a mixed, joint, commotion (Morris 1992; Mack 1990). Once we see that this assumption is a fantasy then we are also able to take account of all the subjects which in the past were often considered to be, somehow, "local" – like gender, or sexuality, or ethnicity – which recent work has shown are crucial determinants of global capitalism. In other words, one could understand "global capitalism" better and, at the same time, realize that not everything can be explained by studying "global capitalism" (Walker 1993).

So now how might we see barometers like illegibility, the space of flows, and time-space compression? To begin with, we can recognize that a globalizing world offers new forms of legibility which in turn can produce new forms of illegibility. For example, electronic telecommunications provide more and new kinds of information for firms and markets to work with, but the sheer weight of information makes interpretation of this information an even more pressing and difficult task (Thrift 1994). The space of flows is revealed as a partial and contingent affair, just like all other human enterprises, which is not abstract nor abstracted but consists of social networks, often of a quite limited size even though they might span the globe. Finally, time-space compression is shown to be something that we have learned to live with, and are constantly finding new ways of living with (for example, through new forms of subjectivity). It might be more accurately thought of as a part of a long history of immutable mobiles that we have learnt to live through and with. After all, each one of us is constructed by these "props," visible and invisible, present and past, as much as we construct them.

For those looking for big answers to the big questions that challenge us now in a globalizing world, this level of provisionality may all seem a bit frustrating. Yet the history of the last hundred years or so suggests that big answers founder when they come up against the messy, contingent world that we are actually landed with. Worse than this, the big answers sometimes become a part of the problem, as their proponents force order on the world by applying ordered force. In other words, the same history suggests that, if we are going to try to clean up the uglier bits of our messy, contingent world, big answers are not the solution. Of course, this means that our actions are likely to be modest and sometimes mistaken, but the fact of recognizing this state of affairs is in itself empowering; it means that the future is open and we can all do something to shape it. There is positive value to be gained in "striving to incorporate the problems of coping with that openness into the practice of politics" (Gilroy 1993: 223). To put it another way, there is nothing definite about the geographies of global change, and we do not have to be definite to change them.

3 Trading Worlds

Peter Dicken

Introduction

One of the most graphic media images of the final days of the last millennium was of the street riots in the US city of Seattle, focused on the meeting of the World Trade Organization (WTO). There, an immensely diverse conglomeration of interest groups – largely, and significantly, organized through the internet – came together to protest about the proposal to initiate a new "round" of international trade negotiations. "After Seattle" there can no longer be any doubt that international trade – as a component of an increasingly contested process of "globalization" – has become an immensely controversial issue. However, as subsequent demonstrations in Washington, Melbourne, and Prague have shown, opinions across the world are not simplistically polarized but, rather, are multiply constructed and articulated by a cacophony of interest groups of widely differentiated ideological and material complexions. Trade – and its effects on jobs, incomes, social groups, the physical environment – has become a *big issue* of the day.

The aims of this chapter are to unravel some of the complexity of the geography of international trade, its underlying processes and the debates surrounding it, especially in the context of the ongoing disagreements on how international trade should be regulated. The discussion will be organized as follows. First, I will provide a brief historical perspective on the evolution of international trade within the world economy before focusing, in the second section, on its contemporary structure and geography. The third section explores the regulatory bases of international trade at both the national and international scales before concluding, in the final section of the chapter, with an exploration of the contentious question of whether international trade is a "good" or a "bad."

International Trade in Historical Perspective

If trade is not quite as old as the hills it is at least as old as the earliest forms of economic specialization. Trade – the exchange of goods and services between producers and consumers – is the direct and inevitable outcome of specialization. Once an individual, a family, or a community is unable, or chooses not, to meet all its material needs through its own efforts then some form of exchange becomes necessary. The two processes – of specialization and of exchange/trade – are mutually constitutive and mutually reinforcing. Increased specialization of production must be matched by the growth of trade so that an individual's or a community's material needs are satisfied. Conversely, as trade grows and as its infrastructure becomes more sophisticated and extensive, a further stimulus to specialization tends to occur. The fundamental significance of trade to material life is reflected in the long and complex history and geography of its regulation. The archeological record shows that maritime-based trading was already in existence around the Mediterranean by 7,000 BC and that some form of regulation of trade was in place at least 2,000 years ago (Braithwaite and Drahos 2000: 175). Of course, for much of human history, most trade was geographically restricted by prevailing means of transport and communication. Trade was overwhelmingly *bimodal*. On the one hand there was *local* trade: short-distance, mainly concerned with basic necessities; the kind of trade focused upon the medieval market towns. On the other hand, there was a very small volume of *long-distance* trade mostly involving trade in luxury items for a very tiny segment of the population.

Although the extensity and intensity of trade were undoubtedly constrained by the costs of, and time involved in, physical movement, they were also constrained by the prevailing fragmented political organization of geographical space. In this regard, the key development was the emergence of the nation-state in sevententh-century Europe. Indeed,

> A major reason why the early nation-states flourished is that they created pacified spaces where trade could flow more freely. It became more possible to travel with goods from one town to another without fear of robbers. Equally importantly, tolls and entry restrictions on the flow of goods between cities were progressively dismantled. Technical standards became more standardized, avoiding many technical barriers to free intranational trade. By the nineteenth century nation-states controlled almost all of the world's territory and trade was mostly free within states . . . States persisted with the same barriers to trade between states as they had dismantled within their borders. In fact, they expanded them. Tariffs – taxes on imports – grew ever higher and became a major source of revenue for states. They were in the grip of mercantilism – the principle of selling more to foreigners than they sell domestically lest a balance of payments deficit leave the state with an inadequate supply of circulating gold or silver. (ibid: 175–6).

International trade expanded dramatically, especially during the nineteenth century, as both the changing technologies and demand growth associated with the process of industrialization created unprecedented flows of materials and finished products at increasing geographical scales. Between 1870 and 1913 world trade grew at an annual average rate of 3.5

percent compared with growth in world production of 2.7 percent per year (Kitson and Michie 1995: 6). Such a differential growth of production and trade is a clear indicator of the increasing international interconnectedness in the late-nineteenth-century world economy. However, the geographical structure of that economy was highly distinctive: it was essentially a *core–periphery* system in which international trade was driven by the core developed economies of northwest Europe and the United States. In this system, the core produced virtually all the world's manufactured products while the periphery acted as both a supplier of raw materials and foodstuffs and as a market for some of the core's manufactured products (Dicken 1998). In both 1876 and 1913, for example, roughly 45 percent of world trade was between developed economies, a little over 50 percent was conducted between developed and developing economies, and only 5 percent was between developing economies (Held, McGrew, Goldblatt, and Perraton 1999: 156). However, the expansionary trajectory of international trade was abruptly reversed in the period between World War I and World War II both by much slower growth and by the increased protectionism that reached its peak in the world depression of 1929–32, while the economic recovery that began in the mid-1930s was rudely interrupted by the outbreak of World War II in 1939. The historical record (see Maddison 1995) shows that the degree of trade integration in the world economy (measured as the ratio of total world exports to total world GDP) increased progressively from around 1820 to 1929 and then was reversed in the period 1929–50. Since 1950, as the next section shows, the upward trajectory resumed, although in a highly irregular temporal pattern.

The Contemporary Map of World Trade

Today's map of world trade – its geographical configuration, the ways in which it is regulated, and the arguments about its costs and benefits – is the outcome of a complex web of political, economic, and technological processes that have evolved since the end of World War II (Dicken 1998). The broad contours of the global economic map that had developed over the previous two hundred years (broadly structured around a core and periphery of national economies) had persisted until the 1939–45 war. In 1938 no less than 71 percent of world manufacturing production was concentrated in just four industrialized countries and almost 90 percent in only 11 countries. This small group of core industrial economies sold two-thirds of its manufactured exports to the periphery and absorbed four-fifths of the periphery's primary products. World War II sliced through this long-established and relatively stable geographical structure so that the world economic system that emerged after 1945 was, to a very great extent, a new beginning. It reflected both the new political realities of the postwar period – particularly the sharp division between the US-led and the Soviet-led systems – and also the harsh economic and social experiences of the 1930s. The kinds of international economic institution devised in the aftermath of war – the IMF, World Bank, and GATT – grew out of these specific circumstances. In this section, two aspects of global trade will be discussed: its temporal volatility and its geographical (re-)configuration. How trade is regulated and the question of its costs and benefits will be explored in subsequent sections of this chapter.

World trade as a roller-coaster

By 1990 the "globalization" of trade (the ratio of exports to GDP) had reached unprecedented levels, a clear indication of the high and growing degree of economic interconnection between countries. The previous peak of such integration in 1929 had been surpassed by the early 1960s. By this time, the average annual growth of world trade was around 8.5 percent; a level never before experienced. Indeed, the 1950s to the early 1970s were universally seen at the time as the "golden age" of world economic growth although, as Webber and Rigby (1996: 6) point out, "the golden age was only partly golden: it was more golden in some places than others, for some people than others." However, as figure 3.1 shows, the subsequent trajectory of world trade – particularly after the 1973 and 1979 oil crises – became extremely volatile. During the 1980s and 1990s, rates of growth were especially variable, with periods of trade recession (1980–2) or relatively slow growth (early 1990s, late 1990s; the latter associated with the East Asian economic crisis) contrasting with extremely high growth (1984, 1988, 1994, 1997).

The changing organization of world trade

This intensification of world trade during the post-World War II period was driven primarily by the very rapid growth of trade in manufactures. Only very recently have services become a significantly traded sector (most services, by their very nature, are not tradable but require face-to-face exchange). Manufacturing came to account for an increasingly greater proportion of total exports from both developed and developing countries, while the relative importance of trade in primary commodities declined overall. By the late 1980s almost 80 percent of developed countries' exports were of manufactures (compared with 70 percent in 1960). The sectoral shift in trade composition was even more marked among developing countries. As a group, the developing countries saw manufacturing exports grow from 20 percent of the total in 1960 to almost 50 percent by the end of the 1980s. Indeed, by the late 1970s the value of manufactures exported from developing countries, for the very first time, exceeded that of food and raw

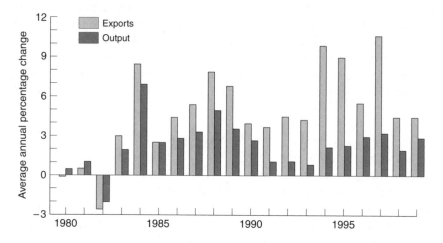

Figure 3.1   The roller-coaster of world trade, 1980–99.

materials. This, as much as any other single indicator, signaled the emergence of a new international division of labor in which manufacturing production was no longer the monopoly of the core economies.

A second shift in the configuration of world trade has been the increased significance, especially for developed economies, of *intra-industry* trade, that is, trade between countries in essentially the same kinds of product rather than in very different products. By far the largest component of all developed countries' trade is now intra-industry trade. For example, in the early 1960s around 45 percent of most developed countries' trade was intra-industry; by 1990 the share had risen to between 70 and 85 percent. Such a trend is far less evident among developing countries, although in some of the newly industrializing economies (NIEs) it is becoming more apparent.

A third significant characteristic of world trade in recent years is directly connected to the growth and proliferation of transnational corporations (TNCs) as the primary shapers of the global economy (Dicken 1998). Two characteristics are especially important. First, TNCs now dominate world trade: the United Nations estimates that around two-thirds of world exports of goods and services are generated by TNCs. Second, a very significant proportion of that trade is *intra-firm* trade; in other words it is trade that takes place across national boundaries but *inside* the boundaries of the firm as transactions between different parts of the same firm. Unlike the kind of trade assumed in international trade theory (and in national trade statistics) intra-firm trade does not take place on an "arm's length" basis in external markets, but on the basis of firms' decisions within their own internal markets. It is estimated that approximately one-third of total world trade is intra-firm trade. For individual economies, the share of their exports tied to the internal markets of TNCs can be very high indeed. In the 1980s, for example, it was calculated that roughly four-fifths of the UK's manufactured exports were intra-firm, either within UK enterprises with overseas operations or within foreign-controlled firms with operations in the UK.

The geographical configuration of world trade

Mapping the geography of world trade is a difficult task simply because, at least theoretically, every country can trade with every other country producing an immensely intricate network of inter-country flows of both imports and exports. In fact, trade flows tend to be channeled into certain dominant and persistent geographical patterns, as figure 3.2 shows. The overall structure is strongly *regionalized* in two specific respects. First, the three mega-regions of North America, Europe and East Asia – the so-called global triad – account for no less than 80 percent of total world merchandise exports. The second dimension of trade regionalization is the extent to which highly intensive trade flows occur *within* each major region. In the late 1990s, according to IMF data, 61 percent of the total exports generated within the European Union were intraregional, that is from and to other EU member states. In comparison, intraregional trade in North America (between the three NAFTA countries of the United States, Canada, and Mexico) was just under 50 percent, while that within East Asia was around 53 percent. Clearly, the vast majority of world trade is focused on these three major regions while substantial parts of the world, notably Africa, parts of South Asia and of Latin America, remain largely peripheralized.

It has become common to attribute such high levels of intraregional trade primarily to political initiatives to create regional economic blocs, such as the EU or NAFTA. No doubt this is part of the explanation and, certainly, political economic integration leads to increased intra-bloc trade flows. But there is a more basic explanation: simple geographical proximity. Trade flows between neighboring countries are facilitated not only by lower transportation costs but also by other less tangible connections enhanced by proximity. So, even where there is no region-wide political trading bloc, a high level of intraregional trade may well develop in circumstances of economic vitality (as, for example, in East Asia). Such intraregional trade flows are greatly enhanced by the decisions of TNCs to create elaborate intraregional production networks within which flows of materials, components, finished and semi-finished products are generated both inside the TNC's own internal networks and between TNCs and their independent suppliers and subcontractors. Such regional organization of a firm's activities – what the former chairman of Sony, Akio Morita, termed *glocalization* – has become increasingly evident in recent years.

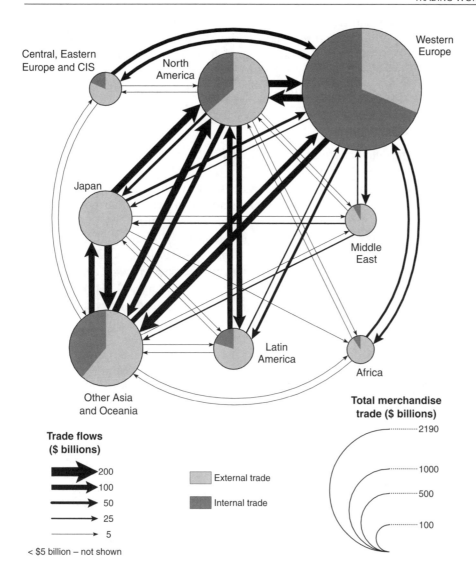

Figure 3.2   The network of world merchandise trade, 1995.

Within this broad regionalized framework of trade relationships the position of individual countries has changed substantially over the past four decades. Table 3.1 is what might be called the "world league table" of merchandise trade in 1963 and in 1999: the top fifteen exporting and importing countries. In both years, the United States followed by Germany were the dominant trading nations in terms of both exports and imports, although their share of world exports declined with the rise, in particular, of Japan

Table 3.1 The world league table of merchandise trade, 1963 and 1999.

Exports					Imports					
	1999		1963			1999		1963		Balance 1999
Country	%	R	%	R	Country	%	R	%	R	$bn
United States	12.4	1	17.4	1	United States	18.0	1	8.6	1	−364.9
Germany	9.6	2	15.6	2	Germany	8.0	2	6.2	2	+67.9
Japan	7.5	3	6.1	5	United Kingdom	5.5	3	4.6	6	−52.3
France	5.3	4	7.0	4	Japan	5.3	4	1.9	13	+108.7
United Kingdom	4.8	5	11.4	3	France	4.9	5	4.7	5	+12.9
Canada	4.2	6	2.6	12	Canada	3.7	6	5.0	4	+18.2
Italy	4.1	7	4.7	6	Italy	3.7	7	4.1	8	+14.8
Netherlands	3.6	8	3.3	9	Netherlands	3.2	8	4.3	7	+15.2
China	3.5	9	–	–	Hong Kong	3.1	9	0.9	–	−6.9
Belg-Luxg	3.3	10	4.3	7	Belg-Luxg	2.9	10	3.4	9	+14.7
Hong Kong	3.1	11	0.9	15	China	2.8	11	–	–	+29.2
South Korea	2.6	12	–	–	Mexico	2.5	12	–	–	−11.5
Mexico	2.4	13	–	–	Spain	2.5	13	1.2	–	−35.6
Taiwan	2.2	14	0.2	–	South Korea	2.0	14	0.3	–	+24.5
Singapore	2.0	15	0.4	–	Taiwan	1.9	15	0.2	–	+10.6
Total	70.6		73.9		Total	70.0		45.4		

Source: based on data in WTO *Annual Reports*, 1996, 2000

and, more recently, of the East Asian NIEs. In fact, the effect of the 1997 financial crisis in East Asia, and of the longer-standing economic recession in Japan, has somewhat reduced these countries' share of world exports. For example, in 1995, Japan's share of world exports was 11.6 percent compared with 7.5 percent in 1999. For the other leading East Asian economies the comparable figures were: Hong Kong (4.4/3.1); Singapore (2.7/2.0); South Korea (3.1/2.6); Taiwan (2.9/2.2). With the exception of Japan, however, these were quite modest declines, while China managed slightly to increase its share of world exports. The final column of table 3.1 shows the trade balance for each of the fifteen countries. Ten of the countries have trade surpluses of varying magnitude; by far the largest being Japan's surplus of $108.7 billion. In contrast, the United States continues to operate at a huge trade deficit of $365 billion. This stark contrast between the United States' trade deficit and the surpluses of Japan and some other, mostly East Asian countries, continues to be a source of major political tension and greatly influences the trade policy stance of the United States in its negotiations with other countries (see below).

Regulating World Trade: From National to International Systems

Historically, virtually all political power centers – from local tribal chiefs and feudal lords, through the city states of classical antiquity and of medieval Europe, to the nation-states that emerged in seventeenth-century Europe, and down to the present day – have erected systems to control or regulate the trade of commodities across their boundaries. *Protectionism*, in one form or another, has been a basic practice from the earliest times. Nationally, the clearest expression of this desire to regulate trade was the principle of mercantilism, a philosophy adopted particularly by the European maritime powers from the seventeenth century. Mercantilism was based on the notion that a nation's wealth and influence depend upon its ability to control its external trade at the expense of its rivals. It is based, in other words, on the assumption that trade is a zero-sum game; that one nation's gain from trade is another's loss and vice versa. Mercantilism as a philosophy was challenged in the early nineteenth century by the economist David Ricardo, who propounded the theory of comparative advantage. This principle states that a country (or any geographical area) should specialize in producing and exporting those products in which it has a comparative, or relative, cost advantage compared with other countries and should import those goods in which it has a comparative disadvantage. Out of such specialization, it is argued, will accrue greater benefit for all. Trade becomes a "win-win" rather than a zero-sum game.

To a large extent, the history of international trade regulation is one of tension between these two philosophies of national protectionism (of some kind and degree) and free trade. Generally speaking, free trade advocates tend to be the dominant economic powers while more protectionist positions tend to be held by weaker economies. Thus, when Britain was the dominant economic power in the late nineteenth century it was the leading advocate of free trade, while the then newly industrializing United States, along with Germany, for example, were pursuing a trade policy inspired by the German economist Friedrich List, which aimed to protect these countries' infant and emerging industries from unbridled free trade until they were robust enough to compete on world markets. After World War II the United States became the unchallenged global hegemonic power and, not surprisingly, the leading advocate of free trade. This has remained its position, although with the seemingly unstoppable economic growth of Japan and other East Asian economies during the 1970s and 1980s (based, to some extent, on Listian policies of economic development) a nascent neomercantilist movement emerged in the United States advocating a more "strategic" trade policy. Nevertheless, the notion of trade based on comparative advantage remains the dominant neoliberal ideology and underpins the structures of world trade regulation that have emerged and evolved since the end of World War II. But it has become increasingly recognized that a country's comparative advantage is not simply "given," as implied in the conventional economic term "factor endowments," but created most notably through the development of skilled human capital and technologies.

Until the late 1940s all trade regulation was *national*. Agreements were sometimes made between countries over specific commodities but these were essentially bilateral in scope. The "new international economic order" created in the aftermath of World War II had as one of its central pillars an *international* regulatory mechanism for trade. This was GATT (the General Agreement on Tariffs and Trade). Initially 23 countries belonged to GATT; today there are around 130 member states in its successor institution, the World Trade Organization (WTO). More countries, most notably China, are anxious to join. Today, more than 90 percent of world merchandise trade is now covered by the WTO.

GATT/WTO represents a *rule-oriented* approach to multilateral trade cooperation. Its fundamental basis is that of *non-discrimination* which incorporates two principles. First, the *most-favored nation* principle states that a trade concession negotiated between two countries must also apply to all other countries within GATT. Second, the *national treatment* rule requires that imported goods be treated in the same way as domestic goods. An important development occurred in 1965 with the inclusion of a *generalized system of preferences*, which granted preferential access of developing-country products to developed country markets (important exceptions were clothing and textiles). Negotiations to liberalize international trade have been conducted in a series of "rounds" which have occurred at irregular intervals. There have been eight such rounds since 1947, the most recent being the protracted and contentious Uruguay Round which began in 1986 and was finally concluded in 1994. GATT is generally regarded as being instrumental in the progressive lowering of tariff barriers to international trade. In 1940, the average tariff on manufactured products was approximately 40 percent; in the mid-1990s it was down to 4 percent. However, GATT was far less successful in constraining the proliferation of non-tariff barriers (such as import quotas).

The Uruguay Round was the most ambitious and wide-ranging of all the GATT rounds. For the first time, it incorporated agriculture, textiles, and clothing into GATT. Special agreements were concluded in services (GATS – the General Agreement on Trade in Services), intellectual property rights (TRIPS – Trade-Related Aspects of Intellectual Property Rights), and trade-related investment (TRIMs – Trade-Related Investment Measures). The major *organizational* development was the incorporation of GATT and the other agreements into a new World Trade Organization (WTO) which came into being in January 1995, almost fifty years after the original proposal to create an International Trade Organization foundered. During that period GATT had continued to be merely a "temporary framework."

However, the Uruguay Round left many issues unresolved. Substantial trade tensions continue to exist between, for example, the United States and the European Union in such commodities as bananas, hormone-treated beef, and GM food. More broadly, attempts by the WTO in late 1999 to initiate a new round of international trade negotiations foundered amid a chaos of internal and external political and ideological differences. Such conflicts bring into sharp focus not only the technical problems of trade regulation but also much deeper antagonisms concerning the whole basis of globalizing trade developments. It is to these issues that we turn in the final section of this chapter.

Is a More Open World Trading System a "Good" Thing or a "Bad" Thing?

The empirical evidence suggests that, in *aggregate* terms, the world economy has grown far faster when the trading system has been open than when it has been restricted. The evidence also suggests that many countries, communities, and individuals have gained very substantially from trade liberalization. From a producer viewpoint, markets have become geographically more extensive and the limitations of a specific domestic market removed. From a consumer viewpoint, there is now a far greater variety of commodities and goods available than ever before, many of them far cheaper in real terms. In the case of food, for example, products formerly available only at certain times of the year – or not available at all – are now available in the local supermarket all year round. All of this is consistent with the conventional economic view that trade that is based upon the principle of comparative advantage – specialization – leads to greater benefits for all. But, as so often, the devil is in the detail.

The short answer to the question posed in the heading of this section is: it depends upon who and where you are in the system. There are losers as well as winners from open trade: a gain from trade at one level of spatial aggregation generally hides losses at a lower level of spatial aggregation. A country as a whole may experience rising aggregate incomes, more jobs, and greater choice of commodities through trade, while at the same time some parts of that same country or some social groups within it experience lower incomes, fewer jobs, and less choice. Exactly the same argument applies at the international scale, as the widening "development gap" between developed and developing countries shows so very clearly. While most developed and some developing countries have benefited from international trade, a significant number of developing countries remain effectively marginalized from any benefits from world trade, primarily because they depend upon just one or two basic commodities in which the terms of trade have deteriorated.

Two major – and related – areas of dispute are currently the focus of attention in the trade and globalization protests. One is to do with the nature and accountability of the international institutions themselves; in the case of trade this is the WTO. Many developing countries feel themselves to be disadvantaged in the decision-making processes of the WTO. This is not so much a matter of formal representation but of the lack of resources available to poorer countries to support their claims within the WTO. This leads to a strong feeling of disempowerment and disadvantage compared with the developed economies, which are seen to dominate the world trade system. Conversely, there is a substantial political lobby within the United States which objects to the WTO being able to pronounce on trade matters affecting the US economy and sees it as an unwarranted interference in domestic affairs.

The second area of current tension relates to notions of "fairness" in international trade. Is "free" trade also "fair" trade? Well, again, it depends on positionality within the system. A *Business Week* poll in the United States in 2000 found that 51 percent of respondents saw themselves as "fair traders" while only 10 percent saw themselves as "free traders." A further 37 percent regarded themselves as "protectionist." But, without question, what the average US citizen regards as "fair" trade would not match that of many people in developing countries who regard themselves as being heavily disadvantaged through various trade barriers. This conflict of interest is especially well illustrated in two of the most contested trade-related issues of the present day: labor standards and the environment. The basic question is: to what extent do international differences in labor standards and regulations (such as the use of child labor, poor health and safety conditions, repression of labor unions and workers' rights) and in environmental standards and regulations (such as industrial pollution, the unsafe use of toxic materials in production processes) distort the trading system and create unfair advantages? In both cases the basic argument is that firms – as well as individual countries – may be able to undercut their competitors by capitalizing on cheap and exploited labor and lax environmental standards.

These two issues were explicitly addressed in the negotiations for NAFTA, in which the United States insisted on the signing of two side agreements to protect its domestic firms from low labor and environmental standards in Mexico. More recently, a group of countries led primarily by the United States but also including some European countries, has made a concerted attempt formally to incorporate the issue of labor standards into the WTO. So far the attempt has failed, partly because not all industrialized countries support it, but also because the developing countries are vehemently opposed. There is no doubt that stark differences do exist in labor standards in different parts of the world and basic workers' rights are denied in many countries. Working conditions, especially in the Export Processing Zones – but not only in these zones – are often appalling. As far as child labor is concerned, the International Labor Organization (ILO) calculates that around 73 million children aged between 10 and 14 years are employed throughout the world: approximately 13 percent of that age group. In Africa, one-quarter of children aged 10–14 are working. If children under 10 are included, as well as young girls working full-time at home, the ILO estimates that there are probably "hundreds of millions" of child workers in the world. However, the ILO also points out that 90 percent of children work in agriculture or linked activities in rural areas and that most are employed within the family rather than for outside employers. Even so, there is substantial evidence that, in many cases, young children are employed by manufacturers (whether in factories or as outworkers) in such industries as garments, footwear, toys, sports goods, artificial flowers, plastic products, and the like. Their wages are a pittance and their working conditions often abysmal.

From the viewpoint of many developing countries, however, there is a strong feeling that the labor standards stance of many developed countries is merely another form of protectionism against their exports and, as such, is an obstacle to their much-needed economic development. There is a suspicion, for example, that at least some of the developed country lobbies are pressing for international agreements on a minimum wage in order to lessen the low labor cost advantages of developing countries. By incorporating such labor standards criteria into the WTO framework, it is believed, developed countries could use trade regulations to enforce such indirect protectionism. There is clearly a basic dilemma for the international community. On the one hand, ethical considerations must be a basic component of international trade agreements; on the other hand, there is a real danger of threatened developed country interest groups using labor standards issues as a device to protect their own commercial interests.

A similar dilemma is central to the other trade-related question: that of the environment. To what extent should variations in environmental standards be incorporated into international trade regulations? Again, there is no doubting the existence of huge differences in the nature, scope, and enforcement of environmental regulations across the world. There is no doubt, either, that the highest incidence of low environmental standards is in developing countries. The existence of such an "environmental gradient" certainly constitutes a stimulus for some firms at least to take advantage of low standards. These may be domestic firms or they may be foreign firms.

But how does the environmental issue relate specifically to international trade? At one level, the problem is exactly the same as that of labor standards. If a country allows lax environmental standards, it is argued, then it should not be able to use what is, in effect, a subsidy on firms located there to be able to sell its products more cheaply on the international market. The question then becomes one of whether the solution lies in using international trade regulations or some other forms of sanction. However, there is an even more extreme position adopted by some environmentalists which is that the pursuit of ever-increasing international trade – which is clearly encouraged by a free trade regime like that of the WTO – should be totally abandoned, not merely regulated. The argument here is basically that *sustainable* development is incompatible with the pursuit of further economic growth and, especially, with an economic system which is based upon very high levels of geographical specialization, since such specialization – as we have seen – inevitably depends upon, and generates, ever-increasing trade in materials and products. One of the bases of this argument is that the energy costs of transporting materials and goods across the world are not taken into account in setting the prices of traded goods and that, in effect, trade is being massively subsidized at a huge short-term and long-term environmental cost.

At the start of the third millennium, therefore, there is no doubt that international trade, as a major component of globalization, has not only become immensely significant economically but also immensely contested politically. The issue facing us all is whether the benefits that have undoubtedly accrued to large parts of the world as a result of the unparalleled growth of international trade offset the costs. But, as we have seen, this is a most complex dilemma because the distribution of these costs and benefits is so uneven. One needs only to look at the incredibly diverse and often contradictory positions taken by the myriad of protest groups against the WTO to realize that, although they may be united at one level, many of their specific goals are contradictory. How, then, should the governance of international trade be organized? Should it continue to be the responsibility of an international body or should it be, once again, the responsibility of individual nation-states? If the latter, then how can the weak be protected? Some form of international rule-based organization seems to be the only sensible way forward. But should it be a reformed WTO or do we need to consider the creation of an entirely new organization? It is a political choice but one which will be very hard to make.

4 From Farming to Agribusiness: Global Agri-Food Networks

Sarah Whatmore

Introduction

Food is a basic requirement of human life, but for most people in the advanced industrial countries of Western Europe, North America, and Australasia today, it has become a taken-for-granted facet of daily consumption. Stacking a trolley in the supermarket is an everyday chore; getting a take-away, a commonplace convenience; eating out, an integral part of many business and leisure routines. These consumer experiences of food are quite profoundly distanced from the social and economic organization of agriculture and the contemporary processes of food production. Milk may still come from cows and apples grow on trees (don't they?) but how does farming, the anchor of commonsense understandings of food production, fit into the creation of oven-ready meals, genetically engineered plants and animals, or synthetic foodstuffs? The prevalent representation of such experiences as the mark of "consumer choice" belies a diminished understanding of, and control over, what it is that we are eating and the social conditions under which it is produced. Moreover, the language of "choice" is at odds with the still widespread realities of food scarcity and uncertainty faced by those without the money-income to secure their basic needs through the market, particularly in the so-called "Third World" and parts of Eastern Europe, but also among the growing numbers living in poverty in the West itself.[1]

These divergent experiences of the political economy of food are intimately connected; bound together in highly industrialized and increasingly globalized networks of institutions, technologies, and products, which have been described collectively as constituting an *agri-food system*. The OECD (Organization for Economic Cooperation and Development), for example, defines this system as "the set of activities and relationships that interact to determine what, how much, by what method and for whom food is

produced and distributed" (OECD 1981). This systemic view of globalization is most readily symbolized by the worldwide presence and cross-cultural potency of such food icons as the McDonald's hamburger or the Pepsi
Cola drink. However, it would be misleading to take these high-profile cases
as in any sense representative of the complex and highly uneven processes
of industrialization and globalization which have been reshaping food production and consumption in the postwar period. This chapter outlines some
of the key features of this restructuring process, focusing on the technological and socioeconomic transformation of food production, its impact on
farming, and the consequent remapping of the geographies of agriculture
emerging from recent research.

Shifting Research Horizons

In the 1980s and 1990s efforts to understand and inform these
transformative processes in terms of what has been called a "*new* political
economy of agriculture" became prominent in agri-food research; a term
reflecting both the changing organizational structure of capitalist agriculture in the postwar period and a refocusing of research questions in geography and other disciplines. As proponents saw it, "the present situation
is one in which the connotations of 'farming' – in particular, rurality and
community, but also other categories that are limited to national economies, nation-states, and national societies – are giving way to vertically
and horizontally integrated production, processing and distribution of
generic inputs for mass marketable foodstuffs" (Friedland 1991: 3–4).
The distinguishing feature of this "new" research agenda was that it took
the analysis of agriculture beyond the farm gate in two directions. The
first has been to look at the wider organization of *capital accumulation* in
the agri-food sector, focusing on the social, economic, and technological
ties between three sets of industrial activities, those of food raising (i.e.,
farming as a rural land-use); agricultural science and technology products
and services to farming (upstream industries); and food processing and
retailing (downstream industries). The second direction has been to look
at the role of the agri-food sector in the wider institutional fabric of *social
regulation* – the focus here being the political and policy processes by
which national and supranational state agencies underpin agricultural
markets by regulating the terms of trade and the food component of wage
costs in the wider economy. In simplified terms, these two research directions stress the dynamics of agri-food sector restructuring generated respectively by the search for profit by private capital, and by state concerns
with securing social order.[2]

Efforts to comprehend these expanded parameters of an agri-food system
beyond the farm have led researchers to adopt a number of unfamiliar terms
which flag a variety of conceptual approaches. Several such terms are in

current circulation in the literature and it seems appropriate to establish some reference points by beginning with a few basic definitions.

One set of approaches focuses on the reorganization of capital accumulation in the agri-food sector, tracing particular agricultural goods through the sequence of processes they undergo before reaching the consumer, and analyzing the social and economic relations within, and between, each stage in this sequence. An early version of this kind of approach adopted the concept of *agribusiness* to mean "the sum total of all operations involved in the manufacture and distribution of farm supplies; the production operations of the farm; storage, processing and distribution of farm commodities and items made from them" (Davis and Goldberg 1957: 3). In practice, this term applies to a particular business configuration, in which a single (usually transnational) corporation coordinates industrial activities in each of these spheres through subsidiary companies; a configuration known as vertical integration, and characteristic of the US agri-food sector in the 1960s and 1970s.

Another influential version of this kind of approach – the idea of agricultural *commodity chains* – emerged in the United States or, more precisely, in California in the late 1970s. It too emphasizes the industrial character of agriculture, treating agricultural products and businesses in the same way as those in, say, the steel or automobile industries.[3] A good example is the study of the Californian lettuce industry by Bill Friedland and his collaborators called *Manufacturing Green Gold* (Friedland, Banton, and Thomas 1981). Despite criticisms about the extent to which such terms can be generalized beyond forms of industrial agriculture particular to certain US contexts, such analytical models have become abstracted and widely used.

A second set of approaches places a quite different emphasis on the regulatory institutions which have evolved to underpin strategies of accumulation in the agri-food sector. There are two related concepts of importance here. The term *agri-food complex* was coined by Harriet Friedmann (1982) to describe the industrial relations of the production and consumption of specific foodstuffs which became dietary standards, such as beef, and canned, frozen, or otherwise "durable foods," in association with distinct *agri-food regimes*. This latter term signifies the regulatory apparatus sustaining world agricultural markets and food prices, as these articulate with the efforts of nation-states to regulate the social conditions of capital accumulation within their borders during particular periods in the development of the capitalist world economy (Friedmann and McMichael 1989). In this chapter we shall be most concerned with the consequences of the decline of the second of two "agri-food regimes" identified by Friedmann, which dates to the period 1945–73 and is associated with what has come to be known as a "Fordist" mode of capital accumulation and regulation.[4]

The concepts of agribusiness and, more particularly, agri-food "chains," "complexes," and "regimes" can be seen as complementary in the sense that they throw light on different aspects of the institutional structure of the contemporary agri-food system. Their composite perspectives are represented in a simplified way in figure 4.1.

Within geography these developments have proved difficult to locate because of established divisions between agricultural geography, with its traditional focus on agricultural land-use (i.e., farming), and industrial geography, which has a strong political-economy tradition but has tended to regard agriculture as an anomalous industrial sector outside its remit. As a result, the approaches and terms outlined above entered geographical research relatively late, in the mid-1980s, with a series of agenda-setting papers from North America (Wallace 1985), the UK (Marsden et al. 1986), and New Zealand (Le Heron 1988). While such approaches can be seen as the core of a reinvigorated agricultural geography, their real significance lies in their contribution to eroding the divisions between agricultural and other industrial geographies, and generating a more creative engagement with agri-food system research in other disciplines (see Marsden et al. 1996).

The Making of an Industrial Agri-Food System

Goodman and Redclift (1991) argue that "to understand accumulation in the agro-food system it is essential to recognize that industrialization has taken a markedly different course from other production systems" (ibid: 90). This distinctiveness is centered on the biological foundations of agricultural production in which the growth and reproduction of plant and animal life that are its products are bound to a series of "natural" processes which have resisted direct and uniform transformation by capitalist relations of production. These biological constraints on accumulation are not fixed but they have made agriculture as a land-based activity unattractive to the direct involvement of industrial capital and its characteristic business forms and divisions of labor. In addition, agriculture has long been seen to face a market handicap in that expenditure on food tends to fall as a proportion of income, as average incomes rise. As a result of these constraints, the capitalist restructuring of agriculture has centered on reducing the dependence of agri-food accumulation on "nature" and increasing the market size and value of agri-food products through a process of industrialization. This strategy has been pursued through the technological modification of biological processes in farming itself, and by valorizing agricultural products "off-farm" in the manufacture of technological farm inputs and the increased processing and packaging of food products after they have left the farm.

An influential study by Goodman, Sorj, and Wilkinson (1987) suggests that the rapid growth of these "off-farm" sectors has taken place through two discontinuous but persistent processes:

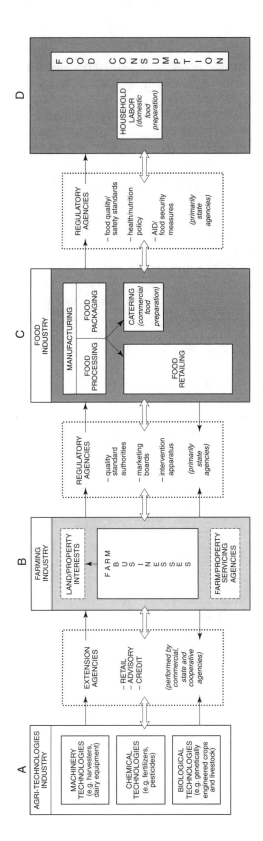

Figure 4.1 An outline of the contemporary agri-food system

- *Appropriationism* – in which elements once integral to the agricultural production process are extracted and transformed into industrial activities and then reincorporated into agriculture as inputs.
- *Substitutionism* – in which agricultural products are first reduced to an industrial input and then replaced by fabricated or synthetic non-agricultural components in food manufacturing (ibid: 2).

Current developments in the application of biotechnologies herald a potentially more radical recomposition of the socioeconomic and spatial organization of the agri-food system. Such genetic and biochemical manipulation techniques fracture the integrity of biologically defined agri-food chains centered on particular types of crop or livestock. They permit, for example, the transfer of genetic features across species barriers as "inputs" into agricultural production, and the transformation of agricultural products into generic raw materials, say starch or protein sources, in the manufacture of industrial foodstuffs. To illustrate some of the implications of these technological processes for the recomposition of the agri-food system, figure 4.2 represents four configurations of the potato commodity chain.

The postwar expansion of these "off-farm" sectors of the agri-food system has seen them become much larger, in terms of market value and employment, and more concentrated, in terms of business and market structure, than farming itself. In the UK, for example, just four companies supply 77 percent of the agri-machinery market and the same number accounts for 50 percent of the food retailing market, compared with over 144,000 units which make up 90 percent of farm product markets (Munton 1992). Contrary to the agribusiness model of agricultural industrialization, corporate agri-food capitals have become *directly* involved in the farm production sector to only a limited extent. Instead their influence has primarily been exerted *indirectly*, through networks of marketing contracts, technical services, and credit arrangements with independent farm businesses. These relations take a number of forms according to the particularities of different commodities, farm business structures, and local political contexts. In concert, they bring into play an important distinction between the corporate profile of today's agri-technology and food industries and the family-based organization of agriculture that prevails, in different forms, in both advanced and developing economies.

The rise to dominance of these "off-farm" sectors of the industrialized agri-food system has been inextricably tied into a complex of state policies and agencies, including public agri-technology research and extension services, agricultural price supports, and subsidized food programs. It is activities and institutions such as these which, in Friedmann's terms, constituted the regulatory apparatus of the second international food regime and played a pivotal role in the globalization of this industrial agri-food system in the postwar era. This regime centered on the consolidation of technological modernization and political corporatism which placed

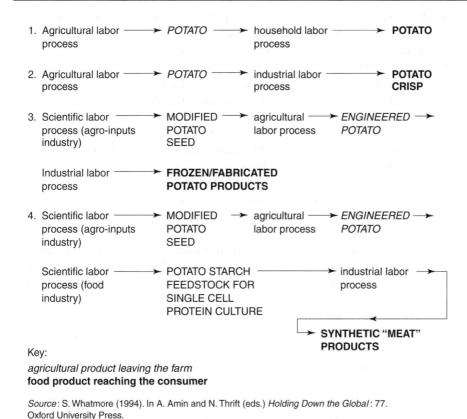

Key:

agricultural product leaving the farm
food product reaching the consumer

Source: S. Whatmore (1994). In A. Amin and N. Thrift (eds.) *Holding Down the Global*: 77. Oxford University Press.

Figure 4.2   Industrialization of the potato commodity chain.

agri-food industries at the heart of agricultural policy-making institutions, first in the US and then, via US technology and food "aid" programs, in Western Europe in the 1950s and the Third World in the 1960s (Goodman and Redclift 1989).

Perhaps the most dramatic and lasting result of the globalization of this agri-industrial system has been the reorientation of much of the food production capacity of developing countries away from staple food crops supplying local markets. Agriculture in these countries has become increasingly tied into Western markets for unseasonal or luxury primary foods, or bulk feed crops for intensive livestock production. Simultaneously, they have been turned into net importers of staple and processed foodstuffs from the US and other Western states (Susman 1989). However, the competitive development of this agri-industrial model in other advanced industrial countries, notably those within the European Community (EC) and Japan, had begun to undermine the hegemony of the US in the international agri-food system by the mid-1970s.

The institutional foundations of this mode of regulating food production and consumption were further destabilized by the escalating fiscal costs of maintaining it, which had come to a head for national governments in all three major trading blocs by the mid-1980s. A final blow came from the growing influence of financial interests in the agri-food sector, through speculative trading in agricultural "futures," and the disciplinary effects of the stockmarket on corporate investment and takeover strategies, which eluded the machinery of regulation (McMichael and Myhre 1991). Cumulatively, these elements have compounded a crisis in the industrial agri-food system that we are still living through today.

The Changing Place of Farming

Making sense of the restructuring of the farming sector of the agri-food system in terms of "commodity chain" and "food regime" approaches is more problematic. Despite their many important insights into the processes of industrialization and globalization, such approaches are inclined to represent them as the logical progression of some inexorable law of motion, popularly translated as "market forces." Their holistic and apparently seamless accounts have limited purchase on the uneven and unstable geography of the integration of farming (and food consumers) into this industrialized agri-food system and all but eclipse the significance of the interests and actions of farmers, and others, in shaping the process and pattern of change. While the global strategies of transnationals in the "off-farm" sectors of this industrial system certainly exercise a powerful influence over the terms on which other actors are drawn into the agricultural restructuring process, they do not determine its outcomes in any straightforward or uniform way (Whatmore and Thorne 1997).

Above all, and despite forecasts of its imminent demise dating back to Marx's time in the mid-nineteenth century, the farming sector retains a highly diverse but distinctive family-based structure. This centers on dynamic, and culturally variable, forms of kinship and household relations in which social divisions by gender and generation are integral to the organization of "family" land, capital, and labor (Whatmore 1991). It also sustains very diverse and locally sedimented farming cultures, political identities, economic strategies, and land-use practices which are active ingredients in the restructuring process. As a result, the articulation between farming and corporate sectors of the agri-food system is a more complex and contested process than the above accounts allow. A good illustration is the tortuous progress of the World Trade Organization negotiations on agricultural trade, resulting from the political embeddedness of farming interests, as well as those of transnational agri-food corporations, in the machinery of agricultural regulation in most nation-states and regional trading blocs.

The restructuring of farm production in the postwar period is marked by one dominant feature: *differentiation* in the extent to which farmers have been integrated into the industrial agri-food system; a feature which also marks trends in food consumption. The leading-edge technological and market relations outlined above, which set the contours for farming in particular commodity sectors of the industrial agri-food system, are realized unevenly. They are concentrated in particular regions and sectors which are characterized by large farm businesses (although many still "family" owned) employing intensive production methods and highly integrated into the global networks of the agri-technology and food industries through the ties of technological dependence, debt, and production contracts. For example, 80 percent of the European Union's agricultural production derives from just 20 percent of agricultural businesses, which are concentrated in agri-industrial "hot spots" such as the Paris Basin, East Anglia, Emilia Romagna, and the southern Netherlands (Commission of the European Community 1988).

Conversely, the same example implies that much of rural Europe, many of its agricultural products, and some 80 percent of its farmers (concentrated in the south) are in some sense "marginal" to these global networks of agri-food accumulation and effectively left out of the "food regime" and "commodity chain" accounts. Similarly, it has been estimated that $10,000 represents the minimum GDP (gross domestic product) per person necessary to "trigger" a consumer transition to the basic ingredients of an "industrial" diet (*The Economist* 1993: 15). The highly uneven social distribution of national income seriously undermines the significance of such estimates but, even on their own terms, it does not take an economist to calculate that those "marginal" to this mode of consumption, particularly in developing countries, exceed the numbers integrated into it. Making sense of the livelihood strategies of households within these rather substantial "margins" of industrialized farm production and food consumption draws attention to the broader significance of social and political processes beyond the scope of frameworks defined in terms of the dynamics of industrial processes or commodities.

From the perspective of these "margins," the globalization of the industrial production and distribution of agri-foodstuffs appears more like a lattice or network of locally embedded nodes in which transnational corporate agri-food capital strategies are variably, rather than uniformly, articulated with farmers (and consumers) through specific regional compromises and institutions, and sectoral regimes (Goodman and Watts 1997). Farmers and consumers deemed "marginal" to this industrial agri-food system occupy its interstices. But rather than seeing such interstices as the "black holes" of a singular, all encompassing, global agri-food system, these tangential spaces might be better understood as indicative of the plurality of agri-food networks which articulate numerous coexisting food production and consumption practices, some maintaining traditional subsistence relations and "informal" market networks, and some bound up with new social movements associated with concerns with the environment, regional political autonomy, cultural identity, and a host of other issues that find expression through food.

Two brief examples must suffice to illustrate this latter point. The first centers on consumer resistance to industrialized food products, such as recent opposition to genetically modified crops and foods in the UK and Europe, and the cultural limits placed on the market strategies of the corporate agri-food industries (Goodman 1999). This takes a number of different forms, but represents an important basis for new political alliances and marketing networks between consumers and farmers in different parts of the world which sustain and promote alternative agri-food practices. The second example centers on the environmental consequences of industrial agriculture, which has become another important focus for the politicization of the agri-food system. This has seen market and regulatory processes in the agricultural sector increasingly reorient towards the production of "environmental goods" in the form of sustainable farming practices that add value to products, for example by certifying their "organic" status, or "conservation landscapes" attractive to visitors and, hence, to the tourist and leisure industries (Murdoch, Marsden, and Banks 2000).

The place of farming amid the globalization of agri-food networks is thus one of growing differentiation. On the one hand, we see a rapid concentration of industrial farming methods and their products onto a smaller number of much larger farm businesses which are bound to the corporate sectors of the agri-food sector by a variety of economic and knowledge ties. On the other, we can trace a diversification of farm livelihood and business strategies that link the rest of the farm sector more closely into wider struggles over the cultural and material resources of the rural environment.

Conclusions

The industrial agri-food system, which has emerged since World War II as an increasingly globalized network, has dramatically altered the contours of the production and consumption of food. Simultaneously, biological time-space rhythms like crop gestation, seasonality, and regional specialization have been disrupted, while the social distance between producers and consumers at the heart of the system and those on its periphery has been magnified. As this account has tried to emphasize, this agri-food system is both uneven and unstable, and is characterized today by a number of overlapping crises. The first is a crisis of *production* which centers on the problems of surplus production and indebtedness among agricultural producers faced with escalating input costs and declining product prices. The second is a crisis of *regulation*, resulting from the growing political and institutional tensions in a policy apparatus which regulates global agri-food trade, on the one hand, and national farm incomes, on the other. The third is a crisis of *legitimation*, which centers on the politicization of concerns about the consequences of industrialized agriculture for food security, food safety, and the farmed environment at national and local levels.

Extending research beyond the farm gate and into other sectors of the agri-food industry and its regulatory framework represents a necessary and important contribution to a remapping of the geography of agriculture. But the accounts that have come to define this new map generate their own silences and blind spots. The challenge now is to make space for a broader canvas of social struggles and political alliances over rural resources and identities and over food health and animal welfare within which this industrial story is couched. We could do worse than to begin by giving greater conceptual and empirical prominence to consumption processes and the cultural politics of food in refashioning the social and environmental relations of agriculture (see, for example, Bell and Valentine 1997).

Notes

1 For example, United Nations data document a growing polarization of income between 1960 and 1990 with a staggering 30 percent of the world's population experiencing hunger, including 10 million people in the USA.
2 This simplified account of the relationship between the dynamics of capital accumulation and social regulation must be placed in the context of a much disputed literature on "regulation theory." A geographer's guide to these debates is provided in Tickell and Peck (1992).
3 A parallel body of work was generated independently in continental Europe, particularly France and Italy, at around the same time using the terminology of agri-food (agri-alimentaire) "filières."
4 This Fordist periodization is not unproblematic, but a critique is beyond the scope of this chapter.

5 Transnational Corporations and Global Divisions of Labor

Richard Wright

Introduction

In the early 1970s, Bulova Watch movements were made in Switzerland. These semi-finished goods were then transported to Pago Pago in American Samoa, where assembly with cases, bands, dials, and other components took place, before they were shipped to the US for final sale. Corporate President Harry Henshel commented on this production geography at the time as follows: "We are able to beat the foreign competition because we are the foreign competition" (quoted in Bluestone and Harrison 1982: 114). How typical is this story? For how many years has international production of this sort taken place? Is this kind of production network stable over space? How do we explain such geography of production? What is the future of the geography of international industrial location? By focusing on the strategies of multinational corporations and the concept of a spatial division of labor at the global scale, this chapter seeks answers to these and other derivative questions.

The transnational corporation (TNC) is the most complex of several archetypes of production unit (Hymer 1979: 146–7). The first and simplest – the workshop – involves a small number of people working together with very little specialization by worker. In the second, the factory, a large concentration of people work together but on separate and specialized linked tasks: this involves a fine division of labor both within the classes of workers directly involved in assembly and between those who plan and those who work. Third, the national corporation comprises many occupations, markets, units, and places within a country. In this type of enterprise, forward integration from single-function firms draws in, say, wholesaling activities, i.e., those activities further along in the sequence of operations. Backward integration (upstream in production to supplying activities) incorporates, for example, mining or wellhead operations. Such vertical linkages produce, at least potentially, multilocation and multimarket firms. It was this type of enterprise that resulted from the US "merger movement" in the early twentieth century.

These corporations evolved, with the development of the market, into a fourth manifestation – the multidivisional corporation, wherein the horizontal integration and subdivision or subcontracting of some operations was based on the introduction of new products or on differentiation within a product line. Accompanied by the creation of a middle level of administration to coordinate activities of a division within a country, management remained to concentrate on strategic planning. When a multidivisional firm's operations span international borders, the firm may be regarded as a transnational corporation. TNCs thus entail the location of a division or part of a division of a firm, or the coordination of some subcontracting operations, in more than one country.

During the last one hundred years international investment and subcontracting by firms has increased steadily to the extent that the TNC at the beginning of the twenty-first century is one of the most important forces creating large-scale shifts in global investment activity. Explanations of the forces driving shifts in foreign direct investment (FDI)[1] include ownership-specific advantages and location-specific factors. The first is generally linked with oligopoly and market power, and frequently associated with the existence of competitors (indigenous or otherwise) who can enter the industry if the TNC does not. While indigenous firms might possess a superior understanding of the local business environment, TNCs entering the market frequently gain advantage through access to finance, credit, and new technology by virtue of their size and oligopoly power. The second explanation is associated with variations in economic and political conditions over space. Market size and competition vary across space, as do national and supranational government policies. Production costs also provide for "locational advantages" such as labor or other factors of production (after Hymer 1979 and Dunning 1981). As it traces the evolution in the geography of these investment patterns, this chapter will show that the explanatory power of these two forces varies considerably over time and space.

In the last twenty years or so, relatively high labor costs plus union activity in technologically advanced nations have influenced the geography of intra- and international investment patterns. As a reaction to these location-specific factors, some corporations have moved or subcontracted manufacturing operations offshore in search of relatively lower-cost, non-unionized labor in the developing nations of the global periphery, creating what has become known as a new international division of labor (NIDL). TNCs control economic activities in more than one country, and therefore can take advantage of geographical differences between countries and regions in factor endowments (including government policies) by virtue of their flexibility – such as their ability to shift resources and operations between locations around the globe (Dicken 1998). According to one group of commentators, "the development of the world economy has increasingly created conditions (forcing the development of the new international division of labor) in which the survival of more and more companies can only be assured through the relocation of production to new industrial sites, where labor power is cheap to buy, abundant, and well disciplined; in short, through the transnational reorganization of production" (Frobel, Heinrich, and Kreye 1980: 15). The overall goal of this chapter is to evaluate this thesis in the context of the evolution of the transnational corporation.

The Evolution of Transnational Corporations

Although the number of transnational networks of production accelerated after World War II, the first signs of transnational enterprise originated around the turn of the century, mainly in the United States. Improvements in communications and transportation at this time, coupled with massive increases in scale economies in industrial production, provided the foundation for the initial development of transnational operations (Vernon 1992). By 1914, US firms had invested $2.5 billion in other countries (Hymer 1979: 209). The United States, along with France, Germany, and Britain, as origin countries, accounted for 87 percent of total global FDI in 1914. As recipients, the total was much smaller – less than 30 percent. Most pre-World War I FDI targeted countries in Asia and Latin America (Dunning 1983).

Two interwoven strategies accounted for this geography of investment. One, associated with resource-based investments, developed backward linkages to supply home producers with primary products in the form of either raw materials or food. For example, the financial decisions by leading European firms to invest in countries with which their home governments had colonial relations explain part of this geography. The other strategy, associated with market orientation, occurred mainly in the form of manufacturing investment in other developed countries. US-based companies made most of the early moves in establishing bridgeheads overseas. Companies like Ford, General Motors, and General Electric, for example, started foreign affiliates to establish footholds in the European market. They were not alone, however, in seeking to develop international operations: Fiat, for example, opened manufacturing branches in Austria in 1907, the United States in 1909, and Russia in 1912 (Fridenson 1986). Merck pharmaceuticals, which grew out of an apothecary shop founded in 1654 near Frankfurt, set up an affiliate in the United States in 1887. Bosch had factories producing automobile parts in France and Britain in the early years of the twentieth century (Hertner 1986).

As late as 1939 the geography of global FDI had changed little from a quarter of a century earlier. Developing countries in Latin America and Asia received almost 67 percent of total global foreign investment. Twenty-one years later, however, technologically advanced nations received two-thirds of global FDI. US companies accounted for almost half of the total international investment (Dunning, Cantwell, and Corley 1986). Although global investment stagnated during the 1930s and most of the 1940s, tremendous changes in the rate and direction of flows took place in the 1950s.

Part of the explanation for this shift is associated with the rapid revival of the European and Japanese economies. A fuller explanation of the shift in the destination of global FDI requires, as in all debates about the structure of international operations, a consideration of "ownership" advantages as well as "locational" advantages. Before World War II, state regulatory mechanisms restricted ownership advantages by US firms in Western Europe – especially the UK. The war changed all that. The war devastated the European economy, providing opportunities for investment that simply were absent before 1939. Even before the war, many countries in Europe had economies with significant structural problems and most of the economies of the larger industrialized countries were operating at capacity. The devastation of the war exacerbated these weaknesses. Also, markets loosened after 1945. For example, Roosevelt continually lobbied Western European heads of state throughout the early 1940s for the end of discriminatory tariffs, especially in the UK and its colonial economies. The establishment of the 1944 Bretton Woods Accords contributed to US industries moving into Europe. The Bretton Woods agreement established the General Agreement on Tariffs and Trade (GATT), the International Monetary Fund (IMF), and made the US central in currency exchange. Each nation was responsible for keeping its currency within a tight range. To effect this, each country bought or sold its own currency in foreign exchange markets. The US consistently ran large deficits in its balance of international payments, forcing foreign central banks to buy excess dollars with their own currencies, providing US TNCs (and other players) with European currencies. These were invested in the UK, France, West Germany, and elsewhere in Western Europe (Bluestone and Harrison 1982).

Vernon (1992) notes additionally that a "follow-the-leader" strategy drew firms to set up operations offshore as a hedge against threats posed by rivals: once a corporation in an oligopoly began production in Europe, other US-based firms in that oligopoly subsequently tended to establish affiliates in the same country. More generally, as Hymer (1979: 210) observed, subsidiaries of TNCs "once established, tend to grow in step with that industry in that country except where interrupted by extraordinary events like war." Decolonization, or the threat of independence by colonized countries, also probably discouraged investment by European-based firms in Latin America, Africa, the Middle East, and Asia, and helped reorient the global geography of FDI away from developing countries and toward (Western) Europe, where new fierce competition was erupting with American affiliates. Last, but not least, the formation of the European Economic Community in the late 1950s helped reorient foreign investment as US transnational corporations sought market access through the establishment of affiliates in Western Europe.

Since 1960 the growth of world FDI has continued apace. Although the proportion of FDI originating in the United States peaked in the 1960s, the volume of US foreign investment grew from a little over $30 billion in 1960 to about $500 billion in the mid-1990s. UK, French, German, and Japanese sources more than matched this rate of growth. Furthermore, over the last forty years transnational corporations have emerged with bases in so-called newly industrialized countries (NICs), such as Singapore, South Korea, and Taiwan. Nevertheless, patterns of TNC transnational investment remain concentrated in the global core, and the TNCs based in the United States are as significant as ever. The evolution and growth of the European Community (EC) has been significant in this regard, as US-based enterprises continue to seek market access in Europe. The development of the Single European Market and the opening up of Eastern European economies will probably affect the size and geography of US investment in Europe.

One of the most noteworthy shifts in global FDI since the mid-1970s has been the change in the position of the United States in terms of the ratio of inward to outward FDI. The United States has always been a destination for FDI, but until recently that inward-bound FDI represented only a quarter or a fifth the size of the FDI emanating from the country. Since about 1980 the USA became a much more prominent destination for global FDI. One result of this trend is that new foreign competition now challenges US firms in their domestic markets. More generally, this widening of the scope of global FDI squeezes profits of some major corporations operating in the global core and firms have sought lower wage labor sites in developing countries in the global periphery to reduce production costs.

The term "new international division of labor" was coined to explain this drift of work from the core to the periphery. The next part of this chapter expands on the reasons why a new international division of labor came about, and then questions its utility in the light of global patterns of investment by TNCs that still tend to favor core-country locations.

The Old and the New International Division of Labor

The term "new international division of labor," popularized by Hymer (1972, 1976) and Frobel, Heinrich, and Kreye (1980), has become common in the discourse on global industrial restructuring. The NIDL refers to a spatial division of labor at the global scale. It can be considered part of a bigger family of labor divisions, which includes social divisions of labor, divisions of labor between production and exchange, and spatial divisions of labor at scales other than the international (see Sayer and Walker 1992). The idea of a new international division of labor opposes that of an old international division of labor (OIDL). Under the OIDL, the global periphery was seen and theorized as the provider of many primary goods and raw materials for processing in "core" countries in Western Europe and North America. In exchange for these materials, the periphery received finished goods manu-factured in the core.

The economic activities in the global core and periphery are changing under the development of the new international division of labor. A process of vertical uncoupling, subdivision, and/or subcontracting of production results in the periphery developing low-skilled, standardized operations such as manufacturing assembly or routine data entry, while the global core re-tains high-skill knowledge- or technology-intensive industries and occupa-tions. Through deskilling labor, and the functional and physical separation of various tasks in the corporation, this process creates "roles" for places in the world economy (cf. Massey 1984).

According to Frobel, Heinrich, and Kreye (1980), countries in the global periphery have transcended their old role, beyond the search for low-cost, union-free labor environments, for three main reasons.

1   The introduction of capital-intensive green-revolution agricultural meth-ods, especially in Southeast Asia, has liberated hundreds of thousands of people from what was largely a subsistence rural life. Much of this labor force is inexpensive and productive by Western standards. More-over, because of the size of this labor supply, transnational corporations (and other companies) can afford to select their employees according to age, skill, and particular disciplinary factors. Notably, the NIDL is espe-cially augmented by gendered labor markets and the underpayment and devaluation of female wage labor (Sayer and Walker 1992).

2   The increasing subdivision of labor processes lends itself directly to the substitution of minimally skilled workers for those who have had more training or education. The fragmentation of productive tasks in manu-facturing has developed to such an extent that the execution of the sim-plified tasks of perhaps a very sophisticated overall process requires only a brief training period. Knowledge functions are therefore extracted from the production process, creating a situation in which skilled workers commanding relatively high wages can be replaced by less expensive semi-skilled or unskilled workers.

3   New, permissive technology extends to the realm of transportation in-novations like containerization, that now present the possibility to pro-duce finished or semi-finished goods at virtually any site in the world. In other words, while some places might be more practical than others, physical space per se is a far less limiting factor in the industrial location calculus than it once was.

The NIDL's basic geography, of the advanced industrialized countries on the one hand and the rest of the world on the other, can be modified in several ways, as much labor-oriented global industrial production has evolved beyond that simple organization. A regional core and periphery has devel-oped within Southeast Asia, for example, with the territorial differentiation of Japan, Taiwan, South Korea, and Singapore in a core from other countries in a new periphery (Lipietz 1986; Donaghu and Barff 1990). In an intranational context, producers seek low-wage locations in peripheries of the global core. Converse, the athletic footwear producer, for example, employs mainly female American Indians (from a tribe unrecognized by the US Bureau of Indian Affairs) to assemble shoes in Lumberton, North Carolina, in the poorest county of one of the lowest-wage states in the country.

TNCs and the NIDL

Joining TNCs and the NIDL almost automatically centers the discussion on an explanatory emphasis of international corporate behavior on the locational advantages of offshore production. More specifically, fusing TNCs and the NIDL focuses attention on the influence of labor on international location. In one respect, this is entirely appropriate. In terms of labor costs, hourly earnings in manufacturing vary hugely across space. Hourly earnings vary from over $12 in certain core countries to less than $1 in some of the poorest developing nations (Dicken 1998: 190). Add the costs of health and social security benefits and the gulf widens even more.

These location-specific advantages associated with labor affect a particular set of economic activities associated with routine, mass-produced, highly differentiated production, such as manufacturing industries like computer electronics assembly, textile and clothing production, and shoe manufacturing, and non-manufacturing activities like routine data entry/processing. For example, Xerox Corporation, based in Rochester NY, manufactures copying machines at various sites around the globe. Large, complex machines, however, are produced exclusively at their New York plant; smaller, simpler machines are assembled in developing nations in Southeast Asia and South America, taking advantage primarily of low-cost labor in the global periphery. Shifts to offshore production sites in Third World countries frequently garner front-page headlines. In the United States, these include the auto plant closures of General Motors in Michigan and their new investments in Mexico, to the shifting of production of Levi's from their factory in San Antonio, Texas, to Costa Rica, and the move of Smith Corona typewriters from rural New York State to northern Mexico.

Expanding the argument to subcontracting relationships, we find similar global patterns of production. Nike, the athletic footwear marketer, used to own manufacturing plants in the United States and the United Kingdom, but presently subcontracts 100 percent of its production capacity to suppliers in South and East Asia. The pattern of Nike's production partnerships has evolved over time, a change powered in part by the geography of labor costs in Asia (Donaghu and Barff 1990). Initially, production of Nike shoes took place in Japan. Soon, subcontracting arrangements diffused to factories in South Korea and Taiwan. Presently, those partnerships are diminishing in importance as labor costs rise and new networks of subcontractors become established in Indonesia, Malaysia, and China. In these new locations, workers involved in shoe production are paid a small fraction of the wage their counterparts make, working for other companies, in the United States (Barff and Austen 1993).

The NIDL and Other Geographies

While many transnational corporations take advantage of low-wage labor in places around the globe, and TNCs play a significant part in the development in the new global division of labor, such a model explains only a portion of TNC operations and their global strategy, and a small portion at best. The geography of core-based TNC investments tends to be in developed countries, because forces driving the international investment decision extend far beyond a singular consideration of labor-cost minimization. Understanding TNC behavior hinges on the issues of the importance of ownership-specific versus location-specific advantages, with labor costs being only a part of the latter. Transnational corporations invest offshore not only to utilize low-wage labor but also for many other reasons such as access to resources and markets, development of import-substituting manufacturing, and so on. At this point, perhaps we should return to Bulova Watch – a small TNC that at the end of the twentieth century no longer produced timepieces in Pago Pago (or the West Indies, or any other developing country) but now has parts production and assembly only in countries in the global core or some of the NICs. Labor cost variation apparently now plays only a small role in Bulova's international locational calculus.

[handwritten margin note: Because new closer to markets buy fancy watches.]

Because of the exclusive focus on labor, other problems occur with the NIDL approach. Schoenberger comments that the NIDL model understates the extent to which unemployment and unstable employment in countries of the global core result from labor-saving technological change: "This omission reinforces the impression that employment gains in one location are made at the expense of jobs in another . . . It also encourages workers in the industrial countries to view low-cost foreign labor as the primary threat to their well-being" (Schoenberger 1988: 116).

Note also that not all developing countries in the global periphery offer the same opportunities for labor exploitation. The geography of transnational corporate foreign investment favors, of course, developed countries primarily, and South America and South and East Asia secondarily. Within the global periphery, transnational corporations invest heavily in locations possessing an Export Processing Zone (EPZ), places where the main lures include not only relatively inexpensive wage labor, but also land, electricity, and water at cheap rates, as well as government grants, tax breaks, tariff/duty reductions, lax pollution controls, and less rigorous health and safety standards in the workplace. Furthermore, even when all of the factors of location are taken into account, including access to primary resources and agricultural products, the continent of Africa is largely excluded from investment consideration. In terms of the NIDL this whole continent plays only a very minor role. For example, the small island of Mauritius has a workforce of 62,000 involved in EPZ activity, which represents about 40 percent of the continent's total EPZ employment (Marshall 1991).

One other criticism centers on the idea at the heart of the NIDL thesis of dividing the world into two parts. Lawson and Klak (1993) argue that the social science literature is pregnant with binary categories, such as new international division of labor/old international division of labor, modern/traditional, developed/underdeveloped, core/periphery, First World/Third World, and North/South. They point out that these binary labels have at least two severe shortcomings. Dualistic theorizing can undermine our making sense of reality. Simply put, new and old international divisions of labor homogenize large areas of the world. Moreover, places become "othered," which creates an unequal power relationship between the namer and object of the labeling. The dualities are not only opposing, but also unequal. In terms of the new international division of labor, Lipietz suggests, in a related argument, that as geographers we should be far more sensitive to spatial variation than the NIDL model implies. "The Third World today appears as a constellation of particular cases with vague regularities constituted out of fragments of the logic of accumulation (which proceeds badly or well depending on local circumstances) and [production] tendencies that rise and fall over a few years" (Lipietz 1986: p. 34).

Conclusion: TNCs and Other Firms

Manufacturing TNCs are not the only global players. Producer service industries also shape and are shaped by new global divisions of labor. High order services, like business consulting and advertising, tend to remain in developed countries; in the last few decades, firms have increasingly moved "back office" (notably data entry) functions to the global periphery (Coffey 1996). Also, the growing efficiency and internationalization of capital markets allows relatively small firms to raise capital in much the same way large firms used to do exclusively. The opening up of foreign markets (via the removal or simplification of legal and tax complexities of international business) allows small and medium-sized companies to operate globally. The use of computers, by narrowing economies of scale in manufacturing and distribution, has made possible inexpensive, small-volume production, further diminishing the advantages of corporate size.

Some might hold that a company like CORPORATION X is an exemplar of a portion of future global economic activity. CORPORATION X is a maker of relatively expensive brand name textile and apparel goods (see Clark 1993). The company is not a transnational corporation, because

CORPORATION X "has not made significant investments or built significant organizational units outside its home base." It is simultaneously a global and a local company: global because of its international sourcing, marketing, and design networks, and local by virtue of its sensitivity to local variation in tastes, fashions, and revenues. Different products are marketed in different places, and experience shows that utilization of local knowledge about place-specific tastes and styles is key to success.

Mass markets around the globe still exist, however, and will continue to do so. It follows that transnational corporations serving mass markets that need to take advantage of the NIDL might go relatively unaffected by niche market servicing. Remember also that new mass markets are emerging, such as China. There should be a continuing place for the TNC and the NIDL in our contemplation of global investment patterns, so long as that understanding is part of a larger theoretical framework accounting for revenue maximizing and other cost-minimization strategies.

Note

1   FDI vs. portfolio investment.

6 Global Change in the World of Organized Labor

Andrew Herod

There are two aspects of contemporary processes playing out at the global scale which are having particular impacts upon workers, both those who belong to labor unions ("organized labor") and those who do not. The first of these is the remaking of the time-space organization of capitalism through the shrinking of the globe brought about by new transportation and tele-communication technologies. The fact that goods can now be made in one part of the world and flown to the other side of the planet in only 24 or 48 hours means that many workers now have to compete globally for work and investment in their particular towns and communities. The second set of changes which have been taking place in the international arena with regard to labor unions are those associated with the end of the Cold War. Specifically, much Cold War rivalry between the United States and the Soviet Union was carried out in the realm of international labor union politics. The collapse of the Soviet Union, though, has dramatically transformed the political environment within which international labor organizations must work. In this chapter, then, I want to outline how international labor organizations are responding to these changes. The chapter is organized into three parts: a brief introductory section outlining the history of international activities by labor organizations; a section which examines how labor organizations are responding to economic processes of globalization; and a section which outlines how the end of the Cold War has impacted international labor politics. A brief conclusion ends the chapter.

Organized Labor and Transnationalism

The history of labor internationalism is, arguably, as old as capitalism itself. Already, in the middle of the nineteenth century, workers and other political activists had begun to form international linkages designed to match the growing linkages of the world economy (Van Holthoon and Van der Linden 1988). Sometimes these were agreements to help provide information and resources between workers in different parts of the world. On other occasions unions in one country actually helped workers in other countries to found unions. Thus, towards the end of the nineteenth century many British union members – especially those working in engineering – traveled to countries like Australia and the United States and helped establish foreign branches of their own unions, a process which helped lay down some of the earliest structures around which international networks of union activists could subsequently develop (Southall 1989). The goal was usually to protect workers' rights (particularly those of workers migrating between countries, which could be done by mutual recognition of union membership cards), to limit the ability of employers to bring in foreign strike-breakers during periods of industrial unrest, and to provide reciprocal financial assistance and to help coordinate labor activities across national boundaries. However, it was not until the last decade of the century that the form of international labor organization which more or less exists today really began to take shape.

There are, primarily, two ways in which workers might consider organizing any international bodies designed to help them pursue their goals: by industry or by geographic territory. Both of these methods of organization have been used in the international arena by unions. With regard to the first of these forms of organization, by the outbreak of World War I workers in different industries had managed to establish some thirty-odd international trade secretariats (ITSs) designed to foster cooperation between unions in particular industries in different countries (Busch 1983; Price 1945; Segal 1953). Industries in which ITSs were established included, among others, boot and shoe workers (1889), miners (1890), printers (1893), metal workers (1893), transport workers (1897), hatters (1900), diamond workers (1905), pottery workers (1905), and even hairdressers (1907) (Windmuller 1980). Obviously, some of these were more powerful and successful than others, but because many had come out of the socialist movements in their respective countries they usually shared similar ideological goals, such as the nationalization of industry and the replacement of the capitalist system with a more humane economic and political system. In practice, though, such ideological goals were usually secondary to more practical matters of providing information and help to workers concerning matters such as wage rates and working conditions.

The second major international entity was organized on a country-by-country, rather than industry-by-industry, basis.[1] This was the International Secretariat of Trade Union Centers, which was founded in 1901. Like the ITSs, this was an organization designed to promote greater labor ties across national boundaries (Sturmthal 1950). However, whereas the ITSs affiliated the unions working in each particular industry – metalworkers' unions in metalworking – in the case of the International Secretariat of Trade

Union Centers it was the various national labor union centers which were the affiliates. Although the ITSs sometimes worked with the Secretariat, the two types of organization were formally independent of one another. However, conflict was about to impact the international labor union movement in momentous ways. First, in 1913 the Secretariat changed its name to the International Federation of Trade Unions (IFTU). Second, the outbreak of World War I in 1914 severely hampered the activities of the ITSs and the IFTU, although some limited efforts were made to continue their work. Third, the Bolshevik revolution in Russia led to the formation of a new international labor union organization, the Red International of Labour Unions (known as the Profintern due to its acronym in Russian). In a portent, perhaps, of things to come, the Profintern criticized severely the IFTU and the ITSs for their lack of revolutionary aims.

Although the ITSs and IFTU continued their work after World War I, the outbreak of World War II effectively ended the activities of both the IFTU and Profintern. Consequently, with the coming of peace, many labor unionists wanted to reestablish international links with confederates in other countries. As a result, in 1946 representatives from many countries' labor union movements (including those of the Soviet Union) came together to found a new international body to replace the old IFTU. This was known as the World Federation of Trade Unions (WFTU). Many of the ITSs were also resurrected. Before long, though, conflict between communist and non-communist elements led the WFTU to split along ideological lines. Communist unions – mostly including those official state unions in the countries of Eastern Europe, together with communist unions in several Western European countries like France and Italy – remained members of the WFTU, whereas the non-communist national centers (such as the American Federation of Labor in the United States) banded together to form a new international entity known as the International Confederation of Free Trade Unions (ICFTU).[2] For their part the ITSs became closely allied with the ICFTU, while the WFTU set up its own version of industry-by-industry organizations equivalent in function to the ITSs, known as Trade Union Internationals (TUIs).

For the most part, this structure of international labor organization has lasted from the late 1940s until today. Although there was a third, less influential, international body constituted by national centers from various countries – this being the religiously oriented Christian labor union international (initially called the CISC, after its acronym in French, but later renamed the World Confederation of Labour) – the ICFTU and the WFTU remained the two main rivals on the world stage. Throughout the period of the Cold War the ICFTU and the WFTU, together with the rival ITSs and TUIs, engaged in conflicts largely determined by political ideology. Thus, for example, in Latin America the two bodies often supported different groups of workers in particular conflicts largely on the basis of their ideological affiliations (see Herod 1997b for examples).

Recent Changes in the Geography of Labor Transnationalism

During the past three decades or so, the world's economy has undergone a tremendous transformation and integration as part of the process commonly referred to as "globalization." Perhaps the most significant aspect of this has been the shrinking of relative distances between places and the growing power of transnational corporations (TNCs) to stretch out their production chains across the planet. Although hampered by ideological divisions, labor unions and workers have tried to develop greater international links between one another as a way of countering the power of globally organized capital. Given that the activities of TNCs have largely, though not completely, been confined to those parts of the globe which have market economies, efforts to develop international links between workers in different countries who might work for the same TNC have primarily involved the ICFTU and, particularly, the various ITSs operating in different industries.

One of the principal goals in trying to develop labor solidarity across international boundaries is to match the geographical organization of TNCs and, more specifically, to limit the ability of TNCs to play workers in different parts of the world against each other by promising one group continued work or investment if they will work for less than workers elsewhere (a process known as "whipsawing"). Through international labor solidarity unions hope to build sufficiently powerful linkages across space so that workers do not succumb to such political and economic pressure. Although such practices are fraught with many difficulties – there are problems in overcoming, for example, cultural and linguistic differences between workers in different parts of the world, in dealing with different legal structures in different countries (particularly with regard to how labor and employment laws operate), and in dealing with fluctuating exchange rates which might make it difficult for unions in one country to provide financial resources to unions in another – there have been a number of very successful efforts to develop such labor solidarity across international boundaries.

For example, in the mid-1980s unions from West Germany pressured executives at the chemical transnational BASF to settle a dispute with workers locked out of one of the company's plants in Louisiana, a campaign which was partially coordinated through the International Federation of Chemical, Energy and General Workers' Unions (ICEF) international trade secretariat (see Bendiner 1987: 62–102 for more on this and other such examples; also, Barnet and Muller 1974: 312–19). Likewise, when the US telecommunications company Sprint fired workers organizing at its facility in San Francisco in 1994, unions in Germany, France, Nicaragua, and Brazil lobbied their own governments to reject efforts by Sprint to buy parts of their national telephone systems unless the company negotiated with its fired US workers (MacShane 1996). When workers at an aluminum plant in West Virginia were locked out in 1990 in a dispute over work rules by the Swiss company which owned the facility, their union (the United Steelworkers of

America) worked closely with its industry's international trade secretariat (the International Metalworkers' Federation, headquartered in Geneva, Switzerland) to develop contacts with metalworkers' unions in other parts of the globe where the company also had operations. These unions succeeded in organizing protests in some 28 countries on five continents, protests which brought the company such bad publicity that it rehired the union workers at its West Virginia plant (see Herod 1995 for more details).

In highlighting such examples, however, we should also recognize that sometimes unions engage in international solidarity campaigns for quite different reasons. In certain instances workers engage in such campaigns as a genuine way of trying to help workers in other countries who might be facing poor working conditions, low wages, or even difficulties in being able to form a union due to opposition from the employers for whom they work. In other cases, though, workers in one country – say the United States – may be interested in helping workers in another country – say Mexico – to organize and increase their wages and better their working conditions only because, in so doing, they make it less likely that a TNC will leave their own particular community to relocate elsewhere. Thus, in both situations groups of workers may be engaging in international solidarity actions but they have very different motives for doing so. Johns (1998) has made an interesting distinction in this regard. She suggests that the first type of solidarity is "transformatory," in that it is designed to challenge the class relations between capital and labor without regard to the consequences of this for particular places. The second type of activity she calls "accommodatory," in that it is designed not to challenge the social relations of capitalism but, instead, to protect certain spaces (the communities in which workers are threatened by capital flight) at the expense of others (those communities which might gain by an influx of relocating capital). In considering workers' practices of international labor solidarity, then, in addition to understanding how they go about such actions we need also to consider their motives and the outcomes of such actions, for these will mean very different things for how the geography of global capitalism is made.

A final issue that I want to consider here with regard to international labor solidarity is the impact of recent telecommunications technologies, particularly things such as the internet and email. As in many other aspects of modern life, such new technologies are changing the way in which many unions conduct business. The internet and email now allow people across the planet to communicate with each other more easily and more quickly (assuming they have access to this technology!) than ever before. Although they may not have adopted such new technologies as quickly as have corporations, unions are nevertheless not immune from their influences and have begun to make use of them in innovative ways (see Lee 1997 for more details). Specifically, such technologies are having at least two important effects on how unions go about developing international linkages in the new global economy. The first of these relates to the speed with which workers in one part of the globe can now contact potential supporters in other parts of the globe and "get the word out" about strikes or other industrial disputes. This, of course, can help them to get support – moral, logistical, and financial – more quickly than before, which may be important in sustaining workers through the early days of a dispute (often the most crucial time in a dispute). Unions have developed numerous information networks involving web pages, bulletin boards, chat rooms, email listservs, and other such devices in their campaigns. They have also been able to develop "cyber-pickets" in a number of high-profile campaigns, encouraging supporters to picket the web pages of various corporations to express their displeasure at the corporation's activities (e.g., by deluging the feedback sections of web pages with negative comments). Thus, for instance, in 1994, using the catch-phrase "to picket, just click it," the United Steelworkers of America (USWA) and the International Federation of Chemical, Energy, Mine and General Workers' Unions (ICEM) trade secretariat, with which the USWA is affiliated, inaugurated a cyber-campaign against the Bridgestone/Firestone tire maker which had fired 2,300 unionized workers at five of its US subsidiaries and replaced them with non-union workers (see Herod 1998a for more details). The unions encouraged members of the public to complain to Bridgestone/Firestone about its activities. In the end, Bridgestone called back to work virtually all of the fired unionized workers.

The second impact of the new technologies of the internet relates to the actual structure of developing international links between workers and unions in different parts of the world. Specifically, for much of the past hundred years or so the model of international organizing adopted by unions was one in which the international trade secretariats and national leaderships of various unions were heavily involved. Usually, the national headquarters of a union or the headquarters of a trade secretariat employed paid "organizers" whose job it was to develop contacts between workers and unions in different parts of a single country or between different countries. Developing links between workers in different places was often, then, a quite hierarchically organized and "professionalized" practice, with workers in one place often having to contact paid staff representatives working in the international relations department of the union's national headquarters, who would in turn contact representatives from an international trade secretariat (e.g., the International Metalworkers' Federation), who would then contact national representatives of the union in a second country, who would finally contact workers in a particular plant or factory. Access to the internet, however, appears to be changing this model of organizing, at least in some cases. Specifically, whereas the older model relied upon a cadre of paid organizers who represented their particular union, a newer model based upon "network activism" appears to be emerging (Waterman 1993). Such labor union activists are often not paid "organizers" or staff members but are, instead, workers and others who are using the internet to make direct worker-to-worker, plant-to-plant contact with their confederates in different parts of the globe. Thus, the "professional labor union organizer" may increasingly be replaced by the "networker," who may be the shopfloor worker who is handy with a computer or the labor sympathizer who can translate documents on the World Wide Web. This has implications both for the structures of international labor activities – the network model is less hierarchical than the older "professional" model – and for issues of union democracy, for it means that international activities may no longer be principally the domain of national and international union organizations.

New Political Geographies of Labor Unionism after the Cold War

The end of the Cold War has brought with it tremendous changes during the past decade in the world of international labor union politics. In particular, the collapse of the Soviet Union and the changes in Eastern Europe have greatly impacted the WFTU. As already mentioned, the WFTU served for most of the post-World War II period as the international labor union body to which many unions of the communist world belonged. Once the Soviet Union lost its grip on Eastern Europe, many of the state-controlled national labor federations which were formerly members of the WFTU disaffiliated and joined the ICFTU. Likewise, many individual unions in particular industries disaffiliated from the WFTU's TUIs and affiliated with the appropriate ITSs. For instance, the International Metalworkers' Federation has signed up formerly communist-controlled metalworkers' unions in Bulgaria, Romania, the Czech and Slovak Republics, Slovenia, Poland, and several other countries of the region (see Herod 2001 for more details). Several unions from Western Europe and North America have also been very active in Eastern Europe. In 1995 the German metalworkers' union IG Metall appointed a coordinator for the Czech Republic, the Slovak Republic, and Slovenia to encourage regional cooperation and mutual assistance between German unions and those in these three countries, while several US unions have conducted seminars with their equivalents in the region. The AFL-CIO has also established offices in Moscow and several other cities in Eastern Europe for the purposes of developing greater links with local labor unions. Moreover, as the post-communist "transition" has unfolded, even some of the unions from the more industrialized countries of Eastern Europe have been active in the other countries of the region. The Czech metalworkers' federation OS KOVO, for example, has conducted training programs and seminars with unions in Bulgaria and Romania, sharing information on the Czech privatization experience. For its part, the much smaller WFTU now draws its membership from a few still-communist unions in Eastern Europe, together with unions in many of the Arab countries, Cuba, and several African, Latin American, and Asian countries.

The unions have also had to adapt to playing a different role than they did during the communist period. Drawing on Leninist ideology, during the communist period labor unions were designed principally to serve as conduits or "transmission belts" by which instructions regarding production were transmitted from central planners to workers in the factories and mines (see Pravda and Ruble 1986 for more details). The unions, then, were primarily supposed to mobilize the workforce for production. Under the type of political and economic system operating in most industrialized Western countries, however, unions play a very different role, one in which they are not appendages of the state but are supposed to represent the interests of workers when dealing with management or the state. This has meant that both workers and union officials have had to learn to play quite different parts in society, something which has often led to confusion and which has taken getting used to. Under the system of central planning, for instance, wages were invariably determined centrally by the planning agencies and so bore little relation to productivity or availability of skills. Now, wage gains are expected to come from hard bargaining on the part of the unions for their members.

The situation in Eastern Europe has been further complicated by the local politics of the post-communist period – specifically, rivalries between communist, former-communist, and anti-communist elements – and by the varied geography of the economic changes associated with transition (for more on the geography of transition, see Pickles and Smith 1998). With regard to the former, when the communist governments of Eastern Europe fell in the early 1990s, the state-controlled unions were impacted in different ways in different countries. In Czechoslovakia, for example, anti-communist elements took over the old communist-controlled unions and began to run them (though in different ways). In Poland, on the other hand, many of the unions remained under the control of communists but were challenged by new, independent unions (such as Solidarity) formed by anti-communist elements. Thus, in many countries there exists a tripartite system of unionization, with workers divided between the new, independent unions, the "reformed" unions (those old communist-controlled unions captured by anti-communist forces), and unions still controlled by communists (even if they go under the names of new political parties today). In some countries there are also a number of Christian labor unions (such as the Krestanska Odborova Koalice – the Christian Trade Union Coalition – in the Czech Republic) which have emerged and affiliated with the Christian-oriented World Confederation of Labor and its versions of international trade secretariats.

This has caused numerous problems (see Herod 1998b for more details). As might be imagined, rivalries between these different unions – which often operate in the same industry in a particular country – are often intense, making collective action to address matters of interest to workers difficult. Indeed, there is frequently greater conflict between different labor unions than there is between collective labor and collective capital. Furthermore, many unions have found it difficult to establish collective bargaining procedures of the sort found in most parts of the capitalist industrial world because, even a decade after the beginning of the supposed "transition" to a market economy, many enterprises are still state-owned.[3] This sometimes leads governments to make a political decision to negotiate with some unions and not with others in a particular industry. Equally, the lack of effective employers' associations means that unions frequently find it difficult to get access to basic information from companies which is necessary for realistic bargaining to occur and/or are sometimes unsure of who it is that they should actually be negotiating with, while the lack of effective procedures and mechanisms for collective bargaining means that talks between union and management often operate on something of an ad hoc basis with few formal guidelines. Finally, the lack of effective employers' associations often means that collective bargaining takes place at the level of individual plants, rather than across whole industries as is common in many other industrialized nations. Combined with the very decentralized structures that many unions, remembering the highly centralized control exerted over them during the communist period, have adopted, such a focus on bargaining at the local level has often made it difficult for the labor movements in some countries to exert economic and political power at the national level.

If the multiplicity of unions, and the rivalries between them, then, have been one set of problems facing workers and their organizations in Eastern Europe, the second set of issues they have had to face relates to the processes of economic restructuring which have been taking place. More particularly, the economic impacts of the transition have been felt in a geographically uneven manner across the region and within countries. For instance, in the case of the Czech Republic much of the foreign investment which has come into the country has been focused on the Prague area, a situation which has led to generally fairly tight labor markets. In other regions of the Czech Republic, such as in the coalmining region of Ostrava, restructuring and privatization of state-owned enterprises has led to high levels of unemployment and opposition to the various "market reforms" pressed by the government. Similar patterns are observable throughout the region. Such opposition has manifested itself both in work stoppages and strikes, and in renewed support for the old communist unions and for the now-restructured and renamed Czech Communist Party. Furthermore, the break-up and privatization of many state enterprises has meant that local labor markets do not operate in the ways in which they did under central planning. For example, under central planning workers tended to be hired at enterprises on the basis of the need to meet *politically* determined production quotas, regardless of the actual labor costs involved. In enterprises working in the market economy, of course, production is now determined to a greater extent by economic forces of the market and enterprise managers usually do their best to keep wages down, a situation which is again causing union officials and workers to have to adapt to new ways of doing things. The economic changes associated with the transition, then, are having dramatic impacts upon workers and labor markets, impacts which are playing out in geographically uneven ways and which are, in turn, spawning an uneven geography of response by unions.

Conclusion

Two major processes of change – the growing economic integration of the globe and the changes associated with the collapse of the Soviet Union – are having significant impacts upon unions and workers across the planet. Unions and their members, though, have not sat idly by as these processes have changed the world around them. Rather, they have adopted new technologies to engage in innovative transnational solidarity campaigns and have been active in shaping the political and economic geography of the transition which is occurring in Eastern Europe. Indeed, through these activities unions are not just responding to processes taking place at the global scale but are proactively shaping these processes – that is to say, they are playing an active role *at the global scale* in remaking the geographies of the contemporary world.

Notes

1   I say "major" because there were several other organizations working in the international arena at this time.
2   The American Federation of Labor (AFL) did not become the American Federation of Labor-Congress of Industrial Organizations (AFL-CIO) until 1955.
3   I say "supposed" transition to a market economy because there is, in fact, a great deal of debate over whether this is actually what is happening. Stark (1996), for example, has argued that a unique type of capitalism is emerging in the region, one shaped by the particular histories and economic and political geographies of Eastern Europe, and that it is incorrect to assume (as many have erroneously done) that the type of economy emerging in the region will simply be a copy of that operating in Western Europe or North America.

7 Trajectories of Development Theory: Capitalism, Socialism, and Beyond

David Slater

Introduction

Ours is an epoch of closures and openings, of chilling conformity and vibrant resistances. The literature of the late twentieth century was pervaded by a sense of departure from the ostensible security of the past. Theoretical debates have been frequently punctuated by the prefix "post" – so we can move, for example, from postmodernism to postcolonialism, or from post-Marxism to post-development. At the same moment, we have been confronted by the signifiers of a radical closure, so that one can encounter the posited "end of history" (Fukuyama 1992). Conversely, in a symptomatic reaction to assertions of endings, we have been encouraged to believe in new beginnings or new affirmations. Critically juxtaposed to the notion of the "*end* of history," it has been suggested that we have been witnessing the emergence of a *new* globalism in human relations, the proliferation of *new* social movements, and the surfacing of a *new* age for democracy. Equally, in the more specific context of development theories, it is possible to discern a connected juxtaposition, whereby ideas concerning the end of development, or "post-development" (Latouche 1993; Rahnema and Bawtree 1997), exist next to the desire to rethink and reproblematize what can be meant by "development" (Nederveen Pieterse 2000; Pretes 1997). Furthermore, the contemporary scene continues to be characterized by the continuing dissemination of an orthodox Western vision of development for the non-West, most clearly reflected in notions of "structural adjustment," deregulation of the economy, privatization, "good governance," and more recently "social capital" (Fine 1999). Such a vision, which is a continuation of an apparently endless Western will to develop the world, has become the subject of an increasingly multi-dimensional critique relating to such questions as democracy, social justice, environmental sustainability, the rights of indigenous peoples, and the geopolitics of knowledge (Coronil 2000; Escobar 1995; Marglin and Marglin 1990; Sachs 1992; Slater 1995; Tucker 1999).

What is now clear is that with the post-1989 dissolution of the Second World, the accelerating tendencies of globalization, and the proliferation of a series of acute social tensions and conflicts, there has also been a resurgence of interest in the state of North–South relations and in the ways in which development is conceived. This resurgence has also been reflected in the domain of geographical scholarship, where contributions by, *inter alia*, Corbridge (1993b), Crush (1995), Peet and Watts (1993b, 1996), Power (1998), Schuurman (1993, 2000), and Simon (1998) have connected development thinking to debt, environmental issues, postcolonialism, and problems of Eurocentrism.

In the present climate of intellectual as well as sociopolitical turmoil, it is even more necessary than before to trace the breaks and connections in the formation of development discourses. And it is not only the history but also the geopolitics of ideas that is so crucial, as I shall seek to demonstrate below. An active politics of memory, carrying on and revivifying what is considered to be significant and relevant, is accompanied by a politics of forgetting that can be used to insinuate the idea of a new truth, the roots of which may actually stretch far back into the past.

In this chapter, I discuss two major perspectives on development, perspectives which are counterposed and essentially incompatible. The first I shall simply refer to as the orthodox Occidental vision, which has been expressed by both modernization theory and neoliberalism. In opposition to such a vision the postwar period has witnessed the rise and decline of a Marxist-inspired theorization of development, which despite a degree of internal heterogeneity, has possessed a certain conceptual and political regularity. In the third and final section of the chapter I shall refer, if only briefly, to the outlines of today's situation within which a critical recasting of development theory has acquired an increasingly global meaning. For the most part, my examples will be taken from the geopolitical encounters between the United States and Latin America. My concentration on the United States relates to two main factors.

- First, it is increasingly crucial to expend more analytical energy on seeking to comprehend the "lone superpower" in its historical, economic, cultural, and geopolitical specificity, since its position in the world at the beginning of the twenty-first century is increasingly pervasive, its impact on all aspects of global affairs being quite profound (Slater and Taylor 1999).
- Second, given the undeniable reality that the United States has been the key imperial power of the twentieth century, it follows that any understanding of the dynamic between geopolitics and development theory must attempt to come to terms with the nature of the relations between the United States and the Third World.

Develop and Rule: A Western Project

It was Enlightenment discourse which originally gave meaning to concepts of the "modern." The West became the universalizing model for societal progress. It was Western civilization, rationality, and order that were proclaimed and bestowed with universal relevance. At the same time, such an enunciation was intimately connected to the creation of a series of oppositions – for example, "civilized versus barbaric nations," or "peoples *with* history and those *without*" – which were reflections of the need to construct a non-West other so as to ground a positive identity for the West itself, as well as providing a legitimizing logic for the will to conquer.

The nurturing of a positive identity for the West found a key expression in the United States of the nineteenth century. Already by the 1850s it was firmly believed that the American Anglo-Saxons were a separate and innately superior people who were destined to bring good government, commercial prosperity, and Christianity to the American continents and to the world. Thomas Jefferson, for example, wrote that England and the United States would be models for "regenerating the condition of man, the sources from which representative government is to flow over the whole earth" (quoted in Horsman 1981: 23). The sense of mission and the ethos of geopolitical predestination were captured in the phrase "Manifest Destiny," which first appeared in the 1840s. This belief in the drive of destiny, embedded in a particular religious conviction, did not remain restricted to the territories of North America, but extended south into Central America, the Caribbean, and subsequently to the whole of the Americas. Furthermore, throughout the nineteenth century Anglo-Saxon assumptions about US civilization being the highest form of civilization in history took firm root, as US attitudes toward other nations and inhabitants came also to be increasingly based on a well-defined racial hierarchy (Berger 1993; Stephanson 1995; Weinberg 1963).

Adherence to a sense of mission, and to Anglo-Saxon supremacy, carried within it a driving desire to bring the posited benefits of a superior way of life to other less fortunate peoples and societies. Frequently the desire to "civilize" and the will to "modernize" another society went together with a belief in the need for order and pacification. At the beginning of the twentieth century, President Theodore Roosevelt, casting his eye southward to the Latin American world – this "weak and chaotic people south of us" – wrote that "it is our duty, when it becomes absolutely inevitable, to police these countries in the interest of order and civilization" (Niess 1990: 76).

The idea of a modernizing and civilizing project, that was justified as part of a wider mission of imperial destiny, was given a practical political realization in a whole series of occupations. Before 1917, and the birth of what came to be seen as the "communist threat" to Western freedom and civilization, the United States occupied and administered the governments of the Dominican Republic (1916–24), Haiti (1915–34), and Nicaragua (1912–

25 and 1926–33), as well as maintaining a Protectorate role over Cuba from 1903 to 1934. In the case of the Dominican Republic, the imposition of development through occupation went together with a five-year guerrilla war against the forces of the US military government, and in other instances too, especially in the Nicaraguan case, there was no absence of resistance. While preserving order and deploying a geopolitical will to power over these other societies, the United States also introduced a series of related social and economic programs that were the precursors of contemporary development projects. There were initiatives to expand education, improve health and sanitation, create constabularies, build public works and communications, establish judicial and penal reforms, take censuses, and improve agriculture. In the case of Cuba, for example, which was occupied and ruled by the United States from 1898 to 1902, public school reformers built a new instructional system with organization and texts imported from Ohio. In 1900 Harvard brought 1,300 Cuban teachers to Cambridge for instruction in US teaching methods, and Protestant evangelists established around 90 schools in Catholic Cuba between 1898 and 1901. At the same time, serious efforts were made to "Americanize" the systems of justice, sanitation, transportation, and trade, while the institutions of the Cuban independence movement – the Liberation Army, the Provisional Government, and the Cuban Revolutionary Party – were disbanded by the US military government.

These geopolitical interventions entailed projects for the modernization and development of other societies. The interventions and penetrations were portrayed and underwritten in terms of order, civilization, and destiny. To develop another society was also to rule over it, and to restore order was part of a wider project of civilizing the Latin South. Moreover, the early part of the twentieth century was also witness to a dramatic expansion of US corporate capital into Latin America. American corporations committed hundreds of millions of dollars to developing mining, agriculture, and the petroleum industry, as well as modernizing electrical and telecommunications systems. US business leaders thought of their industrial civilization as an exportable commodity that could take root in Latin America, provided that business practices could be standardized across the Americas (Salvatore 1998). Nor was this vision only held by American business leaders; the Peruvian president of the 1920s, Augusto Leguía, once declared that if he had his way "Peru would be practically American within ten or fifteen years" (quoted in O'Brien 1999: 109). Naturally, such a vision was not shared across the political spectrum and increasing US corporate penetrations fueled the fires of Latin American nationalisms and growing opposition to US hegemony.

It is important to remember these earlier forms of intervention and representation because in the post-World War II period the growth of modernization theory had key roots in these previous North–South encounters. There were significant discursive continuities, but equally the post-1946 period witnessed the emergence of a series of crucial geopolitical changes.

Since the late 1940s, two related but far from identical discourses of Western development came to be constructed. First, during the 1950s and 1960s, in a time of the Cold War and the coming into being of a whole series of newly independent nation-states, conceptualizations of the modern became central. Modernization theory, as it came to be called, took root in the academic citadels of the West and found expression across a broad spectrum of disciplinary domains. It was multidisciplinary and more multidimensional than the econocentric Marxist analysis of the 1970s. It encompassed questions of economic growth, social institutions, political change, and psychological factors. Its tenets found a home in geography as well as history. Essentially, modernization theory was constructed around three interrelated components: an uncritical vision of the West, largely based on a selective reading of the history of the United States and Britain; a perspective on the non-West, or traditional, societies that ignored their own histories and measured their innate value in terms of their level of Westernization; and an interpretation of the West–non-West encounter which was based on the governing assumption that the non-West could only progress, become developed, throw off its backwardness and traditions, by embracing relations with the West. The posited dichotomy between the "modern" and the "traditional," ideal types of a Weberian vintage, was also replicated within the so-called traditional societies of Africa, Asia, and Latin America. Here, the researcher was encouraged to observe a duality between modern urban centers of growing Western innovation and traditional rural peripheries; "development" would come about, would be engineered, through the diffusion of innovations (capital, technology, entrepreneurship, democratic institutions, and the values of the West).

Such a brief sketch remains unavoidably incomplete. For example, it needs to be stressed that modernization theory went through two phases, especially visible in the political science and sociological literature. In the initial phase, that lasted until the mid-1960s, there was a sense of optimism, a firm belief in the potential success of modernizing and Westernizing the traditional society. If only the premodern society could find ways of accepting and adapting to the spreading tide of modernization, its future long-term development would be assured. But with the rise of radical nationalist movements in the Third World, the onset of the Vietnam War, the Cuban Revolution, and the growth of social protest inside the United States itself, notions of modernization increasingly came to be associated with discussions of political order, societal breakdown, and "diseases of the transition," as ostensibly exemplified by communism. Finally, by the beginning of the 1970s, belief in the universal applicability of the modernizing imperative was clearly on the wane. This can be explained partly as a result of the growing realization in the West that the societies of the developing world were far more heterogeneous and complex than originally depicted by the protagonists of modernization theory, but also because of the upsurge of resistance to the

West's, and especially the United States', project of developmental rule. Defeat in Vietnam, and the earlier "loss" of Cuba to communism were key crystallizing moments for this new *Zeitgeist*. The will to modernize and control the recalcitrant other had been broken, albeit temporarily.

In a second wave of developmental doctrine, frequently considered as neoliberal, and customarily couched in the terminology of structural adjustment, privatization, deregulation, free trade, and market-based development, an apparently new model from the West was prescribed for the Rest, as their model too. Economies had to be opened as never before; state structures, an ostensibly rigid barrier to successful development, had to be rolled back and streamlined; financial discipline was to be strictly imposed; and the logic of the market was to be given full reign for the benefit of all. Initially and fundamentally rooted in earlier currents of economic liberalism, the discourse of the 1980s was amplified to incorporate notions of good governance, fiscal decentralization, participation in development, and the strengthening of civil society. By the late 1990s, governing representations of the economy and the state had been joined to "social capital," which has been deployed to cover notions of trust, networks, associations, and movements. From the early 1980s onwards, the terrain of intervention has been extended and the project of development has become thoroughly global in reach, encompassing both the South and the North. Already, in the early 1990s, an OECD (1992b: 49) report noted that solutions to the domestic problems for which policy-makers in the West have responsibility "are increasingly associated with the economic and institutional functioning of other societies . . . and this creates new scope for mutual understanding and synergy among policy-makers in donor governments as they tackle development as part of achieving a global agenda."

In comparing these two waves of Western development theory, it is important to understand the convergences as well as the points of difference. One important break characteristic of neoliberalism has concerned the treatment of relations between the public and private sectors. In the 1980s the private sector was championed, whereas the public sector, and more specifically the state, has been envisaged as a brake on development, a site of inefficiency, and institutional stagnation. A supreme belief in the benign opacity of market forces, the sanctity of private ownership, and the superiority of achievement-oriented individuals has permeated the orthodox texts of neoliberalist doctrine. In contrast, modernization theory gave greater weight to the nation-state in developing countries, and stressed the importance of a greater degree of balance between the public and private sectors. At this time the influence of Keynesian ideas was still an important factor, and overall the state's role in economic development was seen in a more pragmatic and less doctrinaire light. Equally, however, it needs to be kept in mind that the official discourse of development has its own dynamic. For example, in this particular case, in the wake of growing social inequalities and instabilities, we have seen a shift in the way the state has been depicted, with the World Bank, in its *Annual Development Reports*, moving from an earlier notion of a "minimal state" to a more recent call for an "effective state," where a previously negative portrayal of the state's role in economic development has been replaced by a more balanced interpretation (World Bank 1997; for a critique of the World Bank's concept of an "effective state" see Hildyard and Wilks 1998).

In terms of commonality, both theorizations have shared a belief in the universal relevance of Western models of development. Their points of departure have both drawn on an idealized construction of Occidental history and geopolitics, and their recommendations for development and progress have all assumed that the North–South encounter has been intrinsically beneficial to the South. The ethnocentric universalism of modernization theory provided one of the main targets of the dependency critique of the late 1960s and early 1970s (Slater 1993b) but, in contrast, the neoliberalism of the 1980s and beyond, which has been equally Westocentric, has tended to escape a similar interrogation (for an exception, see Brohman 1996), even though many other elements of the neoliberal "regime of truth" have been subjected to detailed critiques (for a recent geographical text, see Klak 1998).

Overall, the encompassing political climate has clearly favored the reassertion of Occidental hegemony in matters of development and modernization. And it is precisely in this context that we need to stress the fact that it is the political ascendancy of those who believe in the supposed superiority of the orthodox Western model of development, combined with their institutionalized power and enormous resources, that explains the widespread effectiveness of the "development as Westernization" discourse.

The Waning of the Marxist Challenge

The emergence of a Marxist and neo-Marxist challenge to orthodox, or what were customarily referred to as capitalist, views of modernization and development dates from the 1960s, and, with reference to the imperialist countries, especially from 1968. Dependency perspectives, which have been discussed in great length elsewhere (Kay 1989), prepared the ground for a move on to a more clearly inscribed Marxist terrain. The rise of Marxist thought was a generalized phenomenon in the 1970s world of the social sciences, and in the domain of development studies theoretical interpretations of modes of production, unequal exchange, world systems, and class conflicts figured prominently. While primary conceptual importance was given to the relations and forces of production, to capital and wage labor, to the internationalization of capital as a social relation, and to class structures and struggles, in a more directly political language, capitalism was denounced as a system of exploitation that had to be replaced by socialism.

Although it is certainly the case that the Marxist diagnosis of the social and economic issues of development retains a role in the critical geographical and social science literature, its generalized influence has declined. Apart from the impact of the events of 1989 and the disintegration of the erstwhile Second World, the fading intellectual and political influence of Marxist approaches can be related to three interrelated problems.

1   First, there has been a traditional tendency, although now much less marked, to assume that the economy would always be determinant in the last instance. In other words, it was presupposed that the logic of capitalist economic development governed the outcome of social and political processes. In one such reading the nature of the state was read off from the dynamic of this underlying logic. Moreover, it was frequently the case that not only was the economic structure centralized within the overall explanatory frame, but additionally subjects or social actors came to be absorbed within this determining structure. This analytical tendency represents an undiluted example of Marxist economism.

2   Second, in the examination of sociopolitical change, the key subject has always been a class subject, and the class struggle has been interpreted as the defining historical struggle. There have been two interconnected difficulties here. Overall, there has been a failure to analyze the ways in which different forms of social subjectivity come into being. The processes through which individuals in society are constituted as social subjects or agents have been neglected, since the overriding concern has been with class subjects. Instead of viewing the class category as one possible point of arrival in an examination of social subjectivities, it has been taken as a pre-given point of departure. In addition, it has been assumed that in the construction of social consciousness, the point of production is central and determinant. Consequently, the understanding of the heterogeneity and complexity of social consciousness has been severely circumscribed. But also, since the social subject at the point of production has been interpreted as a unified subject, centered around the experiences of the workplace, it has been less possible to begin to comprehend the barriers to mobilization at this site of potential conflict.

The major problem with Marxist class analysis, particularly as it has been deployed in development studies, concerns the failure to theorize subjectivity and identity. This failure is in its turn conditioned by the belief that what classes do is spelled out by their situation in the relations of production, which precedes them causally as well as logically. In Marxism, classes are the agents of the historical process, but its unconscious agents, since it is posited that social being determines consciousness. But it can be more effectively argued that the reproduction of material existence, as well as the constitution of social being, must presuppose thought; they are not prior to it.

3   Third, permeating so much of the Marxist and neo-Marxist canon has been the belief in the existence of a pre-given, privileged social subject,

cast in the role of the historical bearer of the revolutionary rupture from capitalism. Working-class revolution was for so long seen as a species of cure for the diseases of capitalism. But when actual revolutionary breaks occurred, as in a number of peripheral societies, the agents of those ruptures and splits could not be straitjacketed into any category of class belonging. In the cases of the Cuban and Nicaraguan revolutions for instance, at specific historical moments, a variety of social subjects, unified around a particular political horizon, which combined a range of attitudes, feelings, sentiments and desires – around questions of the nation, of the fight against dictatorship, of the need for social justice and equality, of the struggle against US imperialism – came together and took a series of actions that culminated in the moment of revolution. These revolutions, and others in the societies of the South, were not engineered by an insurgent working class but were brought to fruition by an aggregation of forces sharing a common political imagination. And such a sharing was precarious, temporary, and unfixed. What these kinds of revolutionary upheaval demonstrated was the emergence at given moments of a *collective will* to overturn an existing order. Crucial to these processes of change was the fusion of a will to overthrow an established and unjust order with a desire for national dignity and independence.

Further to these three analytical problems of the Marxist reading of development and political change, we have witnessed the institutionalized embodiment of Marxist, and more precisely Marxist-Leninist, ideas, in authoritarian, one-party states. Cuba is a striking example of such a phenomenon. With its one-party system there has been a clear lack of any division of powers within the political space of the state. The notion that the party is a synthesis of society, or the earlier belief dating back to Che Guevara that in the post-revolutionary period there ought to be a total identification between society and government, capture the meaning of a desire for total power. Within such a system, Marxist-Leninist thought was converted into a state doctrine, and used to portray the idea of a society without antagonisms, a society in which the class conflicts of capitalism had been transcended, a society in which political struggle had been successfully concluded. The lack of any institutionalization of the means for the expression of effective difference, or alternative strategies, has sealed into place a relatively inflexible structure that carries within it a reduced potential for constructive renewal. It can be argued of course that such an institutionalization of revolution was perhaps inevitable given the unrelenting hostility of the United States, the continuing blockade of the island, and the permanent specter of destabilization.

On the other side, the development achievements of the Cuban Revolution are well known: transformations in the systems of health and education; impressive improvements in the utilization of agricultural and mineral resources; the installation and extension of public utilities and services; and sharp reductions in the degree of social and economic inequality. The meeting of basic needs was always a key priority of the Cuban Revolution before it became a catch phrase for international organizations. But, unlike the Nicaraguan Revolution of the 1980s, or the Zapatista insurgency in Mexico, post-1994, the Cuban process has not been characterized by any attempts to create new ways of combining different democratic practices. The longevity of the US embargo and general geopolitical isolation in the Western hemisphere ushered in a protracted reliance on the former Soviet-bloc countries. With the dissolution of the Soviet bloc, the drastic decline in trading and financial ties with Russia, and the continuance of US hostility, expressed in the Helms-Burton Act of 1996, Cuba's political future has become far more precarious. The lack of political plurality and the absence of institutional channels for the expression of difference have thrown a long shadow over the basic-needs achievements of the Revolution. For many years the debate on Cuba was polarized between the uncritical supporters of revolutionary change and those who obdurately refused to countenance any positive evaluation of Cuban development. One of the contemporary dilemmas takes shape around the following problem: how will it be possible to preserve those social and economic achievements of the post-revolutionary period, while at the same time opening up the political system to a variety of different currents? In the Cuban case not only is the primacy of the political so apparent but equally the future trajectory of the island's development will be intimately connected to the changing impact of geopolitical circumstances and specifically, at the beginning of 2001, to the future strategy adopted by the incoming Washington administration.

Development and the Geopolitics of Knowledge

As traditional Marxist perspectives have faded in significance, the more recent critical currents of theory have been particularly concerned with questions of identity, difference, subjectivity, knowledge, and power. In development studies, and as exemplified in the contributions of Escobar (1995), Ferguson (1994), Rahnema and Bawtree (1997), and Tucker (1997), attempts have been made to construct a radical critique of the discourse of "development," seen as a hegemonic form of representation of the Third World. This new critical current carries within it certain shared assumptions and concerns. Prominent among these are: an interest in local knowledge and cultures as bases for redefining representation and societal values; a critical stance with respect to the established discourse of development knowledge, as produced and disseminated by international organizations; and the defense and promotion of indigenous grassroots movements.

One of the encouraging and stimulating facets of today's critical research is the opening up of a series of interconnected pathways of analysis. Encouraged and enriched by the influence of feminist theories which have established not only the significance of gender, but also the centrality of questions of identity and difference, new work on the possible meanings of development has increasingly come to include a key gender dimension (Marchand and Parpart 1995). In the connected analysis of new social movements and their relevance for rethinking the purposes and ends of development in neoliberal times, the role of women in these struggles has been paramount, just as the investigations carried out by women social scientists and activists have been so fundamental to the theorization of social movements (Alvarez, Dagnino, and Escobar 1998; Radcliffe and Westwood 1993).

Environmental issues, and the movements that have emerged to counter the damaging effects of conventional development projects, have received increasing attention and have been linked to the relevance of indigenous knowledge for the protection and sustainability of vital resources. When considering environmental politics, the links with the relevance of indigenous movements leads to a broader discussion of the objectives of development within which a challenging of the orthodox Western definition of knowledge for development can take place (see Routledge, chapter 20, this volume). At the same time, as environmental issues are also global issues, the framing of the agenda and the selection of priorities directly connects with the geopolitics of North–South relations. While in the North, for example, there has been a tendency to concentrate on the rural aspects of environmental policy, in the South more urgent issues relate to the *urban* nature of environmental deterioration. Also the attempts by the North to define the environmental agenda and monitor the policies of Third World governments evoke crucial problems of ethics and international justice.

The processes of self-reflexivity, of stretching out toward new themes of dialogue, learning, and rethinking, can well be regarded as a positive and enabling element of the postmodern sensibility. In the North there is the need to reinvent ourselves as other, which can be set in the context of taking historic responsibility for the social locations from which our speech and actions issue. The process of this reinvention requires the will and desire to learn from the South, not in a romantic or uncritical vein, fueled by an unconscious sense of culpability, but as a way of better understanding the North itself, and with it the South. The life of the mind does not begin and end inside the Occident; the enclosure of Occidental thought needs to be fissured and broken open to other currents of thinking and reflection. If we are to develop a genuine global expansion of knowledge and understanding, the West's self-enclosure within ethnocentric standards will have to be transcended. Conventional development theory and practice has been one expression of power over another society and economy – a reflection of a belief in the West's manifest destiny. The Marxist challenge sought to bring into being another form of power based on socialist principles, but here too there has been a tendency to underwrite a privileged position for the West's standards and meanings of development. In today's discussion of the significance and dispositions of development, the politics of the production and deployment of knowledge has become an increasingly pivotal question. In an increasingly global world, the geopolitics of knowledge and power is a theme for which geographers can make an important analytical contribution, and, in the field of development studies, a critical geopolitics can begin by considering the dynamic of power relations involved in the past and present of the North–South divide.

Part II
Geopolitical Change

Introduction to Part II:
After the Cold War

Among the categories of global change covered in this volume, the geopolitical seems, at first, to be quite distinctive in its pattern of global shift. In the 1990s nobody doubted that recent geopolitical changes had fundamentally altered the meaning of world politics, but a decade earlier even thinking of such a likelihood would have been deemed fanciful. During the 1980s, while other scholars were reporting critical changes such as the end of Fordism, the Aids pandemic, the postmodern celebration of cultural diversity, and the depletion of the ozone layer, international political scholars were confronted with the seemingly unchanging politics of the Cold War to study. The pattern of international relations appeared frozen into a bi-polar world of USA and friends versus USSR and friends, where the only changes were ones of degree: "freezes" and "thaws," in the jargon of the time. In the early 1980s President Reagan, viewing the USSR as the "evil empire," led the creation of the so-called "second cold war." This was to be the last "freeze" phase and was a reaction to US suspicions of the previous "thaw" phase, the *détente* of the 1970s. But whatever the political "temperature," the basic pattern of conflict was not changed, although as the conflict intensified the world was periodically made a more frightening place. The "thaw" that followed the coming to power of Gorbachev in the USSR in 1985 was interpreted by contemporaries, quite reasonably, as merely a second *détente*. As late as 1987 Edward Thompson could write about the Cold War seeming to be "an immutable fact of geography," with Europe "divided into two blocs which are stuck into postures of 'deterrence' for evermore." With our immense power of hindsight we know that all was soon to change. The policies developed by Gorbachev, perestroika (restructuring) and glasnost (openness), did not create a new phase of the Cold War but set in train processes that were to bring about its demise.

Geopolitical Transition

The political upheavals between 1989 and 1991 constitute a geopolitical transition (Taylor 1993b) from one world order, the Cold War, to another that, despite early US proclamation of a "new world order," has yet to be clearly defined. These transitions separate one relatively stable pattern of international politics, a geopolitical world order, from another. Despite the fact that the changes are very large, with old friends becoming new enemies and vice versa, the change-round is typically very rapid. Hence their hallmark is surprise as the political world is "turned upside down" in just one or two years. All this is true of 1989 and its aftermaths. The end of communism as a force in Europe (1989), the reunification of Germany (1990), and the demise of the USSR (1991) are each massive world events which together have completely changed contemporary world politics. Post-Cold War geopolitics are much more fluid and unpredictable and, in the 1990s, have been final confirmation of the general world ambience of great social changes out of control. The geopolitical-transition pattern of change – stability followed by very rapid alteration – can be viewed as a political time-lag, a delayed response to the reduction in world economic growth rates that was about two decades old by 1989. During such difficult phases of the world economy we can expect additional competition between states as they strive to protect their economies from the general malaise. In the rich core countries policies are devised to export economic problems to each other and down the economic hierarchy. In the process, the periphery is economically devastated; but it is in between, in what are termed semi-periphery countries, that economic pressures have their greatest impact on politics (Chase-Dunn 1989). The 1980s witnessed the collapse of semi-peripheral Latin American military dictatorships as the precursor to the collapse of semi-peripheral communist Eastern Europe, including the USSR. It is the peculiar role of the latter state, as political superpower but with a semi-peripheral economy, that precipitated a geopolitical transition deriving from the common travails of the semi-periphery.

This interpretation of the 1989 revolutions and their aftermaths shows that the rapid changes of geopolitical transition can be related to more gradual shifts typical in other areas of social change. There is much more to geopolitical change than alternating world orders. There have been important worldwide and gradual political changes throughout recent decades that are at least as important as the 1989–91 geopolitical transition for understanding future geopolitics. For instance, although in Eastern Europe alterations in state functions and the rise of nationalisms had to await 1989, elsewhere key debates about the nature of states and their relationships with national groups had been ongoing for more than two decades. Cutting the welfare state has been a universal response to the slowdown in economic growth, and this has usually been accompanied by a redefinition of the state's role. As markets have been proclaimed to be the arbiters of income and wealth distribution, the old corporatist state of partnership between government, business, and labor has given way to a far less interventionist state. There has even been a wide-ranging discussion on whether states as we know them are in terminal decline, because they are unable to regulate the contemporary world successfully.

Behind all these changes lies the matter of US relative decline in the world system (Wallerstein 1984). Often interpreted as the latest decline of a world hegemon (dominant in economics and culture as well as politics), since about 1970 other Western states, notably Japan and Germany, have become economically powerful to rival the US in the world market: in the 1990s, however, the United States experienced an unparalleled period of economic growth from 1992 until 2000 (contemporaneous with Bill Clinton's presidency), whereas Germany and Japan experienced economic difficulties – the former because of the costs of unification of former East and West Germany; the latter because of unsustainable banking and other policies. From undisputed leadership after World War II, in the 1970s the USA had to settle for a trilateral pattern of power (USA, Japan, Western Europe) in economic matters. The "second cold war" of the 1980s can be interpreted as the USA reasserting its authority over allies by privileging security, where it remained undisputed leader, over economy. However, this just created a huge budget and trade deficit to exacerbate the general relative decline. The result has been the power anomaly of the 1990s, where the USA is the sole remaining superpower but its economy faces major challenges. Nevertheless we should not overlook the important legacies of US hegemony. In particular, the system of trans-state organization constructed by the US in the 1940s is as salient as ever. If the state is really in terminal decline it is these trans-state institutions centered on the United Nations that are expected to take on the burden of world governance in the future.

New World "Disorder"

The early 1990s were heralded as the beginning of a "new world order." With the end of the Cold War the tri-polar division of the world – the USA and its allies; the USSR and its allies; and the non-aligned countries – disappeared, and there were hopes of a much more unified response to international issues. The Gulf War of 1991 appeared to be the harbinger of this, as a substantial number of countries contributed in a variety of ways to the UN response to Iraq's invasion of Kuwait, whereas others largely accepted the US-led strategy. A few years later, collaboration initially among EU countries with regard to conflict in Yugoslavia, later joined by the USA in fashioning a resolution of the Bosnia-Herzegovina conflicts and then by a NATO-led campaign against Serbia over its policies in Kosovo, saw further international collaboration, with other countries either acquiescing (while expressing some concern) or being convinced to participate in other, nonmilitary ways.

Despite these collaborative efforts, however, alongside others led by the USA as the remaining "superpower" (notably in the attempted resolution of continued conflict in Israel–Palestine), international tensions remained considerable and indicated that the collapse of the Cold War was not going to lead to an easy international peace. There were major concerns about a number of the larger Asian nations. India and Pakistan both tested nuclear weapons, for example, and occasionally raised the temperature of their conflict over Kashmir (which proved unresolvable – even to the extent of not agreeing a post-meeting communiqué – at a much-heralded summit of the two countries' leaders at Agra in July 2001), and China remained largely outside the growing *détente* – regarded with suspicion by the "western powers" and itself suspicious about the attitudes and behavior of others, as exemplified by the collision between a US spy plane and a Chinese fighter in the South China Sea in April 2001. As well as leading to the death of the Chinese pilot, the US plane made an emergency landing in China without verbal permission, and ten days of diplomacy over the return of the US crew were required as politicians sought a form of words that both sides could accept. This diplomatic problem spilled over into the debates regarding the 2008 Olympic Games, which were allocated to Beijing in July 2001 and was widely seen (there at least) as a major political success following the condemnation of China's human rights record after Tiananmen Square in 1989.

Alongside these continuing tensions involving the former superpowers and their major allies, much attention was focused on what became known as "rogue states": those whose leaders (at least, because many were not fully accountable in the usually accepted sense of Western democracies, and their control of information flows restricted popular appreciation of many situations) were considered threats to world (or regional) stability, through their possession (potential if not yet actual) of weapons of mass destruction. North Korea is frequently presented as an exemplar, as too is Iraq: its defeat in the Gulf War and expulsion from Kuwait did not lead to the downfall of Saddam Hussein's government, and was followed by a decade of pressures on his regime to allow inspection of weapons establishments there, regularly accompanied by bombing of strategic targets and continuous monitoring of military activity to "protect" dissident minorities in both the north and south of the country – many of whose members were suffering the impacts of the economic sanctions imposed on the regime by the UN. It should be noted, however, that the "rogue state" agenda is one that is promoted by US security interests as a means of replacing the old communist threat with a new less predictable threat. This is a means of reversing the post-Cold War "peace dividend" and legitimating increased new arms through militarizing space (sometimes known as "Son of Star Wars"). Of course, if rogue states were to be defined as states outside current consensus on international law and its future development (e.g., as indexed by failure to accept and implement UN resolutions or refusal to accept new instruments of global governance such as the Kyoto Agreement on the environment) then Israel and the USA become the exemplar "rogue states."

One of the US-defined "rogue states" that attracted much attention during the 1990s and the first years of the next decade was Serbia, which played a substantial role in the violent dismemberment of the former Republic of Yugoslavia. After Croatia and Slovenia seceded in the early 1990s, achieving both independence and international recognition (the latter very much through German pressure within the EU), the Serbs (under their elected leader, Slobodan Milosevic) stimulated intercommunal strife in the multi-ethnic republic of Bosnia-Herzegovina, which involved ethnic cleansing as significant numbers of Croats, Muslims, and Serbs were forcibly moved in the contests for "ethnically purified territory." This conflict was eventually contained after increased US involvement in the search for a solution – which remains uneasy as the division of Bosnia into two semi-independent "statelets" (a Croat–Muslim Federation and a Serbian homeland) is far from tension-free at a variety of spatial scales. The focus then switched Kosovo, a province within Serbia, where the predominantly Muslim population with strong ethnic ties to Albania was being repressed by the Serbian minority. NATO attempts to resolve this led, after the failure of "peace talks," to its air attacks on Serbian forces – in Kosovo and elsewhere – and retaliatory ethnic cleansing by the Yugoslav/Serb armies which stimulated mass forced migrations of Kosovo Albanians into neighboring Macedonia. Serbia eventually withdrew its forces (though not until after several bombing attacks in its capital, Belgrade, in one of which a major diplomatic incident was caused when the US Air Force bombed the Chinese Embassy) and Kosovo was divided and "occupied" by a number of countries, including Russia, to oversee its reconstruction.

As with Iraq, although the NATO-led Kosovo campaign ended the Serb atrocities there, they did not result in the removal of the Serb leadership. This only occurred a year later, when popular protest after a rigged presidential election led to Milosevic's downfall. But the ethnic tensions in the area did not disappear and Albanian Muslims in the former Yugoslav republic of Macedonia began armed action in support of their case that (despite being represented in the Macedonian government) they were discriminated against. During the Cold War the leaderships of the superpowers were able, through a variety of means, to constrain popular protest, but with the end of their conflict the ability of groups (usually stimulated by leaders with political aspirations) to press their national causes has increased, creating zones of instability in many parts of the world, which other countries increasingly feel impelled to get involved with (just as the USA did with conflicts in Latin American states in previous decades). The situation in the former Yugoslavia is just one exemplar of this – intriguingly, in the same part of the world that political geographers and others in the mid-twentieth century identified as one of the main global shatter-belts.

As both Iraq and former Yugoslavia illustrate, short of all-out war aimed at removing the government in power, it is difficult to change the policies of such "rogue states" and prevent future problems erupting. And yet such all-out wars are off the agenda of the states who set themselves the role of policing the "new world order" – notably the USA – because of potential internal difficulties: the Vietnam War proved to US governments the great difficulties of sustaining popular support for international activity involving deaths of their own soldiers and (possibly) civilians, and it was widely believed during the Kosovan conflict that the Americans at least were unwilling to commit their forces to actions that significantly endangered their lives (by, for example, concentrating on bomber raids by planes that flew above the reach of anti-aircraft fire). Thus they have to contain the "rogue states" without being able to defeat them. The problems with post-Gulf War Iraq illustrate some of the difficulties involved. The UN-backed alliance (led by the US and the UK) was convinced that Iraq under Saddam Hussein was continuing to develop weapons of mass destruction (both nuclear and biological): the end-of-war "peace terms" involved Iraq closing the facilities and UN teams being allowed to inspect the facilities – with economic sanctions to "ensure" Iraq's compliance. But negotiations over the inspections have been fraught, and the sanctions have stayed in place: Iraq is only allowed to sell oil on the world market in return for medicines and foods for its population, but there are claims that the money is being shifted to other purposes, that millions are living in poverty and ill-health as a consequence, and that repression continues (against which, the US and UK continue to bomb Iraqi facilities, largely outside the glare of international media coverage).

One other difficulty facing the "policers" of the "new world order" is that of punishment for those who violate its tenets. Popular forces do remove leaders such as Milosevic, but rarely does their punishment extend beyond loss of office and certain freedoms (of movement, for example). An International War Crimes Tribunal has been established by the UN – at the Hague – but relatively few of its indictments of serious war criminals have led to trials, because individual states have been unwilling to hand-over the individuals to such a body, often claiming that any trials should be internal to the country in which the alleged crimes occurred, and then being unwilling to arrest the individuals involved because of fears of popular unrest that might follow. Thus the two most senior figures in the Bosnian conflict – Radovan Karadic and Radko Mladic – have remained unarrested (if not free) even though UN forces are policing that territory and could undoubtedly track down and arrest them if the political will was behind them. One clear exception to this came in June 2001 when the Serbian government agreed to deport Milosevic to the Hague tribunal (a move opposed by the Yugoslav government) under threat of economic sanctions: a conference the following day released $150 billion of aid for Serbian reconstruction. Whether this is the beginning of a new era of international political regulation remains to be seen: Milosevic's initial defense is to challenge the tribunal's legality, and the long process of seeking to try those accused of international terrorism in the 1989 Lockerbie bombing of a Pan-Am plane suggests immense difficulties. In any case, it is argued that it is only the political and military leaders of defeated countries who are "brought to justice" in this way, and there are many examples of human rights violations by the victors which go unheralded and unpunished. It may be that a few successful cases could act as a deterrent to individuals, but the prospects are not bright – as the case of Chilean ex-President Pinochet illustrates: one of the claims in his defense against extradition for violations of human rights from the UK (where he was a visitor seeking medical aid) to Spain was that heads of government are immune from such prosecution.

The move towards a "new world order" – not as well advanced as when it was first proclaimed by US President George Bush in 1991 – involving the territorial restructuring of some areas of the world along ethno-nationalist lines, sustained by the military threats and even presence of external powers, is general but far from universal. Whereas those powers have been prepared (albeit reluctantly in some cases) to become involved with some conflicts, their response has been far from universal. Twice in the 1990s, for example, Russia faced armed rebellion in the small Caucasian republic of Chechnya, with a brutal response on each occasion. It presented the situation as one of Muslim challenge to Russia's hegemony (and thus an internal Russian affair). Protests from outside were muted and not associated with any threats of either force or other sanctions. To charges of pusillanimity by critics of those policies, leaders of such countries argued that it was not always possible to intervene, but that they did what they could – a situation which is far from new: in 1956, the British–French–Israeli invasion of Egypt was halted by American initiatives, but nothing was done to halt the Russian invasion of Hungary.

Arms for All, Votes for All

Furthermore, there are many countervailing tendencies against the forces of peace within the world. One of these is the armaments industry, which is central to the economies of a number of countries – not only the US and the UK, but also others such as France and Sweden. The UK has regulations in place which prevent the export of arms to countries where they might be used aggressively – including against the state's own citizens – but these are often circumvented, in part by arguments that the arms involved can only be used defensively (a distinction that is difficult to appreciate once they are in a potential aggressor's hands). Many governments wish to sustain their defense industries and also, covertly, the policies of some of those who wish to buy and use their arms (as demonstrated in the UK in the 1990s by the "Arms to Iraq" case when the government was accused of covertly encouraging such sales; and by the US decision in July 2001 not to sign an agreement regarding the production of biological weapons – over which negotiations had taken seven years – because, among other things, the proposed inspection regime would compromise US national security and commercial interests). And there is a massive black market trade in arms – again, often implicitly sustained by governments. Many states spend large proportions of their budgets on defense – 25 percent of GDP in the case of Sri Lanka – and very few are likely to follow New Zealand's 2001 decision to disband much of its air force because it could perceive no future threat, which meant that it would not be needing fighter planes to defend its air space.

The geopolitical map is in considerable flux at the century's start. So too is the political complexion of many of its territories. With the collapse of the Soviet Union and the end of the Cold War in 1989, an American political commentator, Francis Fukuyama, contended that we had reached the end of history – meaning not that time would stand still hereafter, nor that there would be no change, but that the virtual universal adoption of liberal democracy was an "end state" in social evolution/progress: the future will just see improvements to that model. Part of the transition to democracy in many parts of the world (notably in Latin America and Asia) reflects American pressure and the US "success" in the Cold War. American (and also European) aid has been tied to not only the economic policies insisted upon by the IMF and other bodies, but also an insistence on the adoption of democracy (however rudimentary that might be) – and countries which "stray," as both Nigeria and Pakistan did when there were military coups in 1996 and 1999, are sanctioned until they regain the democratic path. The main area where liberal representative democracy has been (re-)introduced in recent decades has been in the former USSR and its Eastern European allies. Among the latter, the desire for stronger ties with Western Europe (through membership of NATO and, eventually, the EU) has stimulated the democratic drive – though intriguingly the changes have been much greater in those countries bordering on the EU than in those former Soviet republics on its southern Asian borders.

Analysts identify three main stages in the democratization process. The first involves the universal provision of basic civil rights – such as the right to a free trial under due process. Then comes universal participation in the political processes associated with representative democracy – such as the freedom of speech, the freedom to campaign and to form political parties, and the freedom to vote in elections where all votes are equal. Finally, there is equalization of power, whereby all individuals have equal opportunities to participate. Few of the countries that have recently adopted democratic forms have proceeded far through the second of those stages, and some may still have much to achieve in the first. Even in the longest-established democracies, however, there is much to be done before the final stage can be completed; political power – strongly associated with economic power in most countries, and in some with social and cultural locations too – is unevenly distributed everywhere. But at the macro-scale at least, that unevenness is being reduced. But challenges to it are very unequal in many parts of the world, where the state (often in association with major transnational companies) can readily respond to and put down protests – even when they are supported by human rights and other campaigners elsewhere, as occurred in Nigeria in the mid-1990s over the exploitation of oil in the country's southeastern regions and the detention and execution of protesters, including an author of world reputation, Ken Saro-Wiwa. (Some of that protest, notably against the oil giant Shell, involved geographers protesting against the Royal Geographical Society's acceptance of donations from Shell.) Popular democracy is difficult to mount and even more difficult to sustain under conditions of major power gradients – it may be even more difficult if more superstates are formed, and power is handed to unaccountable international agencies dominated by a few major players.

September 11, 2001: A New Type of War? A New Geopolitical Transition?

On the morning of September 11, 2001, four commercial airliners were hijacked in the United States, two on flights from Boston and two from Washington DC, all heading for the west coast. The first two were crashed into the twin towers of New York's World Trade Center, causing both to collapse with many thousands of deaths; one of the others was crashed into the Pentagon in Washington, with several hundred deaths, and the other crashed in Pennsylvania, en route to Washington, apparently as the result of passengers and crew-members tackling the hijackers. No organization claimed responsibility for these events, which were immediately associated with organized terrorism – specifically the al Qua'eda network and its chief animateur/sponsor Osama bin Laden, a former Saudi citizen who initially came to prominence during the 1980s anti-Russian resistance in Afghanistan (when he was funded by the USA), who turned against the Americans after the 1991 Gulf War. Al Qua'eda and bin Laden were believed to be responsible for an earlier attempt to destroy New York's World Trade Center in 1993, and for later attacks on the US embassies in Africa and a US warship.

September 11 has forever changed the way we think about risk and vulnerability. Americans, in particular, have generally been immune from the sorts of insecurities and violence – and the devastating poverty – that attend the life experiences of large swaths of humanity in the South. The indelible images of aircraft plunging into the twin Trade Towers, and of the buildings' horrific, vertiginous collapse, has however brought the awful realities of, say, the West Bank and the Balkans to the American heartland, with a terrifying pay off: an anti-globalization and anti-imperialistic resistance movement armed with all the knowledge and weaponry of twenty-first century "cyber-cultural" modernity.

The initial response to this first major act of international terrorism on US soil – certainly the first to cause many deaths – was for President George W. Bush to declare "war on terrorism" and announce the "first war of the twenty-first century." (He also referred to it as a "crusade" and the project was initially named "Ultimate Justice" – both insensitive terms for Muslims; for many, stimulated by the media, the war was against "Muslim fanatics," and there was considerable stereotyping, and some consequential racial violence, which identified all Muslims as Arabs, and all Arabs/Muslims as anti-American.) But, as Bush and members of his administration stressed, this would be a different type of war, since the enemy was not an individual state; the goal was to eliminate the terrorists, wherever they might be, using a variety of means. Hence, as well as a military response there were major efforts in the areas of diplomacy (building a coalition) and finance (closing bank accounts). This three-pronged "attack on terrorism" did suggest a very different type of war with, for instance, potential major losers being Afghanistan (military target as hosts of bin Laden), Israel (diplomatic target as obstacle in Arab coalition-building), and Switzerland (financial target as money launderer through its banking secrecy tradition).

In order to launch this war, the US president adopted a cautious approach, since an immediate military reaction against the terrorist leaders and bases – assuming that they could be located – carried the perils of collateral casualties among innocent civilians and stimulation of more anti-American terrorist activity. Thus for several weeks after the events of September 11 major diplomatic initiatives were taken to build an international coalition against terrorism involving as many countries as possible. In addition to the US allies in NATO (who for the first time used Article 5 of the NATO Treaty, which states that an act of aggression against one member is an act against all members) support was sought in Russia, China, and widely throughout the Arab and Muslim worlds, though without any formal backing through a United Nations Resolution. In this, President Bush was strongly supported by the UK Prime Minister Tony Blair, who took a major role in building the alliance through meetings and contacts with a number of national leaders, and whose Foreign Secretary visited Iran to obtain support there. (This was despite Iran being one of the USA's designated "rogue states," having long been accused of funding and supporting Muslim terrorist organizations operating in Israel, Lebanon, and Palestine.)

Although the coalition-building was around a general campaign against terrorism wherever it was based, most of the focus was on Afghanistan, where bin Laden had his training camps and was believed to be. There was a substantial build-up of military resources around Afghanistan whose leaders admitted that bin Laden was within the country but who, although they had invited him to leave voluntarily, were not prepared either to expel him or simply to hand him over to the United States. By early October the US had produced what several leaders of other countries – including the UK and Pakistan – had agreed was sufficient evidence on which bin Laden could be indicted. The coalition was preparing for attacks in and on Afghanistan, which began on October 7. In addition, the ground was being laid to help the Northern Alliance rebels, holding about 10 percent of Afghanistan, to be part of a replacement of the Taliban regime as a new more "friendly' administration.

One major area of attention, however, was the Israel–Palestine conflict, where a year-long intifada had resulted in hundreds of deaths, and all attempts at ceasefires over the previous year had failed. A considerable proportion of the anti-US sentiment in the Muslim-Arab worlds focused on this conflict, and in particular on US support for Israel. During the weeks after September 11 much diplomatic pressure was put on both sides in the conflict to get first a ceasefire and then peace talks, with the coalition leaders realizing that fragile support for their "war against terrorism" (and in particular their campaign against bin Laden and the Taliban) depended on them achieving a breakthrough in Israel–Palestine. Blair, in an important address (recorded in full in *The Times*, October 3, 2001), devised a carefully constructed even-handed approach. The state of Israel must be given recognition by all, freed from terror, and know that it is accepted as part of the future of the Middle East. The Palestinians must have justice and the chance to prosper in their own land, as equal partners with Israel in that future. However, Bush went even further by referring to the creation of a Palestine state, leading to a serious rift between the US and Israel only partially reconciled subsequently. Clearly September 11 has generated a need for new thinking by the US in its Middle East policies.

The long-term effects of the events of September 11, 2001 can only dimly be foreseen – and all of the coalition leaders talked of the "war against terrorism" being a long, perhaps continuous, war. If this is the case a new Cold War might mark the end of globalization because, according to John Gray in *The Economist* (September 29, 2001):

> The entire view of the world that supported the markets' faith in globalization has melted down . . . Led by the United States, the world's richest states have acted on the assumption that people everywhere want to live as they do. As a result they failed to recognize the deadly mixture of emotions – cultural resentment, the sense of injustice and a genuine rejection of Western modernity – that lies behind the attacks on New York and Washington . . . The ideal of a universal civilization is a recipe for unending conflict, and it is time it was given up.

In line with this view, but from a different political perspective, some saw a heightening of the "clash of civilizations" foreshadowed by Huntington (1998) – as expressed in a statement (subsequently only partially withdrawn) by the Italian prime minister in late September 2001 that Western civilization is superior to that of Islam.

Meanwhile, others sought to distance the terrorists from Islam, which they argued is a peace-loving religion. They were terrorists with no need for an adjectival prefix. Their importance derived from the fact that they were attacking our prosperity, hence the need to promote economic activity. There were already fears of a US-led recession in the world economy, which stockmarket reactions to the events of September 11 exacerbated, accompanied by the difficulties of many airlines and associated tourist businesses consequent on a massive reduction in passenger numbers. Bank rates in North America and Europe were significantly reduced to try to stimulate consumer confidence and expenditure, with Bush and Blair both encouraging citizens to resume normal lives. For Blair,

> Globalization is a fact ... The issue is not how to stop globalization. The issue is how we use the power of community to combine it with justice. If globalization works only for the benefit of the few, then it will fail and will deserve to fail. But if we follow the principles that have served us so well at home – that power, wealth and opportunity must be in the hands of the many, not the few – ... then it will be a force for good and an international movement that we should take pride in leading. Because the alternative to globalization is isolation.

Thus in this view globalization, far from ending, is part of the solution.

There is a key problem associated with such optimistic scenarios: the "why" question is not addressed. Why was America hated so much that the killing of thousands of innocent workers could be seen as a legitimate act by the many supporters of bin Laden in the Middle East and beyond? In the initial shock some commentators even suggested that it was morally unacceptable to seek an explanation for such a horrendous act. Drawing on religious language, the acts were designated as "evil," but more often there has been a retreat into escapist psychopathology dismissing the acts as "mindless" and the actors as "crazy." Blair, in his address to his party's annual conference on October 2, 2001, provided a list of the ills of the world – such as conflict in the Democratic Republic of Congo, intolerance in Zimbabwe, and climate change – but such disparate thinking hardly constitutes an explanation. Bin Laden has himself offered an answer to the why question. He lists the provocations as Israeli settlers on the West Bank, US military in the Arabian Peninsula, and continued US bombing of Iraq. These are coupled to his wider goal of overthrowing every pro-US regime, beginning, in order of succession, with Saudi Arabia, followed by Egypt and Jordan. Such an account is not uninformative, but as critical scholars we also know that the reasons people give for why they do what they do cannot always be taken at face value and are always incomplete and partial.

The starting point for addressing the why question must involve two related arguments. First, why is there so little discussion, as Edward Said has noted, of the US role in the world:

Its direct involvement in the complex reality beyond the two coasts that have for so long kept the rest of the world extremely distant and virtually out of the average American's mind. You'd think that "America" was a sleeping giant rather than a superpower almost constantly at war, or in some sort of conflict, all over the Islamic domains. (*Guardian*, September 13, 2001)

For a decade or so the US body politic has entered a deep narcolepsy, drugged by unprecedented economic growth, by unrivaled imperial power, but an almost surreal concern with the trivia and detritus of capitalist modernity: the 24-year-old billionaires, sperm on the blue dress, stockmarket mania. Thus although the USA has more military bases and personnel located in other countries than any state in history (no, you won't find this in the *Guinness Book of Records*), major commentators such as Robert Kaplan (author of *The Coming Anarchy*) can seriously hold the view that the USA has not been practicing *realpolitik*, using its power for its own national interest. Quite simply, the why question has become unthinkable in a national culture of global imperial power. Generally read as "arrogance" in the rest of the world, the US body politic has been "sleep-walking through history" to disaster.

Second, the fact of the global Islamic revival has to be addressed. In a universe marked by locality and diversity, generalization about Islam is always treacherous, but there surely are important discursive shifts and debates within what Ernest Gellner called the "Qu'ran Belt" of 1.2 billion Muslims. One aspect of this revival is Islamism, a series of movements of educated and urbanized groups, very different from the traditional ulema and sufi brotherhoods. These movements are modern, male-dominated, and seek to reinstitutionalize their conception of Islamic laws, institutions, and other imagined practices of the first Muslims. Like other movements there are a variety of tactics employed, from armed insurrection (Islamic Jihad), to building a parallel civil society (virtually everywhere), to the voting booth (Malaysia). Olivier Roy (1994) has shown that Islamism is concentrated among urban youth caught, as Lubeck and Britts (2001: 6) put it, in the "miasmic webs of multiple postcolonial crises." Political Islam, then, represents a shift in popular consciousness from the early postcolonial secular nationalist to the current Islamic narrative. This Islamism operates at many levels – the global ummah, reform of territorially defined nation-state, the moral economy of the urban neighorhood – and this in part explains its appeal and strength (Lubeck 1999).

Of course, Islamic reformism has a long history dating back to the seventeenth century and this is not the place to trace origins of movements such as Wahabbism and Hasan al Banna and the Muslim Brotherhood. But the point is that within this maelstrom of Muslim debate, radical ideas of the likes of the Egyptian cleric Sayid Qutb have had an appeal; they combined puritanical Islam with a sort of Leninist approach to organization. As Olivier Roy says, Islamists received their training not in religious schools but in colleges and universities "where they rubbed shoulders with Marxist militants whose ideas they borrowed and injected with Quranic terminology" (Roy 1994: 4). The huge increase in urban unemployed graduates provided a fertile ground within which such radical and militant anti-imperialist ideas could draw sustenance.

Islamism, one can say, has seized the imagination of sections of the urban youth broadly construed: it is what Immanuel Wallerstein calls an anti-systemic movement opposing US-led globalization (i.e., it is another sort of "anti-globalization" movement). To grasp Islamism's appeal and dynamics requires an understanding of the crisis of the secular nationalist development project in the Qu'ran Belt, and within the Middle East in particular. As a preliminary analysis, we can identify four powerful vectors that are intersecting to challenge globalization. First, the political economy of the oil boom produced rentier capitalism of a decrepit and undisciplined sort, and a profound sense of moral decay and state delegitimation prompted by the commodity booms and the shock of the new. Second, the vast financial resources that flowed to the Saudi and Gulf states exposed immigrant labor to Wahabbi and other Islamist doctrines and in turn funded global networks of associations and charities. Third, the intersection of the 1990s petro-bust and the IMF/World Bank-led austerity and neoliberal reforms further pulverized many already crippled states, throwing millions into poverty and further eviscerating state services and welfare provision (in which Muslim civic organizations came to play an enormous role as the state contracted and withdrew). And fourth, geopolitics – the effects of the Cold War struggles for which Afghanistan is a paradigmatic case, the US support of Israel and of West Bank settlements, and the collapse of the Soviet-socialist bloc – provided a setting in which Islamist ideas provided an obvious bulwark against US hegemony. The particular confluence of these powerful forces – all saturated with an American presence in the form of oil companies, IMF, foreign investment, and foreign-policy interests – crippled, one might say destroyed, the secular nationalist project which had shallow roots in any case in the region.

By the time this book is published and in your hands, much will have changed, but it is unlikely that the future will be very certain. Perhaps globalization will continue unabated, although that is unlikely: recession seems possible if not probable, and the economic order that will follow it is unclear. Furthermore, many governments are proposing greater internal and external security measures, trading-off hard-won civil liberties and human rights for controls over movement (and, perhaps, the right to intern people without trial indefinitely). A new economic world order may be the consequence of the events of September 11, 2001, as may a new global geopolitical order: tracking and analyzing them will be a major task for the immediate future.

8 Democracy and Human Rights After the Cold War

John Agnew

Introduction

The 2000 American presidential election provides a lesson in the fragility of democracy even in those parts of the world in which many people believe it is safely established. First of all, the winner, George W. Bush, had around half a million fewer votes than his opponent, Al Gore, but won because of the operation of the Electoral College, an institution whereby popular votes are cumulated by state and gives an advantage to the candidate winning more smaller states. The Electoral College is a remnant of the founding of the United States when the political elite was not only suspicious of popular election to national office, but also was forced by its southern members to give the southern states more voice in national affairs than their white populations justified. This was accomplished by including their black slave populations in the calculation of Electoral College votes. Second, the nationwide electoral turnout in the 2000 US presidential election was only a little above 50 percent of the total number of eligible voters (around 70 percent of those fulfilling registration criteria). The dynastic undertones of the choice of major-party candidates might have had something to do with the lack of enthusiasm both candidates engendered during the election campaign and with the final closeness of the vote. One candidate was the eldest son of a recent president, the other was the sitting vice president and the son of a long-serving US senator. Third, the lingering controversy over the counting of votes in the state of Florida that brought the election to a close over a month after election day suggests that a somewhat different meaning now attaches to the phrase beloved of American politicians: "Every vote counts." A significant undercount of votes occurred in counties using punch-hole ballots as the means of casting votes and with elderly and minority populations. Each vote, therefore, does count more than ever, but only if actually counted. Finally, according to one US Supreme Court Justice,

Antonin Scalia, referring to the Florida case, it is not entirely clear that the US Constitution actually mandates the election of the president by popular vote. He suggested that how the president is elected should be left in the hands of the states.

In many countries, American observers have frequently drawn attention to the "messiness" of elections and of democracy more generally, but it is only rarely that they have noted the historical and the contemporary messiness closer to home (Keysshar 2000). The 2000 American presidential election serves to remind us that even where long-established, the electoral basis to democracy and, hence, the claim to rule on a popular basis, is subject to institutional and administrative mediation that may produce outcomes that potentially undermine the democratic claim itself. Neither the US nor any other country, therefore, can define its model of democracy as beyond question or as *the* version that all others ought to follow. Democracy still remains everywhere more of an ideal than a fully realized working reality.

A commitment to democracy as a justifying ideal, and certainly as a set of political practices, is a recent innovation everywhere, dating at most from the late eighteenth century and in many places arising only over the past fifty to twenty years. Democracy has also never just "happened." It has always been the result of struggles by subordinated groups for recognition of their "right" to participate in political rule and in governing their own lives. Democracy's limited geographical spread and its apparent fragility even where it has a long history point to the perpetual tenuousness of its hold (Johnston 1999b). The revolutions that struck the former Soviet Union and Eastern and Central Europe in 1989–92 and brought an end to the Cold War have been widely heralded as signifying a celebratory moment in the worldwide spread of democracy (see, in particular, Fukuyama 1992). But experience in these countries since then, and the problems of democracy and human rights elsewhere around the world, suggest the need to be much more circumspect about democracy's success.

Be this as it may, political regimes of all kinds throughout the world now style themselves democracies, irrespective of whether their modes of leadership recruitment or treatment of their citizens conform to ideals of limited government and popular rule. A claim to democracy bestows legitimacy on modern politics. Rule by the few is justifiable only when their decisions, rules, and policies can be portrayed as representing the "will of the people." This democratic claim to rule is a truly modern phenomenon. It is also intellectually controversial. The majority of political thinkers, from the ancient Greeks to today's postmodernists, have been critical of both the theory and practice of democracy (Pangle 1992). Indeed, the very meaning of the word is itself the subject of dispute, even among its proponents. Liberal democrats tend to see democracy as involving the procedures for recruiting political leaders, the application of a rule of law, and the limitation of government to a "political sphere" largely independent of the workings of the economy (e.g., Wolin 1960). Pluralist democrats tend to champion the task of creating mechanisms for peaceful coexistence between contending religious and cultural traditions (e.g., Berlin 1958). Radical democrats tend to see democracy as potentially participatory and involving the democratization of all spheres of life, from the household and gender relations, on the one hand, to the economy and the political sphere, on the other (e.g., Pateman 1989). They are skeptical of liberal democracy's ability to offer anything other than a choice between competing elites represented by different political parties. Democracy's enemies, be they reactionary, fascist, or communist, tend to see it as invariably either a facade behind which "real" power operates without reference to how political personnel are selected, or as largely irrelevant in a world of rampant social inequalities and struggles for power.

After laying out the main ways in which democracy and human rights have been discussed, this chapter addresses how "democracy" was implicated in the Cold War, the dilemma for democracy posed by increased openness of all territorial states to transnational economic and cultural influences, the challenge to democracy posed by the explosion of "identity politics" based on group affiliations (race, gender, interest group) rather than state citizenship (particularly the rise of regional and ethnic politics), and the specific threats to and major challenges facing democratic achievement and possibility after the Cold War.

Democratic Theory and Practice

Proponents of democracy have differed profoundly over what its practice entails. Dispute has centered on such questions as: What constitutes a "people" capable of self-rule? Is democracy a process for selecting rulers or a set of institutions designed to achieve popular control over rulers? Is there an optimum size of political community for democracy? Is democracy only about deliberation on specific issues or are the institutions and rules of democratic practice also open to deliberation? What are the relative roles of obligation and opposition in democratic deliberation? Can democracies coerce recalcitrant minorities in the name of a majority? Are there social rights (e.g., full participation in management of the workplace, full employment, welfare rights, religious freedom, gender and ethnic equality, etc.) as well as political rights (e.g., voting, safeguards against arbitrary arrest or torture, an independent judiciary, etc.) that should be fostered in a democracy? Is democracy compatible with systems of hierarchical subordination, such as some churches and most large business enterprises?

Modern thinking about democracy developed alongside the expansion of the territorial state in Europe from the sixteenth century on. Democratic theories sought to balance the right of citizenship within a given territory against the regulatory and coercive capacity of the state (Held 1987). The relative weight given to the limitation of arbitrary power versus the extension of participation in government differed, respectively, between those with liberal and those with democratic agendas. Eventually, a syncretic "liberal democracy" provided the most frequently institutionalized way of resolving the dilemma in the form of representative democracy. This intellectual solution grew out of the experience of, and the challenges to, absolute monarchy emanating from the English Revolution (1640–88), the American Declaration of Independence (1776), and the French Revolution (1789). The Anglo-American tradition has undoubtedly dominated thinking about liberal democracy because of the global geopolitical importance that first Britain and then the United States acquired. Without this predominance quite different thinking about democracy might have prevailed.

Perhaps the most important Anglo-American statements of the liberal democratic position can be found in the philosophy of James Madison (1751–1832) and James Mill (1773–1836). From their points of view rulers could be held accountable to the ruled only through electoral choice involving a secret ballot, regular voting, and competition between potential representatives. The question of who would count as a "citizen" was left unresolved. Through often violent struggles, the idea that voting and other citizenship rights should be available to all adults (including women and racial minorities) has been slowly realized as a reliable feature of political life in only a limited number of states. Worldwide, the main achievement of liberal democracy, the periodic selection by the populace of representatives to make political decisions on their behalf, is more often than not either absent or episodic and subject to dramatic reversals. Recent doubtful electoral outcomes in the countries where the struggles have been most successful, such as the United States, are reminders of how readily reversals can occur.

Another approach to resolving the claim of citizenship against that of the state derives from the Marxist tradition. To Karl Marx (1818–83), and those who followed him, the great universal ideals of "liberty, equality, and fraternity" associated with the French Revolution could not be achieved only at the ballot box and in the marketplace. Rather, capitalism deepened economic inequalities, and the state only served to protect the collective interests of the capitalist class against the emancipation of the working class. Only in the wake of capitalism's demise would "true" democracy be possible. In this historical context, Marx envisaged a hierarchy of democratic fora extending from local communes to higher-order entities elected directly from the representatives at the lower level. This pyramidal structure was to inspire the Soviet system which emerged in Russia after the Bolshevik Revolution of 1917. A single party, the Communist Party, came to dominate this structure and effectively undermined the "bottom up" vectoring that inspired it. By appealing to the idea of social rights (full employment, etc.) the Communist Party offered compensation for the absence of political rights and introduced an alternative model to that of liberal or bourgeois democracy: "people's democracy."

In Western Europe and North America after the Great Depression of the 1930s a hybrid "social democracy" offered a third approach to the definition of citizenship rights, in which elected governments provided a basic range of social services by means of progressive taxation as well as expanded political rights. Under conditions of economic growth, such as prevailed there during the 1950s and 1960s, social democracy became widely accepted by most political groupings, however grudgingly. But since the 1970s this approach has been questioned, above all in Britain and the United States, as the "welfare state" has been defined by influential right-wing intellectuals and political leaders as a fiscal burden in an increasingly competitive world economy in which geographical and social inequalities are seen as inevitable or even natural. The fundamental historic distinction between the political left and the political right on attitudes towards inequality (Bobbio 1996) comes into renewed focus as people's democracy has disappeared from the global scene and social democracy faces the twin challenges of economic globalization and state budget cutbacks. This distinction has remained largely outside the purview of most writing on democratic theory except that of advocates of radical democracy who insist that, to be more than merely nominal, democracy requires at least a modicum of socioeconomic equality.

The Democracies of the Cold War

The Cold War that emerged between the United States and the Soviet Union in the late 1940s, and which lasted until 1989, was closely tied to competing claims about democracy and human rights. On one side, the United States was represented by its governments as the ideal-typical liberal democracy or "limited" government, whereas the Soviet Union was represented as an expansionist tyranny. On the other side, the Soviet Union was represented by its governments as an embattled people's democracy encircled by the agents of an expansionist world capitalism. The Cold War, therefore, was a conflict of claims to democracy – a conflict between political discourses – as much as it was anything else.

The sudden collapse of communist rule in the Soviet Union and Eastern Europe, for reasons of economic failure more than lack of living up to Marx's model of political democracy, stimulated a burst of theorizing about the "victory" of liberal democracy over its counterpart. This implies that both models were real rather than abstracted and idealized versions of what actually went on in the United States and the Soviet Union during the Cold War. In this construction, ideological struggle over the "best" form of politics was said to be over and, in one extreme formulation (that of Francis Fukuyama in 1992), history itself was declared to be at its end. More temperate commentators note, however, that in most countries with a history of democratic practice, liberal democracy is in a condition of arrested development. Not only has electoral participation shrunk dramatically, most notoriously in the United States itself, but democratic institutions are seen as corrupted by private interests and are inefficient bureaucratically in ways not seen in North America and Europe since the early years of the twentieth century.

Political discourse that was organized largely around the categories and slogans of the Cold War has been unable as yet to find a satisfactory substitute. For example, the collapse of the Christian Democrat–Communist division in Italy as part of the fall-out from the end of the Cold War has produced an interminable political crisis and increased popular cynicism about liberal democratic politics. Those issues high on the agenda in the US and Europe as alternative foci for political competition – namely foreign immigration, "civilizational" clashes (particularly Islam versus Western secularism), free trade and regional economic integration, and geoeconomic competition – do not provide the same capacity for systemic integration as did the Cold War with its Manichean contest of "good" versus "evil" and competing visions of the essence of democracy.

The essential distinction of political versus social rights to which the two sides in the Cold War drew attention remains unresolved, notwithstanding the end of the Cold War and the efforts to create new "third ways" between state- and market-centered delivery of social services. As the welfare state in Western Europe, Australasia, and North America is challenged on the grounds of its undermining national economic efficiency and competitiveness, the defense of social rights becomes as important an issue as the protection and enhancement of political rights. The influence of the Marxist tradition's emphasis on social rights is being felt in countries where its manifestation as Soviet communism was never of much political significance.

Yet the continuing importance of liberal democracy should not be slighted (Przeworski 1991). During the 1980s and 1990s the practice of liberal democracy spread widely into world regions where its previous hold had been tenuous at best. Almost all of the countries of Latin America and many in Asia and Africa experienced some form of transition to a broadly democratic (civilian–constitutional) regime; though the grip of liberal democracy – regular, fair contested elections, the rule of law, respect for basic human rights, open government – is still tenuous throughout the global "South" of underdeveloped and former colonized countries and in many of the countries of Eastern Europe and the former Soviet Union. In Southern Europe (Spain, Portugal, and Greece) rather more successful transitions to liberal democratic rule have been under way since the 1970s. In South Africa and in Israel–Palestine seemingly intractable conflicts have produced new openings that promise more democratic futures. In many of these settings the transition to democratic rule had to coexist with economic stagnation or decline and the austerity programs imposed by the structural adjustment policies of the International Monetary Fund and the World Bank. What is most hopeful is that this conjuncture has not invariably produced demands for a reversion to authoritarian solutions. In many countries it was the authoritarian regimes that helped bring about the economic disasters that more democratic governments must now try to manage (for example, in Brazil or Indonesia). Unfortunately, in this context social rights tend to be neglected in order to reduce government expenditures and attract international investment. Of course, previous liberal democratic interludes produced reversions to authoritarian governments and this could well happen again, as seems to be the case, for example, in Venezuela.

The Territorial State and Democracy

The debate over democracy's content (the mix of rights and duties) grew alongside the modern territorial state as it evolved in Europe in the nineteenth and early twentieth centuries. The association between people's democracy and social democracy, on the one hand, and the territorial state, on the other, is obviously close: the boundaries of the state usually define the limits of those eligible for the services the state provides. But liberal democracy and the state are also closely related, given liberal democracy's focus on procedures for ensuring the election of state rulers for limited periods of time. Indeed, the sense of its *necessary* association with democracy gives the state much of its normative appeal. In his classic work *Politics and Vision* (1960) Sheldon Wolin put the dilemma bluntly:

> To reject the state [means] denying the central referent of the political, abandoning a whole range of notions and the practices to which they point – citizenship, obligation, general authority . . . Moreover, to exchange society or social groups for the state might turn out to be a doubtful bargain if society should, like the state, prove unable to resist the tide of bureaucratization.

But the territorial state's necessary centrality to world politics is increasingly in question. Businesses are less and less organized solely to exploit local or regional market opportunities. They are increasingly multinational or global in scope. Unlike in Marx's day, the state appears less the shield of capital and more its latest victim. States must now choose whether they represent the interests of territorially circumscribed populations or the interests of businesses that operate globally but which originated within their confines (Reich 1991). In fact, in a more recent work Wolin (1989: 16–17) himself captures the sense of the times:

> Compelled by the fierce demands of international competition to innovate ceaselessly, capitalism resorts to measures that prove socially unsettling and that hasten the very instability that capitalists fear. Plants are relocated or closed; workers find themselves forced to pull up roots and follow the dictates of the labor market; and social spending for programs to lessen the harm wrought by economic "forces" is reduced so as not to imperil capital accumulation. Thus, the exigencies of competition undercut the settled identities of job, skill, and place and the traditional values of family and neighborhood which normally are the vital elements of the culture that sustains collective identity and, ultimately, *state power itself*. [my emphasis].

The state's failure to protect its citizens from enhanced capital mobility undermines its claim to represent their best interests. This "legitimacy crisis" is compounded by the deadening impact on political life of the disruption of social and political institutions (such as local governments and community groups) in which democratic practices have become most strongly established.

In this context, the interest of democratic theorists is now more, as it were, above and below the level of the state: in transnational formations of various sorts and in the variegated groupings within society that are not oriented to the state per se. Transnationalism is the obvious result of trying to match political control with the increasingly globalized workings of the world economy (Held 1991). Numerous questions arise. How can vesting powers in different transnational arrangements (regional groupings of states such as the European Union and international organizations linking sovereign states, e.g., the United Nations) be organized given traditional reliance on notions of state sovereignty to justify political legitimacy? How democratic can and should such arrangements be? Will a transnational order be composed of states or composed of individual persons liberated from lower jurisdictions? What threat to "traditional" cultures and religions is posed by the defense of individual rights implicit in most transnational organizations? How can international intervention in the "internal affairs" of sovereign states, by the UN or other international organizations, be justified by reference to the protection of human rights when it involves violating the boundaries of precisely the polities that have formed such organizations?

The Politics of Difference

There is a growing tension between the aspirations of many globalist programs and the increasing interest of many intellectuals and interest groups in the "politics of difference" involving various local and sectoral (ethnic, gender, behavioral) claims to "group rights" (Dryzek 1996). Postcolonial and communitarian theorists defend the mobilization, empowerment, and education of people as members of social groups whose identities have been stigmatized, repressed, and ignored, or who identify issues (such as the global ecology) that are not readily contained within the territorial confines of state-regulated and state-mandated politics. Feminism and environmentalism are the paradigm examples of the "new social movements." The main contrast is with the class-based and nationalist movements which aimed at taking state power by adopting broad political programs. The new movements conversely aim at changing particular state policies and changing the workings of society through "consciousness raising." They are not interested so much in becoming the rulers of territorial states as in creating spaces for associational life autonomous of state power.

Some groups have proved capable of "crossing the borders" of particular states and have begun the construction of a viable international society beyond the direct control of states. Many of these, such as Greenpeace, Amnesty International, and Oxfam, are concerned with acute environmental, political, and food crises. They have profited from the immediacy of modern televisual representations of such crises. Unlike the Catholic Church or the Communist International, they are not simply out to realize a totalizing vision of the world. They want to save the ozone layer, save women's lives from systematic spousal and family abuse, or expose the routine use of torture by police and military forces.

The "new" politics is more tumultuous than the old and less contained by state boundaries (Benhabib 1996). One question concerns how it will link with older political movements based upon class and nationality. What also remains to be seen is whether it is more democratic. Certainly, groups that were previously marginal to mainstream politics (such as racial minorities, sexual minorities, disabled people) now appear regularly in the political arena in many countries. But one frequently asserted threat to democracy from identity politics comes from the elevation of group rights over those of individual persons. The worry is that standards of conduct will be imposed upon some individuals because their beliefs, behavior, or language are offensive to a particular group. In the United States this concern with "politically correct" behavior has emerged most strongly in relation to so-called First Amendment rights involving the constitutionally guaranteed right to express opinions free of prior restraint. As yet, however, only with respect to community regulations governing pornography and in regard to some universities' codes of speech and sexual conduct have these issues had much actual impact (Hughes 1993).

There is a more general challenge to conventional thinking about and practice of democracy from the "politics of difference." This lies in the valuation of communal or sectoral ties at the expense of participation in territorial communities such as states. When political identities are totally bound up with singular social identities then the search for communalities and common understandings with geographical neighbors who are identified as belonging to other social groups is abandoned. Harry Goulborne (1991) worries, for example, that in postcolonial Britain ethnic–communal identities are beginning to exclude the possibility of the shared political identity upon which democratic politics is premised. Sikhs and Guyanese, for instance, care more about the politics of their homelands than the United Kingdom's. At the same time, "British" identity is eroding as more exclusionary and ethnic conceptions of "Englishness" make for difficulty in including foreign-born immigrants within the territorial community.

Regional and Ethnic Politics

The concern with "difference politics" is that long-decried practices of apartheid and ethnic cleansing will escape the geographical confines in which they have been recently contained (in, respectively, South Africa and Bosnia) to afflict societies all over the world. The collapse of Yugoslavia in a frenzy of intercommunal violence, and murderous civil wars in such former Soviet republics as Georgia and Azerbaijan, suggest one kind of outcome for an overemphasis on the politics of difference. Movements for ethnic, religious, and regional autonomy, however, can argue with some justification that they are claiming their democratic rights (Nairn 1993). One feature of democracy and human rights as they developed in the nineteenth century was their connection to the right of national self-determination. Peoples were claiming their rightful place in history when they campaigned for national unification or separation from large and frequently tyrannical empires. The Greek patriots fighting for their freedom from the Turks, and such heroes as Garibaldi fighting to overcome the partition of Italy into disparate kingdoms and principalities, inspired democrats as much as nationalists. Indeed, nationalism appeared as an ideology of freedom in a way that it no longer does, in theory or practice. At the outset of the twenty-first century it appears as parochial and backward-looking as it once appeared universal and progressive. After the bloodletting and concentration camps of two world wars, nationalism has lost most of its earlier sheen.

A case can be made that, contrary to the idea of ethnic nationalism as a worldwide trend, it is in fact a response to local and regional forces. Prime among these is the collapse of centralized communist (and other) governments which used (and abused) ethnic labels as a tactic of divide-and-rule and to reward certain groups at the expense of others. Other forces would include frustration with the corrupt central governments (as with the Northern League in contemporary Italy), the misalignment of political and cultural boundaries (as in large parts of Africa; see Davidson 1992), and the use of language and religious affiliations as techniques of political mobilization (as in India and Sri Lanka). This is not to minimize the importance of violent ethnic conflicts in threatening democratic politics in many parts of the world. It is simply to suggest that ethnic and religious divisions are not universal in the threat they pose to democracy and human rights.

Four Threats to Democracy

At the world scale, four other trends are much more threatening to the past two centuries of fragile democratic accomplishment. One is the reassertion of centralized authority in such hierarchical structures as transnational corporations and multinational churches (such as the Catholic Church under Pope John Paul II and various Islamic and Christian revivalist movements) which, in their investment and moral decisions respectively, show little sympathy for the free expression of opinion and popular constitution of authority that lie at the heart of the liberal democratic tradition. Corporate managerialism and theocratic authority fundamentally challenge the open process of negotiation and decision that democracy requires. Efficiency for its own (and profit's) sake and conformity to hierarchically constituted authority have always coexisted uneasily with modern democracy. Hierarchical subordination did not die out with the rise of the modern territorial state. Indeed, a case could be made that in some world regions, such as the Middle East, Africa, and South America, churches, corporations, and social networks (such as lineages and chieftainships) are more effective units of rule over people's lives than are states, which frequently fail to deliver on their promises. Yet these other organizations are often quite anti-democratic in their goals and procedures.

A second trend, exemplified above all in Eastern Europe and the former Soviet Union, but by no means restricted to them, is the increasing confusion between democracy and economic liberalization. This confusion is characteristic of a certain American position, developed during the Cold War, that capitalism and democracy are somehow synonymous terms. Although the conduct of open elections is a necessary part of this definitional conflation, open markets are more central to the calculus (Robinson 1996). In this light, democracy is redefined as a market-access regime in which the free flow of the mobile factors of production is the central attribute. Such international organizations as the International Monetary Fund and the World Bank seem to have largely adopted this understanding of "democracy." Little or no attention is paid to how responsive governments are to the demands of their citizens. Rather, the needs of external creditors and potential investors have a higher priority. Sometimes, democracy in its classic liberal democratic sense is seen as a barrier to economic development; something that must wait until economic "take-off" is under way. In fact, the empirical research on this connection is equivocal, suggesting that liberal democracy is probably not the villain it is painted as: sustaining the "outrageous" demands of a citizenry without any sense of the economic "realities." In other words, political rights are not a "luxury" conferred by economic development.

A third trend inimical to the prospering of democracy is the growing social-geographical inequality within and between states. At a global scale there is a widening gap in material well-being between residents of some states, particularly those in sub-Saharan Africa, and those in industrialized countries. Within states, even relatively prosperous ones, regional and social inequalities have been growing once again, having been much reduced in the period 1945–75 by means of various income and regional policies. These inequalities not only challenge the achievement of the great democratic virtues in the French revolutionary tradition (one of which is "equality"), they also prevent equal or equivalent participation in political life. Consequently, some regions, localities, and social groups are disadvantaged in the political process. Poorer regions (e.g., parts of southern Italy, the former South African "homelands" under apartheid) are prone to the depredations of patronage politics. Richer regions use their wealth to buy political advantage. Only a reversal to the "discredited" regional investment and income redistribution policies of the 1960s and 1970s could reverse this trend (Denitch 1992).

Finally, a fourth trend threatening democracy around the world is the posing of human rights as particular to cultures rather than actually or potentially universal in nature. At the 1993 UN Conference on Human Rights a major contention of the representatives of several Asian states, including China and Indonesia, was that the understandings of human rights enshrined in the UN Charter, far from being universal, are the global projection of European understandings of human freedom, the rights of individuals in relation to governments and the duties of states in relation to citizens. From this point of view there can be no such thing as universal human rights. There are only rights *particular* to different cultural traditions. The ideal of democracy as a universal form of rule, therefore, is rendered moot.

There is certainly some truth in this position (for a good discussion, both pro and con, see Charney 1999 and Bell 1999). The modern discourses of democracy are largely European in inspiration. Most of the world's cultures, however, do have democratic "currents" of one sort or another and most "cultures" are now so syncretic or hybrid that claims for national or civilizational particularity do not bear close inspection. The idea of "oriental despotism" is in fact a European one; Asia, Europe's classic Other, has always been more politically complex than portrayed by European intellectuals (Springborg 1992). European imperialism, however, worked its effects in contradictory as well as in devastating ways, none so ironic as the spread of democratic ideals that helped undermine European claims to rule their colonies without popular consent. It is difficult to take seriously the charges of Asian governments against a universal model of democracy (however attenuated in practice) when their own legitimacy rests at least in part on popular consent. Most importantly, the consequences of the cultural critique of democracy appear slighter in practice than at first glance. No government, not even that of China, can expect to find international support for the claim that torture or imprisonment without trial can be justified with reference to any self-respecting "cultural tradition." The growth and spread of an international human rights regime has had at least that effect.

In aspiration, democracy is an inventive set of arrangements for the arousal, expression, and mediation of popular political interests and sentiments. It is a reflection of the staying power and spreading influence of the idea of democracy that the battle over its essential elements is likely to continue indefinitely. Democratic politics does not admit of authoritarian definition. This indeterminacy and openness is precisely what disturbs authoritarians of whatever political hue.

Challenges Facing Democracy after the Cold War

One key area of dispute connects democracy to the increasing geographical scope and pace of contemporary life. Conventional thinking of various stripes sees democracy as flourishing in "small places" where active participation can be nurtured. Without extension into the realms of international relations and global flows, however, the possibilities for real or strong democracy in the twenty-first century seem limited at best. Keeping democracy tied to the territorial state seems, like Soviet communism, to be something of the past in a world of increasing international migration and vastly expanded global economic transactions.

A second key area is how to combine democracy's historic commitments to universal human rights and increasing the equality of material conditions of life with a relativistic mission that allows for "cultural translation," as Judith Butler (1999) calls it, between the meanings of needs, justice, and ownership characteristic of different societies. A place for "sorority/fraternity" finally needs to be found in democratic discourses dominated hitherto by liberty and equality (Taylor 1992). Unfortunately, this is not a task for which we have been well prepared by either democratic theory or practice, even in such a historic stronghold of liberal democracy as the United States.

But perhaps the greatest challenge facing advocates of democracy after the Cold War is to tie together the deterritorialization of economic life with increased attention to economic democracy. The one large area of life where liberal democracy, people's democracy, and social democracy have all failed to make much headway is in bringing a degree of democracy into the management of the economy in general and the workplace in particular. The difficulty has lain in restricting the terms of democratic debate to the "political sphere." Yet the arguments that vital decisions should not be made without popular consent, used by classic democratic theorists to justify political democracy, also apply to the economy. The point here is not to advocate a single form of ownership; that road is well traveled and beckons beyond democracy. But only through building a vigorous popular involvement in economic life can the vision of democratic politics itself be realized (Putnam 1993a). Democratic regulation of economic activities would not only expand the scope of democracy into important areas of life, it would also reinvigorate a political democracy that seems increasingly irrelevant to the problems that most people associate with working and making a living. Today, no one anywhere can afford complacency about either what has been achieved or what remains to be done in making the world as a whole more democratic.

9 The Renaissance of Nationalism

Nuala C. Johnson

Introduction

December 1989 was a month of monumental change for the world political map in general and for Eastern Europe in particular. It precipitated the most significant changes in the map of Europe since World War II. The Brandenburg Gate, dividing East and West Berlin, was opened on December 22 after 28 years of closure and this symbolic act prefaced the unification of East and West Germany in October 1990. The former dissident Czech writer Vaclav Havel became the non-communist leader of Czechoslovakia on December 29, 1989 (later to be divided into the Czech Republic and Slovakia), and Romania's dictatorial leader Ceausescu and his wife Elena were executed on Christmas Day of that year. The initiation of glasnost and perestroika under Gorbachev's leadership of the Soviet Union in the mid-1980s eventually resulted in the break-up of that union, with the establishment of independent states in the Baltic in 1991 and the evolution of the loose confederation of states comprising the Commonwealth of Independent States (CIS).

In long-established states, the 1980s and 1990s also witnessed a rejuvenated sense of national identity and patriotism, culminating in Britain with the Falklands War and in the USA with the Gulf crisis. As Anderson (1983: 12) comments, "the 'end of the era of nationalism,' so long prophesied, is not remotely in sight." As the geopolitical blocs that characterized the Cold War have evaporated, nationalism has reemerged as one of the dominant discourses of recent times. Although global processes appear to be eliding the role of the national, and postmodernism is emphasizing the fractured basis of political and cultural identities, the national state continues to exercise power as a mediating link between the local and the global. Indeed David Harvey (1989: 306) has located the renewed popularity of a nationalist politics in the insecurity generated by capitalist globalization. He claims that "there are abundant signs that localism and nationalism have become stronger precisely because of the quest for the security that place offers in the midst of all the shifting that flexible accumulation implies." Whether this insecurity may be more perceived than real, the appeal of place-based national identities in the face of rapid economic transformations endures.

This chapter will treat nationalism not just as an ideology and practice that has recently experienced a revival (Smith 1986), but also as one of the most enduring ideologies that has structured political life over the last two hundred years. As such, nationalism – the desire to bring cultural and territorial imperatives together – will be analyzed by examining key concepts in the lexicon of nationalist discourse, both the symbolic and the literal, to elucidate some of the ways in which we can forge a better understanding of the changes being "mapped out" in the global political landscape.

Imagined Communities

Although there is a huge literature on nationalism (see Hobsbawm 1990), proffering definitions, typologies, explanations, and case studies of the phenomenon, from a theoretical viewpoint Anderson's claim that the nation is an "imagined community," in which there is an assumed cultural communality in spite of class, or geographical and social distance, continues to be a persuasive framework from which to analyze nationalism. Anderson ([1983] 1991) suggests that this imagining emerged in the context of the rise of capitalism in the sixteenth century, the replacement of religious and dynastic orthodoxies with new political and social formations associated with the Reformation and the Enlightenment. These changes were located in an era when there was an extension of the world capitalist economy, which centered on national states (Wallerstein 1974; Agnew and Knox 1989). The invention of the printing press resulting in the production of cheap books and newspapers, written in the vernacular, enabled geographically dispersed peoples to recognize the existence of others with the same language and to unite culturally as a result. While by the sixteenth century the printing press and its products were experiencing a boom, books were still only available to a small, literate minority. It was increasingly with the introduction of mass education in the nineteenth century that widespread literacy was attained, thus making possible the "mass" imaginings underlying nationalist sentiments (Fishman 1972), and allowing for the break-up of poly-vernacular empires and the unification of localized territorial kingdoms. National newspapers, for instance, provide a daily, visible, and shared image of the nation, while the singing of national anthems facilitates horizontal bonding where "people wholly unknown to each other utter the same verses to the same melody" (Anderson [1983] 1991: 145). This image of unity that transcends class, gender, or "racial" difference is important to the exercise of nation-building.

Nation-Building

The "imaginative discourse" surrounding the nation can assume many forms (Bhabha 1990), variously centered on conceptions of history, collective memory, habitat and folk culture, traditions, poetic spaces, and symbolic landscapes (Smith 1986; Williams and Smith 1983; Hobsbawm 1990; Hobsbawm and Ranger 1986; Cosgrove and Daniels 1988). In the drive to achieve territorial and cultural consolidation, exercises in nation-building are necessary and are constantly subject to renewal and rejuvenation. These can be overt exercises (even coercive), such as a state's education policy, or they can be much more subtle processes, such as the gradual promotion of particular landscapes as representations of national identity. Daniels (1993: 5) points out that "Landscapes, whether focusing on single monuments or framing stretches of scenery provide visible shape; they picture the nation." Cultural legitimacy must be achieved by the national state which, as Gramsci noted more than half a century ago, seeks to "raise the great mass of the population to a particular cultural and moral level" (Gramsci 1971: 258). In Eastern Europe the evidence suggests that "national" cultural legitimacy was not achieved under communism and that these states today are struggling to grapple with diffuse political and cultural allegiances (e.g., Bosnia). Similarly in postcolonial Africa (e.g., Nigeria) the construction of unified nation-states has proved difficult: linguistic, religious and ethnic tensions have persisted. The remainder of this chapter will examine, with examples, the process of nation-building and some of the challenges that have been raised against the nation-state.

Linguistic Imaginings

> Has nationality anything dearer than the speech of its fathers? In its speech resides its whole thought domain, its tradition, history, religion, and basis of life, all its heart and soul . . . With language is created the heart of a people. (Herder 1968 [1783])

The cultural definition of identity has frequently rested, as Herder asserts, on linguistic differentiation. Fishman (1972) contends that there are four main reasons why language is useful and often intrinsic to the nationalist cause. First, functionally it can arouse ideas of a common identity. Second, it forms a link with the past; it can safeguard the "sentiment and behavioral links between the speech community of today and its (real or imaginary) counterparts yesterday and in antiquity" (ibid: 44). Third, language becomes a link with authenticity. It provides a secular source of mass communication in modern society and yet can lay claim to uniqueness. Fourth, a vernacular literature can allow elites to become central to a nationalist movement. The politicization of language requires planning.

The standardization of spelling, grammar and so forth, and a mass education system, achieves a degree of uniformity, at least as far as the written word is concerned. Language planning is crucial for the breaking down of old and the construction of new spatial barriers at the scale of the state. To take an example, the adoption of the Francien dialect of the Paris region as the national dialect evolved through various stages of language planning. The Edict of Villiers-Cottêrets by the Paris authorities in 1539 made Parisian French the official language of the royal domains and the edict was facilitated by the first publication of a French dictionary and French grammar text in 1531. The solidification of this process through the revolutionary period and into modern-day France diminished the importance of other languages and dialects spoken within French territory (e.g., Breton, Flemish, and Occitan) and created a general uniformity within the national boundaries. Where language planning is unsuccessful, tensions between linguistic communities forming a state can lead to a separatist politics.

A classic example of linguistic tensions is to be found in Belgium. Created as a state in 1830, with a merging of Flemish areas in the north and French-speaking lands in the south, Belgium has experienced a series of constitutional and linguistic crises centered around divisions between the Flemish and Francophone communities (Senelle 1989; Frognier, Quevit, and Stenbock 1982). Although the state has endeavored to overcome some of these divisions by moving toward a federal system of government and by the allocation of limited autonomous powers to the four regions making up the state, cultural tensions continue to affect the Belgian political scene, despite Brussels's status as the administrative capital of the transnational European Union. The cultural conflict has a micro-geography as well as a regional geography. This micro-geography has been expressed in the suburbs of Brussels, where Flemings have feared for their linguistic future as Francophone Bruxellois move to the suburb of Overijse. This trend stimulated the local Flemish MP to try to quell Francophone migration legally to protect the linguistic and economic well-being of the Flemish population living there: "it is regional not national politics which attracts the Belgian political talent, the language question dominates and permeates all political questions" (*Independent on Sunday*, October 10, 1993). Similarly the constitutional status of Québec, in the Canadian context, has dominated the politics of that region since the 1980s (Laponce 1984). After the 1993 general election Lucien Bouchard, leader of the main opposition party, the Bloc Québecois, claimed "No longer, will English-speaking Canada be able to pretend the constitutional quagmire is a thing of the past" (quoted in the *Independent*, October 27, 1993). Although historically language served as one of the defining characteristics in nation-building projects, today language is an important cause of cultural conflict within many states where communities speaking minority languages are challenging the hegemony of majority-language cultures (e.g., Spain, France).

In Eastern Europe, since the liberalization of the political regimes, the rights of linguistic minorities have firmly reentered the political agenda. In Slovakia, for instance, the Hungarian minority have claimed that the Slovaks have systematically pursued an assimilation policy of "educating Hungarians into state-patriotism towards the Czechoslovak nation-state, and a campaign for Hungarians to learn Slovak" (Carter 1993: 247).

Postcolonial states have also had to make the choice of "whose" language to use in the context of independence. The revival of a local language can be an important legitimation for independence. In the case of Ireland's independence from Britain, the state declared Irish the national language and instituted a series of cultural and educational policies to maintain and extend its usage (Johnson 1992). In the case of a multilingual state such as India, the adoption of an official language (Hindi) has presented great problems, allowing English to remain the official language of administration and government (Seton-Watson 1977). Not only may the language of the colonizer be entrenched in the social relations of independent states, but the landscape can be named in the fashion of the colonizer. Duncan (1989) has highlighted how after independence the Sri Lankan nationalist party renamed the streets of Kandy in the vernacular, an exercise in "symbolic decolonization." Decolonization, however, is rarely a simple process, and just as the naming of streets, towns, and villages in the language of the colonizer is part of the colonial endeavor, the response of the colonized to negotiate their own cultural space after independence is fraught with difficulties. The place of Russian in the context of the Commonwealth of Independent States is a case in point.

Heroic Pasts

While nationalism has a territorial imperative in the acquisition of space or territory for a "nation" to inhabit, the imagined community of nationhood also has a temporal dimension, in which previous periods of "past glory" are resurrected for present self-aggrandizement (Smith 1986). Hobsbawm and Ranger refer to this process as the "invention of tradition," where the chronology of the historical imagination is eschewed in favor of a focus on particular historical or quasi-historical epochs, when the cultural effervescence of the "nation" was particularly heroic. The geographical and the historical merge in the context of these imaginings as particular places and landscapes become centers for the rituals of collective cultural memory (Hobsbawm and Ranger 1986). McCrone (1992), in the context of Scottish nationalism, argues that the tartanization of Scottish culture involved a selective reading of the "national" past which ignored the real political forces that produced such imaginings. The popularization of the tartan kilt and of tartan patterns associated with different clans corresponds with a period of changing economic fortunes in Scotland in relation to the British and the world economy. Not only is the constitution of Scottish identity place-based, but the articulation of separatist politics is embedded in the histories and geographies of particular regions in Scotland, for instance the Highlands (Agnew 1987).

The compelling appeal of "invented traditions" is not confined to minority groups within larger states, but is also found in well-established majority cultures. The contemporary nostalgia for the past has been tentatively linked to public dismay with the purported anonymity of high modernity, where identities are increasingly being fragmented in favor of more situated and flexible ones (Urry 1990). The abandonment of some older myths and traditions associated with national cultures has seen their replacement with new or reconstructed ones. Tourism and the heritage industry frequently reinforce images of a heroic national past, packaged for public, popular consumption (Hewison 1987; Wright 1985). New technologies, new forms of museum display, and the "theme-parking" of historical narratives may have replaced the older traditions of popular ritual (e.g., parades), but the history purported to be represented through these installations often merely anchors conceptions of national identity on new terrains (Lowenthal 1991). While, in the nineteenth century, national languages, school history textbooks, and religion formed the nexus of national imaginings, in the late twentieth century the heritage industry and the associated historicizing of interior design and architecture play a similar role.

In Britain the resurgence of national feeling over the past two decades is attributed to a number of factors: to the decline of Britain in the world economy as the stockmarkets of New York and Tokyo overtake that of London; to challenges from Scotland, Wales, and Northern Ireland to English cultural hegemony through devolved government; and to immigration from non-European states (Samuel 1989). One reaction to these processes is a reassertion of "national" cultural values. The Prince of Wales, in *A Vision of Britain*, published in 1989, articulated this loss of national prowess in postwar Britain and especially in London. The image envisioned by the prince "is a London restored, re-visioned as landscape, framed by lavish reproductions of eighteenth- and nineteenth-century oil paintings" (Daniels 1993: 11). Using Canaletto's painting of London as a blueprint and looking today at St. Paul's Cathedral and environs, where towers disrupt "the symbolic exchange between St. Paul's and the City, the solid image of a community of interest which framed London's supremacy as the centre of world trade" (ibid: 13), the prince is imagining the city being recast in the image of London as glorious, and by inference Britain as supreme. Daniels illustrates that the cathedral had been subject to various interpretations over the centuries and that it was not always eulogized on architectural or civic grounds, yet the prince's view eschews the historical record for a more heroic view of St. Paul's in the history of the nation. The emergence of royal interest in the architecture of the city corresponds with the revival of royal pageantry in the 1970s and early 1980s. The furore generated in the USA in 1991 by the exhibition "The West as America: Reinterpreting Images of the

Frontier 1820–1920" at the National Museum of American Art, underlines the unwillingness of many, and of state representatives in particular, to have their nation-building mythologies challenged. As Watts (1992b: 116) claims, the exhibition's "unflinching account of the brutality of the frontier – a space that . . . has been ideologically formative in the construction of a particular national identity – did not lie well with the jingoistic and nationalist sentiments rampant on the Hill [Capitol Hill]."

The potency of a heroic history rejuvenated for current political reasons can be particularly skewed in the context of civil strife. The historical memory is continually jogged to assert cultural legitimacy and it is incorporated into the iconography of everyday life. In Northern Ireland, where there has been a failure to establish an agreed history and an agreed interpretation of the past, history is a source of constant dispute. Northern Ireland is not so much an exceptional case but an example of more general processes writ large. Varying interpretations of history are popularly consumed through graffiti, street parades, carpet painting of pavements, and wall murals adorning the housing estates there. While street festivals such as London's Notting Hill Carnival may be celebrations of local cultural resistance to dominant groups (Jackson 1988), in Northern Ireland they are heavily politicized dramas of binary opposition carrying important symbolic and material consequences. For the Loyalist population of Northern Ireland, the historic moment that links present-day political strife with past antecedents is the accession of William of Orange to the crown. Wall murals of "King Billy" and his victories assert for Unionists a legitimation of their current political existence (Rolston 1991), and the painting of such murals has a long history in Northern Ireland (Loftus 1990). The murals are principally "concerned with the entrenchment of existing structures and beliefs" (Jarman 1992: 161) and these beliefs rely upon a particular view of the past. The annual celebration of July 12 in Northern Ireland's commemorative calendar and the annual disputes surrounding the routing of Orange parades (e.g., at Drumcree) underscore the significance of space in the representation of cultural memory.

Mythical figures as well as historical ones can also articulate conceptions of national identity (Smith 1986). In Belfast, members of the Ulster Defence Association (UDA) have appropriated the figure of Cuchulain in their wall murals. Traditionally Cuchulain represented "early valour, miraculous feats, generosity, self-sacrifice, beauty and loyalty – evoked an archaic epoch of nobility and liberty, in which the full potential of Irishmen was realized" (ibid: 195). The use of Cuchulain in Loyalist wall murals is an attempt, according to Rolston, "to retrieve a history beyond 1690, and the Battle of the Boyne" (quoted in *The Irish Times*, January 19, 1993). Rather than viewing this mythical figure as a Gael, Loyalists have redefined him as one defending Ulster from the Gaelic queen Maeve. Ironically, then, the interpretations awarded to "golden ages" vary across political communities and create tensions in the discursive practice of myth-making.

While heroic histories form an important part in the establishment of an "imagined community," these pasts are also intimately linked with heroic people in nationalist discourse, and these people are embodied in monuments and statuary that adorn the towns and cities of national states.

Monuments and Nationalism

If nationalism appropriates periods of the past to represent its continuity, it also appropriates historical persons, events, and allegorical figures to reinforce its cultural existence. In the nineteenth century public statuary had firmly entered the public domain (Johnson 1994). Monuments dedicated to important figures in the nation's history began to emerge on a large scale, and such a process was fraught with dissension and disturbance. In Eastern Europe historic figures associated with the evolution of these states are being gradually replaced. Statues of Lenin and Stalin have been systematically removed from the public sphere in towns and cities of the former Eastern bloc – "Moscow has set up a Commission on Cultural Heritage to deal with the statues . . . Few [of the 123 Lenins] will be kept; most will go to the Museum of Totalitarian Art" (quoted in Bonifice and Fowler 1993: 126).

The erection of public monuments often is an intrinsic part of the nation-building process. As Mosse (1975: 8) effectively posits, "The national monument as a means of self-expression served to anchor national myths and symbols in the consciousness of the people, and some have retained their effectiveness to the present day." They commemorate real historical figures such as political leaders, writers, adventurers, and military leaders. Statues also commemorate war or are used to personify for the national community abstract concepts such as justice and liberty (Warner 1985). Statues articulate in a material and ideological fashion the collective memories of a nation's past, but the choice of "whose" heroes and "which" events to commemorate reveals the process by which groups achieve hegemonic positions within a state. The geography of monuments articulates a hierarchy of the sites of memory within a nation and the relative power of different groups.

While the nation-state remembers its founding heroes, it also commemorates its wars – they can be wars of independence, international conflict, or even civil strife. The war memorial varies in iconography depending on whether it is commemorating defeat or victory. Smith (1986: 206) notes that "Creating nations is a recurrent activity, which has to be renewed periodically," and the construction of public monuments is one way in which this renewal and reinterpretation of the national past is articulated.

War memorials that personify particular military leaders tend to be heroic in proportion and may use iconography from previous eras to convey the strength and national importance attached to an individual. The Nelson Column, in London's Trafalgar Square, adopts a design drawn from the iconography of ancient Roman imperial victories (Mace 1976). While it took a long period finally to execute the design of the column and the ancillary figures surrounding the monument, the square itself has frequently been the focus of political protest in London. As a site for lobbying collective memory, Trafalgar Square has simultaneously played an ambiguous role as a center of public protest where the national state has been challenged (ibid).

Memorials to World Wars I and II vary considerably in scale and iconography. They can be simple commemorative plaques placed in towns and villages listing local people who were killed, or they can be colossal national monuments located in prestige positions within capital cities (e.g., the Cenotaph in Whitehall, London). Annual pilgrimages to these memorials and the laying of wreaths at their feet reinforces the states' recognition of the importance of war and the losses it incurs, but these occasions also enable the mass of the population to participate openly in a national event (Mosse 1975). The tone of the inscriptions on war memorials replicates those found on headstones in cemeteries, but unlike the latter they tend to conceal the class and gender divisions of the people they represent. In this sense they function as "nationalized" monuments.

Not all war memorials are heroic in their symbolism, nor is popular support for their construction universal. Where the support for war is contested and ambiguous, the building of a memorial can highlight some of the underlying fissures within the national community. The case of the Vietnam Memorial in Washington is a notable example. As a result of the public's antipathy to America's involvement in Vietnam the conception of a monument proved difficult and raised several issues for the collective memory (Wagner-Pacifini and Schwartz 1991). Rather than being a unifying process, the debates surrounding the construction of a memorial reflected in the popular consciousness the dissension from and ambiguity toward the state's policy in Vietnam. The eventual design of the monument conspicuously deviated from other memorials. First the architect, chosen through public tender, was a female student of Chinese–American descent. She was not an architect with an established record, she was female, and the choice of someone with her ancestry was questioned given that the memorial was commemorating a war which took place in Southeast Asia (Sturken 1991). The politics underlying the commission reflected differing attitudes toward the war. Although the positioning of the monument on the Mall in Washington reinforced it as a national icon and part of the nation's history, the iconography of the design deviated radically from that of other memorials in the capital. The listing of each individual soldier killed in the conflict, and represented in chronological order rather than in order of rank, made it non-hierarchical in conception. Both these features offered an interpretation of the war which treated each casualty as equal in significance and which suggested that the war was not a heroic event in the nation's history (Sturken 1991; Wagner-Pacifini and Schwartz 1991). The monument functioned at both an allegorical and a literal level, and in nationalist discourse it served to heal the wounds of a nation in mourning, a nation that did not offer unequivocal support for the state's actions. In contrast to more heroic depictions of war the iconography and ritual associated with visiting the monument centered the story of Vietnam on individual suffering within the broader framework of a national army. It differs therefore from memorials to the "unknown soldier" where the collective loss is embedded in the anonymity of a dead soldier. The design of the Vietnam Memorial did not satisfy all interests in the US and a more conventional war memorial was erected beside the "wailing" wall (Sturken 1991). Three bronze statues of soldiers, two white and one black, framed the wall in a more orthodox form, masculine, heroic, and anonymous. The recent addition of a new monument dedicated to the women/nurses in Vietnam underlines the gendered division of labor in a war context, and reclaims a more active role for women in the war effort.

The nationalist discourse in which war memorials are conceived is confirmed by the fact that they rarely acknowledge the loss experienced by the "enemy." They are not memorials to all those lost in war, but are interpreted in terms of "national" losses and "national" geopolitical considerations. War memorials commemorate "our" dead not just "the" victims of conflict.

For separatist groups within larger states statuary also articulates the divisions within the national polity, and competing public monuments can be erected to reflect this division. Pierre Nora argues that there are dominant and dominated sites of memory. The dominant "spectacular and triumphant, imposing, and generally imposed – either by a national authority or by an established interest . . . The second are places of refuge, sanctuaries of spontaneous devotion and silent pilgrimage" (quoted in Hung 1991: 107). In China the symbolic meaning of Tiananmen Square and its monuments encompasses the ongoing history of China itself and the competing political ideologies that the state has experienced. The inscription on the monument to the People's Heroes links "separate historical phases into a continuum" (ibid: 99) from the older revolutions of the 1840s to the new revolution of the 1940s. The protest by students in Tiananmen Square and their erection of a statue of the Goddess of Democracy "signified consecutive stages in a pursuit for a visual symbolic of the new public" (ibid: 109). The suppression of that protest and the removal of the monument confirms the hegemony of the state over sections of the population within its territory (Hershkovitz 1993).

Gender and Nationalism

To a large degree analyses of nationalism have been presented as gender-neutral discourses. The imperatives of creating a national "imagined community" have excluded a discussion of gender cleavages, where the desire to create a national imagined community disguises other sources of identity such as class, gender, and "race". Yet, as Warner (1985) points out, pictorial representations of nations have generally been expressed through female allegorical figures: Britannia, Marianne, and Lady Liberty are all examples. Female iconography frequently features in public commissions "because the language of female allegory suits the voices of those in command" (ibid: 37). That the female body has been used to personify concepts such as justice, equality, and liberty is ironic given women's lack of access to such freedoms, especially as nations emerge. Today "feminine" characteristics continue to be ascribed to powerful female political figures. In the case of Margaret Thatcher, where the label "Iron Lady" was used to highlight her tough, resolute, and determinedly "masculine" approach to politics, she was simultaneously depicted as mother, housewife, and her *father's* daughter (ibid).

In nationalist discourse men are active agents and women are typically passive onlookers. Although feminist critiques and histories of nationalist movements have emphasized the role of women revolutionaries as active, determined participants in particular contexts (Ward 1983), dominant theories of nationalism continue to ignore the ways in which gender relations inform conceptions of national identity. The ways in which women's voices are frequently silenced in nation-building projects, despite their role in the achievement of independence, underlines the conventional hegemony of the male voice. That women's role as homemakers was enshrined in the 1937 Irish constitution aptly demonstrates this silencing process. A consideration of the relationship between women and the state in the construction of national citizenship reveals some of the ways in which nation-building practices view men's and women's roles differently. Women's role in the development of national consciousness has often taken specific routes. First as the biological reproducers of a nation's population, encouragement to reproduce often links nationalistic arguments with a duty to motherhood. In Croatia, for example, abortion was abolished by the nationalist party in 1992 for fear that a falling birthrate would threaten the stability of the new nation. Second, women have been regarded as reproducing the boundaries of the national group and thus inter-ethnic marriage has been periodically discouraged because it might dilute the ethnic purity of nation. In Nazi Germany extreme forms of this racialized way of thinking existed, although antipathy towards intermarriage still finds expression across a range of cultural contexts. Finally, as the principal socializers of small children women have been accorded the responsibility of acting as the ideological purveyors of a nation's culture through transmitting it to their young offspring in songs, fairy-tales, and so forth. Thus although, in Eugène Delacroix's painting *Liberty Leading the People*, a female allegory is pointing the path to freedom in revolutionary France (Agulhon 1981), in analyses of nationalism women are seldom leading but loyally following their male protagonists. While the achievement of political independence in no way necessarily leads to the emancipation of women (Kandiyoti 1991), the dearth of studies that deal with gender relations and nationalism makes generalizations difficult.

Nationalism and Territory

While issues of language, symbolic landscape, cultural icon, and gender all play a role in the articulation of a nationalist politics, they also serve to delimit, solidify, or negotiate the boundaries of national space. The marking of the territorial expanse of the nation has been crucial both to the definition and disputation of national borders. Connecting historical processes of nation-building with a specific delineated territory has been central to the geography of nationalism. Anderson (1988: 24) noted this when he claimed that "The nation's unique history is embodied in the nation's unique piece of territory – its 'homeland' . . . The time has passed but the space is still there." Confirming the link between a specific people and a place has been hotly disputed, however, as the cultural geography of places rarely represents an ethnically homogeneous piece of land. Consequently the demarcation of national territory has been fraught with difficulties which at times has resulted in the most violent of territorial disputes. The redrawing of Europe's political map after World War I highlighted the problems of attempting to draw boundaries that would "contain" individual ethnic groups. In the instance of the Federal Republic of Yugoslavia, created in 1919, the multi-ethnic character of the new state survived while a powerful, centralized form of government maintained control. However, the collapse of Yugoslav government in the 1980s and Serbs' desire for territorial control over the region precipitated violent confrontation, especially in Bosnia-Herzegovina which comprised an ethnically diverse set of regions containing Croats, Serbs, and Muslims. The multi-layering of different linguistic, religious, and ethnic groupings in the region, when confronted with a dominant group which sought to achieve ethnic purity within the region through the "cleansing" of those deemed culturally different, led to the most ferocious and bitter dispute over territory, citizenship, and human rights. At the regional and local scale bitter confrontation ensued and as clashes between the different ethnic groups escalated into wholesale war the international community intervened in an effort to resolve the territorial dispute and cease the fighting (Johnston 1999a). The territorial dimension of nationalism cannot be underestimated in any discussions of the building of an "imagined community" of nationhood. The occupation of, and control over, space and the delineation of boundaries has been the source of many regional, national, and international conflicts.

Conclusion

At the beginning of the twenty-first century the decline of the nation-state as the basic structure of global political organization would appear to be nowhere in sight. While local and global processes are increasingly challenging the national state, and the notion of "multiculturalism" is gaining some ground, the evidence suggests that political and cultural identities are still articulated broadly within a national "imagined community."

This chapter has emphasized the ways in which the "imaginings" that consolidate the nation-state occur and can change through time. Whether the basis of nationalist imaginings be linguistic, historical, or symbolic (or combined), the global restructuring that has taken place since the end of the Cold War appears to have raised nationalist discourse more profoundly than ever on the global political stage.

10 Global Regulation and Trans-State Organization

Susan M. Roberts

Introduction

When, in 1994, I wrote this chapter for the first edition of this book, I am sure that the political geography of trans-state organization was far from gripping stuff for most readers. Since then, so much has changed. The politics of global regulation and trans-state organization have become key issues mobilizing thousands around the world. The now infamous "Battle in Seattle" that accompanied the 1999 World Trade Organization meetings marked the dawn of a new era in public awareness of the power of trans-state institutions currently regulating the global economy. Since Seattle, there have been high-profile protests accompanying almost every meeting of any institution formally or informally associated with regulating global affairs. Many people in many different places, often with vastly different agendas, have become concerned and angry about the shape global regulation is taking. Indeed, in this chapter I argue that currently we are witnessing and participating in the establishment of a new order of global regulation. Struggles over the nature and operations of trans-state regulatory institutions are bringing into focus the question of how regulation at the global level may or may not function in a democratic way, given the huge imbalances in the political and economic power of different states, and given the growing salience of internationalized capital – for example in the form of transnational corporations.

The World Trade Organization (WTO) is a most significant trans-state organization, but it is far from being the only trans-state organization engaged in global regulation. There are many organizations set up to regulate particular aspects of trans-state activity. For example, since 1875 there has been an organization to coordinate and regulate international mail services (now the Universal Postal Union). Since World War II (1939–45) though, attempts have been made to establish an institutional framework of organizations that would regulate key areas of international economic and political relations. The resultant institutions served to regulate and in a sense consolidate the postwar geopolitical and geoeconomic order. The Cold War, of course, left its mark on these institutions, as did the political independence of many former colonies in Africa and Asia in the 1960s especially. Now, with the end of the Cold War geopolitical order, the enduring and widening economic gap between rich states and many former colonies, and the further internationalization of capital, the postwar institutions face new geopolitical landscapes and new challenges.

Given the significance of the post-World War II regulatory framework, this chapter will briefly survey its main features. It will also consider how this framework is being transformed at present. The nature of the changes underway, as well as some of the enduring aspects of trans-state organization, are explored through examinations of two recent and significant new institutions of global regulation – one well known, the other less so. Ongoing efforts to set up an International Criminal Court (ICC) highlight some of the tensions between states and the operationalization of a "global rule of law," while the WTO stands as an institutional crystallization of neoliberalism at the global scale that many see as a serious threat.

Global Regulation and Trans-State Organization

This part of the chapter will focus on trans-state organizations that seek to regulate some aspect or aspects of the contemporary international system and the states therein "from above." The phrase "from above" refers to the arena of regulation at a scale higher than that of the state. Regulation is a term that is used in a number of different ways (see Jessop 1990). For those in the so-called "Regulation School" capitalism is understood not only in economic terms as a "regime of accumulation" but also in social terms as a corresponding "mode of regulation" which comprises "institutional forms, procedures and habits, which either coerce or persuade private agents to conform to its [the regime of accumulation's] schemas [of reproduction]" (Lipietz 1987: 33). Here the concept of the "mode of regulation" has a meaning that encompasses formal legal systems of rules, but extends to include elements of the regime of accumulation, such as social processes and cultural norms which also order and "make regular" capitalism's inherently unstable course. In this chapter the case study is focused on a type of formal regulation, but within the context of the wider mode of regulation. Modes of regulation are never completed wholes but are contingent and provisional arrangements – themselves contested and implicated in social and political struggles (see Jessop 1990: 170). Modes of regulation are therefore dynamic: seen as changeable over time and space – usually configured as a series of national regulatory spaces (Clark 1992). In this chapter we will see that a contradiction is that the mode of trans-state regulation is constituted in large part by the actions of states.

Even though there have been recent and quite dramatic changes in the nature of the mode of global regulation, these still have to be seen in the context of the establishment of an institutional regulatory framework as World War II came to a close, and so we will concentrate on developments of the last 50-plus years. Given this focus, three reminders are in order.

First, it is important to remember that trans-state organization did not begin in the mid-1940s. Indeed, those who were involved in setting up the post-World War II system of global regulation were well aware of the failings of the institutions set up after World War I (1914–18), notably the League of Nations.

Thus, secondly, it should be remembered that even though there was undoubtedly the beginning of a new era in 1944–5 the slate was not wiped clean. Just as the emerging new mode of global regulation being contested today is being built upon parts of the preceding Cold War world order, so too, many elements of the institutions set up as World War II came to a close grew out of previously existing arrangements. For example, certain organs of the United Nations (UN) have their origins prior to the founding of the UN itself in 1945 (see table 10.1).

Thirdly, it would be a mistake to think that once institutions of trans-state organization have been set up they remain unchanged. Rather, the four major postwar institutions (the UN, the World Bank, the International Monetary Fund, and the General Agreement on Tariffs and Trade) have not existed unproblematically. They have changed their character and operations in several significant ways. The three major developments in the postwar world that affected trans-state regulatory institutions have been (1) the internationalization of capital; (2) geopolitical shifts such as the independence of many former colonies and the transition of many formerly socialist countries since 1989; and (3) the recent and ongoing organized protest against key institutions. All three of these factors have been significant in rendering problematic the positions and roles of post-World War II trans-state organizations.

Having noted these points, we can now turn to a summary of the formation of the regulatory framework set up as World War II ended. The framework resulted from several meetings held by representatives of the countries that would emerge as victors in the war. From July 1 to 22, 1944, representatives of 45 countries met at a ski resort in Bretton Woods, New Hampshire, USA, to discuss the regulation of the international economy after the war. The debates revolved around competing proposals for the regulation of the postwar world economy put forward by John Maynard Keynes (UK) and by Harry Dexter White (USA). At Bretton Woods it was the White proposals that triumphed – a point which is looked upon as signaling the end of British hegemony and thus confirming US hegemony in the world of 1944–5. Two significant trans-state organizations were established at Bretton Woods – formally coming into existence in 1945. These were the International Monetary Fund (IMF) and the International Bank for Reconstruction and Development (IBRD) – together with the International Finance

Corporation (IFC) and the International Development Agency (IDA) – known as the World Bank.

From August 21 to October 7, 1944, another series of meetings was held at Dumbarton Oaks, Washington DC, USA, with delegates from China, UK, the USA, and the USSR. The conference built upon the Atlantic Charter – a declaration signed by UK Prime Minister Churchill and US President Roosevelt in 1941 – which established political and economic priorities for the postwar world. The Charter called for an international system based on security to guarantee free trade and economic growth. The meetings resulted in a draft of the United Nations Charter. This document then formed the basis for a larger conference (with delegates from 51 countries) held from April 25 to June 26, 1945 at San Francisco, USA. By October 24, 1945, enough states had ratified the Charter and the United Nations officially came into existence.

In 1944 and 1945, then, the Allied powers had set up the key institutions for governing the postwar world in political and economic terms. The commitment to international organization has to be seen in the light of events after World War I. The memories of terrible economic depression, a periodically chaotic international financial system, territorial aggression on the part of certain states, and other factors precipitating a second brutal "world" war, were uppermost. It was envisaged that international organizations could and should be created to prevent such events occurring again.

Subsequently, this postwar framework was refined. A key development was the General Agreement on Tariffs and Trade (GATT), set up in Geneva, Switzerland, in 1947. GATT was intended to be a stop-gap arrangement to ensure open trade through a series of bilateral tariff concessions written into a final agreement while negotiations for the International Trade Organization (ITO) were underway. The ITO's charter had been agreed upon at a UN conference in Havana, Cuba in 1948. The ITO was intended to be the third pillar (together with the IMF and the World Bank) of the international institutional regulatory structure. However, the ITO was never set up because debates in the US over protectionism versus free trade were won by the protectionists and the US Senate declined to ratify the ITO Charter. This was not to be an isolated demonstration of American ambivalence about its economic role as hegemon. However, American participation in GATT did not require Senate ratification and GATT served as the *de facto* regulator of international trade for nearly fifty years.

GATT was both a treaty and an international organization and was infamous for its cumbersome and complicated negotiations that occurred in eight lengthy rounds. The last round, the Uruguay Round, concluded only after eight long years of dispute and disagreement. Despite its difficulties, GATT persisted as the organizing framework for world trade until it was superseded by the World Trade Organization (WTO) in 1995. The WTO is examined in more detail below.

A most significant institution of the postwar international institutional mode of regulation remains the United Nations. The UN currently has about 180 member countries and is organized (as stated in its Charter) around six major internal bodies or "principal organs." The two major bodies are the General Assembly and the Security Council. All UN countries belong to the General Assembly and each gets one vote therein. The General Assembly discusses a range of matters – and often holds special sessions to consider particular issues (for example on Palestine in 1947 and 1948; on the economic situation in Africa in 1986). The Security Council consists of five permanent and ten non-permanent members, each with one vote. The non-permanent members are elected by a two-thirds majority of the General Assembly and serve a two-year term. The permanent members are the victors of World War II: France, Russia, UK, the US, and China. The Security Council of the UN is the body that concerns itself most directly with maintaining peace and security. It is the Security Council, for example, that calls on UN members' militaries for involvement in "peacekeeping operations." There is disagreement over the relative powers and different structure of the General Assembly and the Security Council. The selected membership of the Security Council, arguably the more powerful of the two organs, is contested by many Third World leaders and by those who argue that it does not reflect the present world. Why, for example, should France and the UK each be permanent members? Why not one membership for the EU as a whole instead?

The Economic and Social Council reports to the General Assembly and is concerned with issues of human well-being – notably development and human rights. However, the role of the Economic and Social Council is notorious for being vaguely (un)defined in the UN Charter. The Economic and Social Council is composed of members elected by the General Assembly. The regional economic commissions (such as ECLAC – the Economic Commission for Latin America and the Caribbean) as well as other specialized commissions (such as those on human rights and the status of women) operate under the auspices of the Economic and Social Council.

The International Court of Justice (or World Court) is one of the "principal organs" of the UN according to the Charter, although it is somewhat independent. This feature is evident in the location of the Court in the Hague, Netherlands – a location that is also a legacy of the 1899 and 1907 Hague Conventions establishing the Permanent Court of Arbitration – a precursor of the present Court. All members of the UN are "parties to the Statute of the Court" (Turner 2001). The judges (15) are elected in separate elections in the General Assembly and the Security Council. The Court has not escaped controversy: it has been criticized for having a very light caseload, for being slow and expensive, and for being hampered by having to secure consent of all parties (states) involved. The Court, unlike the ICC examined below, does not try individuals; it deals only with disputes between states.

The fourth organ of the UN in the Charter is the Secretariat, which is the office of the Secretary-General. The Secretary-General is the chief

Table 10.1 Trans-state organizations formally related to the UN.

NAME	Date Estab.	Purpose	HQ Location	Members
IAEA International Atomic Energy Agency	1957		Vienna, Austria	130
ILO International Labor Organization	1919	Formation and promotion of international standards through International Labor Conventions	Geneva, Switzerland	160
FAO Food and Agricultural Organization	1943	To increase food production, raise nutrition and eliminate hunger	Rome, Italy	160
UNESCO United Nations Educational, Scientific and Cultural Organization	1946	Promoting collaboration between countries through education, culture, and science	Paris, France	164
WHO World Health Organization	1946	"The attainment of all peoples of the highest possible level of health"	Geneva, Switzerland	
IMF International Monetary Fund	1945	To promote international monetary cooperation; exchange rate stability and expansion of international trade	Washington DC, USA	167
IBRD (WORLD BANK) International Bank for Reconstruction and Development	1946	To provide funds and technical assistance to facilitate economic development	Washington DC USA	
IDA International Development Association	1960	Lending agency for projects in poorest countries	Administered by World Bank	
IFC International Finance Corporation	1956	Financial and other assistance to private enterprises	Administered by World Bank	143
ICAO International Civil Aviation Organization	1947	Coordinate safety standards, navigation, and promotes international cooperation	Montreal, Canada	173

Organization	Year	Purpose	Location	Members
ITU International Telecommunications Union	1932 (merger of earlier organizations)	Promote international cooperation in telecommunications	Geneva, Switzerland	166
UPU Universal Postal Union	1875 (as General Postal Union)	International cooperation between postal services for reciprocal exchange of mail	Berne, Switzerland	168
WMO World Meteorological Organization	1951	Encourage coordination of collection and transmission of meteorological information	Geneva, Switzerland	168
IMO International Maritime Organization	1959	Through Conventions to encourage cooperation regarding merchant shipping (especially safety, pollution, and traffic)	London, UK	137
GATT General Agreement on Tariffs and Trade	1948	Multilateral treaty negotiated in "rounds" to guard against protectionism	Geneva, Switzerland	103 contracting parties; 29 de facto
WIPO World Intellectual Property Organization	1974 (took over earlier agency [BIRPI] estab. in 1893)	Through Conventions to protect intellectual property	Geneva, Switzerland	131
IFAD International Fund for Agricultural Development	1977	Agricultural and rural development projects in poor countries	Rome, Italy	147
UNIDO United Nations Industrial Development Organization	1986	Promote industrial development in poor countries	Vienna, Austria	

administrative officer of the UN and is appointed for a five-year term. The present Secretary-General, Kofi Anaan, was appointed in 1997 and is the UN's seventh. The Secretariat staff of over 25,000 persons performs all the day-to-day administrative operations of the UN.

Until November 1, 1994 there was an additional functioning organ: the Trusteeship Council. This was set up to supervise and monitor several territories known as "Trust Territories" which were not fully self-governing. There have been 11 Trust Territories since 1946, but when the last remaining Trust Territory (the US-administered Republic of Belau [Palau] in the Pacific) became independent in October 1994 the Council suspended operation.

In addition to the "principal organs" several of the UN's programs are important actors on the global scene. These include the UNDP (the United Nations Development Program); UNICEF (the United Nations Children's Fund); UNFPA (the UN Population Fund) and the UN World Food Program. Another significant UN body is the UN High Commission of Refugees. This organization was established in 1951 by the General Assembly to provide protection for refugees and to seek asylum for refugees and lasting solutions for refugee populations. In addition, agencies such as the Food and Agriculture Organization (FAO), the UN Educational, Scientific and Cultural Organization (UNESCO), and the International Labor Office (ILO), are technically autonomous but work closely with the UN. The full list is to be found in Table 10.1. It is significant that the WTO is not an agency of the UN and is not a part of the wider UN system.

The UN and related bodies, including the World Bank and the IMF, are deeply contested institutions. There is considerable and varied criticism of their roles, workings, and objectives. Many of the criticisms derive from experiences of those living in the so-called Third World – in formerly colonized countries such as those that became independent in the 1960s and 1970s and were incorporated into the already-formed structures of trans-state organization. The most pervasive criticism faced by each of these institutions is that they reflect and reinforce the highly uneven global distribution of political and economic might. Indeed, the locations of the headquarters of each of the bodies listed in table 10.1 reflect the political world in 1944–5 and definitely not the contemporary distributions of member states or population. However, more is at stake here than the location of headquarters offices. Many observers argue that institutions such as the IMF are actively used in the interests of rich countries and work against poor countries. The IMF's response to problems of Third World indebtedness has been to advocate and enforce "structural adjustment" policies in many developing countries – thereby subjecting large numbers of people to the rigors of "economic reform" which have often included worsening standards of living for the majority (see Popke 1994). Powerful countries are never subject to the same "discipline" by the IMF. The UN itself is also under fire – both literal and metaphoric – as it faces a new world, less defined by the simple oppositions of the Cold War. The "success" of UN operations in Somalia and the former Yugoslavia has been contested from many standpoints.

The UN is an example of a global-scale trans-state organization attempting to adjust and shape the so-called new world order. There are many other trans-state organizations that act to regulate in some way but that are less inclusive than the UN. Regional economic alliances such as the Association of South East Asian Nations (ASEAN) or the European Union (EU) are becoming increasingly important elements in trans-state regulation at the sub-global level. Military strategic alliances such as the North Atlantic Treaty Organization (NATO) remain significant despite the challenges of post-Cold War political realignment. Such organizations may not be global regulators but are significant trans-state organizations.

Let us now turn to the two case studies in order to examine some of the issues surrounding current attempts to reform and remake the institutions of global regulation. First, we will consider the recently re-energized attempts to set up an International Criminal Court (ICC) as a permanent court for dealing with war crimes and crimes against humanity. The ICC was originally proposed in the UN General Assembly and its formation is taking place in association with the UN. Debates over the ICC – especially in the US – serve to highlight some of the tensions states face in setting up trans-state regulatory authorities. After considering the ICC we will turn to the WTO. The WTO stands apart from the UN system and perhaps represents a new breed of neoliberal trans-state organization, a type of regulatory institution that has heightened certain concerns over the nature of global regulation in general.

Case Studies of Global Regulation from Above

The ICC

The first case study of global regulation from above illustrates how any attempt at trans-state organization has to be built upon the political–geographic architecture of territorial states even as it might challenge that architecture. The long-discussed International Criminal Court (ICC) came a big step closer to actually being established when, in July 1998, 160 UN member countries agreed to formal procedures for setting up the Court. Since then, sharp debates over the relations between universal human rights and national sovereignty have swirled around the ICC.

After World War II, there were international tribunals held in Nuremberg and in Tokyo to prosecute war criminals. However, since then, even though there has been no shortage of war crimes perpetrated, from Cambodia to Uganda to El Salvador, there has been no permanent trans-state organization set up to deal with war crimes, genocide, and "crimes against humanity" (such as systematic extermination of civilians). The UN has held discussions on establishing such an organization since 1948, but plans lay more or less dormant through the Cold War period. Recently, though, the UN Security Council established ad hoc tribunals to deal with crimes committed in the former Yugoslavia (1993) and in Rwanda (1994). The experience with setting up these two ad hoc tribunals, as well as the failure to set up tribunals in other cases (such as Cambodia, East Timor, Guatemala, Iraq, and so on) led to renewed international interest in setting up a permanent tribunal – the ICC. This, combined with European states' experiences setting up the European Human Rights Convention under the EU within the context of a more general rise in the discourse of human rights, provided the impetus for a renewal of interest in the ICC. After several Preparatory Conferences, a UN conference was held in Rome in 1998 where a statute establishing the ICC was adopted. The statute has been signed by over 130 states but must be formally ratified by at least sixty states before the court can be established. It will be based in the Hague, but will be able to hold trials elsewhere.

The difficulties encountered in setting up the ICC speak to the geopolitics of trans-state organization. For example, it is significant that while 120 states signed the statute in Rome in 1998, seven did not. The seven states in opposition were Libya, Yemen, Iraq, Qatar, China, Israel, and the US. The US has had an ambivalent relation to the ICC. As a purported champion of universal human rights, and as a supporter of the ad hoc tribunals of the 1990s, the US might have been expected to be in favor of the establishment of the ICC. However, the Rome statute made clear some implications of a permanent ICC that the US found troublesome. In particular, members of the US military were concerned that US service personnel would be vulnerable to prosecution. Could US troop deployments across the globe be counted as "crimes of aggression" under the ICC's mandate? As hegemon, the US continues to deploy a large number of troops around the world. Given that there has been considerable discussion aimed at clarifying exactly what is meant by the term "crimes of aggression," the US military probably correctly fears military personnel would be open to prosecution for some US actions, such as the 1989 invasion of Panama. The US position is further complicated by concern on the part of its ally Israel regarding potential ICC prosecution for actions it has undertaken in the occupied territories. Moreover, while the ad hoc tribunals were set up under the auspices of the Security Council (wherein the US has veto power), the permanent ICC would not fall under Security Council control and would not rely on the Security Council for the referral of cases. Indeed even non-state actors such as nongovernmental organizations, or victims or their families, will be able to refer cases to the ICC. These geopolitical implications have prompted the powerful US Senate Armed Services Committee to actively oppose the establishment of the ICC. This makes ratification by the US Senate very unlikely in the near future, despite the fact that the US finally signed the ICC treaty on the last possible day (December 31, 2000) in one of President Clinton's last actions before he left office.

The US political administration's dilemmas over the ICC illustrate some of the difficulties associated with any attempt to set up trans-state regulatory structures and practices. In this case the US does not wish to be seen as backing away from its adopted role as beacon and champion of human rights around the globe, but on the other hand fears that its military actions abroad will subject US citizens to criminal prosecution. In addition, ratifying the treaty would put the US in line with all its major allies, except Israel. Taking a more exceptionalist path and not supporting the ICC would please only one ally – Israel. As one commentator has noted, "At the heart of this debate is the unavoidable question of how much sovereignty the United States is willing to sacrifice to aid in the fortification of a global rule of law" (Frye 1999: 10).

The US position on the ICC is illustrative of how trans-state organization is a form of globalization that like all others operates through the actions of states, even as it affects in different ways the integrity and power of states. The WTO was set up entirely outside of the UN system. While the Bretton Woods institutions (World Bank and IMF) are formally connected with the UN system, they have largely gone their own ways (in contrast, and often in conflict with the UNDP or the FAO, for example). The WTO does not have origins in the UN system, and as such represents a different type of trans-state organization from the ICC.

The WTO

The WTO has been, since its establishment on January 1, 1995, the major trans-state organization regulating international trade. The WTO is significantly different from GATT in three respects. First, the WTO has a broader purview – it systematically regulates trade in services and issues to do with intellectual property (traded inventions, creations, and designs), in addition to trade in goods. Secondly, the WTO is an actual trans-state institution with states as "members" (of a formal organization) rather than "parties" (to an agreement) as they were under GATT. Thus the WTO agreements go through formal processes of ratification in member states' parliaments – whereas under GATT no formal ratification was necessary. Thirdly, at the beginning (1948) GATT involved only 23 states, while the WTO's original members were the 128 countries that had signed GATT in 1994. The WTO currently has over 140 members and has been, from its beginning, more or less global in scope.

The WTO is structured differently from the UN. In the organization's hierarchy, the Ministerial Conference is at the head. The Ministerial Conference is convened irregularly but is supposed to meet at least once every two years and ministers come from every member state. The day-to-day business of the WTO is conducted by the General Council, which also acts as the Dispute Settlement Body and the Trade Policy Review Body. While GATT as an organization has gone, the actual agreements live on and are administered by the WTO. There are three separate Councils which oversee the three most important international agreements that the WTO administers. The Goods Council oversees an updated General Agreement on Tariffs and Trade (GATT) covering trade in goods; the Services Council is responsible for the General Agreement on Trade in Services (GATS); and the TRIPS Council is in charge of the Agreement on Trade-Related Aspects of Intellectual Property (TRIPS). There are a host of committees and working groups in addition to these more formal bodies. The WTO is managed by its Secretariat, headed by the Director General, and located in Geneva, Switzerland. The present Director General is Mike Moore of New Zealand. He was appointed in 1999 and will serve until 2002 when Supachai Panitchpakdi of Thailand is scheduled to take over.

The WTO states that its three main purposes are: foremost, to "help trade flow as freely as possible"; to be the forum for trade negotiations – aimed at "further liberalization"; and to facilitate settlement of trade disputes. The WTO's purposes are congruent with now-dominant neoliberal economic doctrine. Neoliberal doctrine combines a classical approach to trade that stresses the benefits of free trade on the one hand, with a neoclassical emphasis on relying upon the unimpeded operation of the price mechanism – that is, a so-called "free market," on the other. These tenets reinforce the WTO's stance on liberalization (of trade and of markets), which is that it is desirable and necessary for global prosperity. This is, of course, congruent with the positions of the Bretton Woods institutions – the IMF and the World Bank – but in conflict with those of some UN agencies, not to mention with the national development policies of many states.

The WTO has been criticized on many fronts. First, the neoliberal premises upon which the organization is based are criticized. It has historically been the general case that the world's strongest states have pushed for free trade and worked against protectionism on the part of other countries. However, many poorer countries see the mandatory elimination of trade barriers as a threat to nascent industrial development which may benefit from being "incubated" for a period before it is expected to be globally competitive. Indeed, the emphasis of the third WTO ministerial meeting in Seattle was on setting up and "selling" the so-called "Development Round" of trade negotiations. While few developing countries would like no checks on protectionism (indeed they would like to target in particular the measures in place in rich countries to protect key sectors, such as agriculture), many are unwilling to give up the option of strategically employing protectionist tools as part of a development strategy. The Newly Industrialized Countries (NICs) of Southeast and East Asia are often cited as examples of countries that engendered national development only through such means – means that are directly counter to WTO strictures. It is argued that WTO-enforced liberalization effectively compromises national autonomy in matters of economic and social policy.

Second, the push for liberalization is seen by many as being most clearly in the interests not just of rich countries in general, but of transnational capital. The interests of transnational corporations may often be in line with those of their home states but may perhaps be better understood as the interests of a transnational capitalist class, strongly represented in the rich countries of the world, but with elements in just about every country (Van der Pijl 1998). Many of the strongest criticisms of the WTO have argued that the WTO seeks to create a world trading system in which corporations are granted more rights than people (and certainly than the environment). (For one example of these arguments see Wallach, Sforza, and Nader (2000).)

Third, in its actual operations the WTO is charged with being undemocratic and lacking in transparency. Certainly, many poorer countries lack the resources to successfully represent their interests in trade disputes taken to the WTO.

Trans-state organizations have been targets of criticism and protest for as long as they have been around. The World Bank's big dam projects are a notorious case in point. The IMF's structural adjustment programs have been met with street protests on more than one occasion (Walton and Seddon 1994). Recently, though, the nature of the protests and criticism, their targets, and their geography, have changed. The actions of the WTO and the actions of protesters against the WTO have prompted much greater public awareness and concern in many countries as to the nature of trans-state regulatory organizations. Heightened interest in formal organizations such as the WTO has been associated with more popular critical awareness of other, less formal, institutions that serve as quasi-regulators (in the Regulation School sense) of global relations. For example, the agenda-setting claims of the very undemocratic, very Euro-American and very corporate World Economic Forum (WEF) have also come in for scrutiny. The WEF began in 1971 as a kind of club for corporate business owners and executives in Europe with annual meetings in Davos, Switzerland. More recently, the WEF has served to bring together corporate leaders with invited academics and politicians and has taken on a high-profile role in identifying potential (and assumedly temporary) pitfalls entailed in unabashed neoliberalism on a global scale, all the while protecting transnational corporate interests. Even while noting the recent toning down of pure neoliberal rhetoric from the WEF, critics and protesters remain deeply worried that a profoundly unrepresentative and clearly class-aligned elite group appears to have been granted legitimacy, and to have effects upon, international policy-setting and associated trans-state regulatory frameworks.

Trans-State Organization "From Below"

Many of the activist groups that are engaged in seeking changes in the trans-state regulatory framework are themselves networked and organized transnationally. Indeed, another way of conceiving of trans-state regulation is to focus on organizations that are international but not supranational. That is, those groups focusing on aspects of the WTO for instance, are themselves examples of a wider array of trans-state organizational efforts that are not aimed at formally regulating states from above, but rather, aimed at uniting groups in various different countries around common causes. In addition to international confederations of trade unions and labor groups, alliances of labor, environmental organizations, and a wide variety of other grassroots social movements are being constantly constructed and reconstructed. Such groups are sub-state in their constituent elements (as opposed to supra-state) and thus represent trans-state organizations of a quite different nature from that of groups such as the UN or the WTO. Examples of well-known trans-state organizations of this type might include Greenpeace – the international environmental activist group – and Amnesty International – an organization aimed at monitoring human rights and fighting human rights abuses across the globe – but there are thousands of others.

In many cases these sorts of transnational networks are challenging and fighting aspects of the emerging mode of global regulation. Rejecting the premises of neoliberalism, and seeking more community-centered approaches, these sorts of efforts are diverse and not without contradiction. However, what tends to unite such networks is a common concern with the current mode of global regulation. As Brecher, Brown, Childs, and Cutler (1993) note:

> Globalization from below, in contrast to globalization from above, aims to restore to communities the power to nurture their environments; to enhance the access of ordinary people to the resources they need; to democratize local, national, and transnational political institutions; and to impose pacification on conflicting power centers. (Ibid: xv)

Such efforts are examples of a type of trans-state organization not captured in the formal structures of the postwar regulatory regime, but that nonetheless seek to influence and even regulate the activities of states, capital, and trans-state organizations themselves.

Conclusions

The contemporary world is characterized by a series of spatial disjunctures. The formal regulatory agencies of the post-1945 world – such as the UN – have been challenged by recent political shifts and ongoing processes of economic globalization. All these changes have been uneven and have occurred within the context of a deeply unequal world of power relations between populations and between the states in which they live and around which the trans-state organizations are built. The organizations set up as World War II came to an end are under pressure and strain – in particular the UN in its role as peacekeeper.

The case of the ICC, and in particular of US ambivalence towards its establishment, shows that the nature of the trans-state regulatory framework is not that it is supra-state or contra-state. Although at first glance this might appear as a transfer to the trans-state arena of some powers that were previously vested in states, the ICC case indicates that global regulation is itself built out of the strategies and needs of particular states. Trans-state organizations are not extra-state organizations. They do not mark the eclipse of the state. The formal structures of the post-1945 world have been creations of states and reflect the unequal power relations between states. The world political map of states does not tell the whole story of political organization, but it depicts a most significant element.

Further, as we see in the case of the WTO and its opposition, attempts at establishing a mode of regulation at the global level are not just a matter of inter-state politics. Civil-society groups of many kinds are uniting their diverse struggles around seeking more democratic and equitable ways of regulating trans-state flows and relations. The current mode of regulation at the global level is a site of serious contradiction and political conflict, involving capital, states, and social groups, and trans-state organizations charged with regulating at the global level are both the fora for and the targets of such struggles.

11 The Rise of the Workfare State

Joe Painter

Introduction

This chapter considers the global geography of the state, and particularly of that form of the state that developed in the major industrialized countries during the twentieth century: the welfare state. At the beginning of the last century the state played little part in the social welfare of its citizens. Then, gradually, through popular campaigns and the rise of new schools of economic thought, such as Keynesianism, the state took on more and more social functions in industrialized societies. This growth accelerated after World War II and was the primary impetus behind the overall growth in the size and range of state activities. A political consensus developed in many countries and it seemed for a time as if a healthy middle way had been found between the brutalities of unfettered capitalism on the one hand and totalitarian state socialism on the other.

The optimism was short lived, however, and by the early 1980s the combination of economic stagnation and the rise of populist right-wing ideologies undermined the welfare state and ushered in a period of retrenchment and restructuring that continues to this day. Initially much was made of attempts to "roll back the frontiers of the state." More recently there has been a shift in emphasis away from debates about the quantitative aspects of state activity to concern with its qualitative characteristics. At the forefront of these shifts has been the rise of a new focus on work as the best mechanism for delivering social goods to deprived groups in society. These policies take a number of forms, but can be grouped together under the heading of "workfare" rather than "welfare." In what follows I review these developments, but also examine some contemporary alternatives.

State Formation in the Twentieth Century

In the process of the development of human societies, states are comparatively recent phenomena. The *modern* state, organized on clearly bounded territorial lines and with a claim to authority distinct from the person of a monarch, is more recent still, dating from the seventeenth century and confined initially to part of Europe. The growth in the number of such states to the point at which they account for the entire land surface of the globe (with the partial exception of Antarctica) is one of the most significant features of modernity.

This historical perspective provides important clues about how we should understand the process of state development. First, states are not inevitable features of human existence. Organized societies existed for many thousands of years without them. Today's patchwork of states is thus not a natural and inevitable order, but the product of social processes. Second, the forms and functions of states vary through time and across space. Third, explanations of state restructuring and change must be historically and geographically sensitive.

From its emergence the state has been the focus of both hopes and fears. Throughout the complex processes of the formation of states around the globe, a wide variety of social groups has turned to state institutions as potential sources of progress. Of course, what counts as progress has varied immensely across time and space and depends on the interests of the group concerned. Whatever the "progressive" aim, however, the state has most often been the means by which the appropriate strategy has been prosecuted. Throughout the modern age, until the recent rise of the transnational corporation, states have been seen as the only institutions with the resources and organization capable of producing widespread, deliberate socioeconomic changes.

Yet it is precisely the state's control over resources and organization which simultaneously generates its reactionary side. The turmoil bequeathed to Europe by World War I saw the rise of fascist states in Germany, Spain, Italy, and Portugal, and of authoritarian state socialism in Russia. It has often been assumed by liberals that authoritarian or totalitarian states are exceptional: an essentially distinct and aberrant form of the state which has no implications for the development of "normal" liberal democratic states. The sociologist Anthony Giddens has suggested that, on the contrary, a tendency to totalitarianism is inherent in the modern state, constituting one of its defining features. This derives, according to Giddens, from the concentration of "administrative power" (Giddens 1985: 172–81) in the state apparatus generated by its control over resources and organization.

"Administrative power" refers to the ability of states to monitor their territories and populations. The development of technologies of surveillance and information storage, from the invention of writing to today's sophisticated microelectronics, has seen states insert themselves ever more thoroughly into the activities of their inhabitants. The state is an increasing presence: regulating and sanctioning; forbidding and allowing; monitoring and recording. Birth, marriage, and death; employment and unemployment; collective assembly, politics and expression; business and commerce; sex and relationships; religion and culture: fewer and fewer social activities remain wholly private. The administrative gaze of the modern state is increasingly "panoptic" and, Giddens insists, thus tends inherently toward totalitarianism.

Stuart Hall (1984) suggests that the emergence of unambiguously authoritarian states of left and right in the 1930s marked the beginning of a more critical attitude to the state. The power of the state was not always seen as malign in itself, but the potential for abuse had been demonstrated. Yet, following World War II the state was again the focus of hope. In Latin America and in the new countries of the now decolonizing continents of Asia and Africa it was to be the mechanism for economic "development" (Corbridge 1993b). In China, home to a quarter of the world's people, the new communist state promised to sweep away injustice and establish a peasant-oriented socialist society. And in Western Europe new "welfare states" were to be the generators of reconstruction and modernization and the providers of a guaranteed minimum quality of life.

The Corporate Welfare State

The welfare state is popularly thought of as a *part* of the state: those state institutions which provide "welfare" services. Social scientists often prefer to think of the welfare state as a particular *type* of state, in which the state guarantees (in theory) a minimum standard of welfare "from cradle to grave" for its citizens. These guarantees form one defining feature of a particular type of state and an important source of its claim to legitimacy. Seeing the welfare state as a specific type of state is helpful, because it shifts the focus away from the changing *quantity* of welfare services provided and onto the changing *role* of welfare provision. If the welfare state is in crisis today, this is not because state welfare provision has declined in quantitative terms (in fact it has increased), but it may be because welfare provision is no longer so central to state strategies and state survival.

Nonetheless, quantitative measures are important. They show just how limited and unusual welfare states are. The Organization for Economic Cooperation and Development (OECD) is made up of all the major industrialized capitalist countries. Figure 11.1 shows per capita public expenditure at purchasing power parity on health care – a major component of welfare states. There is clearly a massive disparity in public health care provision between the OECD, which accounts for just 19 percent of the world's population, and the "developing" countries, which account for 79 percent.

Welfare states are thus a rather unusual, limited phenomena, in terms both of time (the twentieth century) and space (Western Europe, the United States, Australia, New Zealand, Canada and, rather differently, Japan). Nonetheless, the inclusion of a chapter on them in this book on *global* change is justified as they have global significance. First, their development

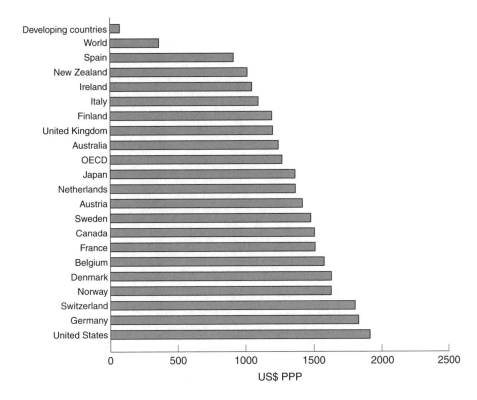

Figure 11.1   Per capita public expenditure on health, ca. 1998.

and subsequent problems have had a number of knock-on effects on the rest of the world, and have also partly depended on global economic and political relations. Secondly, the idea and ideal of the welfare state has met with approval around the world.

Its advent seemed to herald the end of a socioeconomic system based on the impoverishment of the majority for the benefit of a few. Where individuals and their families were unable to meet the expense of fundamental goods and services, the state would step in to make up the difference. This would either be through direct provision (of schools, hospitals, and so on), usually free or heavily subsidized, or through cash benefits to enable goods like food and clothes to be paid for.

Not only did the welfare state mitigate many of the social problems previously experienced, but it also had important economic and political effects. Economically, it was seen as partly an investment, by providing a better educated and healthier workforce. It also helped to iron out the economic instability which had dogged many countries, because it guaranteed a minimum level of consumption regardless of the level of economic activity. Previously, the absence of a "social wage" had left economies prone to crises of underconsumption, in which an economic downturn could turn into a slump as increases in unemployment reduced the demand for goods and services, which generated further unemployment, and so on. By placing a "floor" under the level of popular consumption, the downward cycle could be broken.

Politically, the welfare state was seen as a middle way between the "extremes" of unfettered capitalism and Soviet-style state socialism. It acted as a kind of "truce" between the forces of capital and labor, allowing the continuation of private investment and profit-making, while ameliorating some of the social consequences thereof. In many cases the institutional representatives of capital and labor were given a formal role in the formulation of policy, a strategy known as corporatism (hence corporate welfare state). In Britain, for example, the Trades Union Congress (TUC) and the Confederation of British Industry (CBI) were both involved, with the government, in the National Economic Development Council. This fitted well with the economic doctrines of John Maynard Keynes which were applied at the same time, and this form of the state is often known as the Keynesian welfare state. In many industrialized countries a consensus was established, at least between political parties and elites, that the state should play a significant social welfare role, and the 30 years from 1945 to 1975 are often referred to as the postwar "consensus" or "settlement."

The form of this "settlement" varied significantly from state to state, as did the relationship between welfare provision and the wider social system. The nature and causes of the differences are detailed by Esping-Andersen (1990) and summarized by Johnston (1993). Despite these variations, welfare states were highly successful. This success can be measured in a number of ways. The "welfare output" of states is not easy to assess in quantitative terms. This is partly because welfare is to some extent subjective and partly because it is difficult to determine how much the observable changes are due to state activity, and how much they are a product of a general increase in economic prosperity. Tables 11.1 and 11.2 show the coverage of the West European population by state insurance schemes for healthcare and unemployment respectively. Figures are given for 1925, some fifteen years before the major growth of European welfare states, and for 1975, just before the emergence of criticisms of that growth. They show that there were dramatic increases in state support for the sick and jobless. Table 11.3 gives data for one key indicator of well-being, infant mortality. The rapid decrease in infant mortality certainly had a lot to do with improved nutrition and hygiene as well as state healthcare, but nutrition and hygiene were themselves improved by state policies.

A final, albeit ambiguous, measure of success is the growth of inputs to the welfare state in terms of the proportion of national resources controlled by governments. Table 11.4 shows the expansion in public expenditure in a range of countries between 1930 and 1975. This measure is ambiguous since it is not self-evident that increasing resources lead to increased welfare outcomes. However, these figures do measure the *political* success of welfare states.

Table 11.1   Active members of medical benefits insurance schemes as a percentage of the labor force, selected European countries.

Country	1925	1975
Austria	47	88
Belgium	28	96
Denmark	99	100
France	21	94
Germany	57	72
Ireland	37	71
Italy	6	91
Norway	54	100
Sweden	29	100
Switzerland	50	100
United Kingdom	79	100

Source: Flora (1983: 460)

Table 11.2   Unemployment insurance: members as a percentage of the labor force, selected European countries.

Country	1925	1975
Austria	34	65
Belgium	18	67
Denmark	18	41
France	0	65
Ireland	20	71
Italy	19	52
Norway	4	82
Switzerland	8	29
United Kingdom	57	73

Source: Flora (1983: 460)

Problems for the Corporate Welfare State

Notwithstanding these successes and the enormous benefits that the development of welfare states brought for large sections of the population in industrialized capitalist countries, they also had problems and limitations.

For a start, welfare benefits were distributed in highly socially unequal

Table 11.3 Infant mortality (deaths under one year of age per 1,000 live births), selected countries.

Country	1925	1969	1993
Austria	119	25	7
Belgium	100	22	8
Denmark	80	15	6
England and Wales	75	18	9
France	95	20	7*
Germany	105	23	6
Ireland	68	21	6
Italy	119	30	7
Norway	50	14	5
Switzerland	58	15	6
United States	68	18	7

*1991
Note: The figures for the United States are for white babies. Mortality among non-white babies is 60–100 percent higher in each case.
Source: Mitchell (1975: 130–2; 1983: 130–2; 1998b: 125–6; 1998a: 89)

Table 11.4 General government expenditure as a percentage of GDP, selected European countries.

Country	1930	1975
Denmark	13.5	52.7
France	22.1	38.7
Germany	29.4	47.9
Ireland	20.8	50.2
Norway	17.4	49.3
Sweden	14.0	51.4
Switzerland	17.4	27.2
United Kingdom	24.7	49.9

Note: Data for France relate to 1929 and 1971 respectively. Data for Sweden relate to 1930 and 1974 respectively.
Source: Flora (1983: various pages)

ways. As noted above, many countries were excluded from these developments. However, even within materially rich countries welfare provision was often universal only in theory. "Advanced" societies are marked by social divisions among which three of the most important are divisions of

Table 11.5   The social distribution of expenditure on public services in the UK.

Ratio of per capita expenditure on richest 20 percent of population to per capita expenditure on poorest 20 percent

Favoring the poor	
Council housing	0.3
Equal	
Primary education	0.9
Compulsory secondary education	0.9
Favoring the rich	
National Health Service	1.4
Further education (16+)	1.8
Non-university higher education	3.5
Bus subsidies	3.7
Universities	5.4
Tax subsidies to owner-occupiers	6.8
Rail subsidies	9.8

Source: after Goodin and Le Grand (1987: 92)

gender, ethnicity, and wealth or class. These divisions are reflected in the distribution of the benefits produced by welfare states.

Work by Goodin and Le Grand (1987) on the UK welfare state shows that most of its benefits actually go to the middle class, rather than the working class (table 11.5). This raises the question of why the welfare state finds such strong support among the working class. However, the evidence from Goodin and Le Grand does not suggest that poorer sections of society have made no gains from the welfare state, only that they have not gained as much as the rich. Furthermore, popular support is discriminating. In the UK, changes to the National Health Service are much more politically sensitive than changes to universities.

Welfare benefits are also differentially distributed according to gender and ethnicity (Pierson 1991; Williams 1989). Most welfare states depend on a "subsidy" of their *public* welfare provision in the form of unpaid *domestic* welfare provision, performed mostly by women. For example, there is considerable variety in the extent to which welfare states provide public childcare for those under school age. However, comprehensive provision is very rare indeed and in some cases there is virtually no provision at all. Similarly, caring for the sick is often done unpaid by female family members as a complement to, or in place of, public healthcare. It is also usually women who undertake the paid caring work *within* the welfare state, often for wages lower than those typically paid to men. Members of minority ethnic groups also often suffer from discrimination in welfare provision. Immigrants are frequently permitted to enter their host country only on condition that they will not make claims on the social security system. In some countries minority ethnic groups are denied full citizenship rights and required to become "guest workers." This makes it much easier for the state to exclude them from social benefits, particularly on retirement, when they may be repatriated. Even where there is no overt discrimination of these kinds, welfare provision may still be racist. Formally equal rights may produce highly unequal outcomes in practice, where differences of language, culture, and custom lead to unintentional exclusion from welfare provision. In the UK, for example, documents and claim forms written in English have in the past led to those who speak other languages failing to receive all their entitlements. Similarly, the failure of Western health services to take account of different attitudes to healthcare among patients from non-Western cultures has led to members of some minority ethnic groups losing out.

The corporate welfare state thus brought substantial benefits to the middle classes and to sections of the working class, but it did so principally in the interests of men in general and white men in particular. In its operations in practice, the postwar settlement seems to have had substantial social limitations. However, a number of writers have begun to suggest that there are also political and economic limits to the operation of the welfare state even in its present, socially discriminatory form.

It is over fifty years since the publication of the Beveridge Report in Britain, the document which outlined the basis of the British welfare state, and the political and economic situation now looks very different. It has become clear that Keynesian economics and the welfare state were not permanent solutions to the difficulties and contradictions of capitalist development. During the 1980s and 1990s many governments attempted to reduce (or to limit the growth of) public services. Since the establishment of welfare states their costs have increased, both in real terms and relative to countries' overall production (see table 11.6).

In addition, Keynesian policies have proved incapable in the long run of maintaining full employment, which has led to the reemergence of the social problems that welfare states were set up to solve. State resources have been squeezed, as cash benefit payments have increased and tax receipts have been undermined. Writers on both the left and the right of the political spectrum have argued that the welfare state is in crisis. Increasingly governments blame forces beyond their control for economic problems, particularly economic processes operating at the global scale. In a world of dramatically increased economic, political, and environmental interdependence, some writers have even argued that the state itself is a spent force and is in terminal decline. However, evidence about changes in the welfare state point to a restructuring and reorganization of the state and its roles, rather

Table 11.6 Government consumption and total expenditure, selected
industrialized capitalist countries.

Country	GNP per capita (US$)	General government consumption (% GNP)			Government expenditure (% GNP)		
	1991	1965	1991	1998	1972	ca.1991	1997
Spain	12,450	7	16	16	19.8	34.0	36.8
United Kingdom	16,550	17	21	21	32.7	38.2	41.7
France	20,380	13	18	19	32.5	43.7	46.6
United States	22,240	17	18	16	19.4	25.3	21.7
Germany	23,650	15	18	20	24.2	32.5	33.4
Sweden	25,110	18	27	26	28.0	44.2	44.3
Japan	26,930	8	9	n/a	12.7	15.8	n/a

Note: Data for Germany refer to the former Federal Republic for years up to and including
1991 and to the unified Germany thereafter.
Source: World Bank (1986, 1993, 1999)

than to its demise. The interpretation of these changes is one of the major
tasks facing social scientists at the beginning of the twenty-first century.

Retrenchment and Reform

While there can be no doubt that the welfare state is changing, it is not yet
clear precisely where the changes will lead. We are in a period of experimentation and innovation. A number of competing models are being developed, although it is likely that not all of them will prove sustainable in the
medium term. It is also likely that future developments will be spatially
highly uneven, producing new, more complex, geographies of welfare and
social policy. In this situation the job of the analyst is not to predict with
certainty, but to look for trends and tendencies, and to outline possible
futures and their social and economic implications.

Workfare

One of the most important of the current trends is the emergence of
"workfare" as part of the mainstream of social policy-making, particularly
in the United States, Canada, and the UK. The idea of workfare has been
around a long time. In Victorian Britain, workhouses, which required

inmates to work in exchange for board and lodging, were one of the few means of public support for the very poor. During the depression of the 1930s public work schemes were established in the United States. These also required welfare recipients to work in exchange for food. With the construction of welfare states in the industrialized countries after World War II, workfare fell out of favor. It was widely seen as degrading and stigmatizing. In the context of a broadly social-democratic political consensus, unemployment and poverty were generally regarded not as individual failings but as social problems. Poor and unemployed people were no longer seen as feckless indolents who needed the discipline of workfare, but as equal citizens entitled to a minimum standard of living even when the economic system failed them. The proponents of workfare never entirely disappeared (particularly in the United States), but they dwindled to a small minority on the extreme right of the political spectrum, with no real influence on policy.

In recent years, however, workfare has come back onto the agenda with a vengeance. There are several reasons for this. First, the growth of mass, long-term unemployment placed existing structures of welfare provision under both financial and political strain. Second, the rise of the new right since the 1980s as a major political force saw a renewed emphasis on the obligations, as well as (or instead of) the rights, of welfare recipients, and the development of a discourse of opposition to so-called "welfare dependency." Third, the growth of global neoliberalism as an international policy regime put pressure on national governments to restrain the growth of public expenditure and to pursue fiscal and social policies that were acceptable to multinational capital and international financial markets.

The new workfare is rather broader in scope than the classical food-for-work schemes of the 1930s. Its origins can be traced to a series of policy initiatives in the United States that began under the Reagan administrations of the 1980s. One important feature was the significance of local- and state- (as opposed to federal-) level experiments in states such as Wisconsin and Massachusetts (Peck 2001). The so-called "Wisconsin model" became an important reference point in the subsequent development of workfare, while in Massachusetts, the 1990s saw a new Republican governor shift the state's welfare system from one of the most liberal to one of the most punitive in America (ibid: 129–67). The new workfare involves a wide and disparate range of schemes to encourage work and discourage welfare, the defining feature of which is a degree of compulsion. There is a strong emphasis on moving people off welfare and into work, education, or training, through positive incentives if possible, but through mandatory enforcement where necessary. According to Jamie Peck, one of the foremost writers on workfare, there are three key features of this emerging policy tendency:

- Individually workfarism is associated with *mandatory* program participation and behavioral modification, in contrast to the welfarist pattern of entitlement-based systems and voluntary program participation.

- Organizationally, workfarism involves a *systemic* orientation towards work, labor-force attachment, and the deterrence of welfare claims, displacing welfareism's bureaucratic logic of eligibility-based claims processing and benefit delivery with a more insistent focus on deflecting claimants into the labor market.

- Functionally, workfarism implies an ascendancy of *active* labor-market inclusion over passive labor-market exclusion, as workfarism seeks to push the poor into the labor market, or hold them in a persistently unstable state close to it, rather than sanctioning limited non-participation in wage labor in the way of welfare systems. (Ibid: 12)

Or, more pithily, "workfare is not about creating jobs for people that don't have them; it is about creating workers for jobs that nobody wants" (ibid: 6). Peck is at pains to emphasize that workfare is not yet firmly established as a coherent social policy as a comprehensive replacement to the welfare state. However, workfare developments to date do represent a fundamental break with many of the principles that underpinned the postwar welfare state. While in quantitative terms governments are spending more than ever on social provision of various kinds, in qualitative terms there has been a fundamental shift throughout the industrialized world towards a more individualistic system with a greater emphasis on obligations than rights and a concern to tailor social policies to what governments perceive to be the requirements of global economic competitiveness, as a recent report in the British newspaper the *Guardian* reveals: "Tony Blair heralded the end of the traditional welfare state yesterday when he said that everyone would in future be increasingly expected to help themselves . . . After a second Labour term, people would have to stand on their own two feet rather than automatically turn to the State if they fell on hard times or into ill-health" (Watson, Baldwin, and Webster 2001).

A social Europe?

Peck's emphasis on the unevenness of change is important, not least because it should caution against over-generalizing from the mainly Anglo-American experience that he documents. Elsewhere, particularly in continental Europe, the social-democratic tradition remains an important influence. Indeed, in geopolitical terms the "European project" can be seen partly as an attempt to demonstrate that there is a viable alternative to the out-and-out neoliberalism of recent US and British policy-making. European integration has been a contradictory process. On the one hand, there has been considerable neoliberal emphasis on competitiveness and the construction of a single integrated economic space in which the geographical mobility of all the factors of production would eventually be unimpeded by policy differences or international borders. On the other hand there is a parallel concern with the construction of a "social Europe" in which "social cohesion," European citizenship, and social solidarity are to be promoted and social exclusion, racism, and xenophobia are to be challenged and reduced.

There are many important differences between the welfare policies of different European countries. For example, Budge and Newton (1997) identify four major models in the postwar period: the German model (concerned with status maintenance), the British model (concerned with minimum security), the Scandinavian model (which combined status maintenance and minimum security), and the communist model (in which the state was responsible for the material welfare of all members of society). In one sense, therefore, there has never been a single "European social model." On the other hand, throughout much of the second half of the twentieth century many West European states have sought to define a middle way between state socialism (the "communist model" mentioned above) and unbridled capitalism. Typically this has involved historically high levels of public welfare expenditure, based on notions of social solidarity and the benefits to society as a whole of collective provision rather than free-market individualism. Supporters of such provision have argued that all citizens should have a right to a minimum standard of living, regardless of the cyclical behavior of the economy. Moreover, social inequality, it is suggested, should be reduced to prevent the emergence of deprived groups with little stake in society and few incentives to contribute to the common good, and this can only be achieved through collective action. At the same time, a degree of universalism in welfare provision is important as without it welfare recipients would become stigmatized while wealthier groups, whose taxes and social security contributions fund the system, would have little incentive to continue supporting a system from which they gained few benefits. Taken together, these principles underpinned much social policy-making in Europe for thirty or forty years after World War II.

The rise of global neoliberalism threatens the ability of individual states to pursue social policies that are markedly more generous or more costly than those of other countries with which they are in economic competition. At the same time, there is significant popular political pressure in Europe to protect welfare services and to ensure that welfare state restructuring does not lead to greater social and spatial inequality. Within the European Union there is the additional concern that the completion of the single market will exacerbate such inequality, at least in the short term, unless countervailing policies are put in place. There is thus an important political struggle taking place over the future of social policy in Europe and the extent to which the institutions of the European Union should take a lead in this area.

The Maastricht Treaty on European Union, which was agreed in 1991 and entered into force in 1993, originally included a Social Chapter. However, the British Conservative government at the time refused to sign the Treaty with the Social Chapter included, so it was taken out as a separate protocol signed by the governments of all the other member states. (The incoming Labour government signed the protocol in 1997.) Much of the activity undertaken in relation to the Social Chapter/Protocol has been concerned with the promotion of employment and reduction in unemployment. Very little substantive progress has been made on issues such as the harmonization of social security systems between the member states. Although the European Commission has produced a definition of the European social model as a "unique blend of economic well-being, social cohesiveness and high overall quality of life," European social policy remains one area where national differences are very much to the fore.

The social economy and the shadow state

While a unified European welfare regime is unlikely to emerge in the short or medium term, there are numerous local experiments with alternatives to conventional systems of welfare provision. Moreover, these are by no means limited to Europe. Indeed many of the most innovative have been developed in Canada and the USA, as well as in many poorer countries in Asia, Africa, and Latin America beyond the boundaries of the traditional welfare states. Much recent attention has focused on the potential of the so-called "social economy." Although they are privately managed, organizations in the social economy represent a "third sector" between mainstream capitalist businesses and the public sector. Their proponents argue that they are able to combine an economic role (more usually associated with profit-seeking companies) with social functions (previously the domain, principally, of the state). Drawing on Borzaga and Maiello (1998), Amin, Cameron, and Hudson (1999) identify four defining features of organizations in the social economy:

1   They are *private in nature* even if they have some public sector involvement.
2   They have a high degree of *managerial autonomy* from other public or private bodies.
3   They *produce and sell services of a collective interest.*
4   Although they can take any legal form they should include such features as
    (a)   nonprofit status;
    (b)   user, worker, and community participation in management;
    (c)   a democratic management structure.

The definition of the social economy is therefore fairly broad and can encompass, for example, food cooperatives, credit unions, housing associations, community businesses, environmental service organizations, and so on. Their social welfare functions can come about in two ways. First, they can be providers of social and welfare services to the public. Thus a food cooperative might enhance access to cheap, good-quality food in a deprived community. Second, they may be able to offer employment, work experience, and other opportunities for socially useful activity to people who are otherwise excluded from the formal labor market. Much faith has been placed in the social economy by policy-makers seeking an alternative both to bureaucratic and costly forms of state welfare delivery and to market-driven solutions that have been associated with increased social inequality and deprivation. So far, however, research suggests that the impact of the social economy remains limited and ambiguous. At best it provides interesting experiments that prefigure some bold shifts in welfare policy. At worst it is effectively another mechanism for enforcing the logic of workfare discussed above.

The social economy overlaps to some extent with the "shadow state," a term popularized by the geographer Jennifer Wolch to refer to the growth in the number and influence of voluntary organizations in the provision of public services. Wolch argues that shadow state organizations are ambiguous. Because they often receive substantial public funding for undertaking para-state functions they are heavily regulated by the state. They are thus less able to engage in campaigning or political work, or advocacy on behalf of their users. They also lack the formal democratic accountability of services provided by elected governments. On the other hand, they can represent an alternative to the bureaucratic and in practice often rather *un*accountable state services and can open up new fora for participation and engage in innovation and more flexible and user-centered forms of provision (Wolch 1990). In 2001 the notion of the shadow state was given an added twist by the proposals of the incoming Bush administration in the USA to use "faith-based organizations" (i.e., religious groups) to deliver federal welfare programs. Like other voluntary organizations before them, these associations are divided about the merits of accepting federal funding, some fearing that it will compromise their freedom of action.

Conclusion

In this chapter I have sought to show that the state is neither a natural nor a neutral phenomenon. It is always a historical and social construction, the product and medium of struggles between competing political ideologies and policy regimes. The rise of neoliberalism has brought with it greater instrumentalism in social policy and return to a pre-welfare state emphasis on the obligations as well as the rights of citizenship. Many have mourned the passing of the postwar welfare state, seeing the new workfare regimes as harsher, more disciplinary, and targeted too much on the supply side of the jobs market while ignoring the problems posed by the collapse in the demand for semi-skilled and unskilled labor. At the same time, however, it should not be forgotten that the traditional welfare state was too often inflexible, bureaucratic, and the generator of perverse incentives at the micro-economic scale. It may be that some of the alternatives mentioned here, including the possibility of a coherent European social policy and the innovative potential offered by experiments in the social economy, can provide a route map to a new form of welfare state avoiding the paternalism and bureaucracy of the past and the excesses of the workfarist approaches of the present.

12 Post-Cold War Geopolitics: Contrasting Superpowers in a World of Global Dangers

Gearóid Ó Tuathail

It was just past midnight inside a secret bunker south of Moscow when the alarm went off. The bunker was the control center for a fleet of early warning satellites orbiting above the United States. A signal from one of the satellites indicated that Russia was under missile attack from the continental United States. Soon the electronic screens indicated five intercontinental ballistic missiles in the atmosphere rocketing their way to the Russian heartland. With only minutes to make a retaliatory strike, the lieutenant colonel in charge made a gut-level decision. He decided the early-warning signals were false and that no surprise American missile attack was underway. Fortunately for the millions of people living in the United States, Russia, and surrounding countries, he made the correct decision.

This incident is not fictional. It happened on September 26, 1983 at the height of a particularly tense period in US–Soviet relations, just weeks after the downing of a Korean Air Lines passenger jet by the Soviet Air Force. A subsequent investigation revealed that one of the Soviet satellites had mistaken the sun's reflection off the top of some clouds for hostile missile launches. The officer who made the gut-level decision was initially praised, then investigated for not following procedure, and finally allowed to continue working without recognition or reward until his retirement. He now lives, like most Russian pensioners, on meager and erratic payments from a once modern state that is now mired in poverty, corruption, and decay (Hoffman 1999b).

The Cold War era was a remarkably dangerous time in human history when two of the most powerful states on earth threatened each other with thousands of nuclear warheads. Ever since the development of atomic weapons by an enormous state-funded technoscientific project based in the United

States, the political leaders of one, then a few, and then more and more states have had the capacity to completely destroy rival states quickly and easily compared to the wars of the past. A 1-megaton nuclear warhead placed on a rocket could be delivered thousands of miles in a matter of minutes. Upon impact it would completely destroy 50 square miles of any major world city in seconds and poison the surrounding region with radio-activity for decades. Intercontinental ballistic missiles, silent nuclear sub-marines, and supersonic stealth bombers were all developed to help deliver this destructive capability to enemy states with speed, reliability, and cer-tainty. The United States, the Soviet Union, and other states designed, built, and tested thousands of nuclear weapons of all shapes and sizes in the name of "national security." However, "national security" in the nuclear missile age really meant permanent national insecurity, for no amount of nuclear weapons could purchase protection and defense from other nuclear weap-ons. After Hiroshima, the world was living in a qualitatively new world of technoscientific terror. The best the two superpowers could do was to guar-antee they could destroy each other if they ever attacked each other. Having the superpowers acknowledge this "balance-of-terror" was not easy, but they eventually conceded that nuclear weapons had deterrence value and little else. Behind the doctrine of deterrence was a grim condition called "mutually assured destruction," MAD for short.

What brought the superpower leaders to embrace deterrence in the 1970s were a number of military confrontations involving their allies that nearly degenerated into nuclear war. On three separate occasions in the 1960s – over the Berlin crisis of August 1961, the introduction of Soviet missiles into Cuba in October 1962, and in the Middle East in June 1967 – super-power confrontations brought each side perilously close to the brink of nuclear war despite the desire of both sides to avoid the catastrophe such a war would represent. That the superpowers did not stumble into a thermo-nuclear war was, in part, a product of luck. Reflecting decades later, US Secretary of State Robert McNamara noted that "we came within a hairbreadth of nuclear war without realizing it . . . It is no credit to us that we missed nuclear war – at least we had to be lucky as well as wise" (McNamara quoted in Schell 1998: 47). In the late 1970s and early 1980s, as superpower relations degenerated once again, other geopolitical crises created the conditions for confrontations, miscalculations, and blunders. Fortunately, as in the 1960s, the world's geopolitical luck held and the dan-ger of a thermonuclear exchange between the superpowers was averted. But it could easily have been different.

To many, the collapse of the Soviet Union and the end of communist rule in Russia signaled a new beginning in world affairs, an era of promise be-yond the shadow of nuclear war. The world seems a much safer place with an absence of an overriding ideological confrontation between two heavily armed and hostile superpowers. This chapter takes a more skeptical view. The technoscientific terror born at the end of World War II and developed during the Cold War persists in the post-Cold War era. Weapons of mass destruction and the infrastructures necessary to manufacture them – nuclear weapons complexes, biological and chemical weapons factories and facilities – continue to haunt world politics. As the twenty-first century begins, eight states possess approximately 32,000 nuclear bombs with 50,000 megatons of destructive energy. This global arsenal is equivalent to about 416,000 Hiroshima-sized bombs and is more than enough to destroy the world human beings have created on planet earth (Cirincione 2000: 2). The capability to build biological and chemical weapons, the poor state's atomic bomb, is known to many more states and, in a new development, to non-state actors also. After the Gulf War United Nations inspectors in Iraq, for example, discovered that Saddam Hussein's regime had assembled hundreds of weapons filled with VX and sarin nerve gas and two dozen other biological agents. The manufacture and subsequent release of sarin gas on the Tokyo subway by the Aum Shinrikyo cult revealed that states no longer have exclusive control over weapons of mass destruction. The suicide terrorist attacks on the twin towers of the World Trade Center and the Pentagon demonstrate that even ordinary technologies, like jet airplanes loaded with fuel, can produce extraordinary death and destruction. The threat posed by weapons of mass destruction is one of the most immediate and pressing challenges to the common security of humanity in the twenty-first century. It is a threat that cannot be reduced to state-centric terms, to "us" versus "them." Rather, it is a threat embedded within the very technoscientific systems of production and destruction developed by the imperfectly modern superpowers and their allies since World War II. To understand this threat as a defining feature of the geopolitics of the post-Cold War world, we need first to discuss the general meaning of geopolitics and how geographers can study the subject in a critical manner. We then turn to consider the contrasting modernity of the superpowers today, to specifying the contemporary geopolitical condition more generally, and to briefly discussing the debate over the meaning of "national security" in the twenty-first century.

Geopolitics, Critical Geopolitics, and Geopolitical Discourses

Geopolitics is the study of the geographical dimensions of world politics, most especially the struggles for power by states with worldwide reach and power projection capabilities. As a form of knowledge, geopolitics has its origins in the late nineteenth century within the academic institutions and military academies of states that were or aspired to become "Great Powers." Geopolitics was a problem-solving form of discourse about interstate politics dedicated to serving the leaders of the state. It sought to educate state leaders about the struggle for power in world affairs and how to conduct statecraft and organize military resources to secure more power and influence for their state. In the twentieth century it developed more popular variants that strived to inform the state's population about the nature of world affairs, which states were the supposed enemies of "their state," and what types of threat these states posed to their welfare and survival. As the century progressed, the term "geopolitics" gradually came to define the knowledge used by leaders and ordinary citizens to make sense of the game of power politics across the world (Sharp 2000). What is significant about geopolitics, above all, is that it is the form of knowledge and reasoning favored by the most powerful forces in a state: coalitions of politicians, military institutions, defense contractors, research scientists, and others with a vested interest and commitment to a state-centric and Darwinian survival-of-the-strongest vision of world politics. Geopolitics is not a language of the poor but of the powerful.

The operation of this traditional form of geopolitics, which we will term "orthodox geopolitics," has been challenged in recent decades by an alternative approach to world affairs called critical geopolitics. Critical geopolitics challenges the state-centrism and Darwinian philosophy of orthodox geopolitics. Instead of operating from the perspective of powerful institutions and groups within dominant states, it articulates the perspectives and arguments of a transnational coalition of peace movements, human rights activists, and environmental organizations. Critical geopolitics seeks to challenge how orthodox geopolitics presents the world as "us" and "them," and how it defines "national security threats" in terms of military threats from other states and outlaw groups. It rejects the ethnocentric and chauvinist geopolitics of "us versus them" in favor of a more complex vision of world politics characterized by states dominated by power structures and technological systems that threaten the conditions of habitation and survival on the planet as a whole. Critical geopolitics, in other words, rejects state-centric reasoning and questions the monopoly of the powerful over the definition of "national security" (Ó Tuathail 1996).

Critical geopolitics uses four distinct concepts to analyze the history of geopolitics:

1 Geopolitical world order, the distribution of power, and the configuration of alliances across the world political map. Geopolitical world orders are characterized by a hegemonic state and its allies, which are usually under challenge by an alliance of less powerful states.
2 Techno-territorial complexes, the assemblages of technologies of communication, transportation, and warfare that condition and shape world strategic space. In compressing space and time, techno-territorial complexes influence the relationship between defense and offense in warfare and help shape the practice of geopolitical power.
3 Geopolitical economy, the geopolitical order governing economic production, trade and consumption of goods across the world, and the geoecological consequences of this order.

Table 12.1 Three distinctive "geopolitical worlds."

Geopolitical "worlds"	Geopolitical world order	Techno-territorial complex	Geopolitical economy	Geopolitical discourse
Inter-imperialist rivalry, 1870–1945	Conflict between the "great powers" over their relative power and position across the globe.	The British imperial navy and the emergent naval and airpower complexes of other states. The electrical, telegraph, radio, telephone, and radar networks.	Competing forms of national capitalist modernization with an emphasis on self-sufficiency and obtaining protected markets. Pollution ignored.	Imperialist discourse represents world politics in terms of racial divisions and civilizational hierarchies, justifying the rule of the strong over the weak.
Cold War, 1945–91	Conflict between the superpowers over their spheres of influence, ideology, and relative power position across the globe.	The nuclear weapons complexes of the superpowers. The communications and entertainment networks of the West.	Dominated by a geopolitically organized form of liberal capitalist modernization in the West and Soviet-style socialist modernization in the Eastern bloc. More extensive and toxic forms of pollution ignored.	Cold War discourse represents world politics as a worldwide struggle between, in Western terms, a "free world" and a "totalitarian world." In Soviet terms, the struggle was between Western "imperialist states" and liberation under communism.
Post-Cold War, 1991 to present	Relative predominance of the United States and unpredictable challenges to its power, influence, and symbols across the globe. Persistent regional antagonisms.	The aging nuclear weapons complexes of the superpowers. The internet, wireless communication and pervasive computer-controlled infostructures. The catastrophic potential of accidents and information system crashes.	Neoliberal globalization based on an ideological commitment to unregulated markets, privatization, and the virtues of advanced technological systems. Embedded corruption across large parts of the world. Pollution and toxicity too encroaching to fully ignore. Growing energy crises.	"Global dangers" discourse represents world politics as characterized by a range of borderlessness threats. Debates over which threats are the most pressing and how states should pursue "national security."

4 Geopolitical discourse, the rhetorical and symbolic forms of reasoning used by powerful coalitions within dominant states to explain world politics and justify the exercise of power by their own state. Geopolitical discourses are shifting cultural and political explanatory systems used by state leaders to give meaning to their actions and justify them in the eyes of the public.

Taken together these concepts help us delimit the contours of geopolitical power and conflict that have marked the twentieth century (Ó Tuathail, Dalby, and Routledge 1998). They help us specify the political geographic structures of "geopolitical worlds"as they have come together after a general war and subsequently developed until a new crisis or war changes the order of power. Over the last century and a quarter we can identify three distinctive "geopolitical worlds": a world of imperialist rivalry between the "Great Powers" that produced two cataclysmic worldwide wars, a Cold War world of superpower rivalry and ideological competition across the world's major geographic regions that fortunately ended relatively peacefully, and a contemporary post-Cold War world that is slowly being defined by the technoscientific dangers that characterize it (see table 12.1).

Giving definition and meaning to these worlds are the geopolitical discourses used by the hegemonic state and that used by the leading challenger to that state and its system of alliances. A number of generalizations can be made about these discourses. First, as already noted, these discourses are discourses championed by coalitions of powerful interest groups within the dominant state and across allied states. These coalitions are complex but they conventionally feature an "iron triangle" of conservative politicians, military institutions, and powerful corporations in a state. The conservative politicians normally articulate an exclusivist conception of "the nation" and celebrate its history as a history of "national exceptionalism" and greatness. Marginalized by this discourse are "minority groups" within the state. One clear example of this is the US state's decision to test its nuclear weapons in the American desert on lands claimed and lived upon by native American nations. What was home to these groups was represented by the dominant white Euroamerican nation as "wasteland" and converted into the Nevada Test Site (Kuletz 1998). The US state subsequently conducted numerous atmospheric nuclear explosions upon this site, doing the same upon the homelands of marginalized Pacific islanders.

Second, these discourses seek to monopolize the definition and interpretation of the threats faced by the "nation-state." Geopolitical discourses are discourses of danger that specify a parade of threats powerful interest groups consider important. This discourse defines the meaning of "national security" and, most importantly as far as defense contracting corporations are concerned, sets the agenda for the state spending necessary in order to address these threats. That this definition of "national security" is question-

able is evident from a consideration of the environmental legacy of the nuclear weapons complexes created by the superpowers. The largest polluter
in the United States is the US state, most specifically the "national security"
departments of Defense and Energy. The facilities created by the US state to
manufacture nuclear weapons are some of the most toxic places on the
North American continent; sites like the Hanford Nuclear Reservation in
Washington state and Rocky Flats in Colorado (Hevly and Findlay 1998).
In Russia the environmental legacy of weapons production across a network of ten closed nuclear cities is even worse. All locations suffer from
dangerous levels of radioactive contamination. One location near the formerly secret city of Chelyabinsk-65 (now Ozersk) has been termed the most
polluted spot on earth, for one can receive a fatal dose of radiation there in
less than an hour (Athanasiou 1996: 120). Producing "national security"
by poisoning places with radioactivity that lasts tens of thousands of years,
not to mention exposing workers and communities close to these facilities
to deadly toxins and genetic damage, raises questions about just how "national security" is defined.

Third, geopolitical discourses are frequently simplified spatial visions of
world affairs that organize the complex political struggles across the globe
into abstract conceptual categories and geographic zones. During the Cold
War, world politics was given meaning by Western geopoliticians by the
claim that the Soviet Union was an inherently expansionist empire that sought
to achieve world domination by spreading the creed of communism (Dalby
1992). Western geopolitics, as a consequence, became a somewhat para-
noid discourse that saw a "worldwide communist conspiracy" everywhere
it looked. In the 1980s, for example, the struggle of Nelson Mandela's African National Congress to end apartheid in South Africa, the fight of ordinary Filipinos against dictatorship, and the movement of Central American
peasants for social justice were all interpreted by the Reagan administration
as examples of "worldwide communism" rather than as the diverse place-
specific struggles for justice that they were. Geopolitical discourses, in other
words, are frequently conspiracy discourses in which self-generated anxieties are projected onto externalized foreign others and rendered as colossal
threats organized on a worldwide scale to the very existence and "way of
life" of the virtuous "nation." A characteristic of the operation of Nazi and
Stalinist discourses, this form of reasoning was also found in the West on
occasions of crisis during the Cold War. In the last decade new forms of
geopolitical reasoning have emerged around scenarios of civilizational clashes
and threats from global terrorist networks (Huntington 1998; Weaver 2000).
The catastrophic terrorist attacks of September 11, 2001 have generated a
strongly moral and religious geopolitical discourse that envisions a permanent war between "virtuous civilized states" and "barbarian networks of
global terrorists" and those that harbor them. The failed state of Afghanistan, however, is a poor and absurd substitute for the USSR's Cold War
role as the territorial home of "evil."

The Hyperpower versus the Demodernizing Power

Perhaps the most striking feature of the post-Cold War era is the contrasting contemporary condition of the Cold War superpowers. During the Clinton years the United States enjoyed the longest economic expansion in the state's history. Technological developments opened up new domains of economic activity, like wireless communication and e-business. The stockmarket reached record highs while unemployment reached record lows. The United States was also the unquestioned military power in the world, the "sole remaining superpower" according to some, though others, like the French Foreign Minister, found the term "superpower" inadequate and spoke instead of the "hyperpower" of the United States. Though US military spending declined from its Cold War highs in the mid-1990s, it still dwarfs that of the rest of the world. Today US military spending is on the rise again. The $305.4 billion US military budget request for 2001, for example, is more than five times the size of the current Russian military budget, the second largest military spender.[1] It is more than twenty-two times as large as the combined spending of the seven countries traditionally identified by the Pentagon as "rogue states" (Cuba, Iran, Iraq, Libya, North Korea, Sudan, and Syria) (Center for Defense Information 2000). Remarkably, despite this overwhelming military superiority and lack of a clearly defined territorial state enemy, the United States is committed to continue increasing its levels of military spending well into the twenty-first century. The reasons for this are largely domestic and political, with little relation to any realistic assessment of the external threats it faces. The US military-industrial complex has been reluctant to adjust to the end of the Cold War. Unlike Russia, where economic crisis has forced painful change, the US military bureaucracy has been remarkably successful in resisting any serious reorganization of its structure, mission, and force. Some force reforms are now underway but an entrenched "iron triangle" of military bureaucrats, defense contractors, and conservative politicians wields enormous power in determining the US defense budget. It is not unusual for powerful politicians funded by defense contractors to add items to the US defense budget not even requested by the Pentagon, principally because these items are made in the constituency of these politicians.

   The situation across the former territories of the Soviet Union could not be more different. Instead of economic expansion, the various independent states that emerged from the collapse of the Soviet empire have suffered severe economic contractions and crises. Moving to a market economy after decades of state-directed collective production and planning was always going to be difficult. In practice, this so-called "transition" has been a disaster for the vast majority of the peoples of the former Soviet Union. GNP has fallen by at least half in Russia since the end of the Cold War, while three-quarters of the population have seen their living standards plunge to a condition of impoverishment or near-impoverishment. Some

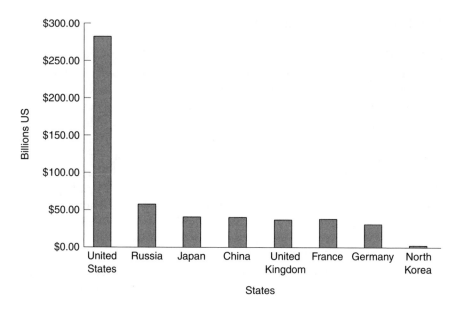

Figure 12.1   Military expenditures, 1999.

50 percent of Russians live below the official poverty line of $30–35 a month and probably another 25 percent are very near to it (Cohen 2000: 49). The neoliberal dream of Russia's transition to "market capitalism" has become the nightmare of transmutation into the "crony capitalism" of oligarchic domination and mafia rule. A hasty and ill-conceived 1995 plan to mortgage the "commanding heights" of the Russian economy for private financial loans from oligarchic controlled banks, a so-called "loans for shares" program supported by Western economists, provided an occasion for corruption, theft, and misappropriation of state assets on a grand scale. Under the rule of Boris Yeltsin, a small cabal of Russian oligarchs quickly gained control over the collective assets of the state and used these to accumulate vast personal fortunes in overseas banks. Control over the rich state oil and natural gas sector was acquired by private interests, as was control over state broadcasting and media networks. Coopting the government of Yeltsin and silencing opposition to their conduct, the oligarchs used their newly acquired power to finance the re-election of Yeltsin and secure their political position close to the center of power. Under Yeltsin's handpicked successor, Vladimir Putin, they have consolidated this power, occupying some government positions themselves and placing proxies in other influential positions of power (Wolosky 2000). Complementing this "top-down" corruption is the "bottom-up" corruption of local mafia groups in cities across Russia who operate through bribes and kickbacks in an alliance with local politicians, state officials, and the law (Handelman 1995).

The wealth of the few has come at the expense of the many. Russian state tax receipts are meager and state institutions are reeling from a generalized funding crisis. Legions of state employees and beneficiaries – teachers, doctors, nurses, postal clerks, planners, factory workers, research scientists, professors, and retirees – are suffering through erratic payment of their wages and benefits. Inflation and currency devaluation have reduced these payments to a pittance. Essential infrastructures of modern life have lost decades of investment and are barely functioning. Though Russia is overwhelmingly urban, it is estimated that three out of four people there now grow their own food (Cohen 2000: 42). One indication of the collective impact of the contemporary Russian depression is that life expectancy fell six times over the last decade of the twentieth century to an average of 65.9 years for both men and women in 2000, about ten years less than the United States and on a par with levels in Guatemala (Wines 2000).

The Russian military has been dramatically affected by Russia's multiple crises. Budgets have been slashed, equipment is aging, and its various branches are grappling with crises of mission and morale. Like others in the society, high-ranking generals have sought to exploit their positions for personal gain, selling state equipment to arms merchants and abusing their power over conscripts to enrich themselves (Odom 1998). The army performed poorly in the first Chechnya war of 1994–6 and only marginally better in the second war of 1999–2001 – a politically inspired war to elect Putin – while being guilty of widespread human rights abuses. Despite the thousands of deaths, population displacements, and recapture of Grozny and surrounding territory, this conflict is still not over. Other regional challenges to the central power and authority of Moscow in the Russian Federation have emerged as local communities and regional bosses eke out survival strategies amid economic depression and institutional collapse (Nunn and Stulberg 2000).

According to some, Russia's "economic and social disintegration has been so great that it has led to the unprecedented demodernization of a twentieth-century country." Russia has "dropped out of the community of developed nations." The political struggles between different factions to monopolize and strip the assets of the state in the 1990s have resulted in the "collapse of modern life" across the country (Cohen 2000: 41). Yet, despite the starkness of the contrast between the "hyperpower" of the United States and the "demodernization" of Russia, the Russian state remains a "superpower" in one crucial respect. It still controls more than enough nuclear weapons to destroy the United States and all of its allies. Table 12.2 contains estimates of the strategic (long-range) and non-strategic (short- and medium-range) nuclear weapons currently controlled by the world's nuclear powers.

The nuclear arsenals of the United States and Russia are governed by the 1991 Strategic Arms Reduction Treaty (START I). Both states subsequently negotiated a START II which promises to reduce warheads to 3,000–3,500

Table 12.2   World nuclear arsenals, 2000.

Country	Suspected strategic nuclear weapons	Suspected non-strategic nuclear weapons	Suspected total nuclear weapons
China	290	120	410
France	482	0	482
India	60+?	0	60+?
Israel	100+?	0	100+?
Pakistan	15–25?	0	15–25?
Russia	6,000	6,000–13,000	12,000–19,000
United Kingdom	100	100	200
United States	7,300	4,700–11,700	12,000–19,000

Source: Center for Defense Information (2000)

each by the end of 2007. After seven years of delay this treaty was finally passed by the Russian Duma in 2000. A START III to reduce levels even further is promised but uncertain. Bilateral arms control agreements, however, assume two equivalent and modern functioning states. With the Russian state, economy, and technological infrastructure disintegrating, the nature of the "Russian threat" has changed quite dramatically for the West. Instead of Russian state strength being a threat, the weakness of Russia as a state is now a pressing source of danger. There are four distinct nuclear dangers in Russia today:

1   The danger of an accidental nuclear war caused by an early warning system accident (Blair 1993). In 1995 the launch of a Norwegian scientific rocket triggered yet another false alarm in this system that reached Yeltsin for a possible retaliatory response. Russia's early warning system is now so decayed that Moscow is unable to detect US intercontinental ballistic missile launches for at least seven hours a day and can no longer see missiles fired from US submarines. At most, only four of Russia's 21 early warning satellites are still working (Hoffman 1999a). This techno-territorial "blindness" only fosters anxiety and danger, especially with the nuclear forces of both superpowers remaining on hair-trigger alert status.

2   The danger of "nuclear proliferation" caused by the illegal commercial sale of Russian nuclear warheads to independent parties or states. With economic times so desperate, the possibility of black-market sales of Russian nuclear technology, material, and expertise is considerable (Allison et al. 1996).

3   The danger of nuclear blackmail and chaos caused by current and future "civil wars" inside the Russian Federation. For the first time in

history, a fully nuclearized state is confronted with significant levels of internal political instability.

4   The danger posed by accidents or catastrophic failures within Russia's nuclear power systems. None of the Soviet-era reactors at electrical power plants or on naval submarines are considered safe by Western standards. The accidental sinking of the nuclear-powered submarine *Kursk* in August 2000 is hardly likely to be the last technological disaster for Russia and the former Soviet states, technological failures that first came to the world's attention with the Chernobyl explosion in 1986.

Problems also exist with the continued manufacture and storage of biological and chemical weapons (Alibek 1999). What all of these dangers have in common is that they are threats to everyone. The threat is as great to the Russian state and peoples as it is to the surrounding states, to the former Soviet republics, Western Europe, and to the United States. The nature of the contemporary "Russian threat," in other words, is a distinct departure from orthodox geopolitical thinking. Rather than being a territorial threat posed by one state to another, these contemporary threats are dangers arising from a disintegrating technoscientific modernity that imperils all surrounding states and the planet in general. These are dangers that "know no borders" for they are produced by the normal and routine (mal)functioning of complex technoscientific systems. As the world learnt when Chernobyl exploded, radioactivity does not respect national borders. It does not have any national allegiance or ideological preference. Undetectable by the human senses, if it is released it travels with the prevailing weather patterns through the atmosphere, raining down toxic fallout on those in its path with consequences that last across generations. Unlike previous wars and disasters, a nuclear explosion would pollute the gene pool of a whole people and generate victims years after any catastrophe.

In a move unthinkable during the Cold War, the United States Senate in 1992 acknowledged the dramatic shift in the nature of the "Russian threat" by funding the Cooperative Threat Reduction Program to aid disarmament and denuclearization initiatives across the former Soviet Union. The program has had some significant successes, including helping Ukraine, Belorus, and Kazakhstan move nuclear warheads from their territory to locations in Russia. Funded until at least 2006, it currently aims to accelerate the elimination of Russian missiles, bombers, submarines, and land-based missile launchers to meet START requirements, improve the safety, security, control, and accounting of Russia's nuclear warheads, end Russia's production of weapons-grade plutonium, and build a storage facility for the tons of fissile material from Russia's dismantled nuclear warheads. This relatively small program is guided by security thinking that departs in noteworthy ways from orthodox geopolitics in order to address the common security challenges of the contemporary geopolitical condition.

The Contemporary Geopolitical Condition: "World Risk Society"

Dangers from accidents, technological failures, and systems vulnerabilities, as well as the environmental challenges posed by deadly substances like plutonium, are not confined to Russia. Rather, the situation there is symptomatic of a much broader feature of the contemporary geopolitical condition. This condition is defined by the struggles of varied imperfectly modern states to address, adapt, and adjust to the multiple consequences and impacts of technoscientific modernization. Everyday life in modern states is secured, surrounded, and sustained by complex technoscientific systems – carbon fuel energy, global transportation and telecommunications webs, capitalist relations of production and consumption, biochemical industries – that deliver short-term "progress," "development," and "growth" but also long-term dangers to human health and the ecosystems that sustain life on the planet. The normalized and taken-for-granted functioning of ever more complex and pervasive formations of technoscientific modernization has produced a range of "manufactured uncertainties" at the very heart of modernity; many, like nuclear energy, hazardous chemicals, genetic engineering, and agro-industrial food systems, with catastrophic potential, either from "normal accidents" or terrorist attacks. This condition has been termed "world risk society" (Beck 1999). It is a condition marked by the globalization and proliferation of potentially catastrophic risks produced not only by the decay and disintegration of the modern, as found in Russia, but also by the successes and excesses of an uncritical embrace of technoscientific modernization, as found in the United States.

The desire of national security managers in powerful states is to control and contain potential threats and dangers. The vexing feature of technoscientific modernization and globalization for them is that it is producing "global dangers" that cannot be controlled and contained by national security institutions. "Global dangers" are threats that know no borders. These can be divided into

- borderless socioenvironmental threats like AIDS and BSE/CJD, acid rain and toxic chemicals, global warming and rising sea levels;
- borderless politicoeconomic threats like transnational crime and narcotrafficking, cyberattacks and global terrorism (Lake 2000);
- borderless catastrophic threats like nuclear energy accidents and proliferating weapons of mass destruction.

"Global dangers" are produced not by warring states but by the regular and taken-for-granted operation of technoscientific modernization and capitalist globalization as they expand and deepen our dependence on complex production systems, fossil fuels, information networks, and technoscientific processes and products. The contemporary geopolitical condition is characterized by the "boomerang effect" of technoscientific progress. That to which we attribute our prosperity and security is also that which threatens us with infrastructural vulnerabilities, systemic failures, environmental degradation, and a range of potential catastrophes. "Global dangers" can be both fast and slow: they range from dramatic explosions in nuclear power plants or sky-scrapers to the slow-motion poisoning of the planetary ecosystem by indus-trial toxins. Because of their spectacular nature, the media tends to focus on the former to the neglect of the latter. Political leaders tend to neglect slow threats by thinking only in terms of the next election cycle. This bias towards short-term thinking makes it exceedingly difficult to develop public policy to address the long-term problems of advanced modernity.

"Global Dangers" and Geopolitical Discourse

"Global dangers" are systemic contradictions in technoscientific modernity that require conceptualization and a sustained coordinated policy response at the global level. Unlike orthodox geopolitics, the enemy is not "out there" but the deep technoscientific modernity that envelops the advanced world. Some state leaders have grown to appreciate that "national security" can only be achieved through mutual security systems at the global level (Gore 1992). International regulatory accords and agreements like the Non-Proliferation Treaty, the Comprehensive Test Ban Treaty, or the Kyoto Ac-cords on the reduction of greenhouse gases articulate a vision of security that recognizes that individual state security is best obtained through col-lective common security. Put differently, no one state can be secure without all states having a shared measure of security. This is hardly a new idea, but it is one made all the more relevant by the "global dangers" that define the post-Cold War world. According to this reasoning, for example, the West's long-term security is best assured by helping Russia to overcome its eco-nomic depression and technoscientific disintegration through programs like the Cooperative Threat Reduction Program.

Orthodox geopolitical discourse, however, refuses to accept this analysis of the contemporary geopolitical condition and persists in defining "na-tional security" in state-centric and territorial terms. Its proponents speak of "global dangers" but interpret them narrowly as dangers posed to "us" by being (mis)used by "them." The rhetoric of "global dangers," in other words, is folded back into an orthodox geopolitical discourse in which a virtuous internal homeland must be secured from a threatening evil foreign power. The world is still primarily defined in terms of threatening "rogue states," "international terrorists," and "mad men" who pose a threat to the "Western way of life." "Security" and "defense" for "the nation" are to be obtained through institutionalizing a "national security state" at home, bombing these enemies abroad, and deploying even "bigger and better" technoscientific military systems.

It was reasoning of this type that lead the Republican-dominated US Senate to reject ratification of the Comprehensive Test Ban Treaty in 1999 and oppose the Kyoto Accords, acts that make the United States a "rogue state" in the eyes of peace activists, environmentalists, and some states. Another example of the persistent power of orthodox geopolitical thinking is the powerful coalition within the United States pushing the expenditure of over $60 billion to construct a National Missiles Defense system. This "Star Wars" system promises to shoot down nuclear missiles launched by "rogue states" at the territory of the United States. Its deployment has a formidable political momentum even though there is no solid scientific evidence that the system will ever work as intended. In supporting even more spending than that proposed by the Clinton administration, George W. Bush noted that "one of the things we Republicans stand for is to use our technologies in research and development to the point where we can bring certainty into an uncertain world" (Bush 2000). The statement reveals the uncritical faith many in America place in technological solutions to geopolitical problems, indeed in technology as a means of salvation and deliverance more generally (Noble 1999). Bush's position reveals the profound disjuncture between orthodox geopolitical discourse, with its clear distinction between "us" and "them," and the contemporary geopolitical condition, with its borderless technoscientific dangers. In a world where technoscientific modernization has created systems and structures with catastrophic potential and global dangers that know no borders, absolute security and "certainty" for states is not possible. Threats from states come from their own vulnerable and polluting technoscientific systems as much as from foreign powers. Yet, rather than acknowledge this and restructure their modernity on safer and more sustainable grounds, the quest for absolute security and salvation via technoscience persists. Deployment of the National Missile Defense system may see the US break the Anti-Ballistic Missile Treaty, undermine the deterrent doctrine of "mutually assured destruction," and, as a consequence, produce greater levels of insecurity among the world's major states. If this is the case, then twenty-first century geopolitics will end up a lot like twentieth-century geopolitics, which is not an appealing prospect.

NOTE

1   The 2001 request was submitted by the Clinton administration and marks a significant increase over the $293.283 billion budget in 2000. President George W. Bush has promised to increase military expenditures even further. Translated into dollars at the prevailing market rate Russia's official defense budget for 2000 amounted to $5 billion, roughly equivalent to the defense expenditures of Singapore and less than that of Argentina or Sweden. The International Institute of Strategic Studies (2000: 119) estimates the purchasing power parity of the rouble as five times that of the US dollar (i.e., the materials 1 rouble will

buy in Russia would cost $5 in the US). This, together with their estimate of military-related expenditures not part of the official defense budget, led them to the calculation of $57 billion for Russia's military expenditures in 1999. In contrast to President Bush, President Putin has announced significant military expenditure cutbacks so Russia's figures are likely to drop below $50 billion.

Part III
Geosocial Change

Introduction to Part III:
People in Turmoil

Economic, political, and cultural changes interact with themselves and with other aspects of social organization, such as the structuring of social life. As the world's population grows, for example, so do the demands for food, with consequences for environmental use: as realization of environmental constraints develops, so there may be calls for restraints on population growth, with clear consequences for the restructuring of social relations. Within societies, too, individuals and groups challenge the positions to which they are ascribed. The result of all these interactions is geosocial change, a people in turmoil, whose main features include population growth and mobility.

Population Growth

The world's population grew very rapidly throughout the twentieth century. Many commentators, especially those in the "developed world" where growth has been relatively slow in recent decades, have argued that unless the increase is very substantially slowed the earth's natural support systems will collapse. Extensive birth-control programs have been promoted in many countries; in China, such a program was accompanied by legislation which penalized married couples having more than one child (the highly controversial "Single Child Law").

Many birth control programs have been at least partially successful, and the number of children born to fertile mothers is declining. Some argue that this is a consequence not of the programs and propaganda per se, but rather of the perceived material benefits that flow from smaller families in many societies, plus the alternative lifestyles offered by education, especially to women (Todd 1987): these are thought to have stimulated declining birth rates in the "developed world" during the twentieth century, and sustain arguments for promoting economic development globally. Nevertheless, so large is the current female population in the child-bearing years that growth will almost certainly continue for several decades yet, albeit at reducing rates. However, there are increasingly two global demographic regimes: one is the low-fertility regimes, in which growth is slow and in which populations are aging (indeed some countries have actually adopted pronatalist policies); the other is the high-fertility regimes in much of the Third World which account for 95 percent of all births each year.

The earth's growing population through consumption alone will therefore place increased pressures on the environment during the coming decades. But growth, with its consequences for food production and distribution, is not the only population characteristic which is generating turmoil at the present time.

Hunger, Disease, and Structural Violence

In a world characterized by much political and military strife, many deaths resulting from behavioral (i.e., intentional) violence are recorded each year. Such premature deaths are small in number, however, compared with those which result from what is known as structural violence.

A capitalist economy is strongly characterized by its class structure, across which the benefits of wealth-creation are very unevenly distributed. Those who receive most can live in better housing conditions, consume many more than the minimum number of calories, vitamins, and other substances needed for daily sustenance, and obtain access to better systems of healthcare. As a consequence, they tend to live longer.

Variations in life expectancy related to class position occur in all countries, and also between countries. The average life expectancy at birth is much greater in Japan (78.6 years in 1990) than in Sierra Leone (42.0 years), for example, because many more Japanese than Sierra Leoneans are in the higher socioeconomic classes. These differences can be expressed in the concept of "lost (or stolen) years." The difference between the highest and lowest life-expectancy figures in those countries is 36.6 years, which will be lost by (stolen from) the average Sierra Leonean born in 1990. If the Japanese can live that long then their West African counterparts should be able to as well. Unequal life expectancies involve inflicting structural violence on the latter, through their position in the power structure that accompanies the map of uneven development.

The concept of lost/stolen years allows measurement of the amount of annual structural violence. If Japan, with the highest life expectancy at birth of any country, represents what is possible, then every child born in Sierra Leone in 1990 is going to live 36.6 years less than possible according to current societal organization: about 200,000 were born there then, producing a total of 7.3 million lost (stolen) years in that small country alone, for just one year's birth cohort. The extent of structural (perhaps better described as silent) violence is thus rapidly indicated: it is many times greater than the extent of behavioral violence (Johnston, Taylor, and O'Loughlin

1987). Interestingly, one of the world's leading economists – Nobel laureate Amartya Sen – has argued that famines are very unlikely to occur in democratically organized countries: where power is shared, so too are entitlements.

Much structural violence occurs in the early months and years of life: infants are the most likely to die because of malnutrition, unsanitary housing conditions, and poor healthcare (and many of their mothers die soon after childbirth for the same reasons). Thus variations in infant mortality rates, within and between countries, provide excellent indicators of the extent of structural violence. Amartya Sen has shown how there is a strong gender dimension to these mortalities in some countries (especially China, South Asia, and parts of the Middle East). Expected sex-ratios are highly skewed – for example over 120 men for every 100 women in China — which is a product of the differing economic and political opportunities, and forms of discrimination, faced by women. Much has been done in recent decades to reduce these variations, through programs of healthcare, housing investment, food provision, and education designed to bring rates down to the "developed world" levels. But they remain wide (from 7 per thousand live births in Hong Kong in 1989, for example, to 173 per thousand in Angola and Mozambique), and the gap between some countries is narrowing very slowly, if at all. The same is the case within countries, even "developed" countries such as the UK and the USA, where there are stark differences between the life chances of those born into the "underclass" and of those born into prosperity (Mingione 1993); those differences are aggravated by the reduction of state welfare provision and the increased reliance on market mechanisms, from which the poor are largely excluded. The problems of hunger and survival will therefore continue to present daily concerns (if not crises) to many, perhaps a majority, of the earth's population. Substantial achievements have been recorded in the control of some killer diseases, but others – including new ones such as AIDS – remain virulent. The goal of a high life expectancy for all, wherever they are born and whatever their parents' backgrounds, remains a very distant prospect. The AIDS pandemic has thrown this issue into bold relief: some African countries such as Zimbabwe and South Africa have HIV rates of 30 percent and more. AIDS has raised awareness of not only the relations between poverty, gender, and sexually transmitted disease, but also the role of transnational corporations in restricting the supply of cheaply available drugs to impoverished Third World constituencies. As new figures on HIV appear for China and South Asia, the short- to medium-term implications appear to be almost overwhelming.

Mobility and Conflict

As demonstrated in earlier chapters, we live in an increasingly mobile world, with goods and information being rapidly – even instantaneously in the latter case – shifted around. People are more mobile, too, responding to the push factors impelling them away from some areas and to the pull factors attracting them elsewhere.

With more people on the move, new problems are created in the places that attract them. Immigrants can be very desirable for a country, especially one experiencing labor shortages. There are many examples of workers being imported – whether to work on the land (as in the sugar-cane plantations of Queensland, Fiji, Trinidad, and Natal), in factories (as in many Western European countries), or in service occupations (as with the UK's National Health Service in the 1950s and 1960s and Filipino maids in much of Western Europe in the 1980s). Some are allowed to remain and can be joined by dependants, but others are sent home once they are jobless – which was the basis of South Africa's Bantustan policy under apartheid. Tensions may arise when immigrants are allowed to remain as permanent residents: they are readily targetable because they are identifiably culturally different. Tension often increases during times of economic difficulty, when immigrants can be regarded as threats to their hosts' jobs and social positions. There have been several national referenda in Switzerland on limiting the number of foreigners allowed to remain there, for example, and many countries have experienced inter-ethnic strife – often aggravated by xenophobic political movements. Refugees, of whom there are increasing numbers throughout the world largely due to famine and xenophobic nationalism, are particular targets for such movements. In Australia, for example, a country built on migration, a political movement achieved some electoral success at the turn of the century with its arguments that Asian immigrants should be repatriated and welfare state benefits withdrawn from Aborigines; several Western European countries (notably, but not only, Austria, France, and Italy) have similarly experienced substantial anti-immigrant movements upon which politicians have been able to capitalize electorally. (In Austria, for example, the success of the Freedom Party at federal elections in 1999 led to its members joining the cabinet, with most other EU states – led by the then presidency, Portugal – ostracizing Austrian representatives at meetings for some time thereafter; in the UK, racial tensions in several cities were inflamed by right-wing groups who were able to win significant support at the June 2001 general election in some towns with large Asian populations, such as Burnley and Oldham.)

Greater population mobility assists the rapid dissemination of contagious diseases. Migrations and tourism increase contacts globally and increase the probabilities of epidemics – for animals as well as humans. This stimulates calls for careful regulation of such movements in order to protect the health status of some, usually relatively privileged, populations. Freedom of movement, increasingly promoted as part of a "new world economic order," is thus threatened by calls to defend "national interests" against the possible depredations of too many newcomers.

Difference

Every society has norms which underpin its social relations; these include (often implicit but some explicit) definitions of acceptable roles and behavior for individuals and micro-social organizations, such as household structures. These norms invariably sanction unequal power relations between groups within society. Such inequalities may be challenged and altered, though usually only after substantial struggle. Often, however, the inequality survives long after it has been explicitly (i.e., legally) removed, as illustrated by the extremely slow and continuing implementation of racial equality after the United States' Civil War more than 130 years ago. Social structures in most of the "developed" world incorporate unequal power relations between the genders which permeate all aspects of economic, social, and political life. There have been many organized challenges to patriarchy, but "equal opportunities" have only recently been built into legal norms: even so, feminist movements have yet to achieve anything like full equality – neither have minority racial groups in most societies.

Gender is only one of the characteristics used to define an individual's position in the structure of social relations: race and sexual orientation are others. Individuals and groups in these various positions are developing their own social movements with which to challenge their unequal status. Their goal is to build new, emancipated societies which have broken free from those dominated by white males in nuclear families comprising a husband, wife, and two children. Many alternative family and household structures are being created, as society becomes more flexible and individual relationships become more fluid.

These social movements are but one set of examples of a society in turmoil. Economic globalization and rapid political change, interacting with continued population growth and a striving for material well-being, are contributing to a global society whose various components are in considerable flux.

13 Population Crises: From the Global to the Local

Elspeth Graham and Paul Boyle

In 1968, *The Population Bomb* warned of impending disaster if the population was not brought under control. Then the fuse was burning; now the population bomb has detonated. (Ehrlich and Ehrlich 1990: 9)

The material conditions of life will continue to get better for most people, in most countries, most of the time, indefinitely. (Simon 1995: 642)

Introduction

The idea that population growth threatens human welfare was promoted in an essay written over two centuries ago by Thomas Robert Malthus. His hypothesis that the power of population is infinitely greater than the power of the earth to produce the means of subsistence, though criticized even during his lifetime, surfaced again in the 1960s when it was reframed within the global thinking that accompanied space exploration. Images of the earth as a fragile planet, coupled with the historically unprecedented growth in world population at the time, renewed fears that a population crisis was imminent. For the neo-Malthusians, the presence of war, famine, and disease on the planet were enough to sound the alarm bells, as these were exactly the "positive checks" to population growth that Malthus had predicted. Yet over the past forty years an alternative voice has emerged in the population debate challenging this interpretation. After all, the causes of war, disease, and even famine are not clearly or straightforwardly rooted in the pressure of population on resources. There have been subtle changes in the focus of debate since the 1960s, but the passage of time has not produced consensus. Indeed there can be few other areas of academic endeavor where such extreme and contradictory views persist. Yet it is of utmost importance to break the deadlock, since we must know whether there is a global population crisis before we can sensibly determine what to do about it.

In 1999 the global population reached a total of 6 billion people. It had doubled in less than 40 years and is growing at around 1.3 percent per annum. This means that some 78 million people had been added to the world in the previous twelve months. According to the United Nations it is now likely that the global population will reach 8.9 billion by the year 2050 and is expected to stabilize at just above 10 billion people after 2200. Can planet Earth sustain this unprecedented number of human inhabitants? The figures are widely accepted but their implications, as the quotations above demonstrate, are hotly disputed. If you agree with Ehrlich and Ehrlich (1990) then the global population crisis is already upon us, whereas if you agree with Simon (1995) there is no crisis. The problem we are all faced with is how to decide between such diametrically opposed views.

The Ehrlichs, like other neo-Malthusians, emphasize the rapid growth of the global population during the last century and the apparent failure of the supply of a basic resource – food – to keep up with such rapidly rising demand. As the twentieth century dawned, the total population of the planet had not yet reached 2 billion; as it ended, this had more than tripled. Put more dramatically, nearly half of those alive today are under 25 years old. Moreover, according to the neo-Malthusians, the planet is already failing to sustain this burgeoning population. The message is uncompromising: we must act to curb population growth or disaster, on a very large scale, will strike – always assuming, of course, that it is not already too late to avoid global catastrophe. It is hardly surprising, then, that the neo-Malthusians have been dubbed "doomsayers."

In contrast, Simon's perspective can be described as Cornucopian, putting faith in the Earth as a horn of plenty. This provides a far more optimistic forecast of what the future might hold, a forecast which contradicts almost every claim that the neo-Malthusians make. The evidence used to support this optimism is based on long-term trends in a variety of aspects of human welfare. Whatever the rapidity of population growth, if life expectancy is generally increasing, standards of living improving, raw materials coming down in price, and the environment becoming cleaner, there can be no crisis – current or imminent. And the evidence, say the optimists, is precisely that all of these good things are happening. Further, since 1976 when Simon first accused the pessimists of peddling "false, bad news," the optimists are even more optimistic than before as they amass further evidence of global improvement. The message here is comforting: we are doing well, and we will be just fine in the future as long as we continue on our current path. Simon saw this as realism rather than optimism, concluding that "it is realistic to forecast improving long-run material trends for humanity, forever" (Simon 1995: 651). A future of infinite improvement is a reassuring thought, but would we be totally irresponsible to embrace it?

It is both difficult to answer this last question and important to do so, for our future may depend upon it. We cannot know how best to act unless we first understand what we are facing. Geographers, given their longstanding concern both with population questions and with environment or resource issues, have much to contribute to this understanding despite their noted reluctance to become involved (Findlay and Hoy 2000). This chapter, therefore, reviews the evidence and ways of thinking which have dominated the global population debate in the hope of stimulating interest. Moreover, we believe that the current dimensions of the debate lack a vital geographical understanding and we call upon recent evidence from geographical research to argue that current frameworks of global analysis are deficient precisely because they ignore geographical diversities. In doing so, we wish to move the debate from the global to the local, while recognizing the interrelationships between the two. Adding a geographical dimension further increases the complexities of the central issue by raising the possibility that we are facing multiple population crises, and we do not pretend that a resolution is within reach. Nor is it possible, within the confines of a single chapter, to be comprehensive in our coverage of these complexities. Nevertheless, the question of whether current population growth threatens human welfare is too important for geographers to ignore.

Population and Resources: Reviewing the Evidence

We begin by reviewing the basics of the debate. First, how accurate are population figures and forecasts, such as those produced by the United Nations? If we cannot be confident in these figures and forecasts, surely we need to be cautious about the implications we draw from them. And, second, what estimates do we have regarding global resources, both current and future, and their potential depletion by the ever-growing global population? Forecasting future resource availability is no less difficult a task than forecasting future population growth, for it too depends crucially on a few key assumptions. Yet, unless we can derive reasonable estimates of global population and global resources, we have no secure evidence on which to base either pessimistic or optimistic conclusions about how many people planet Earth might support.

Table 13.1 summarizes the current state of the global population according to figures provided by the United Nations. The difference between the crude birth and death rates translates into the annual growth rate of just over 1.3 percent, or a net addition each year of more people than currently reside in the United Kingdom. In absolute terms, this growth seems frighteningly large, despite a gradual fall in the rate of growth over the 1990s. Not surprisingly, the dramatic announcement by the United Nations Secretary General that, at 6.24 a.m. London time on October 12, 1999, the 6 billionth person on the planet was born attracted widespread attention from the media. So, too, did the birth of Astha ("faith" in Hindi) on May 11, 2000 in Delhi, which put India in an exclusive club with China as the only countries with populations exceeding 1 billion. The reason for encouraging this media attention, as far as the United Nations was concerned, was to draw attention to the rapid population growth that the globe is experiencing and will continue to experience for some time to come. Of course, it would be foolish to assume that these landmark figures were actually reached by either of these births because the national censuses and population registers from which they are derived vary in their accuracy. Dyson (1996) notes that in 1988 the United Nations estimated that the population of Guinea was 5.4 million in 1980. In 1994 the United Nations revised this estimate downwards to 4.5 million – a difference of 17 percent. If we are uncertain of the population that is alive today, how then can we predict, with any accuracy, what the population will be in the future?

One approach to predicting the future global population starts from an examination of its past increases which pessimists commonly characterize as "exponential" growth. This mathematical expression is used to describe the consistent doubling of the earth's population, with the implication that, unless controlled, global population will continue to grow in this manner. A common portrait of "exponential" growth tracks global population over a long period of time, as in figure 13.1. Following a long period of gradual population increase, the global population has risen remarkably over the last few hundred years and, depicted like this, there appears to be no suggestion that the population could ever begin to fall. In mathematical terms, the shape of exponential curves can vary, but the key point is that the percentage growth of a population growing exponentially will be constant as time passes. Put another way, the population is assumed to double each time a fixed number of years passes. In 1830 Malthus estimated that the population of the United States was doubling every 25 years, correcting for the contribution of immigration, while in many European countries the doubling time was closer to 50 years. In fact, most countries for which there are reliable figures do not experience constant doubling times and this was recognized as far back as the end of the nineteenth century (Cannan 1895). For the globe as a whole, an exponential model of population growth would have failed to predict the acceleration of population growth during the twen-

Table 13.1 Global demographic indicators, 1995–2000.

Area	Annual growth rate (%)	Crude birth rate (per 1000)	Crude death rate (per 1000)	Total rate (per woman)	Life expectancy at birth	Infant mortality rate (per 1000)	Mortality under 5 (per 1000)
World	1.3	22	9	2.7	65	57	79
More developed regions	0.3	11	10	1.6	75	9	11
Less developed regions	1.6	25	9	3.0	63	63	87
Least developed regions	2.4	39	15	5.1	51	99	156

Source: United Nations World Population (1998)

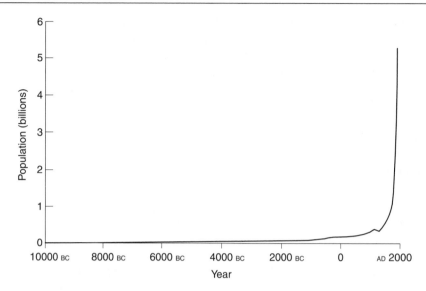

Figure 13.1   Estimated global population.

tieth century up to 1970 (an exponential curve assumes constant, rather than accelerating, growth) and the decelerating population growth that has occurred since the 1970s. Thus, while this simple model is quite reasonable for short-term forecasts over a few decades, it is not useful for long-term projections. The globe's population is not growing exponentially.

The United Nations' current forecast, mentioned above, is that the global population will reach 8.9 billion by the year 2050. This is their medium variant projection which assumes that the global population will stabilize at around 10.8 billion in 2150. However, this is obviously based on assumptions about future fertility and mortality rates and, of these, fertility is the most difficult to predict (although the HIV/AIDS pandemic could have a drastic impact on future mortality rates). Global fertility rates have fallen from 5.1 children per woman in 1960 to around 2.7 in 1998 and the medium variant projection assumes that these rates will continue to fall to a global average of about 2 children per woman. However, if this rate fell to an average of 1.6 children per woman, the predicted population in 2150 would be only 3.6 billion (low variant projection), while, if it remained at 2.6 children, the high variant projection would be a future global population of 27 billion (figure 13.2). Given that there is a difference of 23.4 billion between these two extremes, we are left questioning how useful these predictions are. Moreover, the further into the future we attempt to forecast, the wider the gap between low and high variant projections will be.

Even if we were confident in our knowledge of current and future global population size, drawing implications about whether or not we are facing a

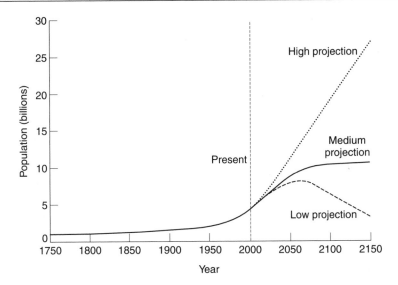

Figure 13.2   Projected global population.

global population crisis is less than straightforward. Malthus also provided predictions of food supply. Unlike the population, which he expected to increase exponentially, doubling every 25 years, he assumed that food supply would increase only arithmetically – hence his apocalyptic prediction that population numbers would be controlled by war, famine, and disease unless preventive measures reducing fertility were adopted. As it turned out, his forecast of the future growth of food supply was rapidly discredited as new agricultural technologies were introduced. Nevertheless, it has proved extremely difficult to estimate reliably even the current global supply of food, far less its future growth.

How much food is there to sustain our current 6 billion people? In addition to the uncertainties of national agricultural censuses, it is not even obvious what needs to be included in the calculation. As Dyson (1996) points out, statistics on agricultural output vary greatly from year to year, influenced by numerous factors including crop prices and weather conditions. The UN Food and Agriculture Organization (FAO) publishes annual information about food production for virtually all countries, but some of these are based on estimates of the area of available land rather than on indices of food production. Such figures derived from land use may be unreliable because arable land may be used for grass or left fallow – 8 million hectares of fallow land were recorded as arable in the United States in the late 1980s, for example (Grigg 1993). In some countries, where conditions allow, crops are sown twice in a year. In others they are sown only once. Thus, there is no general relationship between cropped area and food output. Also, in many developing countries, systems for collecting information on agricultural production are non-existent and, while figures on livestock may exist, the output of milk or meat is not usually monitored. Further, accuracy demands that only net agricultural output is measured, since part of crop harvests may not be destined for human consumption but rather used for seed for the following year, fed to livestock, or used in industrial applications, as oilseeds commonly are. Finally, different countries produce different foods and these may be difficult to compare – instead of converting all foods to standard calorific units, physical outputs weighted by local prices are commonly used. In either case, deriving an aggregate global figure for food supply is beset with problems and reported totals are less than secure.

In fact, in order to decide whether or not we are facing a global crisis, estimates of planetary resources need to be much more comprehensive and must include water, energy, and other vital requirements. There is, too, a complex interplay between these various resources. Focusing on the example of food at least demonstrates that there are numerous problems in producing such estimates. Further, even if we accept estimates of both population and resources as reasonable, predicting a global future also requires some specification of how the two are related. The additional problems of modeling this relationship can be illustrated by considering another model of population growth based on the logistic curve, which introduces the notion of carrying capacity (figure 13.3). The definition of carrying capacity is crucial, as the shape of the logistic curve is dependent on the estimated carrying capacity. The most common starting point for measuring carrying capacity is Penck's (1925; cited in Cohen 1995) formula for the maximum supportable population:

$$\text{Global population} = \frac{\textit{productive area} \times \textit{production per unit of area}}{\textit{average nutritional requirement of a single person}}$$

If we assume that the earth's area is fixed and that there is a maximum level of productivity that this land can achieve, it follows that there must be a limit to the supportable population because the nutritional requirements of people cannot reach zero. There are several limitations to this approach, not least that it considers only food production when population densities could be limited by water, soil, cultural views about space, and so on.

Pearl and Gould (1936) used this model to forecast the global population, expecting it to reach an upper limit of approximately 2.6 billion – in fact this was reached during the 1950s and the population of the planet is now over twice this size. The reliability of this approach is thus debatable. Unlike the exponential curve, which has a constant relative growth rate, the logistic curve has a constantly falling growth rate making it even less useful for predicting the accelerating growth during the first 70 years of the twentieth century. Similar to the exponential model, the logistic curve may provide reasonable estimates of the global population in the short term (in fact the estimates from both models are often extremely similar for short periods) but is far less accurate for long-term projections (Monro 1993). Many more complex methods for forecasting global populations have been advocated (e.g., Rogerson 1997). However, all share this limitation.

According to available figures, global food output increased by 3.1 percent per annum in the 1950s, by 2.7 percent in the 1960s, and by 2.5 percent in the 1970s and 1980s. A comparison between these growth rates and those for the global population suggests that population has indeed increased more rapidly than food supplies since the 1950s (Grigg 1993). Yet this, in itself, is not enough to demonstrate impending ecological catastrophe at a global scale. Even if we take a leap of faith and accept that we have reasonably accurate estimates of both population and food supply, and that growth in the former has recently been greater than the increase in the latter, we would need further evidence to justify predictions of disaster. It could be, for example, that the uneven distribution of food (or, more generally, resources) across the globe is a more potent cause of malnutrition and hardship than any ceiling to what the planet can supply.

Understanding the Problem is Not a Piece of Cake

We have outlined some of the major problems associated with accumulating evidence about global population and global resources, and the even greater difficulties of forecasting what might happen to each in the future. The note of caution is salutary, but it would be foolish to abandon all at-

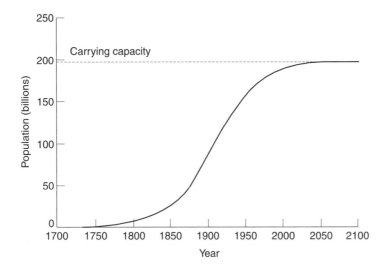

Figure 13.3 The logistic curve.

tempts to understand the "global population crisis" in the face of such uncertainties. The key question is whether the growth of the global population is already, or soon will be, unsustainable given the resources of the planet. And the answers of the main protagonists in the debate depend much more on the conceptual frameworks they employ to understand the relationship between population and resources than is generally recognized.

Malthusian thinking has dominated, and arguably still does dominate, both academic and popular views on the subject. In its cruder versions, the Earth's resources are seen as finite – like a cake. After a certain point, adding more mouths to feed means a smaller slice for all and an unsustainable future. Yet this takes no account of, among other things, the constantly developing impact of technology on our human capacity to produce resources. Indeed, in the twenty years between 1970 and 1990, one estimate suggests that the total food available *per person* in the world rose by around 11 percent (Lappé, Collins, and Rosset 1998: 61). Thus technological innovation must, at the very least, increase the size of the cake, making it possible for the planet to support a larger population than before. In contemporary discussions of the so-called "carrying capacity" of the Earth, this much is recognized. To the neo-Malthusians, increases in carrying capacity brought about by new technologies at best delay the inevitable crisis of population and at worst do irreparable ecological damage. The uncertainties surrounding the precise impacts of such technologies perhaps explain the very different estimates, offered in the last decade, of how many people the planet can support. These range from 2.8 billion (assuming a full-but-healthy diet for all) to a high of 44 billion people in an alternative scenario! Cohen (1995) provides a detailed review of more than sixty such estimates and the assumptions on which they were based. The most striking feature is, again, the lack of consensus.

The hugely different conclusions reached by those who try to estimate the planet's carrying capacity should give pause for thought. However, it is not the difficulties of deciding on the appropriate calorie intake or sources (many of the higher estimates assume a vegetarian diet) to incorporate in the calculation that are of primary interest here, but rather the very notion of applying ideas from non-human ecology to human populations. Humans, as Cohen points out, are not homogeneous in their resource requirements. Environments vary, with different limiting factors characterizing different times and places, and allowance must be made for adaptive economic, technological, and cultural responses. Thus using an analogy such as fruit flies in a jar, as the neo-Malthusians sometimes do, to describe the relationship between human populations and their resources for sustaining life, is misleading. Humans, unlike fruit flies, have the capacity to make choices about their own fertility and about how to manage the resources at their disposal. Their social mores and economic and political institutions shape these choices in complex ways. And with the capacity for choice comes the moral responsibility to make the right choices. Ignoring these dimensions of human life while at the same time emphasizing the need to take control of human population growth seems to us a fundamental contradiction underlying the neo-Malthusian position. However, it would be wrong to conclude that there must, therefore, be no crisis. It remains possible that there are ecological limits to the number of people the planet can support.

Two of the most fundamental resources needed to support a human life are food and water. So, are we already running out of these essentials? The Cornucopians think not. The world's food production gains in the last fifty years have indeed been impressive. Global food output more than doubled but was produced using approximately the same land base as in 1960 (Avery 1995). Despite some decline in annual production per head since the mid-1980s, the world today produces enough grain to provide every human being on the planet with 3,500 calories a day, a more than adequate calorie intake. Even the more cautious commentators conclude from this evidence that there is no question, at a global scale, of population having outrun food production (Grigg 1993). Nor, it seems, is food production globally likely to be limited by the availability of agricultural land. There is one resource, however, which could effectively limit food production in the future, and that is water. Currently, some 70 percent of the global demand for water comes from agriculture (World Resources Institute 2000a). If the pessimists are right and we are in danger of depleting the planet's freshwater sources to such an extent that demand for water significantly outstrips supply, then irrigated crop production may well be the first sector to suffer. With over one-third of total agricultural production coming from the 16 percent of all cropland that is irrigated, the future water needs of agriculture represent a more serious supply challenge than foreseeable difficulties with land or soil (Dyson 1996). Last year, the Worldwatch Institute reported that underground aquifers are diminishing on every continent and that scores of countries are facing shortages as water tables fall and wells dry up (Brown 2000). However, although there are already important constraints on the supply of water relative to demand in some parts of the world, other areas are water rich. This situation is reflected in a recent analysis of water vulnerability published by the World Resources Institute (figure 13.4). The uneven distribution of freshwater availability across the globe is a cause for concern, but not evidence of an imminent global shortage. The complexities of assessing future availability of water for agriculture are no less than for forecasting future food production, and Smil (1994: 270) points out that considerable efficiency gains can be made by lessening the water losses currently incurred through distribution, seepage and evaporation, since these commonly amount to 50–60 percent of carried water. Of course, some parts of the world, especially in sub-Saharan Africa, are likely to remain water poor and this situation may worsen due to climate warming. The extent of their ability to institute significant efficiency gains will, however, be an important determinant of future water availability locally and is related more to economic and political considerations than to water supply per se.

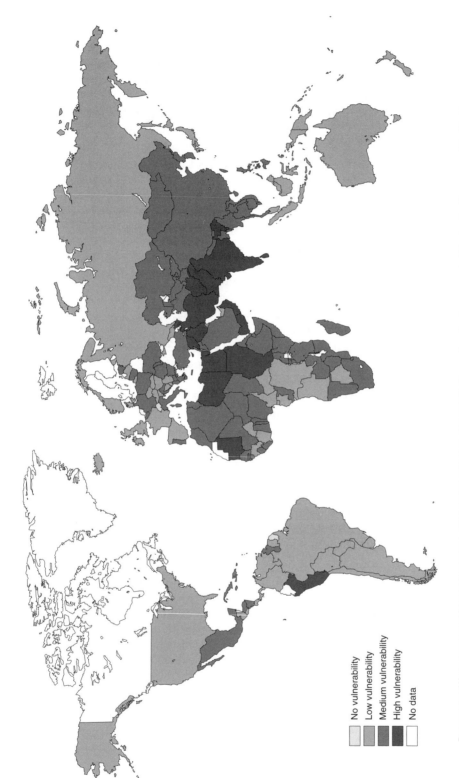

	No vulnerability
	Low vulnerability
	Medium vulnerability
	High vulnerability
	No data

Data source: Water resources Vulnerability Index I, 1995 published in P. Raskin (1997) *Comprehensive Assessment of Freshwater Resources of the World*, Stockholm Environmental Institute: Stockholm, Sweden.

Figure 13.4 National vulnerability to water scarcity.

So far we have been concerned with global futures. Both the doomsayers and the optimists seem content to use national or regional examples selectively to support their global messages, but few appear to take global geography seriously. Our message is that current human welfare depends crucially on intra-global, as well as intranational, relationships of power and dependency which favor some but impoverish others and that changing these relationships could contribute to future gains in human welfare irrespective of any realistic global ecological limits to population growth. Providing a comprehensive defense of this contentious claim is beyond the scope of the present chapter, but we will develop the argument by discussing some of its dimensions.

A Problem of/for Geography?

If population growth is problematic then it must, by definition, be the cause of negative consequences. The ultimate negative consequence is an involuntary reduction in population size produced by an increase in human mortality – the Malthusian scenario. There are several different ways in which this might be brought about and it is interesting to note in passing a shift in emphasis in neo-Malthusian thinking, with environmental damage playing a larger role in the hypothesized causal pathway between population growth and increased mortality (e.g., Meyers 1994). Nevertheless, poverty, malnutrition, and famine remain important intermediate factors. Two major questions arise: first, is poverty caused by population growth and, second, how is poverty related to famine? If population growth were the root cause of poverty, most Western European countries would have been poorer by the end of the nineteenth century than they were at its start, yet all saw unprecedented accumulations of national wealth. Of course, poverty was not eliminated and some sections of the population continued to suffer deprivation and early death. However, the European experience suggests both that the circumstances in which population growth occurs fundamentally determine its consequences and that poverty is socially produced by the system of resource distribution within populations. In other words, we cannot understand the relationships between population growth, poverty, and mortality without considering the details of the contexts in which they occur.

Following the Brandt Commission Report (1980), the geographical patterning of the haves and have-nots across the globe is commonly referred to as a North–South divide. In some areas the population produces more food than it can consume and consumes more food than it needs to maintain health and vitality. At the same time, populations in other areas experience high levels of malnutrition and even starvation. Globally, food aid programs do little or nothing to counteract this maldistribution of resources and it would be naive to think that the problems of malnourished populations could easily be solved by redirecting stockpiles of food in

Europe and North America. More pertinent is to ask why such disparities have arisen and persist.

The most extreme conditions of food shortage result in the famines which periodically ravage the poorest countries, especially in Africa. Ethiopia, for example, experienced a "great famine" in 1984/5 when, it has been estimated, up to a million people died. And from 1998 onwards, when the rains again failed in some areas of that country, a new alert was sounded in the international press. In April 2000 the *Guardian* carried a special report on Ethiopia under the headline "A doubled population: and no aid." Although the report recognized that "Famines are complex things since parts of Ethiopia produce a grain surplus," the implication of the headline is that rapid population growth is a major underlying cause. This conclusion is not supported by the academic literature on famine. In the most influential of recent studies, for example, famine and hunger are defined in terms of entitlements over basic necessities (Sen 1981; Dreze and Sen 1989). Those people who are most food insecure, and hence most vulnerable to severe hunger and starvation, are those with the lowest entitlements and entitlements are socially distributed. Thus, as Swift (1989) points out, not all poor people are equally vulnerable and it is not necessarily the poorest who are at greatest risk. Watts and Bohle (1993) add a geographical dimension to this analysis by considering the space of vulnerability. They treat vulnerability as "a multilayered and multidimensional social space defined by the determinate political, economic and institutional capabilities of people in specific places at specific times" (ibid: 46). Their analysis of the causal structure of vulnerability treats population growth as only one among a number of mechanisms through which vulnerability can increase and not as a root cause of famine and hunger. Two conclusions can be drawn from this literature. First, considerations of the command over food, powerlessness, and the social relations of production are central to any understanding of the relationships between population and resources. And second, the shape and form of vulnerability will vary according to local (geographical), as well as historical, circumstances. Two brief studies based on recent geographical literature lend some support to these conclusions.

Study 1: Population growth and deforestation

The neo-Malthusian hypothesis that population growth leads to environmental damage and hence to reductions in the number of people that can be supported has recently been challenged by a number of detailed studies of population–forest relationships. Conway, Keshav, and Shrestha (2000) examine local practices at the forested frontier of Nepal and note that population growth can be accompanied by environmental enhancement and recovery as well as destruction. They argue that if the skills, perceptions, knowledge, and aspirations of indigenous settlers are recognized, improvements in forest management could be achieved by identifying strategic approaches suited to the particular local context. Leach and Fairhead (2000) also criticize neo-Malthusian thinking, and global analyses more generally, for neglecting crucial questions of social and ecological specificity and history. Their study of the relationships between people and forests in West Africa questions the forest statistics themselves and emphasizes processes, previously ignored, by which populations increase forest cover. In both Ghana and Guinea, significant deforestation is assumed to have begun towards the end of the nineteenth century, yet this is no better than a guess because there are no forest-cover surveys for either country which might lend credence to such a claim. Indeed, the descriptive evidence that we do have, albeit flimsy, points to the early presence of farmed and savannah areas within the forest zone, undermining the use of forest zone area as a surrogate for the extent of primary forests frequently employed as a baseline for determining amounts of deforestation. Further, field research in the Kissidougou prefecture of Guinea has revealed that forest is generally advancing into savannah, rather than vice versa, with today's forest mosaic reflecting institutionally shaped farming practices. Similarly, in the Wenchi district of Ghana the long-established practice among landowners of protecting and encouraging forest trees in their field intersects is being threatened not by local population growth but by the activities of loggers backed by the state Forest Department.

The value of local studies such as those outlined above is that they allow a "bottom-up" approach to the understanding of the complex relationships between populations and their resources which questions many of the simplified assumptions underlying global analyses. Forest cover changes in response to the interactions between populations and diverse institutional and policy arrangements, local social values, and dynamic ecologies. Where policies are ill-conceived in relation to the realities of particular localities, they not only fail to achieve their goal but lead to the impoverishment of local land users. It becomes obvious, then, that population growth ought not to shoulder the burden of responsibility in circumstances where other factors are centrally implicated. Nor, since there are examples of more people leading to more forest, can it be assumed that a growth in population is always detrimental to local environments, even in areas of fragile economy and ecology.

Study 2: Local impacts of global policies

National institutions and policies elicit differential responses within countries according to the diversity of local geographies but, increasingly, economic, social, and environmental policies within the poorest countries of the world are subject to the direction of international agencies. Ghana, for example, is one of many African countries whose economies are being managed within

Structural Adjustment Programs (SAPs) sponsored by the World Bank and International Monetary Fund. Indeed, Ghana has been hailed by the sponsoring institutions as the most successful case of structural adjustment in Africa. A closer look at the local impacts of these global policies, however, suggests a more mixed blessing. Konadu-Agyemang (2000) examines socioeconomic disparities within Ghana and finds a pattern of uneven development both interregionally and between urban and rural areas. SAPs invariably demand institutional changes that reduce the role of the state in the economy and allow market forces to determine access to goods and services. The resultant reductions in government expenditure on social services, such as education and health, increase the vulnerability of some population groups, especially those living on the margins. Thus, because resources are redistributed, policies which appear to be advantageous according to macroeconomic indicators may nevertheless result in increases in poverty and risk of premature death for some, irrespective of levels of population growth.

Global relations of power are also encapsulated in the level of indebtedness of poorer nations to the institutions of international finance. Ghana is fairly typical with a total debt that has more than quadrupled since 1980. External debt amounted to 95 percent of the country's gross domestic product in 1995 and Konadu-Agyemang (2000) argues that debt servicing is diverting resources from local needs on a massive scale. The debt crisis has become a subject of international concern as various pressure groups lobby Western national governments and global organizations to support the canceling of Third World debt and thus contribute to the eradication of poverty.

Ironically, some UN policies appear to be predicated on Malthusian thinking, promoting a simplistic view of the relationship between population growth and poverty by emphasizing the need to control fertility in the Third World. In this context, family planning measures have been advocated by the West as the solution to the "global population crisis." Not all developing countries accept this analysis, however, and reaction to declarations made at high-profile global conferences on population, such as the UN-sponsored International Conference on Population and Development held in Cairo in September 1994, has varied according to national policy contexts – even between rather similar countries. A comparison of two developing countries – Pakistan and Algeria – demonstrates the variety of factors which influence national-level responses (Lee and Walt 1995).

The close alliance between Pakistan and the West following independence in 1947 was reflected in its membership of the Southeast Asian Treaty Organization and the Central Treaty Organization, which were US-sponsored military alliances. Dependent upon US economic and military aid, it was one of the first countries to implement official policies to slow population growth. President Ayub Khan introduced the National Family Planning scheme which attracted substantial investment from abroad (US$68 million annually between 1966 and 1968) and this relationship further resulted in US$480 million of economic aid during this period. Pakistan maintained good relations with the West throughout the 1970s and foreign aid grew as a proportion of GNP but, in fact, domestic politics forced the government to withdraw its support for family planning. Ayub Khan had been strongly associated with family planning measures and it was inevitable that the newly elected government would be less enthusiastic. Hence, resources were diverted from family planning to expanding social services, defense, and public administration.

In contrast, commitment to family planning was always much weaker in Algeria, which was strategically aligned with the Soviet Union. Western ideas about the need to limit population growth were regarded as being neo-imperialist and intended to weaken, rather than strengthen, developing countries. In the 1970s President Boumedienne became renowned for his dismissal of birth control as a solution to problems of development, championing instead changes to global economic relations in favor of developing countries. His aim was to construct an economy that would satisfy Algeria's cultural and economic needs and he argued that economic development would lead to a decline in fertility. As the revenue from oil rose in this period, and economic conditions improved, it seemed even less likely that family planning would assume importance within the national policy agenda.

By the time of the UN Conference on Population in Mexico in 1984, there was, on the whole, increasing acceptance among developing countries of the need to reduce population growth through family planning; the fact that policies were beginning to work in China and elsewhere encouraged this change of views. For Pakistan, though, while rising foreign debt was becoming increasingly serious, the Soviet invasion of Afghanistan in 1979 meant that the country was strategically located and could rely on Western aid to continue. Unlike many other developing countries whose foreign aid was commonly tied to family planning policies, Pakistan was less constrained and spending on debt repayment and defense rose while fertility control remained low on the agenda. In Algeria, economic and political circumstances also changed. Revenue from oil had declined and the increasing economic crises made the country more reliant on Western aid. Without the strategic location that benefited Pakistan, Algeria was forced to allow aid provision to be linked to domestic policies and in 1983 the first official population policy was implemented.

The Cairo conference of 1994 saw a revised global consensus on population policies which emphasized a more holistic approach to family planning, including a stronger focus on women's reproductive health and more general well-being. Prime Minister Benazir Bhutto of Pakistan was supportive of this approach, despite a strong minority of critics, arguing that donor communities should provide more assistance for such policies. This view was much the same as that held by Algeria, where family planning policies continued to be promoted, and the government also criticized donor countries for not investing enough in these policies. Whether the apparent consensus marked a growing convergence of views or had been won by keeping commitments vague is a matter of debate. However, the comparison between Pakistan and Algeria, and other similar comparisons (Lee and Walt 1995), illustrate the different trajectories of national population agendas according to specific political, economic, and cultural contexts which inevitably influence the reaction to powerful global agendas. The demands of the donor countries also vary and the case of Pakistan is a good example of a country that received aid with fewer "strings attached" because of its strategic location. Overall, global initiatives can be influential but it would be naive to expect these to be universally accepted. Finally, we must not forget that even when nations do "sign up" to family planning policies, there will be considerable variations in local responses to these messages within these nations.

Three conclusions can be drawn from the examples of the local impacts of global policies outlined above. First, the realities of life for particular people in particular places are subject to a complex set of interactions between the global and national policy agendas, as well as being influenced by the specifics of political, economic, social, cultural, and ecological contexts. Thus global geographies and relations of power play a part in the production, as well as the resolution of local crises. Second, poverty and other Malthusian ills of life are not caused simply by population growth putting pressure on local resources. Where internationally sponsored SAPs have exacerbated uneven development within poorer countries, as in Ghana, they too are implicated in the production of poverty. Third, the ambivalence of national governments towards global population agendas is hardly surprising since proposed solutions to the problems of development tend to be one-dimensional and insensitive to local contexts. By isolating population from other factors, certain international agencies implicitly adopt a Malthusian framework which, we believe, ignores geopolitical realities and increasingly lacks credibility.

Future Challenges: Making the Right Choices

It should now be apparent that we consider the "crisis" to be neither global nor primarily caused by population increase. Reducing population growth may well be in the long-term interest of poorer nations and poorer people (Shrestha and Patterson 1990), but it will not solve global problems of poverty which have more diverse roots. Both the neo-Malthusians and the Cornucopians underestimate the complexity of these problems because they ignore diverse histories and geographies. In fact, processes associated with globalization impact differentially according to local circumstances. Thus our understanding of the consequences of population growth must move from the global to the local, for it is in local contexts that crises such as famines occur. When we adopt a "bottom-up" approach, it becomes obvious that populations, the environment, economics, and culture all interact *jointly*. It is useless to imagine that population interacts with the environment while economics and culture have no effect and are unaffected (Cohen 1995). Such interactions also vary over time and, most importantly, between places. There is, then, no global *population* crisis, for population growth cannot be isolated as the root cause of the many problems of human welfare we currently face. Yet nor is the happy optimism of the Cornucopians a well-founded response. Poverty remains a momentous problem. The World Bank (2000) estimates that 2.8 billion of the world's 6 billion people currently live on less than $2 a day, and 1.2 billion on less than $1 a day. Up to the end of 1999 the AIDS epidemic had taken or was directly threatening the lives of an estimated 50 million people in Africa (UNAIDS and WHO 1999). Understanding the complexity of these problems is an interdisciplinary enterprise to which geographers, especially population geographers, could make a greater contribution than they currently 2do. By drawing attention to the contingencies of place, we could reframe the "population" debate to focus on the interactions among global processes and local relationships between populations and their environments.

14 Global Change and Patterns of Death and Disease

John Eyles

Introduction

For half a century, most countries have achieved impressive progress in their health conditions. Yet the causes of ill-health do not stand still – humanity, very progress changes them. The past decade has witnessed a profound transformation in the challenges to global health; persistent problems have been joined by new scourges in a world that is ever more complex and interdependent. The idea that the health of every nation depends on the health of all others is not an empty piety but an epidemiological fact.

Thus wrote a group of distinguished health scientists in the *Lancet* in 1997 (Al-Mazrou et al. 1997). There can be no more eloquent comment on the role of global change in shaping patterns of disability, disease, and death. This chapter examines some of those changes with respect to their impact on the burden of ill-health and disease. It first outlines the present-day global burden of disease before examining more closely three interrelated dimensions: globalization and infectious diseases; global environmental change and the potential for changing patterns of death and disease, illustrated by climate change and global warming; and population movements and the diffusion of illness, exemplified by both forced migrations and global travel.

In its 1998 annual report, the World Health Organization (WHO 1998a) noted that the world's population had more than doubled since 1955. The report notes significant progress in many areas. For example, in 1955 40 percent of all deaths were among children under 5: this had fallen to 21 percent of all deaths in 1995 – or, put differently, the under-5 mortality rate fell from 210 per 1,000 live births to 78 from 1955 to 1995. Yet there remain important concerns. In 1997, for example, one-third of global deaths were due to infectious and parasitic diseases, with acute lower respiratory infections, tuberculosis, diarrhea, HIV/AIDS, and malaria being the major causes. Furthermore, while life expectancy continues to increase globally, about 300 million people live in sixteen countries where life expectancy actually decreased between 1975 and 1995. Such findings led WHO (1999a) to write about "the double burden" in its next annual report, by which it means on the one hand the rising age of death and the resulting epidemics of non-communicable diseases and injuries. Globally, non-communicable diseases account for 43 percent of disability-adjusted life years (DALYs), with one DALY representing one lost year of healthy life. In high-income countries the estimates rise to 81 percent, with neuropsychiatric conditions, cardiovascular diseases, and cancers being primarily implicated. On the other hand, the double burden refers to the avoidable burden of disease and malnutrition that the world's disadvantaged populations continue to bear, particularly infectious diseases and complications of childbirth exacerbated by too little food and too few resources for effective use of vaccines and other control mechanisms: "These conditions are primarily concentrated in the poorest countries, and within those countries they disproportionately afflict populations that are living in poverty" (WHO 1999a: 19). For example, it is estimated that poor females are 15 times more likely to die before their fifth birthday than their non-poor counterparts in Malaysia. Furthermore, as figure 14.1 shows, treatable conditions contribute to the burden of illness in low- and middle-income countries. Yet some of these conditions – particularly malaria, HIV/AIDS, and tuberculosis – may be worsened by microbial diffusion and evolution. As Harrison and Lederberg (1998) note, poor prescribing practices or poor patient compliance with treatment have led to the development of strains of *Mycobacterium tuberculosis* which are resistant to available drugs. There is an enhanced problem with malaria: *Plasmodium* parasites, which cause the disease, have become resistant to anti-malarial drugs and the disease vectors, *Anopheles* mosquitoes, to insecticides. These microbial changes are significant for the global burden of disease. The cases discussed in the next section, however, concentrate primarily on the avoidable burden of disease – demonstrated graphically in figure 14.2 with the relationships between infant morbidity rates and income – and the societal forces that relate to this burden.

The Burden of Infectious Diseases and Globalization

Globalization remains largely a complex set of forces that is seen to be inevitable and (in the long run) beneficial. Temin (1999) writes historically of globalization, the integration of the world economy, and how it has ebbed and flowed in terms of the flows of goods, people, and capital over the twentieth century. He points to the growth in merchandise exports as a share of GDP in most countries surveyed between 1913 and 1992. Furthermore, the World Bank (2000) notes that throughout the 1990s trade in goods and services has grown twice as fast as global GDP, with the share attributable to

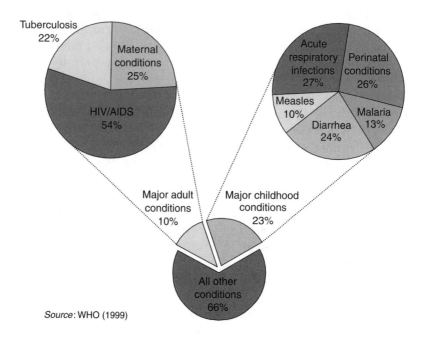

Source: WHO (1999)

Figure 14.1   DALYs attributable to conditions in the unfinished agenda in low-
and middle-income countries, estimates for 1998.

developing countries climbing from 23 to 29 percent of total world trade.
And the Bank notes that financial growth to assist this trade has led to serious
structural problems, including increased debt loads, for many countries and
regions. Indeed, there is a debate on whether the structural adjustment policies of the IMF contributed to these problems and consequently to the
immiseration of the poor, directly through reductions in income and indirectly through reductions in health and social service provisions (see Botchwey
et al. 1998; Collier and Gunning 1999). Naiman and Watkins (1999) have
little doubt on the impact of these policies on African countries. Such nations
subject to these policies saw their per capita incomes decline, along with expenditures on healthcare, education, and sanitation: in Zimbabwe, for example, healthcare expenditure declined as a share of the budget from 6.4 to 4.3
percent and as a share of GDP from 3.1 to 2.1 percent between 1990–1 and
1995–6, despite its failure to meet IMF reduction targets.

As the State of the World Forum noted, the benefits of globalization are
not distributed equally (Global Policy Forum 2000). Globalization has serious medical consequences with the emergence and reemergence of infectious diseases. And with these health impacts can be seen the effects of other
aspects of global change – urbanization, technological innovation in agriculture, threats to environmental integrity, and so on. In fact, in their commentary on infectious diseases being "back with a vengeance" in India,
Chatterjee et al. (1999) conclude that

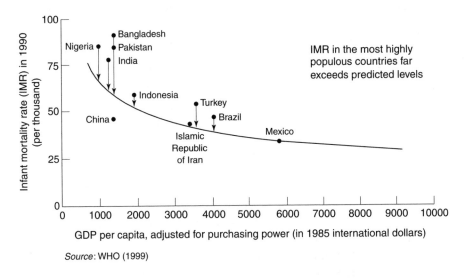

Figure 14.2 Infant mortality rate related to income.

the reasons for the increase of individual diseases are different but some common factors can be discerned. Our inability to tackle preventive healthcare from a broader perspective, the focus on top-down technological and medical solutions and the non-existence of a surveillance system are some important factors. Poverty, the lack of provision of basic amenities of shelter, clothing, food, and education, are closely linked to the infectious disease scenario. Our modes of development, technology-intensive agriculture, dams, dykes, industries, deforestation, migration, and increasing urbanization have an adverse impact on infectious diseases. Most importantly, health has never been a major political issue for the people or for the government.

Infectious diseases are then the lenses through which to examine some of the health consequences of globalization. We focus on the six diseases that cause 90 percent of infectious disease deaths (WHO 1999b): pneumonia, tuberculosis, diarrheal diseases, malaria, measles, and HIV/AIDS. Their impact will be described and then their patterns and distributions linked to factors implicated in global change. Before summing the major causes of death, let us note that infectious diseases cause much disability and illness. For example, lymphatic filariasis is second only to mental illness as the world's leading cause of long-term disability. It results in enlargement of limbs and damage to internal organs and results from a mosquito-borne disease involving infection with parasitic worms. It affects about 120 million people but 1 in 6 of the world's population are at risk. Schistosomiasis, a debilitating disease spread by water snails in stagnant water, affects 200 million people, causing chronic urinary tract disease; it can be spread to new areas through dam and irrigation projects. River blindness affects over

85 million people in Africa, Latin America, and the Middle East. It is a parasitic disease transmitted by blackfly (WHO 1999b). Finally, dengue and dengue hemorrhagic fever are mosquito-borne infections and WHO (1998b) estimates that over two-fifths of the world's population is at risk from these diseases. There have been explosive outbreaks, for example, in Brazil in 1998 (see also Pinteiro and Corber 1997); dengue is now a greater threat because of rapid rises in urban populations, especially if household water storage is common and where solid waste disposal services are inadequate.

Pneumonia is the deadliest of the acute respiratory infections. It affects children in particular, although childhood deaths from pneumonia are rare in industrial nations. It affects particularly children with low birth weight and those weakened by malnutrition and other diseases. Murray and Lopez (1996), in their massive study of the global burden of disease, calculate that malnutrition is responsible for 11.7 percent of all deaths worldwide and 15.9 percent of disability-adjusted life years (see table 14.1 for the relative contribution of ten selected risk factors). For malnutrition, the contribution rises to 31.9 percent of total deaths for sub-Saharan Africa (32.7 percent of disability-adjusted life years). As FAO (2000) points out, malnutrition remains significant in many countries of Southeast Asia and Africa: in Mali nearly half the population was malnourished in 1996–8, a 30 percent increase in the proportion from 1979–81. It has become a significant fact of life (and disease) in the republics of the former Soviet Union, with almost a third of the populations of Azerbaijan and Tajikistan undernourished in 1996–8. In the previous year's report, FAO (1999) notes the devastating impact that poor nutritional status can have on the health status of children. In much of West Africa, for example, one-fifth of all children die before the age of five: in Nigeria this increases to almost one-third. It should be noted that pneumonia is an important but not the sole cause of death of such children; diarrheal diseases claim nearly 2 million children a year under five. The burden is particularly high in deprived areas with poor sanitation, unsafe drinking water, and inadequate hygiene, and some countries and regions have epidemics of such diarrheal diseases as dysentry and cholera. Cholera is endemic in the delta of the Ganges and Meghna rivers in Bangladesh, the delta of the Irrawaddy and Salween rivers in Burma, and in the coastal areas of Indonesia and Africa. Measles is often associated with diarrheal diseases as well as pneumonia and malnutrition and remains a major childhood killer.

Tuberculosis (TB) kills 2 million people each year. WHO (2000a) estimates that between 2000 and 2020 nearly 1 billion people will be newly infected, 200 million people will get sick, and 35 million will die from TB if there are no changes in controls. The threat has been enhanced by its association with HIV/AIDS and the emergence of multi-drug resistant TB: one-third of the world's population is currently infected with the TB bacillus, with Southeast Asia bearing the biggest burden. Also worrying are

Table 14.1 Global burden of disease and injury attributable to selected risk factors, 1990.

Risk factor	Deaths (thousands)	As % of total deaths	YLLs[1] (thousands)	As % of total YLLs	YLDs[2] (thousands)	As % of total YLDs	DALYs[3] (thousands)	As % of total DALYs
Malnutrition	5,881	11.7	199,486	22.0	20,089	4.2	219,575	15.9
Poor water supply, sanitation, and personal and domestic hygiene	2,668	5.3	85,520	9.4	7,872	1.7	93,392	6.8
Unsafe sex	1,095	2.2	27,602	3.0	21,100	4.5	48,702	3.5
Tobacco	3,038	6.0	26,217	2.9	9,965	2.1	36,182	2.6
Alcohol	774	1.5	19,287	2.1	28,400	6.0	47,687	3.5
Occupation	1,129	2.2	22,493	2.5	15,394	3.3	37,887	2.7
Hypertension	2,918	5.8	17,665	1.9	1,411	0.3	19,076	1.4
Physical inactivity	1,991	3.9	11,353	1.3	2,300	0.5	13,653	1.0
Illicit drugs	100	0.2	2,634	0.3	5,834	1.2	8,467	0.6
Air pollution	568	1.1	5,625	0.6	1,630	0.3	7,254	0.5

[1] YLLs = Years of life lost
[2] YLDs = Years lived with a disability
[3] DALYs = Disability-adjusted life years
Source: Murray and Lopez (1996)

outbreaks in places that have seen years of decline in TB rates such as Eastern Europe. The social and economic changes in Russia have had other significant health impacts: from 1990 to 1999, for example, mortality increased by 22 percent and morbidity by 3 percent, with a reduction in life expectancy from 66.5 to 64.1 years (Strukova et al. 2000). A demographic crisis (see DaVanzo and Adamson 1997) has been compounded by the massive reemergence of infectious diseases – not only TB but also diphtheria, for which a large population of susceptible adults, decreased childhood immunization, suboptimal socioeconomic conditions, and high population movement may be implicated (see Vitet and Wharton 1998). Incidence rose from 1.25 cases per 100,000 in 1991 to 26.41 in 1994 in Russia. A massive vaccination program was needed and had by 1996 succeeded in reducing the incidence rate to 9.08 cases per 100,000, still more than seven times higher than the 1991 level.

Malaria is the world's most important tropical parasitic disease, killing over 1 million people a year. Malaria's death toll far exceeds the mortality rate from AIDS: it kills a child every 30 seconds and is a public health problem in over 90 countries with 40 percent of the world's population (WHO 1998c). Yet more than 90 percent of all malaria cases are in sub-Saharan Africa, where it particularly affects poor people in rural areas and its economic cost has been calculated at $2 billion (1997 estimate). Its reach is, however, spreading. Malaria is specific to the ecosystem which breeds it. As climate and weather change, as new dams, canals, and irrigation channels are dug and as people migrate, different reservoirs for mosquitoes to breed are created – in airports, industrial sites, agricultural projects, city waste dumps, and so on. These varieties of reservoir make control difficult, in that adaptation of techniques is required. As with all infectious disease killers, global forces result in the development and spread of malaria with specific local impacts. Indeed, global trade and travel and forced migrations of displaced people and refugees can help spread malaria and TB.

Over 34 million people are living with HIV/AIDS worldwide, with sub-Saharan Africa being the worst affected, having about two-thirds of all cases (UNAIDS 2000; Konrad Adenauer Foundation 2000). By early 1999, 11 million people had already died of AIDS there – as many as were transported during the slave trade (Shell 2000). Life expectancy gains across Africa have been reversed. In Botswana, for example, life expectancy at birth has fallen from 70 to about 50 years. Sibanda (2000) charts the dramatic increase in the numbers of people living with HIV and those dying of AIDS-related diseases in Zimbabwe since the early 1990s. He writes of a nation in pain and of an epidemic out of control. It is estimated that 1 in 4 adults are living with HIV, although there are within-country variations. Blood samples taken from pregnant women at maternity clinics show between one-fifth and one-half of all pregnant women tested positive for HIV. Most transmission occurs through either unprotected heterosexual intercourse or from an infected mother to a fetus. It is argued that the AIDS epidemic has turned back the clock on development. Sibanda discusses the factors that may be responsible for this state of affairs, such as the cultural context of sex and reproduction, emphasizing masculinity and the importance of bearing children; denial, complacency, and policy failure in the early years of the disease; the role of religion and traditional healers that mitigates against the use of condoms and medical treatments; trade and long-distance bus routes to assist the diffusion process; and economic conditions which often force married couples to live apart, potentially encouraging the use of prostitutes. The economic structural adjustment policies of the 1990s have reinforced these conditions by driving many people further into poverty. Furthermore, HIV/AIDS may be seen as one of the four major pandemics of human history. As figure 14.3 shows, it is the one with the greatest potential for devastation not only in Africa but globally. The spread of HIV/AIDS remains significant in sub-Saharan Africa (see table 14.2), but its spectacular growth rates are now found in Eastern Europe and East Asia and the Pacific. For example, UNAIDS (2000) reports that the number of diagnosed HIV infections in Ukraine rose from virtually zero in 1995 to around 20,000 a year from 1996 onwards, about 80 percent of them in injecting drug users. Yet, as the President of the World Bank perceptively noted: "Many of us used to think of AIDS as a health issue. We were wrong . . . Across Africa, AIDS is turning back the clock on development. Nothing we have seen is a greater challenge to the peace and stability of African societies than the epidemic of AIDS . . . We face a major development crisis, and more than that, a security crisis. For without economic and social hope we will not have peace, and AIDS surely undermines both" (Wolfenson 2000).

With reference to all infectious diseases, then US Surgeon General David Satcher (1998) said

> Worldwide, infectious and parasitic diseases remain the leading cause of death. These deaths disproportionately affect the developing countries of the world, with the most vulnerable segment of the population being children under the age of five years. Moreover, the factors that contribute to the resurgence of these diseases – including global travel, the globalization of the food supply, population growth and urbanization, ecological and climate changes and the evolving drug-resistant microbes – show no sign of abatement.

Morse (1995) has summarized these factors seen as responsible for the emergence of infectious diseases, with respect to both the infections themselves and the broad processes of change. Table 14.3 provides recent examples of emerging viral, bacterial, and parasitic infections, and table 14.4, summarizing much work from the US Institute of Medicine and Centers for Disease Control, highlights specific factors and specific diseases. It is important to note how many dimensions of globalization are implicated, from changes in food technology and processing to population migration and environ-

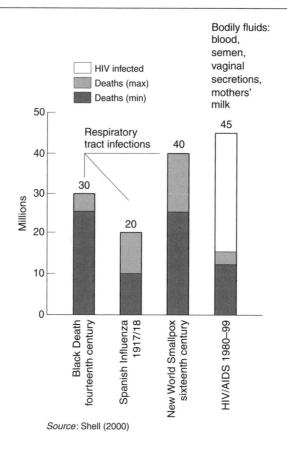

Figure 14.3   The big four pandemics.

mental forces. While all are important, the next two sections will highlight briefly two elements: environmental changes and population movements and their relationships to changing disease patterns.

Environmental Change and Patterns of Disease

We have already noted how some diseases are associated with environmental change, specifically that which occurs with economic or agricultural development or changes in land-use patterns (see also World Resources Institute 1996). Human activities can change ecological balances, potentially leading to new diseases. For example, when agricultural activity shifted to the US midwest in the twentieth century, northeastern forests regenerated and the deer population rebounded but the predator wolf did not; the consequent abundance of deer in the northeast resulted in increased incidence of the

Table 14.2   Spread of HIV/AIDS, 1998.

	Number of AIDS cases (1000s)	Rate of growth 1996–8 (%)
North America	890	18.7
Latin America	1,400	7.7
Caribbean	330	22.2
Sub-Saharan Africa	22,500	60.7
North Africa and the Middle East	210	5.0
Western Europe	500	−2.0
Eastern Europe and Central Asia	270	440
South and Southeast Asia	6,700	28.8
East Asia and Pacific	560	460
Australia and New Zealand	12	−7.7

Source: Mills (2000)

tick-borne Lyme disease (see Levins et al. 1994). Furthermore, with increased urbanization and inadequate waste treatment or removal, litter and garbage proliferate, resulting in increases in the population of rats and mice. These conditions have been associated with hantavirus which may result in hemorrhagic fever, abdominal pain, or respiratory infection depending on the viral strain. Over 100,000 cases have been reported in China and the disease is also found in Asia, much of Europe, and more recently North America (see Leduc et al. 1993).

Uncollected waste is but one hazard in urban neighborhoods. Inadequate water supplies, air pollution, and microbial food contamination are some of the other hazards (see, for example, McGranaham 1993). Lack of access to basic water and sanitation facilities partly explains the profile of causes of death in cities like Accra, with diarrheal diseases, malaria, and measles having significant impacts (see World Resources Institute 1996): in fact, Accra is a microcosm of the traditional hazards identified as threatening human health and prosperity associated with a lack of development (WHO 1997) – they relate to poverty, lack of access to safe drinking water, inadequate basic sanitation in the household and community; indoor air pollution from cooking and using biomass fuel; and inadequate solid waste disposal. "Modern hazards" are associated with unsustainable development practices and include water pollution from populated areas, industry, and intensive agriculture; urban air pollution from vehicular traffic, coal power stations, and industry; climate change; stratopheric ozone pollution; and transboundary pollution. These two types of hazard join together and conspire to ensure the environmental conditions for the development and contaminant disease patterns particularly in the developing world.

Table 14.3   Recent examples of emerging infections and probable factors in their emergence.

Infection or agent	Factor(s) contributing to emergence
Viral	
Argentine, Bolivian hemorrhagic fever	Changes in agriculture favoring rodent host
Bovine spongiform encephalopathy (cattle)	Changing in rendering processes
Dengue, dengue hemorrhagic fever	Transportation, travel, and migration; urbanization
Ebola, Marburg	Unknown (in Europe and the United States, importation of monkeys)
Hantaviruses	Ecological or environmental changes increasing contact with rodent hosts
Hepatitis B, C	Transfusions, organ transplants, contaminated hypodermic apparatus, sexual transmission, vertical spread from infected mother to child
HIV	Migration to cities and travel; after introduction, sexual transmission, vertical spread from infected mother to child, contaminated hypodermic apparatus (including during intravenous drug use), transfusions, organ transplants
HTLV	Contaminated hypodermic apparatus, other
Influenza (pandemic)	Possibly pig–duck agriculture, facilitating reassortment of avian and mammalian influenza viruses*
Lassa fever	Urbanization favoring rodent host, increasing exposure (usually in homes)
Rift Valley fever	Dam building, agriculture, irrigation; possibly change in virulence or pathogenicity of virus
Yellow fever (in "new" areas)	Conditions favoring mosquito vector
Bacterial	
Brazilian purpuric fever (Haemophilus influenzae, biotype aegyptius)	Possibly new strain
Cholera	In recent epidemic in South America, probably introduced from Asia by ship, with spread facilitated by reduced water chlorination; a new strain (type 0139) from Asia recently disseminated by travel (similarly to past introductions of classic cholera)
Helicobacter pylori	Probably long widespread, now recognized (associated with gastric ulcers, possibly other gastrointestinal disease)

Hemolytic uremic syndrome (*Escherichia coli 0157:H7*)	Mass food processing technology allowing contamination of meat
Legionella (legionnaires' disease)	Cooling and plumbing systems (organism grows in biofilms that form on water storage tanks and in stagnant plumbing)
Lyme borreliosis (*Borrelia burgdorferi*)	Reforestation around homes and other conditions favoring tick vector and deer (a secondary reservoir host)
Streptococcus, group A (invasive; necrotizing)	Uncertain
Toxic shock syndrome (*Staphylococcus aureus*)	Ultra-absorbency tampons

Parasitic

Cryptosporidium, other waterborne pathogens	Contaminated surface water, faulty water purification
Malaria (in "new" areas)	Travel or migration
Schistosomiasis	Dam building

* Reappearances of influenza are due to two distinct mechanisms: annual or biennial epidemics involving new variants due to antigenic drift (point mutations, primarily in the gene for the surface protein, hemagglutinin) and pandemic strains arising from antigenic shift (genetic reassortment, generally between avian and mammalian influenza strains). *Source*: Morse (1995)

In some respects the environmental changes identified thus far, while significant, impact at the local and regional levels. The World Bank (2000) identifies global issues, mainly related to rapid population growth and associated trends in urbanization and industrialization. Not only is the scale of the effect global, so too, it is argued, must be the response: "Climate change, the loss of biodiversity and other issues related to the global commons are slowly being recognized as problems that the community of nations must take on collectively" (World Bank 2000: 40). The Bank notes that some three-quarters of bird species and perhaps one-quarter of mammals are threatened with extinction caused largely by modern farming techniques, deforestation, and the destruction of wetland and ocean habitats. The direct impacts of this loss of biodiversity come primarily through changes to ecosystems and ecological balance.

Climate change may have more explicit impacts on human health and disease burdens. The World Bank (2000) suggests that such change is occurring at unprecedented rates as large quantities of carbon dioxide, methane, and other greenhouse gases continue to be released into the atmosphere, this being confirmed by IPCC (2001) summaries of the scientific evidence. Climate influences many of the key features of health: temperature extremes and violent weather events; the geographical range of disease organisms and vectors; the quantity of air, food, and water; and the stability of ecosys-

Table 14.4 Factors in infectious disease emergence.

Factor	Examples of specific factors	Examples of diseases
Ecological changes (including those due to economic development and land use)	Agriculture; dams, changes in water ecosystems; deforestation/reforestation; flood/drought; famine; climate change	Schistosomiasis (dams); Rift Valley fever (dams, irrigation); Argentine hemorrhagic fever (agriculture); Hantaan (Korean hemorrhagic fever) (agriculture); hantavirus pulmonary syndrome, southwestern US, 1993 (weather anomalies)
Human demographics behavior	Societal events: population growth and migration (movement from rural areas to cities); war or civil conflict; urban decay; sexual behavior; intravenous drug use; use of high-density facilities	Introduction of HIV; spread of dengue; spread of HIV and other sexually transmitted diseases
International travel and commerce	Worldwide movement of goods and people; air travel	"Airport" malaria; dissemination of mosquito vectors; ratborne hantaviruses; introduction of cholera into South America; dissemination of 0139 V. cholerae
Technology and industry	Globalization of food supplies; changes in food processing and packaging; organ or tissue transplantation; drugs causing immunosuppression; widespread use of antibiotics	Hemolytic uremic syndrome (E. coli contamination of hamburger meat), bovine spongiform encephalopathy; transfusion-associated hepatitis (hepatis B, C), opportunistic infections in immunosuppressed patients, Creutzfeldt-Jakob disease from contaminated batches of human growth hormone (medical technology)
Microbial adaptation and change	Microbial evolution, response to selection in environment	Antibiotic-resistant bacteria, "antigenic drift" in influenza virus
Breakdown in public health measures	Curtailment or reduction in prevention programs; inadequate sanitation and vector control measures	Resurgence of tuberculosis in the United States; cholera in refugee camps in Africa; resurgence of diphtheria in the former Soviet Union

Source: Morse (1995)

tems. These broad political effects and the association of climate with other driving forces of global environmental change (such as population dynamics, urbanization, production and consumption patterns, economic development) mean that identifying the health effects of climate change is difficult. Figure 14.4 shows how climate change can affect health. Direct effects are primarily altered rates of health, and cold-related events and the death tolls attributed to heat stress can be surprisingly high. In Chicago in July 1995 heat stress was implicated in the deaths of 726 people during a four-day heatwave (Anon 1995; World Resources Institute 1998). Table 14.5 lists the possible health implications of climate change for an industrialized country like Canada, with perhaps the greatest direct impacts being the contribution of greenhouse gases to smog and air pollution and hence respiratory diseases.

If it is difficult to work out the climate change signal in the health consequences of air pollution, it is even more so with respect to its indirect impacts. WHO (2000b) has tried to estimate the likelihood of changes for major tropical diseases (see table 14.6). Especially significant are the potential effects of the changing distribution of the mosquito vector for malaria and dengue as both diseases can potentially affect over 2 billion people worldwide. The situation is however complex, as Patuetne (2000) notes and discusses for impacts on the US. Furthermore, most of Argentina currently lies just south of the zone in which malaria occurs, but if rainfall increases in central Argentina, as climate change models predict, the mosquito might be able to extend southwards to the pampas and savanna regions, introducing malaria to these areas. On the other hand, northwestern Argentina, where malaria mosquitoes are now present, might become less conducive to mosquito survival and malarial outbreaks (World Resources Institute 1998). Globally, Martens et al. (1997) predict that the increase in the epidemic potential of malaria at 12–27 percent, and dengue transmission may increase by between 31 and 47 percent in terms of its epidemic potential. Githeko et al. (2000) point to the potential regional impacts of many vector-borne diseases with climate change. There does, however, remain disagreement on whether the health impacts of climate change are real and/or significant (see, for example, ACSH 1997), in part a result of what are seen as the causes of climate change – the levels of activity in the world economy – and the fact that any attempt to mitigate its effects are not only costly (between 1 and 2 percent of annual GDP of countries like the US) but a challenge to the global economic system itself. We return to this conundrum in our concluding remarks.

Population Movements and Patterns of Disease

The movement of people has been an important mode for the spread of disease. Daily and Ehrlich (1995) point to the many historical examples,

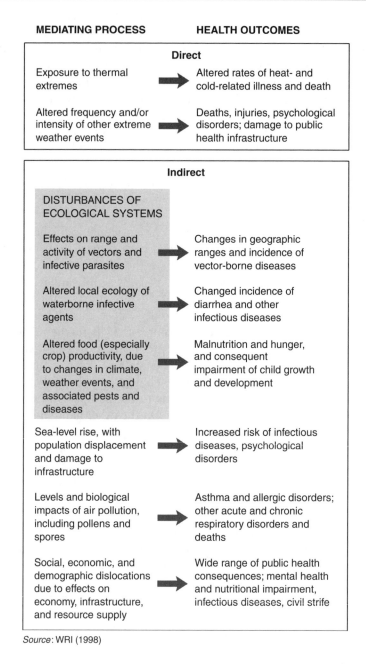

Figure 14.4 Direct and indirect health impacts of climate change.

Table 14.5   Health implications of climate change for Canada.

Temperature and weather changes	
Thermal extremes	Heat-related and cold-related illness
Urban heat island effect	
Air pollution (smog)	Respiratory diseases
Frequent extreme weather events	Deaths, injuries, psychological disorders, stressed public health infrastructure
Ecosystem changes	
Wider distribution of disease vectors and pathogens they carry	Increased risk of vector-borne diseases
Freshwater ecological changes	Increased risk of waterborne diseases
Coastal ecosystem changes	Depleted fish stocks, algal blooms
Proliferation of weeds, grasses, pollens	Allergenic diseases
Food security	
Crop disruption by floods, droughts	Food shortages, malnutrition, perhaps famine
Proliferation of pests, plant diseases	
Sea-level rise	
Displaced populations	Refugee health problems – infectious diseases, psychological disorders, social pathology, etc. associated with overcrowding, civil strife, violent armed conflicts
Social, economic disruptions, resource depletion, decayed infrastructures	Over-extended public health services
	Scarcity of commodities, inflation
Stratospheric ozone attenuation	Skin cancer, cataracts, immune system damage
	Impaired agricultural productivity

Source: Last (1998), based on IPCC (1996), WHO (1996), Patz (1998)

such as cholera to Europe and North America in the 1800s by the movements of traders and soldiers. Such spread can be serendipitous; they cite colonial Uganda where an agricultural officer introduced the shrub *Lantana camera* for use as an ornamental hedge. It also provided a moist habitat for tsetse flies with a concomitant increase in sleeping sickness. Indeed it is through the movement of people for trade, administration, and migration (voluntary or forced) that the threat of disease is brought home to areas of the world that seem immune from, say, tropical diseases. In itself, modern transport may play a vital role in the spread of diseases by helping move animals that are potential vectors or disease reservoirs around the world. For example, West Nile (WN) virus may have reached North America

Table 14.6 Major tropical vector-borne diseases and the likelihood of change with climate change.

Disease	Likelihood of change with climate change	Vector	Present distribution	People at risk (millions)
Malaria	+++	Mosquito	Tropics/subtropics	2,020
Schistosomiasis	++	Water snail	Tropics/subtropics	600
Leishmaniasis	++	Phlebotomine sandfly	Asia/southern Europe/Africa/Americas	350
American trypanosomiasis (Changas disease)	+	Triatomine bug	Central and South America	100
African trypanosomiasis (sleeping sickness)	+	Tsetse fly	Tropical Africa	55
Lymphatic filariasis	+	Mosquito	Tropics/subtropics	1,100
Dengue	++	Mosquito	All tropical countries	2,500–3,000
Onchocerciasis (river blindness)	+	Blackfly	Africa/Latin America	120
Yellow fever	+	Mosquito	Tropical South America and Africa	–
Dracunculiais (Guinea worm)	?	Crustacean (copepod)	South Asia/Arabian peninsula/Central-West Africa	100

+++ = highly likely
++ = very likely
+ = likely
? = unknown
Source: WHO (2000b)

through an individual returning from an endemic country, the importation of an infected bird, or the accidental importation of an infected mosquito (Health Canada 2000). WN virus can result in a type of Japanese encephalitis. This mosquito-borne viral infection was unknown in Europe until 1996, when more than 500 cases were observed in the Bucharest region of Romania, with high rates of neurological disorder and death (up to 10 percent). It was not detected in the Western hemisphere until 1999 when 56 cases of WN encephalitis were confirmed in New York City, from which 7 died. Platanov et al. (2001) report a particularly virulent outbreak in Volgograd City in Russia; about 480 suspected WN virus cases were found, including 84 cases of acute aseptic meningoencephalitis, 40 of which were fatal. Most of the cases were among the elderly (in Romania, Russia, and the US). These three areas are located near large bodies of water and on bird migration pathways and all had unusually dry summers in the year of the outbreak. Mosquito-borne WN virus fever is, however, endemic in Africa, the Middle East, and Southwest Asia. There may be a complex relationship between mosquitoes, birds, and human activities such as trade and travel that led to industrialized and newly industrialized countries being affected by this apparently tropical disease. Nowhere in the world seems immune to the impact of new and reemerging diseases.

The scale of population movement in today's world exacerbates these problems. Each year between 2 to 3 million people emigrate and there are now more than 130 million people living outside the country of their birth, with this number increasing by about 2 percent a year (World Bank 2000). This mainly represents permanent movement. At the same time there has been an exponential growth in international travel. For example, in 1951 7 million people flew internationally; this had risen to 50 million by 1967 and 500 million by 1993 (see Sadler 1998). This scale of travel has rendered national quarantine strategies ineffective and begins to challenge national public health systems themselves. Darrow et al. (1986) pointed out the early association between international travel and the spread of HIV/AIDS, while there is concern about the increasing number of TB and malaria cases in Canada (CMAJ 2001).

This international movement of people (and goods) can lead to the diffusion of microbial risks (Institute of Medicine 1992). One further element of such movement is the forced migration of significant numbers of people following natural disorder, conflict, or war. The World Bank (2000) records that by 1975, 2.5 million refugees had crossed national borders; by 1995 this had risen to 23 million. There are also some 20 million people displaced internally in their own countries. This increase in refugee numbers is often brought about – in human-induced circumstances – by the other dominant forces of the modern world aside from globalization, "tribalization," or local resistance (usually by force) to globalization and difference. Impacts on health are often psychological or mediated by other factors, such as disruption to agricultural and work practices and to access to food and other basic human necessities. FAO (1999) shows how destruction of productive infrastructure and food distribution in Kosovo led to under-nourishment and increases in incidence of disease. Furthermore, the war in Angola had driven 2 million people from their homes, forcing them into unsanitary, crowded conditions in cities and a reliance on airlifts for food. Given that these are war situations, we can only speculate on the levels and types of disease present. But in many ways the traditional and modern hazards of the environment identified by WHO (1997) have been joined together in a hostile environment. These dispiriting conditions and facts are likely to be repeated with, as the years go by, the only change being the regions and nations which so suffer.

Concluding Remarks

Writing on the fiftieth anniversary of the Universal Declaration of Human Rights, Annas (1998) noted that globalization is a mercantile and ecological fact. It is also a reality in healthcare. He suggests that the challenge for medicine and healthcare is to develop a global language and strategy to improve the health of all the world's citizens. Yet the WHO is criticized as an organization for its lack of action, and public health at the global level is seen in a state of collapse (Garrett 2000). Furthermore, the very fact of globalization, while holding the promise of international cooperation and worldwide solutions to poverty and poor health, is implicated as the main cause of these very problems. As Bloom and Canning (2000) note, the positive correlation between health and income per capita is one of the best-known relationships in international development. Furthermore, the implementation of structural adjustment policies to try to break this relationship by tackling indebtedness and what were seen as the problems of dominant public sectors appear to have worsened the situations in, for example, many African countries. At the same time, the forces of localization (tribalization) are at work not only in what often become war zones, but also in the developed world, as wealthy countries try to insulate themselves through restrictive migration policies and their limited involvement in international agreements, particularly environmental and labor matters. Yet such countries remain committed to free world trade and international travel, which means an openness not only to investments and goods but peoples and disease vectors.

The very facts of globalization – its consequences, such as rapid industrialization, urbanization, population shifts, and changes in agricultural and food processing techniques – and the evolution of micro-organisms abetted by the underfunding of national public health systems, inappropriate prescribing practices, and poor patient compliance, mean that the double burden of disease – the emerging epidemics of cancer, heart disease, and neuropsychiatric conditions – compound the persistent problems of infectious and parasitic diseases, particularly in the developing world, where the compounding health impacts of localization or tribalization are also felt. Globalization has created circumstances for great promise and also great danger. As McNeill (1993: 33–4) comments, "the possibility of really drastic epidemiologic disaster bringing to a halt the modern surge of human population seems to me something we should take very seriously. [We present] a marvellous target for any organism that can adapt itself to invading us."

This potential danger and the worldwide impact of different diseases point to the interconnectedness of not only the world economy, but also of goods, people and ideas, and also of problems and solutions. The economic prosperity and its differential social and geographic distribution resulting from globalization is simultaneously problem and solution. So what is to be done? All is implied – population, social, economic, trade, technology, ethical, and health policy interventions among others. The enormity of the problem of combating disease on a global scale must not be a recipe for inaction. Problems that can be tackled must be (and are being) tackled, e.g., vaccination programs, eradication of the reservoirs of particular disease vectors, the internationalization of public health efforts. But until health is as important as trade and finance in the fora of the world, progress will be severely limited. Health is discussed by the World Bank but not significantly by the G7 or G20 nations. Furthermore, until the breaking of treaties pertaining to health and the global commons carry as significant a penalty as breaking trade treaties, globalization will continue, through its consequences, to kill, maim, and debilitate and ensure that the pursuit of health and happiness will be difficult for most of the world's citizens.

15 Changing Women's Status in a Global Economy

Susan Christopherson

Introduction

One of the arguments made for opening markets and lowering trade barriers through "structural adjustment" or other mechanisms is that these often painful processes will lead to economic development and a higher standard of living. The increasingly thorough analysis of women's status across the world provides a window through which to view the processes related to globalization and to assess the extent to which they achieve their promise. The picture we get is complex, in part because women's status is the product of social as well as economic forces. There is, however, general agreement on some important connections between globalization and women's status. First, the type of development that emphasizes market opening and limitations on public spending to encourage inward investment produces social and economic differentiation and inequality among women. Second, political empowerment of women appears to be related to improvements in their economic and social status. When women's political representation declines, as it has in Eastern European countries, women's economic status deteriorates. Third, with the exception of Muslim countries, there is a strong relationship between a country's overall development and its level of gender development.

This chapter describes national trends in women's status related to the uneven economic development trajectories in the global economy, and examines the role women are playing in globalization processes, including changing production organization and labor migration. While women continue to be disadvantaged relative to men with respect to wages, total work hours, and access to societal resources, they are playing central, critical roles in the processes structuring a global economy. Their continued invisibility within these processes is a result of the measures we use to gauge economic and social contributions; of ideologies that marginalize women as either peripheral to, or victims of, economic change; and of the personal nature of many women's responses to exploitation.

Problems in Measuring Women's Status

Although women's status is typically measured in economic terms, other types of indicators are equally important in providing us with a picture of how women are valued in comparison with men. For example, there is now considerable evidence that gender discrimination has resulted in excess female mortality, especially in the developing world. The number of "missing women" who would be alive were it not for gender discrimination has been estimated at between 60 and 100 million worldwide (Coale 1991; Sen 1990b). These women have been the victims of pre-natal choices to abort females, biased health conditions, unequal nutritional provision, and infanticide. The sex ratio, defined as females per thousand males, has declined almost continuously since 1900. Although the ratio increased marginally (from 930 to 934) between 1971 and 1981, the sex ratio at birth declined again in the 1980s. The continued preference for male children reflected in these statistics is a strong indication of the prospects for women throughout their lives. The problem of the "missing women" also suggests that in determining women's status we need to pay attention to what is missing (such as the ability to move freely in the city and the world; and educational opportunity) as well as to what is present.

When women's status is evaluated specifically in terms of their economic contributions, we again encounter problems with what is and what is not counted. Since the 1970s, feminist economic researchers have focused on this problem and on the reevaluation of the concept of work. Much of their empirical research on work has been intended to deconstruct taken-for-granted notions of what constitutes socially and economically necessary work and who are the workers (Waring 1988). A rich variety of empirical studies has documented the significance of female labor in subsistence production, domestic production, volunteer activities, and in the informal sector. Although these studies have raised questions about, and in some cases changed, accepted definitions of labor and of the labor force, they have had almost no effect on national accounting of work. As a consequence, there is very little systematic information available on a national level that reflects all that we have learned about the work that women do. Even in cases where there has been some attempt across a number of countries to measure work better in, for example, subsistence agriculture, problems have arisen over the reliability of the data. "Once market criteria did not apply, what was considered an economic activity became arbitrary and differences developed among countries regarding . . . what activities were included in national accounts" (Beneria 1992).

Even in the sectors of the economy where women's work is reported, there are serious problems in understanding the extent and nature of their role. Take the case of part-time work, for example. Countries define part-time work and part-time workers in very different ways. In Japan part-time workers are defined by their status in the firm – they do not have "regular" employment contracts and so may be laid off at any time. While at work, however, they may be employed for 40 hours per week or more. In some countries part-time work describes any work that is less than full-time. It may be regularly 39 hours per week. In the United States part-time work is divided into two categories: voluntary part-time work, which describes situations in which workers prefer part-time jobs because they are going to school or have family responsibilities; and involuntary part-time work, which describes situations in which people are working fewer hours than they would prefer because full-time work is not available to them. Despite the significant differences in the definition of part-time work from one country to another, the prevalence of women as part-time workers is taken to mean that they have a loose attachment to the labor market. In some countries, such as Japan, when part-time workers are laid off they are not counted in unemployment statistics. Because of the problems with definitions such as part-time work, feminists have advocated new approaches to understanding how people are included in the labor market. They suggest, for example, that instead of examining part-time work we might more usefully look at the trends toward destandardization of employment contracts relative to national norms. Increased variability has been particularly pronounced in those countries in which service employment dominates (such as the UK, Canada, and the US). Women constitute the majority of workers in non-standard contracts, particularly in part-time jobs but also in other forms of non-standard work, such as temporary work on fixed-term employment contracts, and seasonal work. The explanation for women's positioning at the periphery of the labor market is related to their assigned familial role as care givers to both children and the elderly. It is also attributable to their concentration in industrial sectors and occupations which are not protected by collective bargaining agreements such as those which protect the wages and working conditions of many male workers.

There are indications of progress in accounting for women's work in some sectors and countries. It is widely recognized, for example, that a low rate of female labor-force participation may simply disguise the fact that most women are productively employed in family enterprises. This is true in the developed world, in countries such as Italy, Portugal, Spain, and Ireland, as well as in developing countries. The Dominican Rural Women Study, based on a survey of over 2,100 households, resulted in an estimated labor-force participation rate for rural women of 84 percent compared with a 21 percent rate in the 1981 national census. The difference resulted from a broader definition of women's productive work activities, such as garden cultivation, animal care, and cooking for fieldhands (Beneria 1992). Special studies, such as this one, have demonstrated the disparities between official statistics and actual work effort. As yet, the information they provide has still to be incorporated systematically into national data-gathering efforts in such a way that the full range of women's productive activities becomes visible.

The inadequacy in national income accounts and labor-force statistics has implications beyond the recognition of women's contribution. Inadequate measures of women's work in the wage labor force as well as in more informal economic activities have hampered our ability to understand and interpret how women have affected and been affected by recent transformative processes in the global economy. Women are playing a central role in these transformative processes, including the globalization of manufacturing and the enhanced circulation of capital and information. In some regions and countries that role is visible – women are the majority of workers in the new international manufacturing sector. In other regions and countries, particularly in Latin America and Africa, women's role is concentrated in the informal sector but no less significant. Women in much of the developing world are faced with the primary burden of keeping households afloat in a world economy in which trade policies have produced economic instability and high levels of unemployment. Across the world, changes in international markets have interacted with sexually divided patterns of activity to produce differential effects by sector and by region.

What Do We Know About Women's Status in the Global Economy?

Despite the problems in measuring women's contributions and status, some advances have been made. One attempt to evaluate women's status holistically is represented by the human development indices (HDIs) in the Human Development Report of the United Nations Development Program (1997). The UNDP has developed a gender-sensitive "human development index" which calculates male and female HDIs and an overall gender-sensitive HDI using figures on life expectancy, adult literacy, number of years of schooling, employment levels, and wage rates. The use of a gender-sensitive HDI changes the HDI ranking of many countries. In the most recent index, developed for 146 countries, Canada is ranked first, in both the overall Human Development Index and in the gender-sensitive human development index. What that means is that Canada not only has the highest overall scores on the measures used to develop the index, it also shows the highest level of equality between women and men in achieving those measures. In the gender-sensitive index, Canada is followed by Norway, Sweden, and the US. The United Kingdom ranks 12th in the HDI ranking, after Japan (UNDP 1997: 41)

The developing countries which do well in the HDI ranking include Barbados, the Bahamas, Singapore, and Hong Kong. The bottom five countries are Sierra Leone, Niger, Burkina Faso, Mali, and Ethiopia. Women face double deprivation in these countries: overall human development achievements are low as compared to other countries, and women's are considerably lower than those of men within the countries (ibid: 39). Again, it is important to reflect on the particularly negative effects that globalization processes have had on the African subcontinent in interpreting the rankings.

The UNDP has also developed a gender empowerment measure (GEM), recognizing that control of political and economic resources may lead to higher levels of equality. The index measures women's share of managerial and professional jobs, as well as their representation in parliament. There is more diversity in this index, with the Scandinavian countries ranking highest, Canada ranking 6th, the US 7th, and the UK 20th. Among developing countries, Barbados, Trinidad, and the Bahamas rank highest (ibid: 41).

Using a variety of statistical sources we can also obtain a summary picture of women's status relative to that of men, comparing countries by level of economic development (table 15.1). Women clearly still fall behind men on all measures except life expectancy.

To understand women's role in the global economy, however, we have to go beyond statistics to look at their participation in the processes associated with globalization. Although we could look at gender in a wide range of processes, including the changing relationship between the formal and informal economy, in urbanization, or in the changing role of the state, we focus on two: (1) how women are being incorporated into wage work in the global production network, and (2) women's participation in the increasingly complex patterns of international migration.

Women's Central Role in the Globalization of Manufacturing and Business Services

Between 1953 and 1985 the developed market economies' share of world manufacturing output declined from 72 to 64 percent while that of the developing market economies increased to 11.3 percent. There were also shifts in the share of world manufacturing within the older industrialized countries, with Japan experiencing very high employment growth rates and considerably outpacing the United Kingdom and the US, both of which experienced significant employment losses (Dicken 1998). Women's employment in manufacturing was dramatically affected by these shifts. In those countries losing manufacturing jobs, women lost jobs at a much higher rate than men. In those countries gaining jobs, women constituted the majority of the new workforce. In the United States, for example, between 1973 and 1987 (a period of significant manufacturing job loss) the ratio of goods employment to total employment declined 17 percent for men but nearly 23 percent for women (Bluestone 1990). The high female job losses

Table 15.1   Gender-sensitive Human Development Indices (HDI).

	Gender-sensitive HDI	Female HDI as % of male HDI
Sweden	0.938	96.16
Norway	0.914	93.48
Finland	0.900	94.47
France	0.899	92.72
Denmark	0.879	92.20
Australia	0.879	90.48
New Zealand	0.851	89.95
Canada	0.842	85.73
USA	0.842	86.26
Netherlands	0.835	86.26
Belgium	0.822	86.57
Austria	0.822	86.47
United Kingdom	0.819	85.09
Czechoslovakia	0.810	90.25
Germany	0.796	83.32
Switzerland	0.790	80.92
Italy	0.772	83.82
Japan	0.761	77.56
Portugal	0.708	83.36
Luxembourg	0.695	74.88
Ireland	0.689	74.89
Greece	0.686	76.10
Cyprus	0.659	72.32
Hong Kong	0.649	71.10
Singapore	0.601	70.87
Costa Rica	0.595	70.61
Korea, Rep. of	0.571	65.53
Paraguay	0.566	88.82
Sri Lanka	0.518	79.59
Philippines	0.472	78.67
Swaziland	0.315	68.74
Myanmar	0.285	74.07
Kenya	0.215	58.60

Source: UNDP Human Development Report (1992)

are explained by women's concentration in manufacturing industries such as textiles, apparel, and shoes, which were most subject to international competition and thus to shut-downs and lay-offs.

Studies in Canada confirm that women have been more affected by manufacturing job losses than men. An Ontario study of manufacturing job losses in the 1980s shows that women manufacturing workers who lost jobs had a much more difficult time finding another position. While 62 percent of the men were able to find another job, only 38 percent of the women were able to do so and the time it took to find another job was much longer for women than for men. And while the majority of laid-off men in this study improved their wages in the new job, almost half the women reported lower wages in the new job. The result was an increase in the wage gap. Before the lay-offs the women earned 72 percent of what men earned, but after reemployment the women earned 63 percent of male wages (Cohen 1987).

The overall pattern is even more complex than these figures suggest, however. In the US the female portion of total manufacturing employment has actually risen in the past twenty years because of the growth of female-dominant manufacturing sectors such as electronics, and increased productivity (and lower employment) in male-dominant sectors such as machine tools. The reinstitution of labor-intensive female manufacturing in industrialized countries such as the US has been stimulated by shorter production cycles in some industry segments (which favors production location closer to the market) and by the presence of a large, unregulated female immigrant workforce from Central and Latin America and from Asia. So the picture is not one of wholesale female job loss in manufacturing but, rather, a shift in employment from some regions to others, from some labor-intensive industries to others, and from a native-born to an immigrant workforce.

In the developing countries and emerging high-technology industries we see a mirror pattern. Both in the new industries such as electronics, and in industries such as apparel and shoes, in which labor costs have driven the development of production for export, women constitute the majority of the new workforce. Approximately 80 percent of the total export-industry workforce in developing countries is composed of women (Joekes 1987). While there is evidence that more jobs in export zones are now held by men, women are relegated to the lowest paid jobs (Stearns 2000). These women are frequently in culturally conflicted positions, moving between home, where they are expected to play traditional roles, and the public world of the workplace (Ong 1987).

Women are also a central component in the workforce that has increased with the internationalization of finance and business services. Within the industrialized countries there has been a reordering of regional labor markets in response to the global reorganization of production, distribution, and finance. In those countries and cities that have become the corporate command centers, headquarters functions and business services have expanded dramatically. Labor demand has expanded in sectors such as banking and insurance that have historically employed large numbers of women.

Although the number of women in mid-range managerial and professional positions in finance and business services has increased, women are still excluded from upper management positions. Only 4 percent of upper management positions in European and North American international businesses are filled by women. Even new industries founded on the internet and associated technologies show a wage gap between men and women and a more difficult career progression for women in jobs where who you know is as important as what you know.

In another venue of the "new" economy, women are highly concentrated in the "sweatshops" of the information-based business service economy – call centers. The jobs in these centers demonstrate how new forms of labor segmentation are emerging with the development of the global information economy. Customer segments defined by profit potential are served by different segments of the customer service workforce, divided by education, speech patterns, and other class-related characteristics. Mobility from a lower-profit customer segment to a higher one (and from minimum wages to a living wage) is difficult if not impossible because customer-segmented call centers are located distant from one another.

These examples of the ways in which women are being inserted into the global workforce show that new production methods and technologies do not necessarily produce more egalitarian wages and working conditions. Rather, they suggest that class and gender combined is becoming more important than gender alone as a source of labor segmentation.

International Flows of People and Information: Women's Roles

The internationalization of services is a direct reflection of the central process associated with globalization: the increased international flow of people, commodities, and information. Women are central to these flows, too, as migrants and as information providers and processors.

The largest number of skilled migrants from less developed to more developed economies is in the health services. The reasons for this migration are basically economic, but there are some important differences among skilled migrants. Physicians from less developed economies (almost exclusively male) are frequently trained outside their own countries and have become accustomed to higher levels of treatment and diagnostic facilities than are available to them at home. Fewer professional opportunities are available to them because developing countries need general medical practitioners rather than specialists. Migration is thus an avenue to professional development as well as higher remuneration. For nurses, almost all of whom are female, the situation is quite different. Nurses are trained in their home countries and acquire general rather than specialized skills. In several countries, most notably the Philippines but also Korea and Pakistan, the training of nurses reflects a recognition that nursing graduates are being prepared for service in a world market rather than a national one (Ball 1990). While the nursing migration stream to the US comes primarily from Asia as well as from Ireland and Haiti, a parallel migration stream is developing from Eastern European countries into Western Europe.

The increased demand for this category of skilled female labor is the result of a complex set of developments. As employment opportunities have opened for women in non-traditional occupations, nursing has become less attractive as a professional choice. This unattractiveness has been exacerbated by an extremely flat career hierarchy and generally low pay. In addition, the restructuring of the health sector in some countries has considerably worsened working conditions for nurses, causing many of them to leave the profession. Given this shortage, there is considerable pressure to look to foreign sources for nursing graduates to fill available positions.

The case of nurses suggests some ways in which female migrant labor may become more important in industrialized economies. As women in industrialized countries become more qualified and more specialized, occupational mobility may depend on a willingness to spend periods of time outside their native country. In addition, in economies where the birth rate is low (such as Germany and Italy) and where native-born women are moving into more qualified positions, immigrant women may be used to fill the gap for routine service work. In Western Europe and North America female immigrant labor is increasing in response to a demand for personal services such as childcare, elder care, or housekeeping, to replace services previously provided by women who are now employed. The hiring of immigrant women may also be used as a strategy to reduce the cost of providing routine services in industries such as healthcare.

On the supply side, there is evidence from data on legal and illegal migration from Mexico to the US that women form an increasing proportion of the migrant workforce, a workforce that until recently was almost exclusively male. Many of these women identify their occupation as "manufacturing operative" and have entered the workforce as manufacturing workers in industrial zones. Having obtained basic job skills and confidence, they decide to migrate to higher-wage manufacturing sites in the US. Thus the globalization process comes full circle, with workers as well as commodities entering the markets of industrialized countries.

Another area where women are a prominent and increasing component of economic globalization is in information processing. Although services have conventionally been thought of as market driven, some services such as data preparation and processing are aspects of complex production and distribution processes in both manufacturing and service industries. The location of these activities is influenced by the absence of product transport costs (products are transmitted through telecommunications networks), by labor cost, and skill. Barbados, Jamaica, Ireland, the Philippines, Taiwan, Sri Lanka, and China are major locations for data entry. For US companies the high literacy rates in some countries, particularly Barbados, are strong attractions (Howland 1993). The development of information processing industries in these countries demonstrates that even in developing countries the increased educational attainment of women is directly related to employment growth.

Advances in information processing technologies, particularly the replacement of manual data entry with optical scanners, will change the nature of work in this industry and affect thousands of women workers in both industrialized and developing countries who are currently employed as data-entry operators. As was the case with technological change affecting other female-dominant occupations – librarians, nurses, and computer programmers among others – the expansion of necessary skills and specialization may not translate into better jobs or higher wages for women if they are denied access to opportunities for skill acquisition. Instead, the transformation of the work process and its attendant occupations may produce a workforce in which the majority of the more technically skilled jobs are held by men.

New Divisions in the Global Labor Force

In many conventional respects the prospects for women in the global economy look positive. Female labor-force participation is increasing, women's educational attainment has improved significantly, women's work is, to a greater degree, recognized, and in some countries the wage gap has decreased. Using a standard in which these indicators explain women's status, we would have to say that women's prospects in the global economy are improving. As was noted at the beginning of this chapter, however, status is not static; and economic, political, and technological developments have occurred which put these achievements in a different light. Both technological and organizational innovations have constructed a global workforce composed of a small number of higher-skilled high-wage jobs and a large number of poorly paid service and manufacturing jobs with scant opportunities for upward mobility.

Some women are able to move into the highly compensated, high-skilled jobs, but their mobility may be blocked in various ways. In those countries where labor markets are strongly developed within firms, for example, women are hampered by requirements for continuous employment and geographic mobility. A study of the Japanese information technology industry demonstrates that the career path from programmer to manager is reflected in age cohorts in the various "steps." Of the 20 percent of the software engineers who are female, approximately 40 percent are programmers, 6 percent are systems engineers, and 1 percent are managers. This categorization reflects a dual internal labor market with men in the management track and women in the routine service track. So, one way of responding to the increased demand for skilled workers is to routinize the lowest-level job – that of programmer, where most women are employed – in order to induce labor-saving effects and increase the male labor supply for the more skilled jobs (Christopherson 1991).

In countries where firms rely more on the external labor market, such as the US and UK, another type of dual tracking has evolved. Service firms paying for performance reward those workers able to expend the greatest work effort on behalf of the firm. The result is a "mommy track" for those women who are unable to devote all their waking attention to their jobs.

In both these cases women's increased educational attainment, which in many industrialized countries now exceeds that of men, has not translated into occupational mobility. Instead, new organizational mechanisms are being introduced to use a more generally skilled female labor force more effectively without rewarding them with higher compensation or enhanced job responsibilities. These innovations are most notable in the US, where limited collective bargaining and labor regulation allow firms to design new work organization schemes and compensation systems without any input from workers.

In addition to the customer-segmented call centers described above, another new economy innovation is "broad-banding" – a compensation system that groups different types of jobs in broad bands in order to separate skill acquisition, workload, and increased responsibility from the expectation of higher compensation. This system is designed to make effective use of the external labor market by requiring that new employees enter at the lowest wage level of the band without regard for their skill or experience. Broad-banding is now being introduced in large public and private service organizations predominantly employing women. It may constitute a new form of labor segmentation, confining women to broadly defined job "families" and limiting their opportunities for occupational mobility and increased compensation. The case of broad-banding demonstrates that despite women's increased educational attainment, jobs skills, and work experience, increased equality in the workforce is not assured. Just as women master the rules of the game, the rules may change.

Conclusion

In summarizing the effects of globalization on women, one conclusion we could draw with some certainty is that, in any national context, the situation of the "average" woman is less descriptively accurate than it may have been in the 1960s and 1970s. Possibly the signal characteristic of the period of the 1980s and 1990s was differentiation of markets and within the workforce. With globalization of trade in commodities and increased interaction derived from capital and information flows, women are differentiated from each other in new (as well as old) ways. Class has arguably increased in importance as the variable determining women's economic prospects and status in society.

Another force promoting differentiation among women is the general economic situation in each region. East Asian women, for example, have shared in the increased prosperity of their countries as members of households and because of increased paid work available to them. At the same time, in Africa and Latin America, women's employment prospects have deteriorated because of the structural adjustment measures imposed with the debt crisis. This has been particularly problematic because of the higher incidence in these two regions (than in Asia) of households headed by women, and the greater dependence of children on women's earnings. All these factors suggest that we need to look at women's status as relative but also as constantly evolving.

More broadly, the evidence presented suggests that an analysis of globalization that neglects the role of women is missing a key explanatory ingredient. We need to move our analysis of the gender dimension in globalization out of "ghettoized chapters" such as this one, into the mainstream theoretical discussions about the nature, direction, and ideology of the global economy.

16 Stuck in Place: Children and the Globalization of Social Reproduction

Cindi Katz

There is nothing natural about childhood or what it means to be a child. The two are historically and geographically contingent and always up for grabs. By the late twentieth century both were in flux thanks to the processes of global change that are the focus of this volume. Globalization reconfigures childhood and what it means to be a child by reordering and rescaling relationships between production and reproduction, altering the spaces of everyday life, and tightening the links between certain places while hurling others further away. The introduction to this volume makes the important point that the political, economic, and social relations of the world under contemporary conditions of global change are interstitial. It has become impossible to live with the fiction that "others'" struggles are separable from "our" comforts.

Indeed, one of the figures of the "global child" is as a worker, and the very processes of globalization enable, or more aptly, force "us" to recognize that child as instrumental to "our" comforts. For some, often on university campuses, this has galvanized a robust anti-sweatshop movement. Certain multinational corporations have responded to this movement by raising compensation rates for all workers, providing minimal standards for employment, and providing schools and other on-site facilities for child workers. Yet much of this movement is misguided; missing the essential points that if these children did not work, they and their families might suffer greater harm and that the alternative to employment for most of them is not schooling. Movements to change the horrible truth that "our" comforts rely upon "their" work need to work a broader ground than an opposition to sweatshops alone allows.

A less frequently mourned figure of the child in contemporary globalization debates is the child consumer. Yet this child is as much a part of contemporary processes of global change as the child worker and the two are not separable. Gayatri Chakravorty Spivak (2000) has recently added another global child to the picture, the child investor, who likewise causes little consternation to the (Northern) global imaginary. Why, she asks, is only one child – the child worker – troubling to our imaginations and unsettling to received notions of childhood? If, indeed, childhood is not a time of life that is sacrosanct from the market, then shouldn't all children have the right to "get a piece of the action"? Without confusing the vast differences between children born to substantial stock portfolios and those with few means to economic survival apart from early employment, it is important to recognize the ties that bind them, both literally and metaphorically, and to figure out political strategies that might redress these contradictory relationships. The child consumer is, of course, one of the links. Children are not only a huge and growing "market niche" in the global North, they are increasingly seen as influencing household consumption patterns. Thus, a recent issue of *Sports Illustrated for Kids* featured a two-page foldout advertisement for a minivan (Kapur 1999: 125). Not only do children with their own disposable assets and significant market clout coexist with child workers on the contradictory terrain of globalization, the technocultural circumstances of contemporary globalization have the capacity to render child consumers, child workers, and child investors intelligible to one another, thus heightening the contradictions but making possible a politics that would embrace them all. These issues, however compelling, are not the direct focus of this chapter. I raise them to make clear the complex integrations between and among globalized children, and to forestall the comforts of innocence about their coexistence.

My research has taken a different tack. Rather than counterposing children of such obviously different privilege, I have sought to uncover how the shifts associated with the globalization of capitalist production and the restructuring of the global economy that began in the middle of the 1970s have reworked the grounds in which poor children grow up in two places, rural Sudan and urban United States. This project asks at what scale do we find those "others" and construct "our" selves? The distinctions between comfort and struggle shift and even blur at different geographic scales. In my work, for example, I have argued that contemporary processes of global economic restructuring have had startlingly similar effects on young people in working-class New York City and rural Sudan (Katz 1998a: 2001). While of course these two locales embody great differences of national wealth, power, and development, they also can be imagined together along the lines of what Giuseppe Dematteis has called "fragmented globality," wherein certain parts of the world get bound more tightly as others become more distanced both from their nearby centers and from all that might be associated with the "comforts" and benefits of globalization (Dematteis 2001). Such fragmentation exacerbates differences and inequalities at a range of scales and can increase social polarization within territories that become integrated more closely into global networks.

If we can find *common* experiences between children in places as different as rural Sudan and New York City, what might that tell us about contemporary globalization? In looking at the conditions of young people's everyday lives in both places, I have tried to get at their responses to being marooned in the flux of all that is possible. My intent is to render the toll of "globalization" more sensible and to signal its possible reconfiguration at the hands of those it is most likely to strand. Many of the issues at stake are accessible through the theoretical concept of *social reproduction*. This chapter presents a brief overview of what is entailed in social reproduction, examines some of the ways that globalization has altered its grounds and practices, and connects these changes to children's everyday lives in New York and the Sudanese village I call Howa.

Social Reproduction

Social reproduction is the fleshy, open-ended stuff of everyday life as much as a set of structured practices that unfold in tension with social production. It encompasses daily and long-term reproduction both of the means of production (e.g., factories, equipment, raw materials, and the environment) and the labor power to set them in motion and make them productive. Social reproduction is contingent upon the biological reproduction of the labor force, both on a daily basis and across generations, through the acquisition and allocation of the means of existence, such as food, shelter, clothing, and healthcare. But social reproduction also involves reproducing a differentiated labor force. This differentiated and differently skilled labor force is the fluid outcome of a host of material social practices that vary over time and across space. One of the key struggles of workers in many parts of the world is over what constitutes an "adequately" prepared workforce, so that the bundle of things considered integral to social reproduction, such as educational levels, skills acquisition, health and nutritional status, and workplace rights, not only varies tremendously historically and geographically but is the object of often intense contestation. When, for instance, workers succeed in making the case for certain social benefits such as healthcare and insurance or in-service training, it becomes incumbent upon their employers or the state to provide for such things. Such shifts obviously alter the costs of labor to capital and have redefined the contours of social reproduction and its contents.

But the compass of social reproduction extends well beyond the workplace and has many non-economic aspects. Social reproduction is accomplished through a broad range of sources associated with the household, the state, capital, and what is often called civil society, which includes religious institutions, community organizations, non-governmental organizations (NGOs), and the like. The balance among these varies depending upon location, historical circumstances, class differences, and other factors. In the industrialized countries of Europe and North America during the twentieth century, for example, organized labor and community activists succeeded in getting the state and capital to shoulder broader responsibility for securing social reproduction. These shifts, coupled with those associated with industrialization more generally, altered the role of both the household and civil society in ensuring social reproduction. Both gave way to social reproduction activities provided through the state. For instance, the education of children was increasingly removed from the household to government-sponsored schools, while such institutions as orphanages, almshouses, and settlement houses supported by private charities were displaced by government social welfare programs. Taxation funded state provisions of social reproduction while expansions of social benefits increased the costs of wages for capital.

By the latter part of the twentieth century these relatively expensive labor costs coupled with a number of significant changes in the regulation of international finance and trade that made capital much more mobile, increased competition, and weakening labor organization in the face of employer onslaught, helped to trigger disinvestments in the traditional manufacturing centers as capitalists searched for cheaper sites of production. The costs of social reproduction figured centrally in these decisions to relocate or disinvest, and make clear its importance in what has come to be called globalization. Shifting production elsewhere spurred unemployment and reduced the tax base in abandoned sites largely in older industrial areas, and in many cases altered the particular local commitments of various multinational and transnational corporations. At the same time, and in direct response to widespread recognition of capital's enhanced mobility, municipal, state, and provincial governments in the global North, wary of further capital flight and eager to entice new investments, began to give extensive long-term tax and other breaks to various corporations in an attempt to lure them or tie them down. This practice, among others, reduced the funds available for public spending and increased the burden of individual taxpayers. The tax revolts that were common in the US and many other parts of the industrialized world (and beyond) over the last two decades have been fueled in part by anger and resentment at having to pay for the promises of a set of social relations of production and reproduction under such vastly altered and often inequitable circumstances. Workers, whose struggles through the twentieth century had resulted in broader commitments to and expansions of the social wage, have had to pay twice for the changes wrought by capitalist resistance to paying the higher costs of social reproduction: first in the reduced benefit packages that are commonly listed among the costs of industrial restructuring (for those lucky enough to have benefits and full-time work), and second in the reduced tax base to provide for reproduction socially. For children, particularly poor and working-class children, the price is even higher.

These issues can be counted among the costs of globalization to the industrial North. The fluidity of capital and the increasing porosity of national boundaries enable capitalists and various corporations to take advantage of different social reproduction costs across space, and to mobilize these differences to their continuing advantage so that more and more of the costs of social reproduction are borne by individuals and households, no matter what the location. As political–economic circumstances have grown more and more uneven in the last decade, the gaps between rich and poor people and rich and poor nations have become more marked than ever. In this climate, the liberal promises of earlier decades – associated as much with social welfare as with "development" – have been largely rescinded and even repudiated.

The increasing mobility of capital, the quest for lower production costs, and the decolonization and anti-imperialist movements of the 1960s and 1970s, among many other things, altered the nature of foreign investments in the "Third World" by the latter decades of the twentieth century. Some of the capital that was withdrawn from older industrial areas of the global North was reinvested in production in a number of areas of the South, while some of it flowed "southwards" in the form of "development" initiatives intended to bring increasing numbers of people and goods into the world market and enfold them in capitalist relations of production. By the late 1970s the International Monetary Fund, one of the key organizations regulating international finance, developed and began to impose "structural adjustment programs" (SAPs) upon nations of the global South who were carrying large amounts of debt to northern nations or financial institutions that were not being repaid. SAPs, which have been employed widely throughout the global South, were disciplinary strategies calling for a fairly standard set of measures designed to ensure the debts were serviced if not repaid. Among their most important aspects were the removal of food subsidies (which often caused the price of staple foods to soar), an insistence that cultivation shift from subsistence to cash crops as a means of improving the trade balance of affected countries, and a reduction in various social welfare and "development" programs which were seen as siphoning funds from national coffers that might otherwise be used to repay international debts, incurred over decades of "development" programs. These measures individually and in concert hurt children (Bradshaw et al. 1993)

These questions of social reproduction set the stage for a discussion of what happens to children under conditions of contemporary "globalization." By looking at social reproduction in relation to the shifting nature of investments in the global North and South and linking the two, I am trying to find common grounds in the lives of children growing up in the shadows cast and possibilities created by these local global processes.

Altered Geographies of Social Reproduction

With the globalization of capitalist production, capitalism appears to be increasingly placeless. As capitalist relations of production encroach everywhere, no one place is critical to its survival. But this placelessness is part aura and part alibi, recalling what Donna Haraway called the "god-trick" of objectivity; a way of being everywhere and nowhere at the same time (Haraway 1991: 189ff.). Such is the case with capital. Capital investments are no longer secured *in place* strictly speaking; investments are neither as fixed as they once were nor as committed to any single place. Capital investment is geographical to be sure, and thus secured in many places, but the need for and commitment to any *one* place are largely moot in an era of globalized production. Capitalism's "god-trick" is an issue of great concern because a putatively placeless capitalism unhinges social reproduction from particular forms of production. Producers have fewer commitments to reproducing any particular labor force or conditions of production. Profitability may be greatly enhanced by reneging on social reproduction, but social life and environmental conditions, among other things, suffer.

Neither capitalists who recognize their investments as placeless or unmoored from the specificity and demands of any one place, nor financially strapped governments have much inclination to pay for the spaces in which children come of age, for instance. Yet in these spaces children acquire knowledge and skills to reproduce the society and some version of its social relations of production and reproduction while at the same time constructing identities, cultures, and material social practices of their own. This lack of support is particularly evident in relation to working-class and poor children, whose reproduction is no longer tied so directly to the accumulation of local wealth and capital and whose families have few means to secure support privately. In fact, it is when factoring in the costs of social reproduction that the expendability of this population comes directly to light. Their reproduction, crudely put, may actually represent a loss to capital both in the present and over the longer term because in a "globalized" world with advanced labor-saving technologies and the possibilities of hypermigration, their labor may be unnecessary to the continued workings of the economy. If such a calculus suggests a deeply flawed analysis, it nevertheless drives disinvestments in public environments such as parks and playgrounds, schools and housing for poor people in the global North, and curbs even more basic social investments in the South. This calculus, as much as its crude entailments, is enabled and fostered by capital's fluidity expressed most recently in the globalization of production. Social reproduction, which pivots on the reproduction of people and place, and on the social, political–economic and political–ecologic relations that bind them, is inevitably more rooted.

Children's Everyday Lives I: New York

A recent issue of the conservative New York City tabloid, the *Daily News*, blared the banner headline, "No Place to Play" followed in smaller print with "Many city parks suffer from budget cuts, neglect." A full-color photograph of broken swings at a neighborhood park in Brooklyn accompanied the headline. None of this is news. But how bad things must be that the *News*, an ardent supporter of conservative mayor Rudolf Giuliani and the sort of neoliberal, privatized, and sharply divided city his policies have produced, was covering it. The *News* story reiterated a litany of problems from lack of infrastructural investment in neighborhood parks and playgrounds, an almost complete absence of maintenance, skeletal staffing (whether of gardeners, recreation workers, or maintenance workers), and huge disparities between the sorts of parks most of the city's population relies on for their everyday needs and the small number of "jewels" in the public eye, which are lovingly tended and sustained with private funds (Katz 1998b). The blunt headings of the newspaper say it all, "A question of money," "Glaring inequities," "Shrinking Staff" (Wasserman 2001). But why was this so in 2001 when New York had experienced nearly a decade of unprecedented economic growth and the city's coffers had been unusually full due in large measure to the strength of the stockmarket and its ancillary effects? If the deterioration of New York City's once glorious public parks system began during the recessionary years following the oil shocks of the 1970s, why did it not rebound with the rising economy?

The disinvestment in social reproduction has much to do with the answer. The public spaces of everyday life provide an arena within which children can play in an open-ended or organized way, encounter friends, acquaintances, and new kids serendipitously, and be more in touch with the elements (or at least get their hands dirty, play in the open air, get wet, and learn about plants, birds, gum wrappers, and the like). Open space is a crucial arena of social reproduction, yet these days it gets little attention as such, especially compared with other places routinely associated with social reproduction like schools and the home. This was not always the case. In the early decades of the twentieth century, for instance, parks and playgrounds were recognized as key arenas of social reproduction and treated accordingly. Social reformers, especially those associated with the US Progressive movement, urged park construction in urban areas, and if necessary built and staffed children's play environments as settings for the "Americanization" of a polyglot, largely working-class urban immigrant population (Rainwater 1922; Goodman 1979; Nasaw 1985; Gagen 2000). Municipal governments quickly saw the advantage of providing play and recreation opportunities for working-class people who, living in overcrowded conditions and working long hours under dismal conditions, might find other – political – outlets for their frustrations and anger. Likewise, it was hoped that their children would be incorporated into an American way of life through recreational and cultural activities provided through parks and playgrounds, and have a chance to release childish energy pent up in small living quarters and workplaces which might have exploded elsewhere. The plan was to suck children off the streets, which they vastly preferred, and away from nefarious influences which were thought to be everywhere in immigrant and working-class communities.[1]

Now more private means are deployed to suck children off the streets. Indeed, Denis Wood and Robert Beck (1994) argue that television is a means of privatization for precisely this reason. Likewise, computers and the internet increasingly lure people away from the physical environments of their every-day lives. Yet, of course, the public spaces of everyday lives still matter and these are often neglected. For instance, my colleagues and I worked for years to renovate and redevelop two bleak and broken-down Harlem schoolyards in partnership with the schools and various members of the neighborhood. More than a dozen years later, the two yards languished in roughly the same shape as in 1988 when a parent–teacher group interested in schoolyard improvement at two elementary schools designated as Com-munity Schools (expanded access and hours) sought out our research group on children's environments for assistance. The schools had over $800,000 available from the Board of Education for repaving and refencing the schoolyards, and together we engaged in an intensive participatory design project that had the inputs of students, teachers, community residents – young and old, school administrators, school custodians, and neighborhood leaders. The architectural program we developed (Hart et al. 1992) was the basis of an international design competition for the two yards, and the School Facilities Division of the Board of Education promised to take these designs seriously in their plans for the schoolyards. After much foot dragging on their part, it turned out that there was no money for the improvement of these yards (others in wealthier neighborhoods underwent renovation dur-ing the several-year period in question). Moreover, the schools' principals were informed that the money that was originally theirs for repaving and new fencing was no longer available. After almost a decade of effort, the schoolyards remained little improved until the principal of one of the two schools sought and gained funding from the Manhattan Borough President's Office. In the last couple of years the ball seems to have gotten rolling again and some improvements have been made. This sort of delay – despite com-munity organization and initiative as well as the support of the school administration and staff – is all too common in poor and less powerful parts of the city.

While conventional playgrounds may not be children's first choice in play spaces, multipurpose schoolyards – especially well designed ones – can offer a range of opportunities for children and others to get together. The designs we put forward, for instance, had open space for team sports, raised beds for gardening, a small stage for performances, shaded and planted areas with seating for conversations, contemplation, storytelling, and the like, as well as some play equipment for younger children and lines painted on the ground for a number of games and activities. These schoolyards were a mere two blocks from the northern limits of Central Park, and over the same time period that we were struggling to change the schoolyards,

Central Park was splendidly renovated from bottom to top by the very well funded Central Park Conservancy. I note this coincidence for two reasons. First, most children in this neighborhood were not allowed to visit Central Park without adult supervision and so its pleasures were largely wasted on them, despite its proximity. Second, while the schoolyards were more accessible to neighborhood children either because they were sometimes staffed during non-school hours or because parents could see them from their apartment windows or at the very least have their children within earshot, the schools were not able to find the modest sums required to make their yards more hospitable to neighborhood children and other residents.

The early years of the twenty-first century, as this example suggests, are a far cry from those of the twentieth century. Through the first half of the twentieth century an extensive parks infrastructure was established in New York City, driven in the earlier years by the "playground movement." Parks and playgrounds were often staffed not just with maintenance personnel but with play leaders. Not only were the grounds and equipment in good repair, most playgrounds had non-fixed play equipment such as balls and bats to distribute to children unlikely to be able to afford their own. The relationship of these endeavors to questions of social reproduction was direct. Reformers and city officials viewed such things as team sports, sing-alongs and calisthenics as means of inculcating conforming behavior, respect for authority, obedience to rules, self control and the like; things not coincidentally associated with producing a pliant workforce. New York's industrial economy grew through most of the first two-thirds of the twentieth century, and for much of this time allocations for parks and recreation remained at relatively high levels. By the 1970s, as the processes associated with contemporary globalization began to take hold, and New York lost most of its manufacturing sector and entered a "fiscal crisis," one of the budget areas hardest hit by cuts was open space. Where in mid-century expenditures on parks accounted for about 1.5 percent of the New York City's operating budget, by 2000 the share for parks had dropped to 0.48 percent (Citizens Budget Commission 1991; Wasserman 2001). The disinvestments in public open space by the end of the twentieth century parallel the investments in it at the start of the century. Both processes were deliberate and revolved around the reproduction (or not) of a home-grown workforce.

I realize that this is a rather crude and overly simplified argument. Neither the investments nor the disinvestments in public open space were that straightforward. The decisions around them were inflected by broader questions of race, class, and gender as much as by socially constructed views of children, urban life, the role of government, immigration, and the future, among other things. Nevertheless, economic data suggest a clear relationship between the necessity of producing a differentiated labor force and investments in public spaces of social reproduction by governments and civil society. Following the oil shocks of 1973 and New York's fiscal crisis, there was a stark diminution of funds for parks and recreation. By 1980 the city's parks were in a shambles, and it was then that an altered stance concerning social reproduction came most clearly to light. Starting with New York's most prominent park, mid-town Manhattan's Central Park, public–private partnerships were established to refurbish and tend to some of the city's public spaces. But now the focus was on the landscape – the visible landscape of those with wealth and privilege – rather than the health and well-being of the city's children. Through the aegis of public–private partnerships, the parks seen by tourists, wealthy residents, and office workers have been restored and groomed. They are kept clean, well-maintained, and sponsor cultural programs. They are the minority of the city's parks. In the parts of the city that tourists rarely visit, that most office workers never see, and that high-income people tend to avoid, parks (and schoolyards) are in a woeful state. With only 0.48 percent of the city's annual budget going to the Department of Parks and Recreation compounded by the fact that wealthy and powerful people tend privately to the parks in their neighborhoods, this situation is not likely to change (Katz 1998b).

Where in the past various wealthy and privileged constituencies lobbied for the public provision of open spaces for all (even if for many their intent was to properly socialize young people), the impulse now is essentially to mind their own backyards. Such tendencies mesh with what Neil Smith (1996) calls revanchism, a policy of revenge against those who seem to threaten the social order of the white middle class. Revanchism is associated with neoliberal imperatives made possible in part by the globalization of capitalist production and its loosening of capitalists' commitments to particular places and the production of any specific localized workforce. The disintegrating grounds of everyday life – the broken swings, the filthy patches of dirt that pass for parks or gardens in New York City, and the bulldozing of carefully tended community gardens for the construction of affluent housing not meant to house those in the gardens – are the very palpable local effects of "globalization" for young people and others.

Children's Everyday Lives II: Rural Sudan

In rural Sudan globalization works differently. There, I have worked with young people over the past twenty years trying to understand how various aspects of global economic restructuring have affected them. At first I went to Howa, a village in Arabic-speaking central Sudan, to see how children fared after a state-sponsored development project was established in their village (Katz 1991). I was particularly interested in children's knowledge under these circumstances – I wanted to see what children learned, from whom, and how they used this knowledge in their everyday lives. My project was concerned with the nitty-gritty of social reproduction in a period of rather dramatic local transformation spurred by the impetuses of international "development." What I found was that children were learning about agriculture, forestry, and animal husbandry – the economic mainstays of their community – when the viability of each of these was increasingly becoming compromised as a result of the political–ecological and political–economic changes associated with the project (see Katz 1991 for a detailed discussion of these issues). Thanks to the labor requirements of cultivating cash crops in the agricultural project, which exceeded those of subsistence cultivation; the environmental deterioration provoked by the land clearances necessary for the project; and the increasing importance of cash in the local economy, children worked long hours farming, herding animals, and collecting fuelwood, leaving them proportionately less time to attend school. Yet when young people came of age under these circumstances they were more likely to need the sorts of knowledge they might have acquired in school had they had the chance. Again, the broad forces of globalization – this time in the form of a multilaterally financed and state-sponsored agricultural project – appeared to be displacing selected young people from a viable future. Yet they managed to craft more promising futures for themselves when they grew up than what I had envisaged when they were 10 years old, and this suggests the indeterminate possibilities that inhere in the practices of social reproduction.

Young people in Howa were best prepared to continue the agriculturally based life of their village, but with a fixed number of tenancies in the agricultural project and serious local environmental degradation which made both forestry and animal husbandry difficult, most young people would not be able to succeed on the land. Where I had imagined dramatic increases in migration to the towns in search of work for which most of them were ill-prepared, something else happened. What happened was something I call "time-space expansion," a distinctly spatial strategy that enabled young people to continue doing what they knew how to do even under the vastly altered conditions in which they came of age. In short, while David Harvey (1989) notes that one of the signal effects of advanced capitalism is the "annihilation of space by time" which provokes "time-space compression," that only seems to be the case from a privileged vantage point. For those reaching adulthood in mid-1990s Howa, the world seemed to be getting bigger not smaller. With Sudan suffering from a decade and a half of structural adjustment and a relentless civil war, many young men were only able to survive by continuing the sorts of activities that had sustained their community for over a century – agriculture, animal husbandry, and forestry – but over a vastly expanded terrain. Where previously these activities took place within a 5-kilometer radius of the village, by 1995 animals were grazed more than 100 kilometers from the village, charcoal was produced over 200 kilometers away, and agricultural laborers routinely traveled 50 kilometers for work. In other words, the area of production and social reproduction in Howa was some 1,600 times larger by 1995 than it was in 1985, just so people could stay in place! Many young men also took day or seasonal employment in the towns, and labor migration abroad was on the rise. Nevertheless, by drawing on the considerable knowledge they had acquired as children and deploying it across a wider terrain, young people in Howa crafted survivable futures for themselves despite the hardships suffered by their country and endured by their village. I recognize that such strategies have their limits, but in highlighting this one from Howa I want to signal the vitality, promise, and open-endedness of the material social practices of social reproduction.

Conclusion

The failures and retreats of capitalists from the landscapes of social reproduction, wherein play, social life, social learning, sociality, and shared cultural experiences happen, are part and parcel of globalization. Examining global change through the lives of children allows us to see its perils and possibilities. As accumulation and production have been increasingly organized at a supranational scale, various locations and certain populations lose much of their consequence for capitalists. Those sites that are no longer of interest to capital suffer the effects of what we might call "scale collapsing," where the reorganization of production by the dominant classes pushes upon, abandons, or eliminates the material social practices and spaces associated with reproduction at another scale. Such was the case with the public spaces of New York City. In Sudan, on the other hand, international capital was used to draw the local population more fully into capitalist relations of production and reproduction, and in the process many of the existing spaces and practices of production and reproduction were thrown into disarray. In both places it seemed that global processes enacted on the ground had troubled the geographies of social reproduction, and at a more abstract level those very processes had made the viability of young people's futures there moot.

Under these circumstances the question becomes how adults and children can turn around and make work for young people (or at least not work against them) the conditions associated with globalization that have proven so destructive to the geographies both of social reproduction and of children and childhood as well. How might social reproduction be made to figure in accounts of global change? From the deterioration of agricultural areas, woodlands, and pastures around Howa, to the deterioration of the public spaces of reproduction such as schools, playgrounds, parks, and housing in the contemporary US, the spaces of social reproduction have been under assault in recent years (cf., Katz 1991, 1998a). These conditions are frightfully reminiscent of earlier times and recall how from the Middle Ages until the nineteenth century, "childhood" as such was a luxury that only the upper classes could provide for their children (Ariàs 1962). With the advent of mass schooling, child labor laws, and the reform movement in the industrializing economies of the West, some semblance of childhood was available to children more generally by the end of the nineteenth century. But the messy entailments of globalization seem on the one hand to have led to a retreat from these advances for many and on the other to have produced a fetishization of childhood among more privileged parts of the population. That the promises of "development" failed to deliver such privileged European notions of "childhood" to the "South" and elsewhere is a long and troubled tale that I and others address elsewhere (Katz forthcoming; Roberts 1997; Ruddick 1998). Now, rather than the *time* of childhood being a luxury only for the rich, it may be that the *spaces* of childhood have become a luxury.

Note

1   Playground construction is a contradictory business. In the Reform era as now, most children preferred not to play in playgrounds (although parks are another matter entirely). Standard US playgrounds are after all rather boring children's environments and all but the youngest of children tend to prefer unprogrammed, even dangerous places like streets, empty lots, alleyways. Of course, this was precisely what Reformers were responding to when they called for the construction of playgrounds and schoolyards. To paraphrase the movie, they thought if they "built them, the children would come." Never mind that they didn't or that it was difficult to attract children to playgrounds more than a couple of blocks from their houses, it is the intent of the Reformers that I am interested in here and their beliefs in providing public space for social reproduction (Goodman 1979; Nasaw 1985).

   Nowadays children are more likely to be incorporated into the fabric of American life through consumption or through various exhortations to consume. Again, social reproduction comes into play. The New York City Board of Education – woefully underfunded and behind in computerizing its schools and providing electronic resources for all students – entered a public–private partnership to accelerate the provision of computers throughout the system. Until parental outcry, they seemed to see no problem in the fact that students would log on to a Board of Education website that advertised consumer products – not necessarily educational – that children were invited to purchase through the internet (Wyatt 2000). Has consumership replaced citizenship as the key vehicle for engaging children in a broader matrix of social relations?

17 Race and Globalization

Ruth Wilson Gilmore

Theorizing Racism

While there is no legitimate biological basis for dividing the world into racial groupings, *race* is so fundamental a sociopolitical category that it is impossible to think about any aspect of globalization without focusing on the "fatal coupling of power and difference" (Hall 1992) signified by *racism*.[1] Racism is the state-sanctioned and/or extra-legal production and exploitation of group-differentiated vulnerabilities to premature death, in distinct yet densely interconnected political geographies. Wherever in the world the reader encounters this chapter, she will have some knowledge of racism's everyday and extraordinary violences; she will also be sensible of the widening circulation of cultural, aesthetic, and oppositional practices that subjectively mark the difference race makes. For the purposes of this chapter political economy is primary, because so much of globalization concerns material changes in ordinary people's capacities to make their way in the world. Therefore, by emphasizing racism, the next few pages examine how race is a modality through which political–economic globalization is lived (cf., Hall 1980). A case study of the United States demonstrates how the conjuncture of globalization, legitimate-state limits, and white supremacy reorganizes and contains power through criminalization and imprisonment. These significant political practices, while devised and tested behind the sturdy curtain of racism, have broad national and global articulations – connections not impeded by racialized boundaries (Gilmore 1998; Gordon 1988). The purpose of focusing on the US in this chapter is not to study an "average" much less "original" racism, but rather to consider how fatal couplings of power and difference in one place develop and change. Then we will consider how they connect with, are amplified by, and materially affect, modalities of globalization elsewhere.

Why should race so vex the planet? Variations in humankind can be regarded in many ways, as contemporary genetics demonstrates (Lewontin, Rose, and Kamin 1984). However, the coupling of European colonialism's economic imperatives – expansion, exploitation, inequality – with

European modernity's cultural emphasis on the visible (Berger 1980) produced a powerful political belief that underlies racialization. The belief can be summed up this way: What counts as difference to the eye transparently embodies explanation for other kinds of differences, and exceptions to such embodied explanation reinforce rather than undermine dominant epistemologies of inequality (Gilroy 2000). Geographers from Linnaeus forward have figured centrally in the production of race as an object to be known, in part because historically one of the discipline's motive forces has been to describe the visible world (Livingstone 1992).

To describe is also to produce. While any number of "first contact" texts show that in fact "all cultures are contact cultures" (Williams 1992), the powerful concept of a hierarchy of fixed differences displaced both elite and common knowledges of an alternatively globalized world (Blaut 1994; Lewis 1982; Mudimbe 1988). For example, in the mid-fifteenth century, Azurara, court historian to Henry the Navigator (intellectual and financial author of Europe's African slave trade), noted how many in the first group of human cargo corraled at Lisbon strongly resembled then-contemporary Portuguese; indeed, the captives' sole shared feature was their grievously wept desire to go home (Sanders 1978).

The triumph of hierarchy required coercive and persuasive forces to coalesce in the service of domination (Said 1993). While European militarization constituted the key force that produced and maintained fatally organized couplings of power and difference, Catholic and Protestant missionaries explained and reinforced hierarchical human organization in terms of God-given ineffable processes and eternally guaranteed outcomes (Stannard 1992). National academies – precursors to today's colleges and universities – codified the social world in stringently insulated disciplines which further obscured the world's interconnections (Wallerstein 1989; Bartov 1996).

In the long, murderous twentieth century, geographers used three main frameworks to study race: environmental determinism (see Mitchell 2000), areal differentiation (see Harvey 1969), and social construction (e.g., Jackson and Penrose 1993; Kobayashi and Peake 1994; Gilmore 1999, 2002; Liu 2000). The variety of frameworks, and the fact of transition from one to another, demonstrates both how geography has been deeply implicated in the development of inequality, and how critical disciplinary reconstruction at times seeks to identify and remedy the social effects of intellectual wrongs. In other words, frameworks – or "paradigms" (Kuhn 1996) – are not structures that emerge with spontaneous accuracy in the context of knowledge production. Rather, they are politically and socially as well as empirically contingent and contested explanations for how things work that, once widely adopted, are difficult to disinherit.

Geographers who embraced environmental determinism sought to explain domination and subordination – power and difference in terms of groups' relative life-chances – by reference to the allegedly formative climates and landscapes of conqueror and conquered. The framework assumed, and therefore persistently demonstrated, that inequality is a product of natural rather than sociopolitical capacities; while culture might revise, it can never fully correct (e.g., Huntington 1924). In this view inequality is irremediable, and thus should be exploited or erased. Examples of exploitation and erasure include US and South African apartheid, the Third Reich's "Final Solution," scorched-earth wars against Central American indigenous groupings, and other cleansing schemes.

As if in recognition of environmental determinism's horrifying social consequences, the second framework, areal differentiation, seized the seemingly unbiased tools of the quantitative revolution to map distributions of difference across landscapes. The areal approach featured a mild curiosity toward the political–economic origins of inequalities, by suggesting causes for certain kinds of spatial mismatches or overlays. But in the end, taking race as a given, and development as the proper project for social change, the approach described territorialized objects (people and places as if they were things) rather than sociospatial processes (how people and places came to be organized as they are) (Gilmore 2002).

Inquiry into processes shapes a prevalent critical geographical framework. Neither voluntaristic nor idealistic, social construction refuses to naturalize race, even while recognizing its sociospatial and ideological materiality. At its relational best, the social construction approach considers how racialization is based in the (until recently) underanalyzed production of both masculinity and whiteness (foundation and byproduct of global European hegemony), and how, therefore, race and space are mutually constituted (Ware 1992; Pulido 2000).[2] How do spatially specific relations of power and difference – legal, political, cultural – racialize bodies, groupings, activities, and places? Why are such relations reproducible? For example, how is it that globally dynamic interactions, organized according to liberal theories of individual sovereignty, protection, grievance, and remedy ("human rights"), reconfigure but do not dismantle planetary white male supremacy – as measured by multinational corporate ownership, effective control over finance capital, and national military killing capacity?

While the three approaches span a wide political spectrum, from racist eugenics to anti-racist multiculturalism and beyond, all, at least implicitly, share two assumptions: (1) societies are structured in dominance within and across scales; and (2) race is in some way determinate of sociospatial location (Hall 1980). A way to understand the first point is to think about all the components – or institutions – of a society at any scale, and then ask about differences of power within and between them. Are corporations stronger than labor unions? Do poor families rank equally with wealthy ones? Does education receive the same kind of financial and political support, or command the same attention to demands, as police or the military? Do small food producers enjoy the same protections and opportunities as agribusinesses? Are industrial pollutants and other toxic wastes spread evenly across the landscape? Do those who produce toxins pay to contain them? Are people tried in courts by juries of their peers? Having thought about these kinds of institutional relationships, turn to the second assumption: According to the society's official or commonsense classifications, how does race figure in and between the institutions?[3] While this thought-experiment is only a crude cross-section, the conclusions suggest strongly that – as all the twentieth-century frameworks agree – race, while slippery, is also structural.

But what structures does race make? Let us turn the question inside out, and ask how might fatal couplings of power and difference be globally represented. Any map of modernity's fundamental features – growth, industrialization, articulation, urbanization, and inequality – as measured by wealth, will also map historical–geographical racisms. Such a map is the product of rounds and rounds of globalization, five centuries' movement of people, commodities, and people *as* commodities, along with ideologies and political forms, forever commingled by terror, syncretism, truce, and sometimes love. The cumulative effects of worldwide colonialism, transatlantic slavery, Western hemisphere genocide, and postcolonial imperialism – plus ongoing opposition to these effects – appear today, on any adequate planetary map of the twenty-first century, as power-differerence topographies (e.g., North, South) unified by the ineluctable fatalities attending asymmetrical wealth transfers.

So far the discussion is pitched at a general level of abstraction. Our map of contemporary globalization circulation models (GCMs) is built on the historical geographies of past GCMs, and signifies underlying struggles that indicate global warming of a peculiar kind. Indubitably anthropogenic, the racialized heat of political–economic antagonisms sheds light on the forms of organized abandonment that constitute the other side of globalism's uneven development coin (Smith 1990): structural adjustment, environmental degradation, privatization, genetic modification, land expropriation, forced sterilization, human organ theft, neocolonialism, involuntary and superexploited labor.

At the same time, the realities of racism are not the same everywhere, and represent different practices at different geographical scales – which are connected (or "articulated") in many ways (see, for example, Pred 2000). Within and across scales – respectively configuring nation-states, productive regions, labor markets, communities, households, and bodies (Smith 1992) – anti-racist activism encounters supple enactments and renewals of racialization through law, policy, and legal and illegal practices performed by state and non-state actors. The key point is this: at any scale, racism is not a lagging indicator, an anachronistic drag on an otherwise achievable social equality guaranteed by the impersonal freedom of expanding markets. History is not a long march from premodern racism to postmodern pluralism (Linebaugh and Rediker 2000). Rather, racism's changing same does triple duty: claims of *natural* or *cultural* incommensurabilities secure conditions for reproducing *economic* inequalities, which then validate theories of extra-economic hierarchical difference. In other words, racism functions as a limiting force that pushes disproportionate costs of participating in an increasingly monetized and profit-driven world onto those who, due to the frictions of *political* distance, cannot reach the variable levers of power that might relieve them of those costs.[4]

What is the character of such friction? Why is the cost of mobility so prohibitive for some, especially in the current period that is colloquially characterized by increased – some say *hyper* – mobilities? Race and racism are historical and specific, cumulative and territorially distinct – although distinct does not mean either isolated or unique. But while already-existing *material inequality* shapes political landscapes, the contested grounds are also *ideological*, because how we understand and make sense of the world and ourselves in it shapes how we do what we do (Donald and Hall 1986). In any society, those who dominate produce *normative* primary definitions of human worth through academic study, laws, and the applied activities of medical and other "experts," as well as through schooling, news, entertainment, and other means of mass education (Omi and Winant 1986; Bartov 1996; Chinn 2000). Those who are dominated produce counter-definitions which, except in extraordinary moments of crisis, are structurally secondary to primary definitions. While such counter-definitions might constitute "local" common sense, their representation in the wider ideological field is as sporadically amplified *responses* to regional norms – rather than as the fundamental terms of debate (Hall 1978). On all fronts, then, racism always means struggle. Whether radically revolutionary or minimally reformist, anti-racism is fought from many different kinds of positions, rather than between two teams faced off on a flat, featureless plain. Indeed, organized and unorganized anti-racist struggle is a feature of everyday life, and the development and reproduction of collective oppositional capacities bear opportunity costs which, in a peculiar limit to fiscal metaphor, are hard to transfer *collectively* to other purposes within "already partitioned" political geographies (Smith 1992: 66). Therefore, if, as many activist-theorists note, coercion is expensive (e.g., Fanon 1961), anti-coercion cannot be cheap.

The deepening divide between the hyper-mobile and the friction-fixed produces something that would not surprise Albert Einstein: depending on their sociospatial location in the global political economy, certain people are likely to experience "time-space compression" (Harvey 1989) as time-space *expansion*. We shall now turn to a case study of the United States to see how intensified criminalization and imprisonment constitute such an expansion, and then conclude by considering some global effects of US anti-Black racism. The reader must bear in mind that US racism is not the model but rather the case, and that US racism is not singularly anti-Black; the larger point, then, is to consider both how racism is produced through, and informs the territorial, legal, social, and philosophical organization of a place, and also how racism fatally articulates with other power-difference couplings such that its effects can be amplified beyond a place even if its structures remain particular and local.

Prison and Globalization

Ever since Richard M. Nixon's 1968 campaign for US president on a "law and order" platform, the US has been home to a pulsing moral panic over crime. Between 1980 and 2000 the "law and order" putsch swelled prisons and jails with 1.68 million people, so that today 2,000,000 women, men, boys, and girls live in cages.[5] The US rate of imprisonment is the highest in the world (Gainsborough and Mauer 2000). African-Americans and Latinos comprise two-thirds of the prison population; 7 percent are women of all races. Almost half the prisoners had steady employment before they were arrested, while upwards of 80 percent were at some time represented by state-appointed lawyers for the indigent: in short, as a class, convicts are the working or workless poor. Why did "the law" enmesh so many people so quickly, but delay casting its dragnet for a decade after Nixon's successful bid for the presidency?

The 1938–68 World War II and Cold War military buildup produced a territorial redistribution of wealth from the urban industrialized northeast and north central to the agricultural and resource dominated south and coastal west (Hooks 1991; Markusen et al. 1991). While one urban–rural wealth gap was narrowed by state-funded military development, the equalization of wealth between regions masked deepening inequalities *within* regions as measured in both racial and urban–rural terms (Schulman 1994; Gilmore 1998).

Military Keynesianism characterized the US version of a *welfare* state: the enormous outlays and consequent multipliers for inventing, producing, and staffing *warfare* capacities underwrote modest social protections against calamity and opportunities for advancement. Prior to the military buildup, the New Deal US developed social welfare capacities, the design of which were objects of fierce interregional struggle (Egerton 1995). In concert with the successful political struggle by the Union's most rigorously codified *and* terrorist white supremacist regimes (Ginzburg 1962; McWilliams 1939) to make the south and west principal sites for military agglomeration, the federal government also expanded to the national scale – via the structure of welfare programs – particular racial and gender inequalities.[6] As a result, under the New Deal white people fared better than people of color; women had to apply for individually what men received as entitlements; and urban industrial workers secured limited labor rights denied agricultural and household workers (Gordon 1994; Edid 1994).

The welfare–warfare state (O'Connor 1973 – another way to think of "military Keynesianism") was first and foremost a safety net for the capital class as a whole (Negri 1988) in all major areas: collective investment, labor division and control, comparative regional and sectoral advantage, national consumer market integration, and global reach. Up until 1967–8 the capital class paid high taxes for such extensive insurance (Gilmore forthcoming). But in the mid-1960s the rate of profit, which had climbed for nearly thirty years, began to drop off. Large corporations and banks, anxious about the flattening profit curve, began to agitate forcefully and successfully to reduce their taxes. Capital's tax revolts, fought out in federal and state legislatures, and at the Federal Reserve Bank, provoked the decline of military Keynesianism (Dickens 1996). The primary definers of the system's demise laid responsibility at the door of unruly people of color, rather than in the halls of capital – where overdevelopment of productive capacity weighed against future earnings (Brenner 2001) and therefore demanded a new relation with labor mediated by the state.

The 1968 law and order campaign was part of a successful "southern strategy" aimed at bringing white-supremacist Democrats from *anywhere* into the Republican fold (see, for example, McGirr 2001). Mid-1960s radical activism – both spontaneous and organized – had successfully produced widespread disorder throughout society. The ascendant right used the fact of disorder to persuade voters that the incumbents failed to govern. The claim accurately described objective conditions. But in order to exploit the evidence for political gain, the right had to interpret the turmoil as something they could contain, if elected, using already-existing, unexceptionable capacities: the power to defend the nation against enemies foreign and domestic. And so the *contemporary* US crime problem was born, in the context of solidifying the political incorporation of the militarized south and west into a broadening anti-New Deal conservatism. The disorder that became "crime" had particular urban and racial qualities, and the collective characteristics of activists – whose relative visibility as enemies inversely reflected their structural powerlessness – defined the face of the individual criminal. To deepen its claims, the right assigned the welfare–warfare state's *social* project institutional responsibility for the anxiety and upheaval of the period.

The postwar liberation movement focused in part on extending eligibility to those who had been deliberately excluded from New Deal legislation. While some factions of the civil rights movement worked to bring about simple inclusion, radical African, Latino, Asian, and Native American groupings fought the many ways the state at all scales organized poor people's perpetual dispossession (Jones 1992). Radical white activists both aligned with people of color and launched autonomous attacks against symbols and strongholds of US capitalism, and Euro-American racism and imperialism.

Indeed, growing opposition to the US war in Southeast Asia helped forge one international community of resistance. At the same time, activism against colonialism and apartheid on a world scale found in Black Power a compelling renewal of linkages between "First" and "Third World" Pan African and other liberation struggles (James 1980). Meanwhile, students and workers built and defended barricades from Mexico City to Paris: no sooner had smoke cleared in one place than fires of revolt flared up in another. The more that militant anti-capitalism and international solidarity became everyday features of US *anti-racist activism*, the more vehemently the state and its avatars responded by "individualizing disorder" (Feldman 1991: 109) into singular instances of criminality – that could then be solved via arrest or state-sanctioned killings.

Both institutional and individualized condemnation were essential, because the deadly anti-racist struggle had been nationally televised. Television affected the outlook of ordinary US white people who had to be persuaded that welfare did not help them (it did), and that justice should be measured by punishing individuals rather than via social reconstruction (Gilmore 1991). Thus, the political will for *militarism* remained intact, but the will for *equity* (another way to think about welfare), however weak it had been, yielded to pressure for privatizing or eliminating public – or social – goods and services. In other words, the basic structure of the postwar US racial state (Omi and Winant 1986) has shifted, from welfare–warfare to workfare–warfare, and that shift is the product of, and is producing, a new political as well as economic geography.

The expansion of prison coincides with this fundamental shift, and constitutes a geographical solution to socioeconomic problems, politically organized by the state which is, itself, in the process of radical restructuring. This view brings the complexities and contradictions of globalization to the fore, by showing how already-existing social, political, and economic relations constitute the conditions of possibility (but not inevitability) for ways to solve major problems. In the present case, "major problems" appear, materially and ideologically, as surpluses of finance capital, land, labor, and state capacity that have accumulated from a series of overlapping and interlocking crises stretching across three decades.

In the wake of capital's tax revolt, and the state's first movements toward restructing both capital–labor and international economic relations, the US slipped into the long mid-1970s recession. Inflation consequent to abandonment of the gold standard (Shaikh and Ahmet Tonak 1994) and rising energy costs sent prices skyward, while at the same time steep unemployment deepened the effects of high inflation for workers and their families. Big corporations eliminated jobs and factories in high-wage heavy industries (e.g., auto, steel, rubber), decimating entire regions of the country and emptying cities of wealth and people. Even higher unemployment plagued farmworkers and timber, fishing, mining, and other rural workers. Landowners' revenues did not keep up with the cost of money because of changing production processes and product markets, as well as seemingly "natural" disasters. Defaults displaced both large and smaller farmers and other kinds of rural producers from their devalued lands, with the effect that land and rural industry ownership sped up the century-long tendency to concentrate (Gilmore 1998).

Urban dwellers left cities, looking for new jobs, cheaper housing,[7] or whiter communities, and new suburban residential and industrial districts developed as center-cities crumbled. Those left behind were stuck in space, their mobility hampered by the frictions of diminished political and economic power. As specific labor markets collapsed, entire cohorts of modestly educated men and women – particularly people of color, but also poor white people – lost employment and saw household income drop (see, for example, Waldinger and Bozorgmehr 1996). Meanwhile, international migrants arrived in the US, pushed and pulled across borders by the same forces producing the US cataclysm.

The state's ability to intervene in these displacements was severely constrained by its waning legitimacy to use existing welfare capacities to mitigate crises. However, what withered is not the abstract geopolitical institution called "the state," but rather the shortlived *welfare* partner to the ongoing *warfare* state (Melman 1974). Unabsorbed accumulations from the 1973–7 recession lay the groundwork for additional surpluses idled in the 1981–4 recession, and again in 1990–4, as the furious integration of some worlds produced the terrifying disintegration of others.

Prison Expansion

Many map the new geography according to the gross capital movements we call "globalization." This chapter proposes a different cartographic effort, which is to map the political geography of the contemporary United States by positing at the center the site where state-building is least contested, yet most class based and racialized: the prison. A prison-centered map shows dynamic connections among (1) criminalization, (2) imprisonment, (3) wealth transfer between poor communities, (4) disfranchisement, and (5) migration of state and non-state practices, policies, and capitalist ventures that all depend on carcerality as a basic state-building project. These are all forms of structural adjustment, and have interregional, national, and international consequences. In other words, if economics lies at the base of the prison system, its growth is a function of politics not mechanics.

The political geography of criminal law in the United States is a mosaic of state statutes overlaid by juridically distinct federal law. Although no single lawmaking body determines crimes and their consequences, there are trends that more than 52 legislative bodies have followed and led each other along over the past two decades. The trends center on (1) making previously non-criminal behavior criminal, (2) increasing sentences for old and new crimes, and (3) refiguring minor offenses as major ones. More than 70 percent of new convicts in 1999 were sentenced for non-violent crimes, with drug convictions in the plurality – 30 percent of new state prisoners and 60 percent of all federal prisoners (Gainsborough and Mauer 2000). Even what counts as "violence" has broadened over this period.[8] The summary effect of these trends has been a general convergence toward ineluctable and long prison terms.

The weight of new and harsher laws falls on poor people in general and especially people of color – who are disproportionately poor. Indigenous people, and people of African descent (citizens and immigrants), are the most criminalized groups. Their rate of incarceration climbed steeply over the past twenty years, while economic opportunity for modestly educated people fell drastically and state programs for income guarantees and job creation withered under both Republican and Democratic administrations (Gilmore 1998). Citizen and immigrant Latinos in collapsing primary or insecure secondary labor markets have experienced intensified incarceration; and there has been a steady increase in citizen and immigrant Asian and Pacific Islanders in prison and jail (Waldinger and Bozorgmehr 1996). Finally, at the same time that revisions to federal law have curtailed constitutional protections for non-citizens accused of crimes and for all persons convicted of crimes, immigration law has adopted criminalization as a weapon to control cross-border movement and to disrupt settlement of working people who are non-elite long-distance migrants (Palafox 2001).

Does the lawmaking and prison building fury mean there's more crime? Although data are difficult to compare because of changes in categories, the best estimate for crime as a driving force of prison expansion shows it to account for little more than 10 percent of the increase. Rather, it is a greater propensity to lock people up, as opposed to people's greater propensity to do old or new illegal things, that accounts for about 90 percent of US prison and jail growth since 1980. People who are arrested are more likely now than twenty years ago to be detained pending trial; and those convicted are more likely to be sentenced to prison or jail, and for longer terms than earlier cohorts (Blumstein and Beck 1999).

A counter-intuitive proposition might also help further understanding of why there are so many US residents in prison. The lock-up punishment imperative must be positively correlated with lock-up space. Of course, legislative bodies can make any number of laws requiring prison terms, and they can, in theory, drastically overcrowd prisons and then build new prisons to correct for non-compliance with constitutional, if not international (UN 1976), custody standards . However, if one scrutinizes the temporality of prison growth in California, the largest US state, one sees that lawmaking expanding criminalization followed, rather than led, the historically unprecedented building boom the state embarked on in the early 1980s. And the inception of the building boom followed, rather than led, significant, well-reported, reductions in crime (Gilmore 1998, forthcoming). A similar pattern holds true for the other leading prison state, Texas (Ekland-Olsen 1992; Kaplan, Schiraldi, and Ziedenberg 2000). The new structures are built on surplused land that is no longer a factor in productive activity. Virtually all new prisons have been sited in rural areas, where dominant monopoly or oligopoly capitals have either closed down or, through centralization and/or mechanization, reorganized their participation in the economy.

In search of new prison sites, state prison agencies and private prison entre-preneurs (to whom we shall return) present lock-up facilities as local eco-nomic development drivers. Recent quantitative and qualitative research in the US (Hooks et al. forthcoming; Gilmore 1998, forthcoming) demonstrates that prisons do not produce the promised outcomes for a number of reasons. New prison employees do not live in amenities-starved towns where prisons go, while 60–95 percent of new prison jobs go to outsiders. Prisons have no industrial agglomeration effects. The preponderance of local institutional purchases is for utilities which are usually extra-locally owned. Locally owned retail and service establishments such as restaurants are displaced by multina-tional chains, which drain already scant profits from the locality.

When a prison site is authorized, land values increase amid the euphoria of expected growth, but after construction values drop again. Anticipatory development – particularly new and rehabilitated housing – fails, leaving homeowners (especially the elderly) with their sole asset effectively deval-ued due to increased vacancies. Renters bear higher fixed costs because of hikes during the shortlived construction boom. As a result, prisons can ac-tually intensify local economic bifurcation.

At the same time, prisons produce a local economy dependent on con-stant statehouse politicking to maintain inflows of cash. In one mayor's words: "Beds. We're always lobbying for more beds." "More beds" means more prisoners (Huling 1999). Most prisoners come from urban areas, where the combination of aggressive law-enforcement practices (Bayley 1985) and greater structural strains (Laub 1983) produces higher arrest and convic-tion rates than in rural areas (Gilmore 1998); suburbia is following urban trends (Bureau of Justice Statistics 2000).

The movement of prisoners is, in effect, a wealth transfer between poor communities, and there isn't enough wealth in the sending community to create real economic growth in the receiving community (Huling 2000b; Gilmore forthcoming). Taxes and other benefits that are spatially allocated on a per capita basis count prisoners where they are held, not where they are from (Huling 2000a). When prisoners' families make long trips to visit, they spend scarce but relatively elastic funds in motels and eating establish-ments. Towns disappointed by the lack of prison-induced real growth con-sole themselves with these meager rewards, although modest tax subventions and families' expenditures hardly constitute an income tide to lift ships. Prisons also provide localities with free prisoner labor for public works and beautification, which can displace local low-wage workers.

Global Implications

Throughout the globalizing world, states at all scales are working to reno-vate their ability to be powerful actors in rapidly changing landscapes of accumulation. Already-existing capacities, antagonisms, and agreements are the raw materials of political renovation; embedded in renovation work, then, is the possibility (although by no means *certainty*) that already-existing frictions of distance may be intensified. The rise of prisons in the United States is a potentially prime factor in future "globalization circulation models" because prison-building is state-building at its least contested, and the US is a prime exporter of ideologies and systems. The transfer of social control methods, in times of political economic crisis, is not new. A century ago, Jim Crow, apartheid, racist science, eugenics, and other precursors to twentieth-century hypersegregation, exclusion, and genocide took ideological and material form and globalized in conjunction with technology transfers and dreams of democracy (cf. Blaut 1994).

In the current period the legitimizing growth of state social control apparatuses productively connects with the needs of those who struggle to gain or keep state power. Such political actors (whether parties, corporations, industrial sectors, or other kinds of interest groups makes no difference) are vulnerable to the arguments of private entrepreneurs and public technocrats about how states *should* function in the evolving global arena, when the norm has become neoliberal minimalism. Increased coercive control within jurisdictions is, as we have seen in the US context, one way to manage the effects of organized abandonment. At the same time, the struggle for *international* sovereignty in the context of "postcolonial" globalization can, and often does, feature a rush to institutional conformity – which today includes expanded criminalization, policing, and prisons. As a result, new or renovated state structures are often grounded in the exact same fatal power-difference couplings (e.g., racism, sexism, homophobia) that radical anti-colonial activists fought to expunge from the social order (Fanon 1961; Alexander 1994).

In other words, structural adjustment – most ordinarily associated with shifts in how states intervene in the costs of everyday-life basic-goods subsidies, wage rules, and other benefits – flags not only what states stop doing, but also what states do instead.[9] Policing and lawmaking are internationally articulated, via professional and governmental associations (see, for example, Bayley 1985), and the pressures of international finance capitalists (whether commercial or not-for-profit) seeking to secure predictable returns on investments. In short, while not all countries in the world rush to emulate the United States, the very kinds of state-based contigencies and opportunities that help explain US prison expansion operate elsewhere (see, for example, Huggins 1998; Chevigny 1995).

US prison expansion has other broad effects. While most US prisons and jails are publicly owned and operated, the trend toward public service privatization means firms work hard to turn the deprivation of freedom for 2,000,000 into profit-making opportunities for shareholders. Success rates differ across jursidictions, but privatized market share, currently about 6 percent, grew 25–35 percent each year during the 1990s (Greene 2001; Austin and Coventry 2001). The largest firms doing this work also promote privatization in such disparate places as the United Kingdom, South Africa, and Australia (Sudbury 2000).

Public and private entities package and market prison design, construction, and fund-development; they also advocate particular kinds of prison-space organization and prisoner management techniques. The "security housing unit" (SHU), a hyper-isolation "control unit" cell condemned by international human rights organizations, is widely used in the United States. The US imported the SHU from the former West Germany, which developed it as a death penalty surrogate to destroy the political will and physical bodies of radical activists. The US has both the death penalty and the SHU, and promotes control units abroad (Davis and Gordon 1998). At the end of 2000 more than 10,000 prisoners throughout Turkey participated in a hunger strike to protest spatial reconfiguration from dormitories to cell-based "American"-style prison, with a particular focus on the punitive SHU (Prison Focus 2001).

Exported structures and relationships can take the form of indirect as well as deliberately patterned effects. In addition to the transfer of wealth between poor places, prison produces the political transfer of electoral power through formal disfranchisement of felons. While elections and politics are not identical, the power to vote has been central to struggles for self-determination for people kept from the polls by the frictions of terror and law throughout the world. In the United States Black people fought an entire century (1865–1965) for the vote. As of 1998, there were nearly 4 million felony-disfranchised adults in the country, of whom 1.37 million are of African descent (Fellner and Mauer 1998). The voter effect of criminalization returns the US to the era when white supremacist statutes barred millions from decision-making processes; today, lockout is achieved through lock-up.

The 2000 US presidential election, strangely decided by the Supreme Court rather than voters, was indirectly determined by massive disfranchisement. George W. Bush Jr. won Florida, and therefore the White House and the most powerful job on the planet, by fewer than 500 votes. Yet 204,600 Black Floridians were legally barred from voting; additionally, many others of all races who tried to vote could not because their names appeared on felon lists. Had felons not been disfranchised, candidate Bush would have lost; however, candidate Albert Gore's party shares equal responsibility with Bush's for creating widespread disfranchisement, and could not protest on that front. Thus, the structural effects of racism significantly shape the electoral sphere with ineluctably global consequences for financial (G8), industrial (WTO and GATT), environmental (Kyoto), and warfare (NATO; Star Wars) policies.

Conclusion

As exercised through criminal laws that target certain kinds of people in places disorganized by globalization's adjustments, racism is structural – not individual nor incidental. The sturdy curtain of US racism enables and veils the complex economic, political, and social processes of prison expansion. Through prison expansion and prison export, both US and non-US racist practices can become determining forces in places nominally "free" of white supremacy. Indeed, as with the twentieth, the problem of the twenty-first century is freedom; and racialized lines continue powerfully, although not exclusively, to define freedom's contours and limits.

Acknowledgments

The author thanks Craig Gilmore, Lauren Berlant, Rachel Herzing, and the editors of this volume for their thoughtful criticism of earlier drafts.

Notes

1   Some theorists prefer the plural – racisms – to underscore how there is not a single universal practice. I use the singular because racism, like other forms of violence, tends to produce the same outcomes regardless of technique: premature death and other life-limiting inequalities.

2   The articulations of race and space – as and through multiscalar hierarchies of colonialism, slavery, and other relations of unfreedom – are more evident in some contexts than in others. For some examples of how race becomes both amplified and entrenched, see CCCS (1982) and Mitchell (2000).

3   Susan Christopherson's chapter on gender (chapter 15, this volume) provides an exemplary chart for doing this exercise on a global scale.

4   I use "friction of distance" to theorize the metaphorical and material drag coefficients that differentially impede the movements of people, things, relationships, and ideas across geometric as well as social space. See Isard (1956) for his thoroughly *unmetaphorical* introduction of the term as the regional science's key revision of neoclassical economics.

5   2,000,000 does not include persons detained with or without charge by the US Immigration and Naturalization Service.

6   While there was plenty of racism and sexism outside the south and west, the structure of New Deal social welfare programs equalized across a differentiated landscape a series of perspectives about eligibility, need, and merit that became common sense (see, for example, Mink 1995)

7   About 65 percent of US households are owner-occupied. When the data are broken down by race, we see a different picture: for example, only about 45 percent of Black households are owner-occupied, because of federally mandated racist lending criteria as well as lower-than-average incomes (Massey and Denton 1993; Oliver and Shapiro 1995).

8   The meaning of violence used to define racism in this chapter (see note 1) is far narrower than the meaning of violence used by current lawmakers to expand punishment.

9   Rarely, if ever, does a delegitimated state, or state-fraction, simply disappear.

Part IV
Geocultural Change

Introduction to Part IV:
Modernity, Identity, and Machineries of Meaning

> I am prepared to take off my ski mask if Mexican society will take off its own mask. Both would show their faces but the great difference will be that Marcos always knew what his face was really like, and civil society will wake from a long and hazy dream that "modernity" has imposed on it at the cost of everything and everyone. (Subcomandante Marcos, the masked rebel leader of the Mexican Chiapas rebellion, Zapatist National Liberation Army, January 1994)

On January 1, 1994, the day the North American Free Trade Agreement (NAFTA) sent trade barriers tumbling between Mexico, Canada, and the US, a guerrilla group was seizing half a dozen towns in Chiapas, southern Mexico. Mr. Salinas's economic restructuring policies had widened already huge economic disparities; 40 percent of the Mexican population was said to be poor or extremely poor. According to the Zapatist guerrilla leader, Subcomandante Marcos, NAFTA represented a "death sentence" for the poor.

On the other side of the Atlantic, Radoslav Unkovic, director of the Institute for the Protection of Cultural, Historical, and Natural Heritage in the Bosnian Serb government, was in charge of the current linguistic purge. Even the name "Bosnia" is a target for linguistic cleansing since it has become, in the course of the break-up of Yugoslavia, a symbol of Turkish and Austro-Hungarian oppression. "By eliminating this term," says Unkovic, "we are eliminating the memory and the consequences that stem from it" (*San Francisco Chronicle*, May 2, 1993). Among the casualties, he continues, are those names which "in an ethnic–linguistic sense do not correspond to Serb traditions." Names disappear from maps, street signs, dictionaries, travel directories, and encyclopedias. This erasure – the annihilation of meaning on the basis of a putative tradition – presumably extends to Radoslav Unkovic himself, since his own name derives from the Turkish word "un" meaning forward.

Cultural Globalization of Difference

The horrors of the Balkan wars and the chimerical presence of Subcomandante Marcos pose quite sharply the paradox of the late twentieth century: to wit, the collapse of many forms of existing socialism and the radical shift toward market integration under the auspices of free trade and heavy-handed interventionism by global regulatory institutions such as the World Bank and the IMF. What has been called market triumphalism has not produced the End of History, as Francis Fukuyama (1992) would have us believe – the final triumph, in other words, of liberal democracy. Rather, we have seen the proliferation of militarized ethno-nationalisms and other forms of illiberal politics whether radical Hindus in India, Zulu secessionist movements in South Africa, or the Islamic Front for Salvation in Algeria – what Ben Anderson has properly called the New World Disorder (B. Anderson 1992). Against the backdrop of the instabilities of the nation-state, and the erosion of national economies and identities in the face of unprecedented globalization, the post-Cold War period is both a complex and a dangerous moment (Hall 1991a). In the vacuum created by the demise of Stalinism and the undermining of state-led modernization, new sorts of territorial place-based identities spring up which, when conflated with racial, ethnic, religious, or class-based identifications, can be "one of the most pervasive bases for both progressive political mobilization and reactionary exclusionary politics" (Harvey 1993: 4). Imagined communities of various stripes – Iran-ian Muslims in Houston, Israeli settlers in the West Bank, Punjabi separatists in India, Palestinians in Gaza, fundamentalist Christians in Waco, Texas – all lie at the intersection of, and in some way are products of, local and global forces. Put differently, the late twentieth century witnessed not only radical economic restructuring at the hands of globalized markets and flexible accumulation, but also a globalization of culture – that is to say, publicly meaningful forms – and new patterns of cultural complexity, diffusion, innovation, and polycentralism (Hannertz 1991). Mapping this cultural cosmopolitanism is key to understanding the contours of the New World Disorder and the variety of new social movements and postmodern alternatives to development and modernity.

The specifically "cultural" content of global culture refers to both meaningful social forms and the means by which they are rendered meaningful, recognizing that both forms and means are in some fundamental sense global. Ideas and modes of thought, the forms in which meaning is made accessible, and the way in which the cultural inventory is distributed, have globalized aspects and features (ibid: 6–7). The Muslim diaspora, for example, is worldwide, interconnected through media and global networks (including the hadj) that link disparate communities in Houston, Oxford, and Kuala Lumpur; some of the most active Punjabi nationalists are resident in Canada, an instance of what Ben Anderson calls "fax nationalism." All states participate in the creation of a public culture in which the cultural materials and cultural capitals are drawn from a global system of signification. Adeline Masquelier (1992) illustrates this point in her description of how a Dunlop tire advertisement featuring a European woman was converted into a road siren – a peculiar conflation of local spirit traditions and Western advertising imagery – which spoke to the contradictions experienced by rural people in postcolonial Niger. As people make history they do so within a historically shaped imagination and often with what Taussig (1993) calls mute things that are given human significance. These mute things, however, now have a global reach and distribution, facilitated by the machinery of global media and transnational capital. All countries create their own modernity, as Paz (1992) says, but draw upon local and global meanings to fashion it, while specific constituencies participate in and contest its creation.

The machineries of meaning, and the global media in particular, are instrumental in the distribution and externalization of cultural meanings, and this is a process which is aided and abetted by the extraordinary mobility of people. Cities, and certain world cities such as London, São Paulo, and New York, emerge as centers of extraordinary cultural diversity, as loci in the world system of signs, and as sites of what one might call a critical cultural cosmopolitanism. Even Tokyo, long distinguished by its cultural closure, has working-class concentrations of Chinese, Bangladeshis, Koreans, and Filipinos. But the fluidity of people and culture signals not so much the Frankfurt School nightmare of cultural homogenization – a sort of erasure of difference on the grounds that everyone watches *Ally McBeal* and *Baywatch*, drinks Pepsi, eats a Big Mac, and worries about Michael Jackson's personal life – as much as it signals a destabilization of sites and places, in which hybridity and the refashioning of new identities seem to proliferate: "Not only had the border [between the West and the Rest] been punctured porous by the global market [and capital and migration] . . . but the border has dissolved and expanded to cover the lands it once separated such that all land is borderland" (Taussig 1993: 248–9).

A world of flows and hypermobile capital does not so much produce powerless places and unremitting deterritorialization, but rather results in the endless production of difference with a complex and unstable globality. Cultural complexity stands less as a monument to the final annihilation of space than as a testament to the complex mappings between force fields of local and global provenance.

World Cities

The rapid growth of many world cities at various levels in the hierarchy of economic and political power has been associated with growing inequalities among their populations. The growth of many of those cities has been achieved through immigration, at two levels in the occupational status and reward system. At the highest levels, professional and managerial classes have been attracted to the opportunities offered in the vast concentrations of economic activity, and the associated rewards; and at the lowest levels, migrants (most of them from different origins than the other groups, and a not inconsiderable number of them illegal) have been drawn to the low-paid, insecure service industries. The economic polarization within world cities is thus associated with an ethnic fragmentation, in what some have termed EthniCities. Over one-third of the population of cities like Los Angeles and Sydney were born outside their country of residence, for example; 60 percent of Miami's population reported speaking another language than English at home; less than 45 percent of Auckland's population identified themselves ethnically as New Zealand Europeans (with another 15 percent identifying as New Zealand Maori); and one quarter of all London residents claimed a non-UK ethnic identity (even though many of them were born there). The political significance of such ethnic communities and the demographics of their growth and decline cannot be overestimated. The Spanish-speaking community in the US (and in California in particular) is a case in point. It has been estimated that President Bush will have to double his "Hispanic vote" in order to be re-elected in the next presidential election, while Los Angeles mayoral politics is now inexplicable outside the growing political enfranchisement of what has become the ethnic majority.

Ethnic variety is not new to expanding cities in many parts of the world, of course, but the social and spatial processes operating within them have been changing. Until recent decades the dominant attitude was of assimilation or separation. Over time, the members of most ethnic immigrant groups and their descendants were expected to improve their life chances through educational and economic advancement, and then become culturally and socially assimilated to their new homeland's society as well – with such cultural assimilation (and the removal of differences between hosts and new arrivals) leading to spatial assimilation. Initially, many of the ethnic minority group members were constrained to living in low quality housing areas by a combination of their low incomes, their desire to maintain social and community ties with their co-ethnics, and discrimination against them by the host society (which may have encouraged spatial distancing on both sides). As disadvantages disappeared and familiarity led to reduced discrimination, so the newly assimilated found a wider range of housing areas available to them, and they became spatially dispersed. Some groups were excluded from this process, however, preferring to remain culturally and spatially separate despite improvements in their economic status – as occurred with Jewish populations in many cities. Others continued to suffer disadvantage and discrimination and continued containment in ghetto-like enclaves – as was the case for African-Americans in US cities, as well as for all races under the apartheid policies of South Africa until that regime was discontinued in 1994.

The "melting-pot" process is particularly associated with American cities in the twentieth century, where immigrant assimilation was the desired outcome, and separate residential enclaves were considered temporary phenomena (apart from the situation of African-Americans). By the end of the twentieth century, however, and the increased ethnic complexity of many societies, multiculturalism became the declared public policy. Economic assimilation did not have to lead to cultural assimilation too, and various ethnic groups (if they or at least some of their members wished) retained their cultural identity in a pluralistic situation where no one ethnic group predominated. Whether such groups lived in spatially separated areas was a matter of choice for them; they may choose to maintain their social ties while becoming spatially dispersed (what Zelinsky terms heterolocalism), or they may prefer to concentrate in particular areas to facilitate both formal and informal social interaction. In large cities, in particular, it seems that spatial dispersion has proceeded furthest: there is much less spatial segregation of ethnic minority groups in London than other English cities, for example, and in Auckland than in smaller New Zealand towns.

Voices from the Margin

All of this returns us to territorial and other sociocultural identifications. The effort to erase place, community, and memory in Bosnia by ethnic and linguistic cleansing does not reflect irreducible ethnic hatreds; indeed the history of killings really only began in 1928. For war to occur in the context of the collapse of the Soviet Union, nationalists speaking the language of self-determination had to convince neighbors and friends that they had been killing one another for millennia. The events of the 1980s "turned the narcissisms of minor differences into the monstrous fable that people on either side were genocidal killers" (Ignatieff 1993: 3). It is this rewriting of history, the reinvention and reimagining of cultural traditions, that runs through the Balkans, not eternally fixed identities from some essentialized past. It is the continual play of history, culture, and power in an increasingly globalized world capitalist system.

If the border and the hybrid are the metaphors for late capitalist culture, this is no less the case for the knowledges by which capitalist modernity is interpreted and understood. The challenging of metanarratives and of totalizing histories has come from many quarters (Gregory 1994). But it is perhaps entirely appropriate that the crisis of representation has above all emerged from those voices who in some way stand at the margins – feminisms of various strands have drawn self-consciously from this marginality – and more often than not at the borderland of First and Third Worlds. It is from the vantage point of the postcolonial subject, and from cultural studies more generally, that so much of the ferocious attack on the West has emerged.

The West is itself a difficult and ambiguous category in this regard, since its very existence is rooted in the creation and history of the Rest. Whether the celebration of fragmentation, difference, and marginalization that accompanies much of this critique, and its abandonment of a sense of totality or system, is to be uncritically welcomed is a matter for debate, of course, but the impact of the subaltern critique from within the belly of "big science" and "grand theory" has been unmistakable. Moreover, to take the case of the critical rejection of development as a trope for those parts of the world that are poor and peripheral, the postmodern assault reflects a critical engagement between the ideas of intellectuals and the political practice of new social movements. For some, these popular movements, often emerging from the ashes of Draconian structural adjustment during the 1970s and 1980s, are held aloft as the bearers of new sorts of politics, new sorts of subjectivities, and new sorts of democratic practice, all socially embedded and expressed through local cultural traditions (Escobar 1992a). Whether grassroots movements around dam construction in Brazil, in the barrios of Rio, or in the name of eco-feminism in India do actually contain the potential for feasible alternatives to development, any more than Derridean deconstruction represents the means to refigure human geography, strikes to the heart of contemporary debate over theory and, one should add, *realpolitik* outside the academy. In all of this, culture has returned to center stage. No longer an epiphenomenon or a residue of political economy, the realm of meanings – their construction, contestation, and interpretation – stands at the very center of the reworking of *fin de siècle* modernity (Pred and Watts 1993).

18 Consumption in a Globalizing World

Peter Jackson

From the moment we wake up each day, our lives are characterized by choices over what to wear, what to eat, and how to spend our leisure time. Such choices are, of course, not equally available to everyone and the geography of contemporary consumption includes some stark variations across the globe. In India, for example, less than 8 percent of households own a colour television, fewer than 9 percent have a refrigerator, and less than 5 percent own a washing machine. While it would take a middle-class Indian over two months' earnings to buy a colour television and over eight years' earnings to buy a medium-sized car, a middle-class American could acquire the television set with just over two days' earnings and the car in a little over four months (Shurmer-Smith 2000: 28–32). As these examples show, even at the beginning of the twenty-first century the world is still characterized by stark inequalities of poverty and wealth, of hunger and malnutrition in some places and super-abundance in others, of extravagance and waste amid scarcity and need.

This chapter attempts to make sense of the changing geographies of consumption, not just in terms of such social and spatial inequalities (important though they are). Rather, it approaches the geographies of consumption through a series of debates that require connections to be made between people and places at a variety of scales, from the global to the local. The chapter begins by establishing a theoretical approach to consumption as an intensely practical activity, grounded in the politics of everyday life. It goes on to examine the geography of three specific commodities and some of the issues that surround their consumption. Such an approach takes us beyond the "tyranny of the single site" (Jackson and Thrift 1995) – whether focused on the shopping mall or the world's fair, the department store or the car boot sale – to explore the links that connect people and places through complex chains and circuits whose contours have only recently begun to be traced.

Theorizing Consumption

Though often represented in a very academic language – of simulations and simulacra (Baudrillard 1988) involving complex processes of externalization and sublation (Miller 1987) – consumption is an intensely practical activity, grounded in the mundane realities of everyday life. The fact that such a diverse range of activities as watching television, listening to music, going shopping, surfing the internet, or visiting a pub can be lumped together under the umbrella term "consumption" is part of the problem. The term itself encompasses the purchase of material goods (such as food ingredients, recipe books, and cooking utensils), what we do with those goods once they have been purchased (preparing and eating a meal), as well as the wide range of symbolic meanings that attach to notions of culinary culture (whether "eating in" at home or "eating out" at a restaurant, zapping an individual ready-made meal in a microwave, cooking for the family, or preparing an elaborate meal for a dinner party). The meaning of particular consumption practices clearly varies across time and space, and has a complex relationship with aspects of our personal and social identity. As Bell and Valentine (1997) remind us: we are *where* (and when) we eat as well as *what* we eat. Besides the actual consumption of particular goods, we can also consume their meanings independently from their material form, as when we enjoy an advertisement with no intention of buying the product, or when a bottle of perfume is marketed through words, music, and visual images without any direct reference to our sense of smell.

As all these examples suggest, consumption involves much more than a single, isolated act of purchase: the moment when money is exchanged for a specific commodity, extending "forwards" into cycles of use and re-use and "backwards" into the social relations of production (Jackson 1993). Indeed, the "act" of consumption need not involve any kind of purchase at all, as the practice of "window shopping" clearly demonstrates. Consumption is as much about looking, comparing, and desiring as it is about buying, exchange, and possession (Gregson and Crewe 1997).

As consumers we are engaged, in often complex ways, with a world of ideas and symbolic meanings as well as with specific material goods. We "make sense" of particular commodities through a web of historical associations and geographical knowledges (Cook and Crang 1996). Consumption is a social accomplishment, an active process involving practical and discursive skills (searching for a "bargain," knowing what represents "value for money," assessing "quality" or appropriateness, etc.). And yet the very terms of our engagement as "consumers" imply a lack of agency, signaling a world where consumers respond passively to market forces, as though consumption were an anonymous and disembodied process, unshaped by our classed, gendered, and racialized subjectivities. The recent history of consumption research has therefore seen a move away from semiotic readings of shopping malls and other sites of "heroic consumption" where the subjective meaning of particular environments and commodities is largely inferred from external visual evidence, towards more ethnographic or experiential accounts of how consumers actually engage with particular consumption sites and material goods (cf., Goss 1993; Ley and Olds 1988; Falk and Campbell 1997; Miller et al. 1998).

As these preliminary thoughts suggest, we need a more nuanced geography of consumption, establishing its complex meanings for differentially positioned and embodied consumers (cf., Valentine 1999). For consumption is deeply embedded in the social relations of family life, rooted in ideologies of gender and generation, "race" and class, that are effective across a range of geographical scales from the neighborhood to the nation, from the body to the globe. Conventional views of consumption are often underpinned by a series of unhelpful dualisms that divorce production from consumption, "the economic" from "the cultural," the global from the local, social structure from human agency. Transcending these dualisms – by focusing on networks and associations between people and things, discourses and practices, material goods and their symbolic meanings – is a central theme of current research on the geography of consumption (as reflected in this chapter).

Consumption and Everyday Life

To say that consumption is rooted in everyday life does not mean that it is devoid of any wider moral or political significance. Providing "proper" meals for the family or buying "environmentally friendly" products are examples of everyday practices whose meanings extend far beyond the basic provisioning that is necessary to sustain our lives. Focusing on the geography of everyday life is a good way of extending our understanding of the politics of consumption beyond such self-conscious forms of "ethical consumption" as buying fairly traded coffee or supporting consumer boycotts of goods that are produced under oppressive conditions.

An emphasis on the politics of everyday life also helps to overcome the more pessimistic accounts of recent social change which link the globalization of production with the homogenization of culture. While phrases like "Coca-colonization" and "McDonaldization" have become shorthand ways of describing the flattening out of geographical differentiation across the globe, local cultures of consumption can still thwart the power of even the most global corporations, as demonstrated by advertising campaigns which fail to take such differences into account (Jackson and Taylor 1996). For, as Michael Watts reminds us, while it might appear that everyone now watches Oprah and drinks Coca-Cola:

> Globalization does not so much mark the erasure of place [as] in a curious
> way contribut[ing] to its revitalization . . . Globalization here implies less the
> erosion of place than a sensitivity to how location, identity, and community
> are refashioned in incompletely globalized sites. (Watts 1996: 64–5)

A different view of globalization emerges from those who resist a worldview
where cultural creativity is uniquely located in the West, with the rest of the
world cast in the role of passive consumers. Whether one thinks of Bollywood
movies or Japanese sushi, of Brazilian football stars or cheap clothing from
the Far East, it can now be confidently asserted that

> World systems, regarded especially from the cultural point of view, now emerge
> as much from Bombay, Tokyo, Rio de Janeiro and Hong Kong as they do
> from Los Angeles, New York, London and Paris. (Appadurai and Breckenridge
> 1988: 2)

Though, as David Morley (1992) warns, such a view runs the risk of
confusing consumer agency with consumer power, it highlights the need
for a more subtle understanding of the continued salience of local contexts of consumption, rather than an uncritical emphasis on the power
of global corporations such as McDonald's or Coca-Cola (cf., Gillespie
1995).

In order to trace the politics of everyday consumption in more detail,
from the scale of global corporations to the local level of individual lives,
this chapter focuses on three specific commodities: bananas, sugar, and soap.
These examples encourage us to think in new ways about the politics of
place through circuits and networks of connection, and to reconsider the
scope for consumer agency in a globalizing world.

Commodity Cultures

The fruits of Third World labor

We start with a particularly well-worked example, the humble banana, as a
way of highlighting the connections between "Third World" producers and
"First World" consumers. Though they are often disguised in the commodity form, such connections are readily established, as Susan Willis (1991:
50) outlines:

> From southern California and Florida to the tropical zones of Mexico, Cen-
> tral America, Colombia, and Ecuador, the produce department [of American
> supermarkets] features the fruits of the Third World, whose only acceptable
> attribute of tropicality is color . . .
>   Maintained in a constant bath of refrigerated air, these fruits are incapable
> of producing scents, harboring bugs, growing molds, and becoming decayed.

Air-conditioning is the medium of abstraction which severs the agricultural production of the Third World from the heat of labor and the heat of the marketplace. It swaddles the product in First World antiseptic purity and severs its connection with the site of production.

In this account, consumers are assumed to be entirely unaware of the labor involved in getting the fruit onto the supermarket shelves. Consumption creates "the illusion of difference" (Costa Rican or Colombian bananas are virtually indistinguishable in taste) while the "real difference . . . is located in production" (ibid: 52) .

Cynthia Enloe (1989) paints a more subtle picture of the complex relationships between production and consumption in her study of *Bananas, Beaches and Bases*. For her, too, the banana has a history, though in her case it is a specifically gendered history. In an essay that spans the globe, Enloe goes on to show how bananas originated in Southeast Asia and were carried westwards by traders to become a staple food for Africans on the Guinea Coast by the fifteenth century. Portuguese and Spanish slave-traders introduced the red banana to the West in the eighteenth century, the now more familiar yellow banana only being introduced from the Caribbean in the early nineteenth century. Enloe shows how the banana became an exotic delicacy in the homes of wealthy Bostonians in the late nineteenth century, coming to symbolize America's new global reach at the US Centennial Exhibition in Philadelphia in 1876. The United Fruit Company introduced their cartoon character Chiquita Banana in 1944, effectively creating brand loyalty for a previously unmarked "natural" product. Later adverts played on the "exotic" appeal of Latin-American sex symbol Carmen Miranda, confirming Enloe's argument that "Gender is injected into every Brooke Bond or Lipton tea leaf, every Unilever or Lonrho palm-oil nut, every bucket of Dunlop or Michelin latex, every stalk of Tate & Lyle sugar cane" (ibid: 134). As banana production developed on a global scale, the derisive label "banana republic" came to describe the effects of undue foreign influence over the political independence of plantation economies. Banana production has continued to be a source of political conflict to the present day, as the recent trade wars between American and (British) Commonwealth exporters confirm.

The sweet and sour history of sugar

An even stronger case for the political significance of apparently mundane commodities can be made in the case of sugar – described by Willis (1991: 133) as "the alimentary chemistry of colonialism." For the growth of consumer demand for sugar on both sides of the Atlantic in the eighteenth and nineteenth centuries is inseparable from the history of colonial production, where the plantation mode of cultivation encouraged slavery and oppression on a global scale. The rigors of plantation work are vividly described in Sidney Mintz's study of sugar production in Puerto Rico:

> From a distance, the scene is toylike and wholesome. Up close it is neither. The men sweat freely; the cane chokes off the breeze, and the pace of cutting is awesome. The men's shirts hang loose and drip sweat continuously. The hair of the cane pierces the skin and works its way down the neck. The ground is furrowed and makes footing difficult, and the soil gives off heat like an oven . . . The men of Jauca [where Mintz's ethnography was set] grow drawn in the first two weeks of the harvest . . . It's a way of life that can make menial jobs in the continental United States seem like sinecures. (Mintz 1960: 21)

In a later book, Mintz documented the place of sugar cane in modern world history. The book begins with the following quotation from J. H. Bernardin de Saint Pierre's *Voyage to Isle de France, Isle de Bourbon, The Cape of Good Hope* . . . from 1773:

> I do not know if coffee and sugar are essential to the happiness of Europe, but I know well that these two products have accounted for the unhappiness of two great regions of the world: America has been depopulated so as to have land on which to plant them; Africa has been depopulated so as to have the people to cultivate them. (Quoted in Mintz 1985: frontispiece)

Mintz goes on to demonstrate how the history of the Caribbean, in particular, has always been entangled with the wider world, "caught up in skeins of imperial control, spun in Amsterdam, London, Paris, Madrid, and other European and North American centers of world power" (ibid: xv–xvi). At the time of his first fieldwork in Puerto Rico in the late 1940s, the island had already been producing sugar cane for four centuries, but always for an overseas market:

> . . . whether in Seville, in Boston, or in some other place. Had there been no ready consumers for it elsewhere, such huge quantities of land, labor and capital would never have been funneled into this one curious crop, first domesticated in New Guinea, first processed in India, and first carried to the New World by Columbus. (Ibid: xviii–xix)

The apparently innocent taste for sweetness among Western consumers clearly has global repercussions. Similar arguments about the "outside" history that lies "inside" such commodity cultures are made by Stuart Hall in his memorable deconstruction of the English cup of tea:

> People like me who came to England in the 1950s [from the Caribbean] have been there for centuries . . . I am the sugar at the bottom of the English cup of tea. I am the sweet tooth, the sugar plantations that rotted generations of English children's teeth. There are thousands of others . . . that are . . . the cup of tea itself." (Hall 1992: 48–9).

These "hidden histories" within the commodity form resurface from time to time in small-scale protests which hint at much wider global connections. A recent example is Annie Lovejoy's "stirring@the international festival of the sea" project in Bristol which drew attention to the human costs of slavery. In an event (the International Festival of the Sea) that had otherwise been remarkably silent about the city's involvement in the transatlantic slave trade, Lovejoy's project – described in more detail by Nash (2000) – made available to the public at local cafes and restaurants a series of specially designed sugar packets, each bearing a triangular symbol (to remind consumers of Bristol's role in the "triangular trade" of slavery) and a list of commodities (sugar, tobacco, cocoa, tea, spices, rum, slaves . . .), inviting those who stirred the sugar into their tea or coffee to reflect on the circulation of commodities and human capital that lay at the heart of Britain's maritime trade. Such examples provide a telling critique of more celebratory models of (British) multiculturalism which tend to emphasize social and cultural diversity rather than political and economic inequalities.

Soft-soaping the Empire

Our final example involves similarly far-flung geographies and historical interconnections. The history of soap dates back to Roman times when it was made from soda ash brought over from Egypt and the Far East. Large-scale production followed the discovery of how to produce soda from common salt in 1790. Fats and tallows, imported from Russia and Australia, were also used, later substituted by palm oil from West Africa, coconut oil from the South Sea Islands, Ceylon (Sri Lanka), and the East Indies. By the early nineteenth century giant chemical concerns like Unilever and ICI (Imperial Chemical Industries) were manufacturing soap on a massive scale, exploiting Britain's imperial connections with West Africa.

These historical geographies of production are paralleled in equally complex geographies of consumption. As Timothy Burke's (1996) fascinating study of soap in Zimbabwe demonstrates, consumers played an active role in the commodity's history. Commodities were not simply imposed through colonial domination but were actively desired. The history of soap is a particularly good example of the development of modern consumer culture because of its close connection to the body and to the making of human subjectivities. Its place in postcolonial Zimbabwe must be located within a longer history of colonial ideals about hygiene, domesticity, and manners, connected to wider discourses of "civilization," and specifically associated with the production of cleanliness, health, and racialized bodies. As late as the 1940s, for example, advertisements for Atlas Soaps were couched in the highly racialized language of cleanliness and civilization:

> Nature's greatest gift is perhaps water – and one of civilization's greatest gifts – SOAP. The day of the witch-doctor's craft and all its evils are over. Today educated Africans know that disease is spread by germs – and that germs live in dirt. Be educated – be healthy – keep your body, clothes and house clean by regular washing with soap and water. If you are wise you will make sure it is ATLAS Soap – the Best. There is an Atlas Soap for each and every use. Ask for it by name – ATLAS SOAPS. (Quoted in Burke 1996: 151)

As so often, too, the discourses of race and gender are mutually constitutive, with firms like Lever Brothers (who arrived in Zimbabwe in 1943) advertising Lifebuoy as a strong soap, suited to particularly dirty bodies (and hence connected to discourses of masculinity and blackness), while Lux was pitched at African women, signifying "the definitional essence of glamour and 'smart' living" (ibid: 155). While male consumers were lured by the promise that "Successful Men Use Lifebuoy," African women were reminded that "thousands of beautiful women have won lovely complexions from using Lux Toilet Soap":

> Lux Toilet Soap is pure, you can see that because it is white. Lux Toilet Soap has a rich creamy lather that makes your skin soft and smooth, beautiful to look at. Lux Toilet Soap is the simple secret of beauty. (Ibid: 156)

Burke also demonstrates some of the creative appropriations that these new products were subject to, using toothpaste as a skin medicine, for example, or soap as fish bait. Such dynamic local contexts of consumption need to be emphasized if we are to avoid the image of "Third World" consumers as purely passive victims of Western exploitation or hapless colonial subjects who simply mirrored Western consumer desires.

Establishing such connections lie at the heart of Anne McClintock's (1995) recent study *Imperial Leather*. In sketching the history of soap advertising, McClintock demonstrates the connections between representations of race in distant parts of the British Empire and discourses of domesticity "at home" in Britain: "The first step towards lightening the White Man's Burden is through teaching the virtues of cleanliness" (1899 advert for Pear's Soap). Victorian advertisements show the fetishized commodity's magical ability to whiten black skin and to domesticate man's animal nature (most notably in adverts for Monkey Brand Soap). Ever since Pear's appropriated Sir John Everett Millais's "Bubbles" painting ("A Child's World"), soap advertising has transgressed the boundaries between culture and commerce, most notably in the birth of "soap opera," originally designed to hold the attention of American housewives between successive commercial breaks. McClintock shows how the bourgeois Victorian obsession with cleanliness (at home) was connected to the development of imperial commerce (abroad). Its legacy can still be seen today in popular associations between "foreignness" and ideals of purity and pollution (articulated through racist ideas about "foreign muck" and "dirty foreigners," for example).

Chains, Circuits, and Networks

As the previous examples suggest, the study of commodity culture helps us to connect the geography of consumption back into the social relations of production as well as helping overcome the previously mentioned "tyranny of the single site." These connections can also be explored by tracing the geography of specific commodity chains. There are many examples of such an approach in the current literature, ranging from Brad Weiss's (1996) analysis of the multiple "breaks" and "connections" between the experiential worlds of Tanzanian coffee producers and European consumers, to Paul Robbins's (1999) account of the cultural politics of the Indian meat economy at various points along the commodity chain, from the shrublands of rural Rajasthan to the fast-food markets of urban Delhi.

A particularly good example of the strengths and possible limitations of this kind of approach is Elaine Hartwick's (1998) analysis of the commodity chain involved in the production and consumption of gold. Her discussion begins with an analysis of the images of men and women in jewelry advertisements, tracing the commodity chain back to jewelry factories in Italy and gold mines in South Africa. But the connections do not stop there, as Hartwick traces the supply of labor in South African gold mines to its source in Lesotho where male migrants leave behind "gold widows" whose plight she compares with the affluence depicted in Tiffany's "gold windows" in New York City. Hartwick argues that connecting the two ends of the chain might lead to a new "commercial geography,"' fusing a materialist approach to the geography of production with a more cultural understanding of the symbolic meanings involved in consumption. My own emphasis would be rather different. Rather than seeing commodity chains in such linear terms, as "two ends" of a single strand, I would urge a more holistic understanding of the multiple circuits and complex networks involved in contemporary commodity culture (Jackson 1999).

Robert Goldman's and Stephen Papson's (1998) analysis of the "commodity circuitry" involved in the production and marketing of Nike sports shoes provides a useful guide. While design, research, and development are carried out by Nike Inc. in Beaverton, Oregon, the manufacturing process is subcontracted to Taiwanese and Korean producers. Various trading companies in Japan and the newly industrialized Asian countries handle tariffs, customs, and transportation, while Nike's advertising agency (Weiden & Kennedy) is based in Portland, Oregon. Distribution involves a range of outlets, including franchises such as Footlocker and Footaction as well as the company's own dedicated Niketown stores. Endorsements are sought from international athletes such as the legendary Chicago Bulls basketball player Michael Jordan, while their customer base includes athletes and teenagers across the world. As Goldman's and Papson's attempt to map these links diagrammatically suggests (see figure 18.1), this is no single-stranded "commodity chain" with a clear point of origin and a finite point of sale.

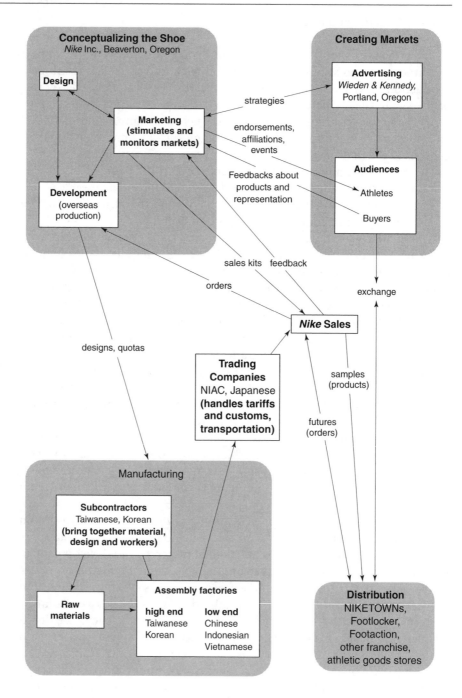

Figure 18.1   "Commodity circuitry": the production and marketing of Nike sports shoes.

Rather, it involves a series of feedback loops and interconnections best captured in the language of (complex) networks and (leaky) circuits.

In their recent work on the home furnishings industry, Deborah Leslie and Suzanne Reimer (1999) make a similar case for "spatializing" commodity chains. They note that different products have very different spatialities (contrasting goods such as clothing and food that are closely articulated with the body, with other goods such as furniture that are closely related to the space of the home). Specific products also occupy a different place in the geographical imagination of consumers in different national contexts (where, for example, consumer knowledge about the environmental origins and domestic production of furniture in Canada contrasts markedly with the much sketchier knowledge of British consumers). Leslie and Reimer endorse the value of a commodity chain approach in helping to bridge the gap between materialist analyses which often ignore the symbolic meaning of goods, and culturalist writings which often fail to engage with questions of poverty, exploitation, and differential opportunities for consumption. But they also have some reservations. Contrasting the "vertical" logic of Fine's and Leopold's (1993) "systems of provision" approach, which emphasizes the ways in which production and consumption are vertically integrated in the provision of specific commodities, with an emphasis on "horizontal" factors (such as gender and place) which operate across commodity-specific systems of provision, Leslie and Reimer emphasize the multiple and shifting connections between sites and the "leakiness" of commodity chains (also noted by Glennie and Thrift 1993). So, for example, there are growing links between the marketing of furniture and clothing, acknowledged in "lifestyle" magazines such as *Elle Decoration* and *Wallpaper*, which suggests that too stark a separation between different commodity forms may be misguided.

In contrast to the finite conception of chains, which have a clear beginning and end, notions of circuits and networks have proved increasingly popular, shifting analysis away from the often futile search for the origin of specific commodities from which all subsequent departures are regarded as increasingly "inauthentic." As Cook's and Crang's (1996) work on culinary culture clearly demonstrates, debates over authenticity and depth are rarely helpful in the case of commodities like food, whose material and symbolic geographies are better approached in terms of juxtaposition and displacement. Their work provides a valuable extension of the commodity chain approach to include the circulation of consumer knowledges and geographical imaginaries, rather than simply "unveiling" the conditions of production that are "masked" in the process of commodification.

Notions of "unveiling" or "unmasking" derive from an understanding of the fetishized nature of capitalist production whereby material objects and social relations are transformed into commodities which are produced for the purpose of exchange. Whether or not one thinks of the commodity as fetish, Thrift argues, "What is certain is that the process of commodification has reached into every nook and cranny of modern life . . . Practically every human activity in Western countries either relies on or has certain commodities associated with it, from births to weddings to funerals, at work or in the home, in peace or war" (Thrift 2000: 96).

Besides commodity chain analysis, recent studies have also drawn on actor–network theory, focusing on the role of human and non-human "actants" in shaping the geographies of production and consumption. So, for example, Sarah Whatmore and Lorraine Thorne (1997) have attempted to demystify the geography of food networks by demonstrating that their global reach "depends upon intricate interweavings of *situated* people, artefacts, codes, and living things and the maintenance of particular tapestries of connection across the world" (ibid: 288; their emphasis). They take the example of fair trade coffee networks and contrast the process of network lengthening (the effects of which are to distance producers from consumers via the operation of a range of what network theorists such as Bruno Latour call "immutable mobiles" such as money, telephones, computers, or gene banks) with processes of network strengthening (involving UK-based fair trade organizations working more closely with Peruvian export cooperatives). Similar arguments have been pursued by Bill Pritchard (2000) in his account of the relational networks involved in the establishment of a manufacturing facility in Thailand by the US-based breakfast cereal company, Kellogg. His research acknowledges the methodological complexities of network-based research, tracing the flow of commodities across space, and the indivisibility of "economic" and "cultural" processes.

Conclusion

Examining a range of specific commodities, this chapter has tried to sketch a series of more complex geographies of consumption than those implied in more standard accounts of globalization. While some have argued that globalization can be understood as a process of increasing homogenization, obliterating the particularities of place – "a social process in which the constraints of geography on social and cultural arrangements recede and in which people become increasingly aware that they are receding" (Waters 1995: 3) – this chapter has been much more skeptical of the claim that economic globalization inevitably leads to cultural homogenization where local differences are simply obliterated. Not only do stark variations in consumption persist at the global scale (as exemplified at the start of this chapter), but local contexts remain important in understanding the consumption of even the most globalized brands. As our case studies of the commodity cultures associated with tropical fruit, sugar, and soap further demonstrate, consumer choices can scarcely be described as "divorced from geography" (as Waters insists in his sociological account of globalization).

Reinstating the links between people and place at a variety of spatial scales, this chapter has attempted to transcend traditional distinctions between production and consumption, the economic and the cultural, material goods and their symbolic meanings. Focusing on the circuits and networks through which such commodity cultures are constituted and reproduced provides a useful way of understanding the political dynamics of contemporary consumption. Such an approach also confirms the value of tracing the social geography of "things in motion" (Appadurai 1986) as a key to understanding the geographies of consumption in a globalizing world.

Acknowledgment

The ideas in this chapter have been developed in collaboration with Phil Crang, Claire Dwyer, and Suman Prinjha through our current research on "Commodity culture and South Asian transnationality" (funded by ESRC award no. L214252031).

19 Understanding Diversity: The Problem of/for "Theory"

Linda McDowell

One day, the philosopher of science Paul Feyerabend, then teaching at the University of California, Berkeley, confessed to a female African-American student his reservations about teaching imperial Western philosophy to representatives from different cultures, who had their own ways of knowing and their own values. She replied: "You teach. We choose." (Interview with Feyerabend, *The Higher*, December 10, 1993)

Introduction

This story is the key to this chapter. In the last two decades or so, a set of challenges to the theoretical basis of disciplinary knowledge posed a major challenge to a discipline like geography, whose *raison d'être* is the explanation of difference and diversity. Critiques of Western science from a variety of positions, that exposed its purportedly universal and objective truths as ethnocentric, masculinist, and historically specific assumptions, seem to many to have left us marooned in a quagmire of relativism in which different ways of knowing have apparently equal claims to validity. An angry debate erupted, ranging across disciplinary boundaries, within and between the social sciences and the humanities, and reaching into the sciences, in which adherents of different perspectives exhorted each other variously to celebrate difference or to hold on to old verities. The bad-tempered arguments within a related discipline, anthropology, give a flavor of the debate and the passions with which each side holds to its arguments for or against notions of rational science and cultural relativism (see, for example, Gellner 1992). Our own discipline, too, has seen fierce exchanges about the nature of theory between, for example, Marxists, feminists, and a number of adherents of various versions of postmodernism.[1] These exchanges took a particularly extreme form, perhaps, in the arguments about the relative significance of cultural practices for "the economic" (Amin and Thrift 2000; Barnes 199; Martin and Sunley 2000; Sayer 199).

I want to argue here, however, that abandoning old certainties does not necessarily entail the whole-hearted embrace of relativism. Recognizing different ways of knowing does not mean abrogating responsibility for distinguishing between them. For teachers, as well as students, one of the main purposes of theory is to permit and to justify such choices. As geographers, we need theoretical perspectives that not only permit the elucidation of the main outlines of difference and diversity, of the contradictory patterns of spatial differentiation in an increasingly complex world, where ever-tighter global interconnections coexist with extreme differences between localities; but perspectives that also allow us to say something about the *significance* of these differences. Thus, I argue here that geographers must develop *Principled Positions* (Squires 1993) as a basis for constructing a *politics* of difference.

In this chapter I shall try to summarize the main features of three sets of theorists of diversity: feminists, poststructuralists, and postcolonial theorists. This is an almost impossible task as the literatures in each area are now huge, both within and without geography. References to other introductory summaries are listed in the bibliography. It is an important task, however, as without a flavor of these debates it is impossible to address the question of relativism that arises if we argue that discourses are geographically and socially specific, or the possibility of distinguishing between different ways of knowing.

What is Theory/What is Theory For?

In debates about the status of knowledge, the term "theory" has a privileged position. A major consequence of the debates is that its definition has changed. A theoretical analysis of a geographic issue consists of a rule-governed activity. These rules establish "what counts as evidence, what constitutes an argument, what appeals are to be made to empirical truth, explanatory adequacy, rhetorical effectiveness, or any other standard of judgment" (Landry and Maclean 1993: 136). Conventionally, this has meant a scientific, or objective, analysis: this is usually what is meant by "theory" in the technical sense and is the definition that informs geographic approaches such as spatial science or Marxist analyses that dominated the discipline in the 1960s and 1970s and still, of course, retain a hold. But since those decades, geographers have increasingly come to recognize that theories themselves are "texts" or narratives that tell a single and particular story, and not others that might have been told (Gregory 1994; Duncan and Ley 1994; Duncan and Gregory 1998). Thus an alternative kind of theoretical analysis has become common in geography: analyses of theories themselves, revealing the ideological assumptions in the work; the contradictions, unvoiced assumptions, the limits to the texts' logic (see Barnes 1992). Scholars from a range of positions – feminist, poststructural, postmodern, and postcolonial – have raised crucial questions about whose point of view is represented in geographic writing. These scholars pointed out the absence of certain voices in geographic narratives – in the main, those of the powerless and dispossessed – arguing against conventional scientific methods that assume a correspondence between "reality" and its representation by geographers, the idea that maps and texts make transparent a singular meaning for the reader. Rather, they insist that knowledge reflects and maintains power relations, that it is partial, contextual, and situated in particular times, places, and circumstances. Representations of these partial truths are produced by authors who are "raced," gendered, and classed beings with a particular way of seeing the world. Further, these socially produced texts have no necessary or fixed meaning; the reader is implicated in the construction of meaning. A reader's own set of assumptions and sociogeographic location affect how texts are read and interpreted.

These arguments are tremendously powerful and potentially liberating. They enabled new questions about power and knowledge to be raised – questions that were particularly pertinent ones for geographers as they problematized the relationship between place and representations, raising new issues about the social construction of geographic knowledge and about our research methods. They are also unsettling, posing difficult questions about the significance of different views of the world. If there is no singular Truth any more, but diverse truths, how, we might ask Feyerabend's student, does she go about choosing *between* them?

To understand the extent of the challenge posed by new theories of diversity, it is helpful to reflect on the history of theoretical endeavor in the discipline. Surveys of the history of geographical explanation tend to give the impression that the replacement of one approach by another is orderly progression in which increasingly sophisticated theories replace earlier and, by implication, discredited perspectives. But this is a false impression. With the possible exception of spatial science – and even this approach never achieved complete dominance – a more accurate picture is one of competing knowledge claims, of interpretive communities with different views of the world and different methodological approaches. Geography, like most social sciences, has a history of multiple perspectives, competing paradigms, and theoretical diversity. However, these diverse ways of seeing the world were regarded as *alternatives*. The adherents of each approach argued that their view was the correct one. Each held to a notion of a single truth embodied in their own view of the world and uniquely revealed by their approach to research – beliefs that are crucial in arguments that geography is "scientific" but are disrupted by the claims by the multiple "others" who have made a more visible appearance in contemporary geographic texts. The voices of women, working-class subjects, "Third World" peoples, of the underclass and transgressors, make different claims – that the singular geographies we have been accustomed to are equally partial; they reflect the view of the world of the powerful rather than a rational, scientific, or objective view of reality as claimed. The theoretical challenge of representing the partial perspectives of multiple "others" rather than what Haraway (1991) has termed "the view from nowhere" is subsumed in part of my title for this chapter: the problem *of* theory. The current problem *for* theory is how to respond to it.

The Problem of Theory: the Status of "Truth/truth"

Science, ideology, knowledge, discourse, narrative: these five words indicate alternative positions in the intellectual landscape of contemporary geography. The first two reflect Enlightenment views of Truth that underpin the liberal and Marxist perspectives in geography, whereas the latter three are common terms in more recent theoretical positions. Whereas Marxist analysts in geography are concerned to demonstrate what Marx himself termed "the economics of untruth" in liberal theories of the world, more recently other geographers, especially those influenced by the French philosopher Foucault, have challenged the scientific basis of Marxist theory in their focus on "the politics of truth" (the term is Foucault's). I shall briefly spell out these differences, but interested readers will find a more thorough discussion in Michele Barrett's *The Politics of Truth* (1991).

The first challenge to the scientific basis of positivist approaches in geography came from the work of Marxist scholars and from socialist feminists, who argued that the work of too many theorists bore little relationship to the reality that it claimed to portray. Positivist spatial science, for example, portrayed a landscape without power, poverty, or political struggles: areas to which Marxist and socialist feminist analysts drew particular attention. Although adherents of Marxian perspectives continue to hold to notions of scientific knowledge and Truth, they argue that positivist science is ideological, justification of the privileges of a class society dressed up as truth. Here we see the idea that knowledge is positional or embedded, a version of the world that embodies a particular justification of the distribution of rewards and privileges. However, it is within that set of knowledges variously labeled postmodernist, poststructuralist, and postcolonialist, and in recent developments in feminist scholarship, that resistance to singular truth claims has been most strenuous. The rest of the chapter briefly outlines these arguments, before turning to the possibility of some sort of accommodation between them and other theoretical perspectives.

Introducing the "Others": Feminism, Poststructuralism, and Postcolonialism

What was it that happened to disrupt theoretical confidence in a singular notion of Truth? What were the challenges to geographers' ways of seeing the world? How were spaces opened up in which ideas about plurality, diversity, multiplicity, difference, disruption, and contradiction challenged the singular emancipatory vision of Western science, that central notion of development and progress that sustained both the liberal/scientific and the Marxist/socialist discourses? The answer lies partly in the political movements developing from the late 1960s onwards, as well as in theoretical shifts in the academy. Within geography a group of committed and radical academics were active not only in class struggles but also in feminist, green, and ecosocialist struggles. (The founding of the journal *Antipode* in 1969 gave these radical geographers a disciplinary standpoint.) Their combination of academic and political work raised questions about the assumptions behind the notions of social justice and progress that informed both liberal-humanist notions of individual equality and socialist versions of the future. A sustained attack was launched from several quarters on the supposedly universal "we" of both these discourses. A choir of voices – women, gay men, colonial subjects, prisoners, and other "deviants" – was raised to protest that the supposed universal "we" of humanist and socialist discourses in fact excluded their specific experiences, based as they were on a particular view of the world: a view through the eyes of the Western, male, bourgeois subject.

Theoretical developments in the critical social sciences coincided with these political movements to necessitate the reevaluation of the privileged status of existing theoretical frameworks, with their commitment to notions of a single universal truth and a unitary, universal human subject (Barrett 1991). Instead of a search for Truth and universal laws, we have seen a number of geographers, especially those working in the social, historical, and cultural areas of the discipline, turning toward the theorization, and in some cases the celebration, of difference and diversity, in opposition to the totalizing tendencies in what have become known as the "grand narratives" of liberal humanism and Marxism. Instead of searching for general laws these geographers insist on the particularity and plurality of knowledge, on analyses of the specificity of its social construction by certain social groups located, in particular, in space–time frameworks. While debates about the situatedness of knowledge have permeated the social sciences and humanities more generally, for geographers they strike a particular chord as they coincide with our long-held interest in difference, diversity, and place-related specificity.

The key players in the displacement of old certainties have, appropriately, been many and various and their arguments overlap. However, I shall give a flavor of their claims under three headings: feminism, poststructuralism, and postcolonialism, indicating the implications for geographical work, although many of the key theoretical texts are outside our discipline.

Feminism

Despite the insistent claims of a parade of grand (and mainly European) male theorists – Michel Foucault, Jacques Derrida, Jacques Lacan, Jean-François Lyotard, Richard Rorty, et al. – some of the earliest attempts to "deconstruct" the humanist subject were made by feminist theorists, "as long ago as 1792 by Mary Wollstonecraft in her demand for women to be included in the entitlements claimed by the 'Rights of Man'" (Soper 1990: 229–30). As innumerable feminists have demonstrated since, Western humanism succeeded in writing women out of the theoretical agenda by constructing the category "Woman" in opposition to "Man," in such a way that women were defined as the "Other," as an absence or a lack, compared with the "One," that is an idealized rational, disembodied subject or individual, that, in fact, embodied masculine (and Western) attributes. Through the construction of a set of binaries that equated masculinity with science, rationality, objectivity, and the public and political arenas of life, mapping femininity onto opposite characteristics – emotion, irrationality, subjectivity, the private and domestic sphere – women were relegated to an arena that apparently needed no theoretical investigation but might be taken for granted as "natural." By naming femininity in opposition to the category "Man," which was assumed to be human, woman was situated in a very different category from that of man (Pateman and Gross 1986). Feminism has convincingly demonstrated how "women's silence and exclusion from struggles over representation have been the condition of possibility for humanist thought: the position of woman has indeed been that of an internal exclusion within Western culture" (Martin 1988: 13).

The early achievements of feminist scholarship involved establishing women's right to inclusion, based on an appeal to liberal notions of social justice. As women, too, were individuals, to deny them the rights and benefits available to men was patently unjust. By the mid-twentieth century, however, it became clear to feminists that a claim for equality within humanist thought was impossible. As the humanist subject was masculine, women would always be excluded. So, it is argued, humanism as such was inherently flawed and must be replaced by an alternative set of knowledge claims (although based on what claims and in what form remains contested and uncertain). However, for feminists, the death of the humanist subject necessarily led to the deconstruction of "woman," as well as man, and a huge expansion of exciting and challenging work that mapped the dimensions of difference in gendered identities and gendered social relations as well as retheorizing the very notion of gender itself.[2]

It is here, in the deconstruction (the challenging and taking apart) of knowledge claims, that recent feminist, poststructuralist, and postmodern theories unite to pose a common challenge to the theoretical basis of conceptions of knowledge, truth, and science: a challenge that is resulting in the "displacement of the problematics of science and ideology, in favor of an analysis of the fundamental *implications* [original emphasis] of power-knowledge and their historical transformations" (Morris 1988: 33). By this, Morris means that we must look at whose interests theories support, how they vary across space, and how they change over time. It is this emphasis on change that perhaps distinguishes poststructuralism from Marxism, but there is also a significant difference in the way in which power relations are theorized, as I shall demonstrate below.

Poststructuralism

The term "power/knowledge" is a reference to the work of Foucault, who, although resistant to the labeling of his work, is generally regarded as a poststructuralist. What is important here is his rejection of the notion of ideology, and a reconceptualization of power relations as complex and unstable, part of the social construction of self rather than an external system "imposed" on people from above. Let us take these claims in turn.

Foucault rejects the concept of ideology on several grounds: first, that it necessitates an opposite notion, that of truth; secondly, that it stands in a secondary position to the material or economic determination of society; and thirdly, that it rests on the humanist notion of an individual subject. Poststructuralists have developed instead the concept of "discourse," a concept that is dependent upon Foucault's redefinition of power. Rather than seeing power as a social and political system in which a repressed group is dominated for the benefit of the oppressors, Foucault (1970, 1977a, 1977b, 1980) argued that language, texts, discourse, as well as social practices, are part of the maintenance of particular power relations in modern societies. He provided what he called a different "grid" for deciphering history – a new way of looking at and analyzing power which focused on the ways in which discourses, and the pleasures and powers they produce, have been deployed in the service of creating and maintaining hierarchical relations in Western societies (Foucault 1978). Thus power relations are maintained not through the false consciousness of the repressed, who will eventually act as a class for themselves, but rather as part of the very construction of identity. Thus an individual subject is constructed through a grid of discourse and practice. In poststructuralist theory there is no notion of a prior or essential moral self, as in liberal and Marxist theory, and so traditional ideas of "truth" and "self" are rejected. As Foucault argued, "power seeps into the very grain of individuals, reaches right into their bodies, permeates their gestures, their posture, what they say, how they learn to live and work with other people" (quoted in Sheridan 1980: 217). Our bodies, identity, and sense of self have no prior existence but are brought into play through language, discursive strategies, and representational practices, and so conventional distinctions between thought and action, between language and practice, are collapsed, as all social actions have a discursive aspect.

While all discourses are changeable, open to contestation because they are social practices, Foucault termed the establishment of an interlocking set of successful or dominant discursive formations a *regime of truth*. The term "truth," however, is used here in a different sense from conventional notions. It is impossible to distinguish between what is true or false by appealing to the "facts" of the matter, as the "facts" themselves are part of a particular discursive formation. The "truth" of a situation is determined by the outcome of struggles between competing discourse.

Postcolonialism

It is within recent work under this heading, linked to Third World feminists' critiques of the ethnocentric biases of Western feminism, that perhaps the most powerful arguments about contested and incompatible knowledges, as well as multiple subject positions, have been advanced. Within Western theory, "women of color" have criticized the racially exclusionary theory and practices of white, middle-class feminism, problematizing assumed commonalities as women. At the international level the theoretical focus is on imperialism rather than racism, and here the most influential work has been that of Said in his arguments about the construction of a colonial "Other" by the West, and the work of the Subaltern Studies group, who are rewriting the history of colonial India from the point of view of peasant insurgents. In both cases, despite his profound Eurocentrism, Foucault is a key inspiration, as he has been for a number of feminists who argue that women's exclusion from struggles over representation within Western culture are another demonstration of Foucault's concept of power (Martin 1988).

Said's *Orientalism* (1978) has probably been the most influential text in the development of new ways of thinking about Western imperialism. Drawing on Foucault's notion of discourse, Said argued that "without examining Orientalism as a discourse one cannot possibly understand the enormously systematic discipline by which European culture was able to manage – and even produce – the Orient politically, sociologically, militarily, ideologically, scientifically, and imaginatively during the post-Enlightenment period" (ibid: 3). Through a range of discursive strategies and social practice, people of the Orient were constructed as the "Other" to the Western "One," in ways such that "European culture gained in strength and identity by setting itself off against the Orient as a sort of surrogate and even underground self" (ibid).

Whereas Said focuses on the discursive construction of the Orient in Western texts, the Subaltern Studies group is interested in the impact of colonialism on native cultures. Spivak (1988) argues that the imposition of colonial rule in India effectively ruptured and silenced indigenous cultures, a fracturing that she terms, after Foucault, "epistemic violence." The colonial subject is unable to answer back in a language that is untouched by imperial contact, and which may only be glimpsed in the interstices of colonial texts. The truly marginalized, the "subaltern" of the group's title and Spivak's best-known paper, is unable to speak at all.

These arguments have interesting implications for how geographers undertaking ethnographic and field-based work in postcolonial societies theorize and analyze the narratives collected from "native informants." The implications of Spivak's arguments are that no one falls outside the field of imperialist discourse. Whether engaged in peasant insurgency or postcolonial critique, the native is forced to speak the language of imperialism. However, rather than resulting in capitulation, Bhabha (1990) has argued that what he terms "hybridization" is the result, a sort of double displacement whereby native mimicry of imperialist discourses embodies cultural and political resistance, disrupting authoritative representations of imperialist power. Thus, rather than either a nostalgic search for lost origins or a desire to re-establish authentic traditions, the postcolonial critic analyzes the ways in which resistance lies in hybridization. In the context of the West, where the huge migration flows related to slavery, imperialism, and global capitalism have resulted in peoples who are of neither First nor Third World, Bhabha's notion of hybridization also seems relevant. Indeed, the sociologists Gilroy (1993) and Hall (1996) have developed interesting parallel arguments about cultural translation. Gilroy, for example, in his work on African-American peoples, argues that a hybrid Black Atlantic culture has developed. Similarly, "women of color" in the US, UK, and elsewhere are exploring the position of women who are the product of international and cultural borders – women of mixed origins, whose identity is constructed out of difference (Anzaldua 1987; Mirza 1997; Sandoval 1991).

In combination, the work summarized under the above three headings altered the subject of geographic research, refocused theoretical and methodological assumptions, and challenged the political practice of "radical" geographers. Many have realized that, in the words of Landry and Maclean (1993: 125–6), "the white male heterosexual worker of the First World industrial nations just won't do any longer as the alternative or insurgent subject of history, the locus of resistance to ruling class dominance." Similarly, geographers expanded their foci. Drawing in particular on qualitative and ethnographic methods, with increasing attention to everyday social relations and to discursive strategies, a number of geographers have turned to detailed analyses of the "other": women in different positions (see my review in McDowell 1999), female travelers (Domash 1991), native peoples and ethnic minorities (Anderson 1992; Radcliffe and Westwood 1996), disaporic, "displaced," and transnational communities (Lawson 1999; Mitchell 1997; Walter 1995), young people of different ethnicities in different locations (Hyams 2000; Skelton and Valentine 1998), the mad (Philo 1989), the sick and less able (Butler and Parr 1999; Dyck 1995), and the perverse (Valentine 1993); and of the transgressive landscapes of desire of "alternative" sexualities (Bell 1994; Bell, Binney, Cream, and Valentine 1994; Bell and Valentine 1995). There has also been a "discursive turn" in which the methods of discursive and textual analysis have become influential in the interpretation of landscapes of power and oppression, as well as social relations in localities. As Peet and Watts (1993a: 248) have suggested, "one of the great merits of the turn to discourse, broadly understood . . . is the demands it makes for nuanced, richly textured empirical work": a type of work admirably illustrated in Watts's book with Pred (Pred and Watts 1993) and in the growing body of rich ethnographic and qualitative research by geographers.

Attention also has been turned onto the discursive strategies of geographers themselves, onto textual analyses of our own writing. Geographic language is recognized as being not simply a medium expressing reality "out there" but as playing an active role in creating social and cultural worlds. Interpretive schemes "in part . . . create the reality that they seek to interpret" (Barnes 1992: 118). Theories "have the imaginative capacity to represent and reconfigure the world like *this* rather than like *that*" (Gregory 1994: 182; original emphasis). In new histories of geography, theoretical perspectives are seen not as alternatives in a struggle for hegemonic status, but as alternative and incompatible interpretive communities. The purpose of theoretical endeavor in our discipline is thus disrupted. Rather than a struggle to establish universalizing claims, the grand and certain aim of grand theory, we (not this time united by theoretical consensus, but a motley collection of geographers with different views of the world) have in common what Gregory terms a "strategy of supplementarity – *dis*-closing what theory closes off" so that "the certainties of theory can be capsized, its confidence interrupted, and its conditional nature reasserted" (ibid: 181–2). The singular geographical imagination so eloquently defined by David Harvey in 1990 has become the partial and plural geographical imaginations of Derek Gregory's (1994) survey of the discipline. Even the most insistent of class analysts – David Harvey – has surveyed the terrain of difference and its association with notions of social justice, although he insists on the continuing dominance of a materialist analysis in which class is the difference that makes the difference (Harvey 1996, 2000). Indeed, there is a noticeable reassessment of the significance of material inequalities in some contemporary theoretical work, both by geographers and other theorists, including feminist scholars (see, for example, the recent books by Phillips (1999) and Segal (1999)). In the penultimate section of this chapter I shall briefly consider contemporary feminist approaches to social justice wherein "difference" theory and class theory find some accommodation.

Problems for Theory

Claims that geographies are multiple, that knowledge is contested, and that space is discontinuous and fluid, a discursively constructed set of relationships that discipline, control, or privilege (Lefebvre 1991), a "plane of contest" (Ashley 1987) rather than a set of fixed regions defined by Cartesian coordinates, led to an exciting set of debates in the "new" geography of the 1990s and the new millennium, but also to a continuing expression of unease. Fears of relativism, enunciated not only by those critical of claims that knowledge is embedded and situated but also by many who generally welcome the challenge to Enlightenment notions, are perhaps the most serious, although it is important to note Barrett's claim that poststructuralist critiques of truth and knowledge are "not so much relativist as highly politicized" (Barrett 1991: 161). Theoretical acceptance of plural and fragmented identities, of multiple and local sites of power, has been enormously empowering for those "others" condemned as handmaidens, helpmeets, "perverse," or slaves of Western "civilization," whether marginalized by the universalist pretensions of humanism, or left at home/outside by Marxists while the "real" struggles take place elsewhere. But questions remain. How, for example, might we judge between competing knowledge claims? On what may we base our political opposition to the oppression and exclusions that are revealed in these new discourses? Are all claims to knowledge and power equally valid?[3]

Many, geographers included (Driver 1992; Gregory 1994), have pointed to a paradox at the heart of the three theoretical positions discussed above: their critique of the exclusion of "others" surely depends for its critical force on acceptance of modern ideals of autonomy, human rights, and dignity. Thus, as Soper (1990: 149) asks, "why concern ourselves with the exclusions from the 'humanist' universal of women, or blacks or gay people, or the insane, or any other oppressed or marginal group, except on the conventional ethical grounds that all human beings are equally entitled to dignity and respect?"

It seems to me that here we must establish ways of criticizing universalistic claims without completely surrendering to particularism. And surely this is a more general example of the problem that geographers, in their focus on the particularity of place, have been grappling with for many years. Geographers have long accepted a form of relativist argument: that the specificities of local socioeconomic structures, differences in cultural attitudes and in ways of living, are part of the explanation for uneven development (Massey 1984). Drawing on this work, in which universal and particular explanations are neither counterposed nor seen as alternatives, we need to recast our general notion of relativism and no longer see it as counterposed to universalistic explanations. Rather, we must see relativism, in the sense of difference, as inescapable. Acceptance that there are different ways of knowing which are historically and geographically specific does not mean that we cannot make judgments about their value. Foucault drew our attention to the ways in which discourses, or speaking positions, are constituted by power relations. Thus oppositional knowledges of the subjugated are inherently and inescapably political, demanding judgment about the validity of different discourses on the basis of political considerations. We might, therefore, accept the multiple claims to power of some groups and reject those of others, in attempts to construct a multiple and inclusive notion of social justice that includes the claims of the non-white, non-masculine subject (Flax 1990; Young 1992). But as Barrett points out, this second move, one that she terms "transformation," is much more difficult than the first, the transgressive deconstruction of dominant binary oppositions where feminism, poststructuralism, and postcolonialism have been so successful. Once the categories of "woman," or "postcolonial subject," have been dismantled, it becomes more difficult to speak on their behalf or, rather, for their own varied voices to be heard. There is a tension between challenging the oppositional construction of others and continuing to make political demands on their/our behalf. A possible way forward is suggested by "Black" theorist Du Bois (1989; see also Gilroy 1993) in his theorization of "double consciousness" that simultaneously allows for the deconstruction of the category "Black" and for the need to assert the truth and authenticity of the Black identity, constructed in common through the racism of dominant social institutions. Biddy Martin (1988) similarly has suggested a "doubled strategy" for feminist struggles, where a political commitment to challenging women's subordination does not necessarily imply a unified theoretical field (see Pollock 1996).

Forms of Difference and Claims of Justice: Misrecognition and Maldistribution

As I noted earlier, the recognition of diversity has unsettled theories of social justice based on liberal or humanist claims of the rights of an individual to equal treatment. The recognition of group difference and the denial of claims for parity from, *inter alia*, women, people of color, and "alternative" sexualities, led to the development of new theories of social justice based on claims for cultural parity and the recognition of multiple forms of identities. An interesting debate arose about the extent to which claims for recognition were distinct and separate from claims for the redistribution of material resources, perhaps seen most clearly in the exchanges between a number of feminist theorists, including Iris Marion Young, Judith Butler, and Nancy Fraser. Here the arguments of Nancy Fraser (1997, 2000) are particularly relevant to this chapter, as in her most recent work she has attempted a theoretical reconciliation between claims of misrecognition and those for redistribution (Fraser 2000). Her impetus lies partly in the form of unease I mentioned earlier: how to judge between claims for cultural recognition in a world in which the nature of these claims seems in recent years to have become increasingly urgent and troublesome in an era of ever-expanding transcultural interaction. However, as Fraser notes, "With the turn of the century, issues of recognition and identity have become even more central, yet many now bear a different charge: from Rwanda to the Balkans, questions of 'identity' have fueled campaigns for ethnic cleansing and even genocide – as well as movements that have mobilized to resist them" (ibid: 107). The insistence on cultural recognition lies not only in exciting movements of hybridizing and plural cultural forms, but also in claims for separation and cultural assertions that oppress many of the people for whom the movements claim to speak.

But the impetus to reassess claims for justice also lies in the increasingly apparent disparity between the relative decline of both theoretical and political claims for egalitarian redistribution and an increasingly unequal world in which the inequalities in income and living standards between the richest and poorest inhabitants within and between nations become ever wider as a consequence of the dominant aggressive form of neoliberal economic globalization. The responses to this growing unease have been complicated. As Fraser notes,

> many have simply washed their hands of "identity politics" – or proposed jettisoning cultural struggles altogether. For some this may mean reprioritizing class over gender, sexuality, "race," and ethnicity. For others, it means resurrecting economism. For others still, it may mean rejecting all minoritarian claims out of hand and insisting upon assimilation to majority norms – in the name of secularism, universalism, or republicanism. (Ibid: 108)

And in contemporary geographical debates it is possible to see all these positions under discussion (see, for example, Harvey (2000) and Smith's (2000) discussion of the significance of class politics). However, it is also possible to discern the outlines of a new agenda in the social sciences, geography included, in which the reintegration of class and identity politics is a key focus. It is an agenda that is, I believe, the key theoretical challenge facing geographers in the new millennium (and one that I enthusiastically endorse: see McDowell 2000). Fraser's work is so provocative here as she insists that cultural difference is both a site of injustice in its own right and is deeply imbricated with economic inequality. British feminist political theorist Anne Phillips (1999) has also recently begun the theoretical reconciliation between different forms of inequality. Their arguments deserve wider recognition among geographers (although they are too complex to effectively summarize here) who endorse the attempt to move beyond the impasse of universalist versus relativist claims, as well as between cultural claims for recognition and economistic claims for redistribution.

Conclusions: Principled Positions

The theoretical and moral implications of a rejection of both universalism and relativism as conventionally defined are, however, complex. It is clear that the exercise of moral responsibility is culturally dependent and temporally specific. But the fact that political choices and everyday decisions are made in relation to a contestable code or set of conventions as to what is "good" or "bad" does not in itself detract from the element of judgment and the assertion of values in our choice. As Haraway (1991b), a profoundly thoughtful theorist, has suggested, we need both to accept irreducible differences and yet hold on to what she terms a "successor science project" in which "projects of finite freedom, adequate material abundance, modest meaning in suffering and limited happiness" (ibid: 187) are the focus of action. And as Gregory persuasively argued in his conclusion to his ambitious summary of the theoretical currents that have washed through our discipline: "If we are to free ourselves from universalizing our own parochialisms, we need to learn how to reach beyond particularities, to speak to larger questions without diminishing the significance of the places and the people to which they are accountable" (Gregory 1994: 205).

Theorizing diversity does not mean abandoning these larger questions and projects. The challenge for geographers is to find ways to respond to them which, through the recognition of the interconnections between cultural difference and material inequalities, enhance theories of social justice.

Notes

1   I avoid the use of the term "postmodern" in the rest of this chapter, preferring instead to use "poststructural" to refer to theoretical claims, and reserving "postmodern" for social and cultural shifts. However, like Rosenau (1992: 3), I recognize that "the terms overlap considerably and are sometimes considered synonymous. Few efforts have been made to distinguish between the two, probably because the difference appears of little consequence."

2   This work is impossible to summarize here, but for introductions to the key debates within feminist geography at different periods see, *inter alia*, Jones, Nast, and Roberts 1997; McDowell 1999; Massey 1994; Rose 1993; WGSG 1984, 1997).

3   Some of the ways in which these questions affect our teaching have been addressed in Crang, Cook, and McDowell.

20 Resisting and Reshaping Destructive Development: Social Movements and Globalizing Networks

Paul Routledge

In contemporary, Western, popular parlance, development means change, but it also implies betterment and advance. The term conveys a sense of optimism and expectation of progress, and of a general improvement in the human condition. Indeed, the fruits of development are potentially manifold, and have included, for example, advances in medicine and education. The project of development and modernization has been particularly directed at countries in the South (i.e., the states of Latin America, Africa (excluding South Africa), and Asia (excluding Japan)) by the advanced capitalist countries. It has also been directed at indigenous peoples within the developed world, such as Native Americans in the United States and Canada. In contrast to the expected improvements that it would engender, the development project has caused widespread environmental damage and disrupted or destroyed the cultures and economies of numerous traditional (subsistence) and tribal communities. In response to this assault, social movements have emerged across the planet to pose challenges and alternatives to the process of development. This chapter will critically examine the nature of the development resist project and discuss some of those social movements which attempt to resist and reshape the forces of development.

Development as Discourse

The word "develop" was first associated with ideas concerning the nature of economic change in the nineteenth century, implying the notion of a society passing through definite evolutionary stages. The term "underdeveloped" began to be used after 1945, referring to lands in which natural resources had been insufficiently developed and exploited, and to economies and societies destined to pass through predictable "stages of development" according to a known model (Williams 1983: 102–4); although as Peet and Watts (1993a) have noted, these notions were already embedded in the colonial development and welfare acts of the French and British colonies during the prewar period.

The term was used by US President Truman in 1949 to describe those areas of the world that had yet to reach the standard of living experienced by those in the West. Truman went on to advocate an era of development whereby the West would provide assistance to "underdeveloped" areas to enable them – via "scientific advances" and "industrial progress" – to improve and grow (Esteva 1992: 6). The construction of the world into developed and underdeveloped areas and the articulation of a particular remedy for this situation (the modernizing project of development) formed the basis of the discourse of development, becoming part of the process by which the "colonial world" was reconfigured into the "developing world" (Peet and Watts 1993a: 232).

Discourse is a field of strategies (statements, views, theories, concepts, and objects of analysis and their interrelations) that create knowledge about something and create differentiations by posing limits on what can be said and by whom (Foucault 1977a, 1977b). The development project has been deployed by many of the Western states through discourses of underdevelopment that include, for example, theories of development, World Bank policies, and the apparatus of development programs. These discourses privilege the West's economic systems, institutions, and policy "experts" at the expense of those of the South, imagining the West as "the transcendental pivot of all analytical reflection" (Slater 1992: 312). Indeed, the development discourse serves to manage, control, and create the South politically, economically, ideologically, and culturally.

Ideologically, the development project is based upon a linear theory of progress and evolution (the meaning of development in the biological sciences) that places the industrialized states at the top of the evolutionary scale ("the developed") and Southern countries at the bottom ("the underdeveloped"). This differentiation, therefore, requires the countries of the South to further evolve – to develop – so that they can reach their full potential.

Economically, the project of development is based upon a program of economic growth, modernization, and industrialization that is seen to be essential to the alleviation of poverty. Since 1945 a plethora of development theories has been proposed variously privileging the role of the market (e.g., Rostow's (1960) modernization theory, and Bauer's (1966) theory of neoclassical economic development), civil society (e.g., Hirschmann's (1958) notion of Weberian modernization), and the state (e.g., those proposed by Mao and Nehru in China and India respectively) in the process of development (Peet and Watts 1993a). While a variety of new approaches to development have emerged since 1980 (see Peet and Watts 1993a), recent recommendations by the United Nations' Brandt and Brundtland Commissions continue to argue that economic growth is the necessary condition for development and the subsequent alleviation of poverty (Sachs 1992; Esteva 1992).

Culturally, the development project defines people partly as poor because they do not participate in the market economy. It emphasizes Western values (of capitalist production, economic growth, rationality, calculability, etc.) and devalues indigenous and traditional systems of knowledge, economy, and culture. By deeming culturally specific economic practices such as subsistence agriculture as backward, development discourse has legitimized the intervention of the processes of modernization into traditional economies (Bandyopadhyay and Shiva 1988).

Politically, the development project was initiated as a response by French and British colonial powers to anti-colonial movements in regions such as Africa. Development became a means by which the perils of independence could be negotiated, and colonialism continued (Peet and Watts 1993a: 232). More recently the development project has served as a means to encourage capitalist market economies, thereby providing conditions under which Western-style "democracy" could flourish. This purpose of the project emerged in the West at a time when the Soviet Union became a superpower competitor to the United States. As Rostow's (1960) "non-communist manifesto" implies, the development project was constructed as a strategy to bring Southern states into the geopolitical orbit of the United States and its allies (Sachs 1992).

In development discourse, then, the South does not represent itself; rather, it is represented by Western academics, experts, professionals, bankers, and government officials. Communication and information technologies such as the mass media, television, and commercial cinema serve to reinforce the stereotypes by which the South is viewed (e.g., poor, non-modern, undemocratic, etc.). The discourse of development serves three purposes. First, it creates problems and abnormalities (such as underdeveloped countries, poverty, etc.) which require treatment and reform. Second, the project of development is professionalized through the creation of development studies courses, the classification of problems, and the formulation of policies. Third, the development project is institutionalized through the establishment of international agencies, governments, national planning bodies, and local-level development agencies, the "professionals" who determine what is best for developing countries (Escobar 1984). Two types of intervention have coevolved with these processes: (1) intervention in the South by transnational corporations and institutions such as the World Bank, the International Monetary Fund (IMF), and the World Trade Organization (WTO); and (2) intervention in the traditional economies and cultures within particular states by national governments which are seen as the only legitimate form of political authority.

Development as Dependence

The extent to which the political economy of the Southern countries is in-fluenced by a global economy dominated by the advanced capitalist coun-tries has been analyzed by a variety of scholars known as the "dependency theorists." The analysis of dependency includes a variety of related theo-ries, including those formulated by the Economic Commission for Latin America (see Love 1980), Frank (1967), Dos Santos (1970), Cardoso (1972), Emmanuel (1972), Amin (1976), and Cardoso and Faletto (1979).

Dependency theory focuses on the unequal economic and political ex-change that takes place between the advanced capitalist countries (the "core") and the countries of the South (the "periphery"). The economies of the periphery are seen as conditioned by, and dependent upon, the develop-ment and expansion of economies in the core. The process of development is seen as selective, reinforcing the accumulation of wealth in the core at the expense of the periphery. Dependency is seen as both the relationship be-tween states – an industrialized core and an impoverished periphery – and also the relationship between groups and classes within states. It is impor-tant to note that while certain broad trends are identified below, regional and intrastate differences exist.

In adopting the Western-defined development project, the countries of the South are confronted with various problems that place them at a disad-vantage in relation to the developed countries. First, the particular eco-nomic and political conditions that enabled the West to industrialize – based upon the domination and exploitation of natural and human re-sources in the West's colonies – are different to those that confront the South. For example, whereas the West was able to support its moderniza-tion upon an expanding resource base (in the colonies), the Southern states are confronted with diminishing resources within their political borders. Second, the process of colonialism integrated the South into a world divi-sion of labor whereby the major function of the colonized economies was the production of raw materials for the European colonizers, and the ex-port of those raw materials for the colonizers' use. This facilitated the industrialization of the economies of the West at the expense of those in the South. It also integrated the economies of the South into a world economy dominated by the industrial centers in the West, placing them in a dependent relationship to the West that has continued, albeit in differ-ent forms, since decolonization.

The development project that fosters the economic dependency of the South on developed countries has several features. First, most of the tech-nology required for development (especially information technology such as satellites, computers, etc.) is produced in the developed countries who determine the price and availability of such technologies to the South. Much of this technology is also capital intensive, whereas traditional and subsist-ence economies are labor intensive. Hence, when adopted as part of the process of modernization, traditional labor is displaced, resulting in unemployment. Second, the South relies upon foreign investment to accelerate the development process, to access new technology, and to gain new markets. The markets are often represented by multinational corporations (MNCs). This results in the MNCs being in a position to exert considerable economic and political influence over Southern economies. For example, many of the countries in the South have pursued export-led economic policies involving, for example, the production of cash crops and manufactured goods for export to developed countries (e.g., the US and the EU) on whose markets they are dependent (Chandra 1992).

Finally, although development is mostly financed from domestic sources, foreign aid is important to the South as an added source of foreign exchange, human resources, and technical assistance. However, various factors, including the interest payments made by Southern countries on the debts incurred by foreign loans, and price fluctuations in commodities sold by Southern countries to the West, mean that a "reverse transfer" of resources has occurred from the former to the latter. For example, since 1983 the Southern states have transferred over US$30 billion in net wealth to the industrialized countries (Franke and Chasin 1992). This process has had enormous ecological, cultural, political, and economic repercussions in the South, since in the past up to half of the loans and credits from international banks have gone to environmentally sensitive projects such as irrigation, forestry, agriculture, and dam construction. For example, between 1983–4 Brazil borrowed US$950 million per year to develop farming for export which necessitated deforestation and human displacement in the Amazon (Bandyopadhyay and Shiva 1988: 1231). Through imposing structural adjustments (such as privatization) upon a country as the condition for awarding loans, such institutions as the World Bank also influence long-term economic policy and not just single projects.

An example of the development project creating dependency is provided by the green revolution which emerged in the 1960s as a development strategy designed to improve the yields of food production in developing countries. The project was capital intensive and relied on the application of expensive new technologies such as pesticides, chemical fertilizers, and high-yield variety (HYV) seeds, which were manufactured by MNCs, and had to be imported from the West. The project was based upon the introduction of HYV monocrops such as rice, maize, and wheat, whose increase in productivity was achieved by undermining the productivity and availability of other locally important crops such as pulses. The project destabilized traditional farming methods, which further rationalized the use of new technologies from the West, and the displacement of traditional foodstuffs by the HYVs. For example, during 1965–6 HYV sorghum was introduced under irrigated conditions in Dharwar district in Karnataka, India. It replaced indigenous sorghum varieties that had been previously cultivated with pulses and oilseeds. Because of its susceptibility to pests, the HYV sorghum required extensive pesticide spraying which destroyed the pest–predator balance in the area. As a result new pests appeared which gradually destroyed the traditional varieties of sorghum. Because sorghum is the main food crop in the region, farmers were compelled to plant HYV sorghum at the expense of indigenous crops so that by 1980–1 no area was sown by traditional crop varieties. Traditional sorghum cultivation involved mixed cropping with other important protein sources such as pulses, and the use of sorghum straw for animal fodder. Its replacement by the HYV monocrop decreased overall foodgrain and fodder yields and availability, and reduced soil fertility (see Shiva 1989: 123–5).

The green revolution increased the South's dependence upon the West in various ways. First, Southern economies were dependent upon foreign banks who provided the necessary credit to farmers to purchase the expensive inputs. Second, the project supplanted traditional agriculture with Western methods, which required the intervention of Western professionals and institutions into the South's agricultural sector. Third, the project made agriculture dependent upon industrial outputs (e.g., fertilizers, pesticides, expert advice, credit, etc.), thus subordinating agriculture to the requirements of industrial growth (Alvares 1992).

Neoliberal Development

Contemporary economic development is guided by the economic principles of neoliberalism and popularly termed "globalization" (e.g., see Herod, Ó Tuathail, and Roberts 1998). The fundamental principle of this doctrine is "economic liberty" for the powerful; that is, that an economy must be free from the social and political "impediments," "fetters," and "restrictions" placed upon it by states trying to regulate in the name of the public interest. These "impediments" – which include national economic regulations, social programs, and class compromises (i.e., national bargaining agreements between employers and trade unions, assuming these are allowed) – are considered barriers to the free flow of trade and capital, and the freedom of transnational corporations to exploit labor and the environment in their best interests. Hence, the doctrine argues that national economies should be deregulated (e.g., through the privatization of state enterprises) in order to promote the allocation of resources by "the market" which, in practice, means by the most powerful. As a result of the power of international organizations like the IMF, the World Bank, and the WTO to enforce the doctrine of neoliberalism upon developing states desperately in need of the finances controlled by these organizations, there has been a drastic reduction in government spending on health, education, welfare, and environmental protection across the world. This has occurred as states strive to reduce inflation and satisfy demands to open their markets to transnational corporations and capital inflows from abroad. Transnational liberalism celebrates capital mobility and "fast capitalism," the decentralization of production away from developed states and the centralization of control of the world economy in the hands of transnational corporations and their allies in key government agencies (particularly those of the seven most powerful countries, the G7), large international banks, and institutions like the World Bank, the IMF, and the WTO. As transnational corporations have striven to become "leaner and meaner" in this highly competitive global environment, they have engaged in massive cost-cutting and "downsizing," reducing the costs of wages, healthcare provisions, and environmental protections in order to make production more competitive.

Transnational liberalism has been institutionalized through various international free trade agreements, such as the North American Free Trade Agreement (NAFTA) between the US, Canada, and Mexico. These agreements are based upon the doctrine that each country and region should produce goods and services in which they have a competitive advantage, and that barriers to trade between countries (such as tariffs) should be reduced. However, such agreements are more concerned with removing the barriers to the movement of capital, to enable transnational corporations to operate without government interference or regulation, and to exploit the "competitive advantage" in cheap labor, lax environmental regulations, and natural resources.

The resulting global competition for jobs and investment has resulted in the pauperization and marginalization of indigenous peoples, women, peasant farmers, and industrial workers, and a reduction in labor, social, and environmental conditions – what Brecher and Costello (1994) term "the race to the bottom" or "downward leveling." In response to these processes numerous social movements have arisen in the South (and in the developed North) to challenge the violence of the development project, and it is to this resistance that I will now turn.

Social Movements: Resisting and Reshaping Destructive Development

In response to the development project, myriad social movements have emerged articulating struggles for cultural, ecological, and economic survival. Although this chapter has focused primarily on countries in the South, it is important to acknowledge that the development project continues to be directed against indigenous peoples in the developed world, where it has also been met with resistance (see figure 20.1). Examples of such movements include the Cree struggle to prevent the Hydro-Quebec dam in James Bay, Canada; Aboriginal struggles against mining in Australia; and Yakima Indian struggles for fishing rights in the United States (Moody 1988).

Social movements articulate resistance on a number of interrelated realms within society, including the economic, cultural, political, and environmental. Economic struggles may also contain political dimensions, political

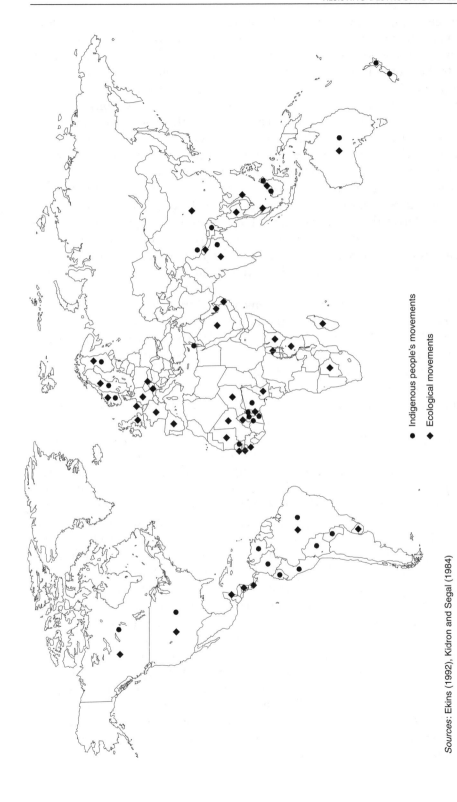

● Indigenous people's movements

◆ Ecological movements

Sources: Ekins (1992), Kidron and Segal (1984)

Figure 20.1   Location of some contemporary non-violent social movements.

struggles may also contain cultural elements, and so on. This multi-dimensionality is indicative of an alternative politics that seeks to create autonomous spaces of action outside of the state arena (Peet and Watts 1993a).

In the economic realm, social movements articulate conflicts over access to productive natural resources such as forests and water that are under threat of exploitation by states and transnational corporations. The economic demands of social movements are not only concerned with a more equitable distribution of resources between competing groups, but are also involved in the creation of new services such as health and education in rural areas (Guha 1989). Indeed, social movements have emerged in many areas, including civil liberties, women's rights, science and health, that are themselves related to problems caused by the development project. For example, during the mid-1980s in the Philippines, farmers, peasants, and scientists joined forces to demand the closure of the green revolution's International Rice Research Institute, which they perceived as being instrumental in the destruction of their traditional farming economy (Alvares 1992).

In the political realm, social movements challenge the state-centered character of the political process, articulating critiques of neoliberal development ideology and of the role of the state. These movements are frequently autonomous of political parties (although some have formed working relationships with political parties or formed alliances with voluntary organizations, non-government organizations, and trades unions). Their goals frequently articulate alternatives to the political process, political parties, the state, and the capture of state power. By articulating concerns of justice and "quality of life," these movements have enlarged the conception of politics to include issues of gender, ethnicity, and the autonomy and dignity of diverse individuals and groups (Guha 1989). For example, the *Ejercito Zapatista Liberacion Nacional* – the EZLN or the Zapatistas – are a predominantly indigenous (Mayan) guerrilla movement that have emerged in Chiapas, Mexico. They have articulated armed and discursive resistance to NAFTA and the Mexican state, exposing the inequities on which economic development in Chiapas is predicated. In their demands for equitable distribution of land, their calls for indigenous rights and ecological preservation (i.e., an end to logging, a program of reforestation, an end to water contamination of the jungle, preservation of remaining virgin forest), they also articulate an economic, ecological, and cultural struggle (Routledge 1998).

In the cultural realm, social movement identities and solidarities are formed, for example, around issues of class, kinship, neighborhood, and the social networks of everyday life. Movement struggles are frequently cultural struggles over material conditions and needs, and over the practices and meanings of everyday life (Escobar 1992a). For example, in Colombia during the 1980s, the Indigenous Authorities Movement sought to reaffirm the integrity of tribal territory, community, and culture against incursions made by the hacienda system (Findji 1992).

In the environmental realm, social movements are involved in struggles to protect local ecological niches, e.g., forests, rivers, and ocean shorelines, from the threats to their environmental integrity through such processes as deforestation (e.g., for logging or cattle grazing purposes) and pollution (e.g., from industrial enterprises). For example, resistance against the Nam Choan Dam in Thailand during the 1980s arose because the construction of the dam threatened to destroy wildlife sanctuaries and riverine forest environments and displace up to 6,000 Karen and Hmong tribal people (Eudey 1988). Whereas environmental struggles in the developed countries tend to concentrate upon "quality of life" issues, in developing countries, movements have often focused on access to economic resources. Such groups articulate an "environmentalism of the poor" (Martinez-Allier 1990), whose fundamental concerns are with the defense of livelihoods and communal access to resources threatened by commodification, state take-overs, and private appropriation (e.g., by national or transnational corporations), and with emancipation from material want and domination by others.

Social movements are by no means homogeneous. A multiplicity of groups including squatter movements, neighborhood groups, human rights organizations, women's associations, indigenous rights groups, self-help movements among the poor and unemployed, youth groups, educational and health associations, and artists' movements are involved in various types of struggle (Corbridge 1991). Many of these struggles take place within the realm of civil society, i.e., those areas of society that are neither part of the processes of material production in the economy nor part of state-funded organizations. For these movements, civil society represents "the domain of struggles, public spaces, and political processes. It comprises the social realm in which the creation of norms, identities, and social relations of domination and resistance are located" (Cohen 1985: 700). Although not exclusively, these movements frequently employ non-violent methods of resistance (see Routledge 1993).

Many of these movements are also place-based and frequently involved in what I term "constructive resistance." That is, not only do these movements articulate dissent (and often non-compliance) with central and state government policies, they also actively seek to articulate and implement alternative development practices. Viewing the state-directed development process as inimical to local tradition and livelihood, many social movements actively affirm local identity, culture, and systems of knowledge as an integral part of their resistance. In doing so, these movements articulate locally based "terrains of resistance" to the dominating discourse of development, expressing their own counter-hegemonic (or anti-development) discourses and practices. Hence particular places become contested terrains between different social groups who assign different meanings and values to those places: between local communities and their traditional lifestyles, and the state (and private corporations) as the implementers of the development project.

The responses of state authorities to social movements vary according to the type of movement resistance and the character of the government involved. When faced with social movement challenges, governmental responses include repression, cooption, cooperation, and accommodation. Repression can range from harassment and physical beatings, to imprisonment, torture, and the killing of activists. For example, in El Salvador during 1983, 73 members of the National Association of Indian People of El Salvador (ANIS), who were involved in struggles for economic rights, were assassinated by the military (Environmental Project of Central America 1987). In concert with, or instead of, repression, governments may attempt to coopt the leadership of oppositional movements through bribes, the offer of government jobs, etc. If neither of these approaches is effective, governments may be forced to accommodate the demands of the social movement, or at least cooperate partially with the social movement by instituting certain reforms. For example, during the late 1980s in the Philippines' capital of Manila, the squatter movement, Sama Sama, was able to negotiate a partnership with the government to plan and implement the urban land reforms that the movement had demanded (Parnell 1992).

As mentioned, social movements can be defensive or assertive (or both) in character, attempting to resist and reshape the development project. Examples of defensive resistance are widespread and include recent struggles in Brazil and Sarawak. In Brazil during the 1980s, numerous conflicts developed over access to, use, and exploitation of the Amazon and its forest and mineral resources. For example, between 1985 and 1987 the number of people involved in land conflicts rose from 566,000 to 1,363,729. These included indigenous tribes, gold miners, rubber tappers, land speculators, and mining and timber concerns. One conflict, in the Amazonian state of Acre, pitted land speculators and ranchers against rubber tappers and subsequently indigenous groups. In order to move the rubber tappers off the land, the ranchers and speculators engaged in the wholesale burning of the tappers' houses and crops, and destruction of the rubber trees. In resistance, a rubber tapper, Chico Mendes, organized the rubber tappers to conduct direct action in the form of *empates*, or stand-offs. When confronted with the destruction of the rubber trees, men, women, and children of the rubber tapper communities began blocking the tree clearers to prevent them from carrying out their work. During the 1980s literally millions of acres were saved by *empates*, and the rubber tappers joined forces with various indigenous groups to form the Forest People's Alliance to resist ranching and logging practices in the area (Hecht and Cockburn 1990).

In Sarawak, between 1963 and 1985, over 2.8 million hectares of forest were logged by commercial timber companies to trade tropical hardwoods on the world market. The massive deforestation led to the destruction of forest wildlife, pollution, and massive soil erosion into the river systems which decreased fish supplies. The forests are the traditional homes of the

Orang Ulu peoples, which include the Penan, Kelabit, and Kayan communities. The Orang Ulu depend entirely on the forest and river resources for their subsistence. In 1987, in response to the logging of their homes, the Penan, Kelabit, and Kayan communities formed human barricades across the logging tracks in an attempt to prevent the further destruction of the forests. Within three months blockade sites had been established along a 150 km length of road in Sarawak. One of the corporations targeted was the Limbang Trading Company, owned by Malaysia's Minister for Environment and Tourism. As a result, the Orang Ulu faced severe police repression. However, their struggle has received support from the Sahabat Alam Malaysia (Friends of the Earth in Malaysia) non-governmental organization, and continues to resist deforestation (Sahabat Alam Malaysia 1987; Ekins 1992).

In their assertive forms of resistance social movements are attempting to reshape the development project by articulating alternative practices at both the local and national levels. For example, the Sarvodaya Shramadana Movement in Sri Lanka is active in 8,000 of Sri Lanka's 23,000 villages (incorporating 3 million of the country's 15 million population). The movement is involved in providing healthcare, education, housing, irrigation, and agricultural resources via the creation of Shramadana Camps. These employ donated labor by the villagers for collective projects, each program being structured into separate autonomous organizations. In Burkina Faso the NAAM Movement has established traditional village organizations of principally young people to establish culturally appropriate agricultural projects adapted to local needs. By 1987 there were over 2,700 NAAM groups in the Yatenga area of the country with approximately 160,000 members (Ekins 1992: 113). Meanwhile, in Kenya, the grassroots Green Belt Movement has conducted reforestation programs, soil conservation practices, and set up nurseries to counter deforestation and provide incomes for local women. By the mid-1980s the movement had established about 600 tree nurseries, had planted about 2,000 green belts of at least 1,000 trees each, and had helped between 2,000–3,000 women earn their own incomes (ibid: 151).

Social movements also combine the strategies of resistance with the articulation of alternatives to development. One of the most celebrated examples is the Narmada Bachao Andolan (Save Narmada Movement, NBA) which is resisting the Narmada river valley project in India. This river, which is regarded as sacred by the Hindu and tribal populations of India, spans the states of Madhya Pradesh, Maharashtra, and Gujarat, and provides water resources for thousands of communities. The project envisages the construction of 30 major dams along the Narmada and its tributaries, as well as an additional 135 medium-sized and 3,000 minor dams. When completed, the project is expected to flood 33,947 acres of forest land, and submerge an estimated 248 towns and villages. According to independent estimates up to 15 million people will be affected by the project, either by being forcibly evicted from their homes and lands as they are submerged, or by having their livelihoods seriously damaged. With two of the major dams already built, opposition to the project has been focused on the Sardar Sarovar and Maheshwar dams – the former funded by the World Bank until 1993, when, after an independent review, the Bank withdrew its funding, and the latter partly funded by transnational corporations such as Siemens and Asia Brown Boveri (ABB). The resistance to the project has been coordinated by the Narmada Bachao Andolan, a network of groups, organizations, and individuals from various parts of India who have demanded the curtailment of the scheme.

The movement's repertoire of protest has included material struggles such as mass demonstrations, road blockades, fasts, public meetings, and disruption of construction activities. While localized protests have occurred along the entire Narmada valley, wider public attention has been drawn to spectacular events such as mass rallies and protest marches. Moreover, in addition to articulating resistance to the impacts of large dams upon the local economy and ecology, the NBA has also argued for, and practiced, sustainable irrigation and development alternatives. While the movement has been almost completely non-violent, its leaders and participants have been harassed, assaulted, and jailed by police. In one tragic event, police opened fire on a demonstration in the Dhule district of Maharashtra, killing a 15-year-old boy. The movement has attracted widespread global support from various environmental groups and non-government organizations such as the London-based Survival International. However, despite the resistance, construction of the dams continues.

The NBA provides a good example of how the different realms of social movement struggle are interrelated. In representing a threat to the ecology of the area surrounding the Narmada river, the construction of the dams also threatens the economic survival of the tribal and peasant peoples who will be evicted from their homes and lands – from which they earn their livelihoods – when the land is submerged. Moreover, these inhabitants have a profound religious connection to the landscape around the Narmada river. This spiritual connection to place – which eviction threatens to sever – intimately informs their customs and practices of everyday life. Hence opposition to the dam also articulates the inhabitants' desire for cultural survival. In addition, many of the villages that border the Narmada are demanding a level of regional autonomy, seeking "our rule in our villages," thereby articulating political demands as well (Gadgil and Guha 1995).

Globalizing Resistance

The recent creation of the WTO and its use by corporations has hastened the extension of previously local struggles to the international level. While the WTO is serving to increase the centralization of global economic policy-making, it has also provided a central object of protest. Local conflicts between citizens, governments, and transnational institutions and corporations have begun to globalize as a result of the increased uniformity of policies and international agreements among governments to implement global sets of rules. This has resulted in the perception of common interests amongst resistance formations to challenge these rules. In addition, the common ground that has begun to be articulated against neoliberalism and its agents has manifested itself in myriad local points of protest (Cleaver 1999). Ironically, such a globalizing resistance has been facilitated by the discourses of globalization, since it has allowed resistances to engage in critical analysis of the present in which no theoretical or political privilege is given, *a priori*, to experience or analysis of any social group or actor about their vision of the future. In addition, globalization has enabled the forging of new political alliances, as different social movements representing different terrains of struggle experience the negative consequences of neoliberalism (Wallgren 1998).

The recent emergence of People's Global Action (PGA) provides a fascinating example of such a process. The PGA owes its genesis to an encounter between international activists and intellectuals organized by the Zapatistas in Chiapas in 1996. In a meeting in Spain the following year that sought to build upon the Zapatista encounter, the idea of a network between different resistance formations was launched by ten social movements, including Movimento Sem Terra (Landless peasants movement) of Brazil, and the Karnataka State Farmer's Union of India. The official "birth" of the PGA was February 1998, its purpose to facilitate the sharing of information between grassroots social movements without the mediation of established non-government organizations. At the 1998 Ministerial Conference of GATT/WTO in Geneva, an alternative conference of groups from Asia, Africa, and Latin America was held under the PGA banner and convened by such groups as Movimento Sem Terra (Brazil), Karnataka State Farmer's Union (India), Movement for the Survival of the Ogoni People (Nigeria), the Peasant Movement (Philippines), the Central Sandinista de Trabajadores (Nicaragua), and the Indigenous Women's Network (North America and the Pacific). From this meeting a manifesto was established which called for direct confrontation with transnational corporations and an end to globalization.

However, the PGA is not an organization. Rather, it represents a *convergence space* of social movements, resistance groups, and individuals from across the world. Its main objectives are (1) inspiring the greatest number of persons, movements, and organizations to act against corporate domination through non-violent civil disobedience and people-oriented constructive actions; (2) offering an instrument for coordination and mutual support at the global level for those resisting corporate rule and the capitalistic development paradigm; and (3) giving more international projection to the struggles against economic liberalization and global capitalism.[1] It was the PGA network, along with other movements, that put out the call for the recent global days of action against capitalism, such as the event actions of June 18, 1999, and the protests against the WTO in Seattle.

Globalizing resistance is all about creating networks: of communication, solidarity, information sharing, and mutual support. The core function of networks is the production, exchange, and strategic use of information. The speed, density, and complexity of such international linkages has grown dramatically in the past twenty years. Cheaper air travel and new electronic communication technologies have speeded up information flows and simplified personal contact among activists (Keck and Sikkink 1998). Indeed, information-age activism is creating what Cleaver (1999: 3) terms a "global electronic fabric of struggle" whereby local and national movements are consciously seeking ways to make their efforts complement those of other organized struggles around similar issues. Certainly, the use of telecommunications has the potential to alter the balance of power in social struggles. This is in part effected by the refusal of social movements to accept the boundaries of communication taken for granted by established systems of domination (e.g., states). Through their use of media vectors, social movements can escape the social confines of territorial space, upon which much of the legitimacy of the state is predicated. Indeed, the globalization of communications provides new opportunities for decentralized political practices, as many social movements increasingly locate their strategies within local and translocal spaces as well as national and transnational spaces (Adams 1996: 419).

For example, the Narmada Bachao Andolan has conducted its resistance simultaneously at several scales. It has grounded its struggle against the dams in the villages along the Narmada valley, mobilizing tribal and cash-crop peasant farmers to resist displacement. The NBA has been able to use its local knowledge of the valley to facilitate communication between disparate communities, and to mobilize tens of thousands of peasants to resist the dams. The NBA has also taken its struggle to non-local terrains, such as the national level – through serving writ petitions to the Supreme Court of India and through participating as a convener for the National Alliance of People's Movements – a coalition of different social movements in India collectively organizing to resist the effects of liberalization upon the Indian economy. The NBA has also forged operational links with various groups outside of India, such as the solidarity group Narmada UK (who recently climbed the Millennium Wheel in London in protest against the Narmada dams) and the International Narmada Campaign, which consisted of a broad alliance of interest groups and NGOs whose terrain of resistance was that of international lobbying against the World Bank's financial support for the largest of the Narmada dams, the Sardar Sarovar (Udall 1997).

What characterizes convergence spaces such as the PGA is a fragmented geography, one that is heterogeneous, fluid, and discontinuous, where the virtual geography of the internet and other media vectors becomes entangled with the materiality of place, local knowledge, and concrete action. It is comprised of myriad grounded material struggles in particular places as well as a globalizing network of alliances that are attempting to share information, support one another, and coordinate various struggles. Hence both the Zapatistas and the Narmada Bachao Andolan engage with the convergence space of the PGA network. Some of the globalizing forms of resistance to neoliberal development may be characterized as (1) *globalized local actions*, which are political initiatives which take place either at the same or different times, in different locations across the globe, in support of particular locally based struggles (such as the various solidarity actions that have taken place around the world in support of the Zapatistas and the Narmada dam struggles), or against particular targets (such as the global day of action against global capitalism on June 18, 1999, when there were over 100 demonstrations in 40 different countries); and (2) *localized global actions*, whereby different social movements and resistance groups coordinate around a particular issue or event in a particular place, such as the November 30, 1999 global day of action against the WTO in Seattle, USA.

Conclusion

Despite its claims to bring prosperity and the alleviation of poverty through economic growth, the development project has caused enormous environmental destruction, and the impoverishment, displacement and, at times, cultural ethnocide of poor and landless peasants, urban workers in both the formal and informal sectors, women, and tribal peoples. In resistance to these processes, social movements have emerged throughout the world, attempting to protect their homes, lands, and cultures. Within the existing culture of domination created by the neoliberal development project, there is a growing realization among many social movements that they should not only resist development projects, but also articulate alternatives to the dominant economic culture. Thus many movements are regenerating traditional health practices, re-embedding learning in local culture, regenerating environmentally sustainable agriculture, and recovering a definition of their own needs and autonomous ways of living that were dismantled and redefined by the development project. In addition, social movements have made visible the particular ideology of development, its inherent injustice and non-sustainability. In doing so they are both resisting the modernization process and articulating alternatives – economic, ecological, political, and cultural. While these movements tend to be active in locally based contexts, the issues that they address – such as ecologically sustainable practices – have both national and international importance. As a result, globalizing resistance networks are emerging – such as the PGA – wherein locally based social movements are attempting to communicate with one another, share information, and provide mutual support. Although only in its nascent stage, such globalizing resistance represents potentially potent material and representational challenges to the exploitation and domination of neoliberal development.

Note

1   The hallmarks of the PGA are: (1) A very clear rejection of the WTO and other trade liberalization agreements (like APEC, the EU, NAFTA, etc.) as active promoters of a socially and environmentally destructive globalization. (2) A rejection of all forms and systems of domination and discrimination including, but not limited to, patriarchy, racism, and religious fundamentalism of all creeds. We embrace the full dignity of all human beings. (3) A confrontational attitude, since we do not think that lobbying can have a major impact in such biased and undemocratic organizations, in which transnational capital is the only real policy-maker. (4) A call to non-violent civil disobedience and the construction of local alternatives by local people, as answers to the action of governments and corporations. (5) An organizational philosophy based on decentralization and autonomy. (See the PGA website at: http://www.agp.org.)

Further Reading

For a comprehensive critique of development discourse from historical and anthropological viewpoints, see W. Sachs (ed.) (1992): *The Development Dictionary* (London: Zed Books). For a wide-ranging discussion of local and global environmental development issues and social movements, see W. Sachs (ed.) (1993): *Global Ecology* (London: Zed Books) and M. Watts and R. Peet (eds.) (1996): *Liberation Ecologies* (London: Routledge). Two books dealing with the effects of development in different cultural contexts are: W. Fisher (ed.) (1997): *Toward Sustainable Development*, which addresses issues of development and resistance in India's Narmada valley; and S. Hecht and A. Cockburn (1990): *The Fate of the Forest* (New York: Harper Collins), which analyzes the destruction of the Amazon rainforest and resistance to it. For a discussion of grassroots initiatives within the context of globalization (including the Zapatistas), see G. Esteva and M. S. Prakash (1998): *Grassroots Postmodernism* (London: Zed Books). For an analysis of globalizing resistance networks, see R. Cohen and S. M. Rai (2000): *Global Social Movements* (London: Athlone Press).

For further information on particular social movements mentioned in this chapter, see the following websites:

Zapatistas: www.eco.utexas.edu/faculty/Cleaver/zapsincyber.html
Narmada Bachao Andolan: www.Narmada.org
People's Global Action: www.agp.org

21 World Cities and the Organization of Global Space

Paul L. Knox

Between 1980 and 2000 the number of city-dwellers worldwide rose by 1.1 billion. Cities now account for almost half the world's population. There are 372 metropolitan areas of a million or more people and 45 with over 5 million. Looking ahead, population projections for 2010 suggest that there will be around 475 cities with a population of a million or more, including about 55 of 5 million or more. Urbanization on this scale is a remarkable geographical phenomenon, the manifestation of one of the most important sets of processes shaping the world's landscapes. From small market towns and fishing ports to megacities of millions of people, the urban areas of the world are the linchpins of human geographies. They have always been a crucial element in spatial organization and the evolution of societies, but today they are more important than ever.

Some cities, though, are more important than others. Ever since the evolution of a world system in the sixteenth century, certain cities – world cities – have played key roles in organizing space beyond their own national boundaries. In the first stages of world-system growth, these key roles involved the organization of trade and the execution of colonial, imperial, and geopolitical strategies. The world cities of the seventeenth century were London, Amsterdam, Antwerp, Genoa, Lisbon, and Venice. In the eighteenth century Paris, Rome, and Vienna also became world cities, while Antwerp and Genoa became less influential. In the nineteenth century Berlin, Chicago, Manchester, New York, and St. Petersburg became world cities, while Venice became less influential. Today, with the globalization of the economy, the key roles of world cities are concerned less with the deployment of imperial power and the orchestration of trade and more with transnational corporate organization, international banking and finance, supranational government, and the work of international agencies. World cities have become the control centers for the flows of information, cultural products, and finance that collectively sustain the economic and cultural globalization of the world. World cities also provide an interface between the global and the local. They contain the economic, cultural, and institutional apparatus that channels national and provincial resources into the global economy and that transmits the impulses of globalization back to national and provincial centers.

Today's world cities are both cause and effect of economic, political, and cultural globalization. They must be seen as the product of the combination of a new international division of labor, of the internationalization of finance, of the global strategies of networks of transnational corporations, and of the proliferation and increasing influence of international non-governmental and inter-governmental organizations (NGOs and IGOs) – all of which have been facilitated by new modes of regulation and by revolutionary process and circulation technologies. At the same time, world cities must be seen as the places and settings through which large regions of the world are articulated into the space of global capital accumulation: centers of economic, cultural, and political authority that give shape and direction to the interdependent forces of economic, political, and cultural globalization. The result is a "smaller" world in which our lives are lived and shaped through the global metropolitanism of "larger" cities. These world cities – no more than a couple of dozen of them altogether – are not necessarily the largest in terms of population, but they are the most capacious in terms of economic and cultural capital and innovation. Embedded within them are the nodal points of a "fast world" of flexible production systems and sophisticated consumption patterns. This fast world currently extends from the world's triadic economic core to its dependent theaters of accumulation in the megacities of semi-peripheral and peripheral world regions, and altogether involves about 900 million of the world's 6 billion population. Its corollary is the "slow world" of catatonic rural settings, declining manufacturing regions, and disadvantaged slums, all of which are increasingly disengaged from the culture and lifestyles of world cities. Yet these slow worlds are not altogether separate from the global metropolitanism of world cities. Both the internal and the external proletariat of world cities contribute capital (economic, human, and cultural) to the cause of global metropolitanism, and in so doing they are unavoidably inscribed into the economic and cultural landscapes of world cities.

Globalization and Urbanization

As we have seen in earlier chapters, globalization is by no means a new phenomenon; nor is it novel that a few major cities play a key role in the capital accumulation circuits of the world system. A globalized infrastructure of unitary nation-states, international agencies and institutions, global forms of communication, a standardized system of global time, international competitions and prizes, and shared notions of citizenship and human rights had all been established by the mid-nineteenth century, with roots stretching back to the sixteenth century. The recent acceleration of globalization and the emergence of a distinctive generation of world cities was grounded in the twentieth-century development of this legacy. What is distinctive about the globalization of the late twentieth century is, first, that there has been a decisive shift in the proportion of the world's economic activity that is transnational in scope (Sassen 1997). At the same time, there has been a decisive shift in the nature and organization of transnational economic activity, with international trade in raw materials and manufactured goods being eclipsed by flows of goods, capital, and information that take place within and between transnational conglomerate corporations (Castells 1996). A third distinctive feature, interdependent with the first two, is the articulation of new worldviews and cultural sensibilities – notably the ecological concern with global resources and environments and the postmodern condition of pluralistic, multicultural, non-hierarchical, and decentered world society (Bauman, 1998). All this adds up to an intensification of global connectedness and the constitution of the world as one place – at least for that portion of the world's population that is in fact tied in to global systems of production and exchange and to global networks of communication and knowledge. For most residents of the fast world there has been a profound redefinition of their roles as producers and consumers, and an equally profound reordering of time and space in social life.

Much of this change has been transacted and mediated through world cities, the nodal points of the multiplicity of linkages and interconnections that sustain the contemporary world economy (Hall 1996; Short and Kim 1999). Yet world cities themselves have to be understood not only as the legacy of past phases of globalization and urbanization, but also as the product of enabling technologies, of the strategies of transnational corporations, and of the responses of local, national, and supranational governments and institutions. In this context we can draw on several aspects of the geoeconomic and geopolitical change described in Parts I and II of this volume:

1   The "New International Division of Labor" (see chapter 5), which has resulted in a locational hierarchy, the top of which is constituted by the concentration of high-level management functions, mostly in major cities of the world's core economies. It has been the expanded management, planning, and control operations of transnational corporations that have formed the nucleus of contemporary world-city formation.

2   New production technologies and advances in telematics that have made for a more variable geometry of economic activity, a faster-paced economic and social environment, and a compression of space and time around the world. The development of these technologies, however, has

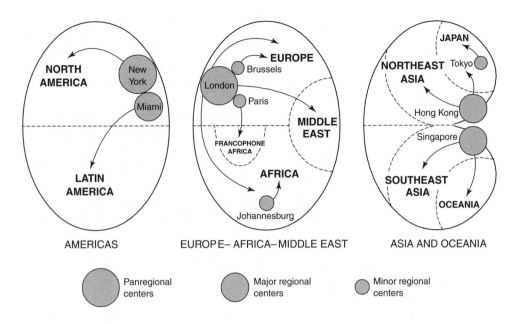

Figure 21.1 Regional world cities and their spheres of influence.

followed patterns of initial economic advantage and geopolitical influence. Three main circuits of traffic (the Americas, Europe/Middle East, and Asia/Oceania) have come to dominate the "space of flows" of the informational economy (figure 21.1; see also Castells 1996; Sklair 1999). Cities with major concentrations of transnational corporate headquarters, of international news, information, and entertainment services, and of international business services – world cities – have naturally succeeded to nodal positions in this framework.

3 The trend toward global neo-Fordism (see chapter 3), which has required more sophisticated, internationalized financial and business services. This in turn has resulted in the renewed importance of major cities as sites not only for management and coordination, but also for servicing, marketing, innovation, the raising and consolidation of investment capital, and the formation of an international property market (Sassen 1999). These trends have been reinforced by the corporate strategies of transnational companies, including joint ventures, alliances, and global networks.

4 A new mode of regulation has been created which features public–private cooperation, selective trade reforms, less restrictive labor laws, and heavy subsidies for telematics, high-tech infrastructure, and science and technology with commercial potential. The ideology of

competitiveness attached to this new mode of regulation has resulted in a distinctive geopolitics of "techno-nationalism" (Petrella 1991), whereby state policies in the leading countries have generally protected the interests of those involved in commercial innovation and corporate control, which has in turn fostered the development of world cities within the core of the world system.

5 A proliferation of transnational, non-government organizations (NGOs) – partly as a result of global geopolitics (see chapter 7), and partly in response to economic globalization. Between 1973 and 2000 the total number of transnational NGOs tripled, from around 2,000 to more than 6,000. With this proliferation there has also been a consolidation and a localization of transnational NGOs in centers of international politics and mediation: London, Paris, New York, Brussels, Strasbourg, Geneva, Vienna, and Helsinki.

It must be acknowledged that these same changes have made for a great deal of economic decentralization: from core economies to semi-peripheral ones, from rustbelts to sunbelts, from metropolitan areas to smaller towns and cities, and from downtowns to suburbs and edge cities. World cities are not so much an exception to this decentralization as they are a consequence and shaper of it. As Amin and Thrift (1992) point out, centeredness is essential within a globalized world economy. First, centers – world cities – are needed for their *authority*: their knowledge structures and their ability to generate and disseminate discourses and collective beliefs relating to economic strategies and business climate. Second, they are needed for their ability to sustain settings of *sociability*, in which key actors can gather information, establish and maintain coalitions, and monitor implicit contracts. Third, they are needed for their ability to foster *innovation*: places where there are sufficient numbers with the specialized knowledge to identify gaps in markets, develop new uses for technologies, and produce innovations; where there is sufficient mass in the early states of innovation; and where social networks provide rapid reactions within a sophisticated market.

World Cities

World cities, then, are nodal points that function as control centers for the interdependent skein of material, financial, and cultural flows which, together, support and sustain globalization. They also provide an interface between the global and the local, containing economic, sociocultural, and institutional settings that facilitate the articulation of regional and metropolitan resources and impulses into globalizing processes while, conversely, mediating the impulses of globalization to local political economies. As such, there are several functional components of world cities:

- They are the sites of most of the leading global markets for commodities, commodity futures, investment capital, foreign exchange, equities, and bonds.
- They are the sites of clusters of specialized, high-order business services, especially those which are international in scope and which are attached to finance, accounting, advertising, property development, and law (Beaverstock, Smith, and Taylor 1999; Leslie 1995; Moulaert and Djellal 1995; Warf 1996).
- They are the sites of concentrations of corporate headquarters – not just of transnational corporations but also of major national firms and of large foreign firms (Godfrey and Zhou 1999).
- They are the sites of concentrations of national and international headquarters of trade and professional associations.
- They are the sites of most of the leading NGOs and IGOs that are international in scope (e.g., the World Health Organization, UNESCO, ILO (International Labor Organization), the Commonwealth Lawyers' Association, the International Federation of Agricultural Producers).
- They are the sites of the most powerful and internationally influential media organizations (including newspapers, magazines, book publishing, satellite television), news and information services (including newswires and on-line information services), and culture industries (including art and design, fashion, film, and television).

There is a great deal of synergy in these various functional components. A city like New York, for example, attracts transnational corporations because it is a center of culture and communications. It attracts specialized business services because it is a center of corporate headquarters and of global markets; and so on. At the same time, different cities fulfill different functions within the world system, making for different emphases and combinations of functional attributes (i.e., differences in the *nature* of world-cityness), as well as for differences in their absolute and relative localization (i.e., differences in the *degree* of world-cityness).

As Short et al. (1996) have noted, the difficulty of obtaining reliable comparative data makes it very difficult to undertake empirical research on these attributes of world cities. Nevertheless, world cities have come to be regarded as settings with distinctive attributes. John Friedmann (1986), writing largely in the context of the New International Division of Labor, hypothesized that world-city formation would result in metropolitan restructuring to accommodate not only the physical settings for concentrations of international activities and their supporting infrastructure, but also the new class fractions and the spatial and class polarization that is consequent upon evolving local labor and housing markets. The linkages between world cities, along with their relationships to processes of globalization, have been subject to rather less attention. World-system theory tends to portray world cities as the "cotter pins" that hold together the global

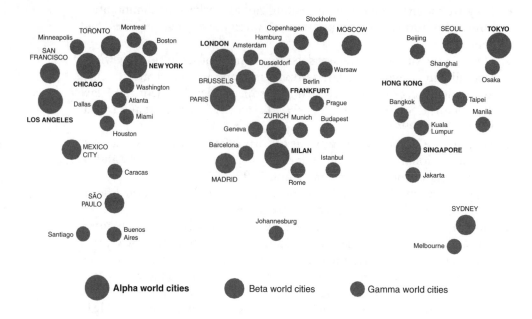

Figure 21.2   The GaWC inventory of world cities.

hierarchy of core, semi-periphery, and periphery. This fits comfortably with the widely held notion of a global hierarchy of world cities (Friedmann 1994). Analyses of key world-city functions (international accountancy, advertising, banking, and legal services) in 122 cities by the University of Loughborough's Globalization and World Cities (GaWC) Research Group suggests a threefold hierarchy (figure 21.2). At the top of the hierarchy are ten "alpha" world cities, each of global significance in all four service areas. Not surprisingly, the cities with the highest scores are London, Paris, New York, and Tokyo. A second tier of ten "beta" world cities, headed by San Francisco, Sydney, Toronto, and Zürich, is of global significance in three of the four key world-city functions. Beneath these in the hierarchy are 35 "gamma" world cities, each of global significance in two of the four key world-city functions; these include Amsterdam, Berlin, Miami, Osaka, Rome, and Washington.

World Cities and Metrocentric Global Cultures

Throughout the urban system represented by this hierarchy, the "transnational practices" of the "transnational producer–service class" necessary to globalization have begun to generate new cultural structures and processes which echo and reverberate through the daily practices and spatial organization of the rest of the "fast" world. The global metropolitanism resulting from these transnational practices is not merely a state of interconnectedness and a shared, materialistic culture-ideology of consumerism (Sklair 1999). It also involves, at various levels, not only cultural homogenization and cultural synchronization, but also cultural proliferation and cultural fragmentation. It involves both the universalization of particularism (i.e., dissolving the traditional boundaries of space and time, the relativism of postmodernity continuously propagating and redefining uniqueness, difference, and otherness) and the particularization of universalism (i.e., crystallizing transnational practices around specific regional, class, gender, and ethnic groups) (Robertson 1992).

This global metropolitanism is closely tied to the time-space compression of the world and the speeding up of production and consumption, of politics, and development. Yet globalization involves much more than the speeding up and spreading out of people's activities. While traditional links to family, neighborhood, region, and nationality are subverted by high-tech, high-speed networks and devices, *quantitative* changes – more decisions, more choices, more mobility, more interaction, more objects, more images – become *qualitative* – new lifestyles, new worldviews. In short, the global metropolitanism of world cities facilitates new processes affecting the construction of identity, new forms of meaning, and new movements that can take on material representations (King 1997; Oncu and Weyland 1997).

Global metropolitanism is, of course, closely tied to the material culture promoted by transnational capitalism: designer products, services, and images targeted at transnational market niches, promoted through international advertising agencies, the motion picture industry, and television series. One common interpretation of this is that it represents the homogenization, universalization, and "Americanization" of global culture through the economic and political hegemony of the United States and of US-based transnational corporations. But though the world may "dream itself" to be American, this means different things to different people, depending on the ways in which American imagery is appropriated and, more often than not, subverted by the mere fact of becoming iconic (Olalquiaga 1992). Furthermore, although American-based transnational capitalism and media may be the indisputable locus of pop-culture mythologies, it is clear that America has difficulty competing with Japanese cars, cameras, and hi-fi systems, Italian design, German engineering, French theory, and British television comedy shows. Meanwhile, it is now clear that the recourse to Orientalism – the Western worldview that developed as a repository for all the exotic differences and otherness repressed or cast out by the West as it sought to construct a coherent identity (Said 1978) – has had to yield to the plural histories, diverse modernities, and alternative moral orders uncovered by globalization.

Rather than suggesting cultural homogenization, then, global metropolitanism invokes a differential and contingent reach, through world cities, that embodies tensions and oppositions rather than convergence and uniformity. McGrew (1992) characterizes these oppositions as follows:

- *Universalism vs. Particularism*: although globalization tends to universalize many spheres of social life (e.g., the iconography of materialistic consumerism, the idea of citizenship, the ideology of the nation-state), it also provokes a sociospatial dialectic in which the social construction of difference and uniqueness results in particularism (e.g., the resurgence of regional and ethnic identities).
- *Homogenization vs. Differentiation*: just as globalization fosters similarity in material culture, institutions, and lifestyles, the "differential of contemporaneity" means that "global" tendencies are articulated and imprinted differentially in response to varying local circumstances.
- *Integration vs. Fragmentation*: the functional integration of labor markets, consumer markets, political institutions, and economic organizations that unites people across traditional political boundaries also gives rise to new cleavages. Labor, for example, has become fragmented along lines of race, gender, age, and region.
- *Centralization vs. Decentralization*: another aspect of the sociospatial dialectic provoked by globalization involves new movements (e.g., localized environmental movements and the "postmodernism of resistance" that seeks to deconstruct modernism: see Foster 1985) in opposition to concentrations of power, information, and knowledge.
- *Juxtaposition vs. Syncretization*: whereas time-space compression and global economic interdependence tend to juxtapose civilizations, lifestyles, and social practices, they can also fuel and reinforce sociocultural prejudices and sharpen sociospatial boundaries.

The global metropolitanism mediated and reproduced by world cities is thus complex, dynamic, and multidimensional. The most direct contribution of world cities to global metropolitanism stems from the critical mass of what Sklair (1999) dubbed the transnational producer–service class, with its "transnational practices" of work and consumption. These are the people who hold international conference calls, who send and receive international faxes and e-mails, who make decisions and transact investments that are transnational in scope, who edit the news, design and market international products, and travel the world for business and pleasure. Theirs is a transcultural environment, one of entanglement, intermixing, and hybridity (Iyer 2000; Welsch 1999).

World cities not only represent their workplaces but are the proscenia for their materialistic, cosmopolitan lifestyles, the crucibles of their narratives, myths, and transnational sensibilities. These new sensibilities are, in turn, adopted by the mass-market consumers of the fast world. The lingua franca of this populist dimension of global metropolitanism is the patois of soap operas and comedy series; its dress code and worldview are taken from music videos and the sports pages, its politics from cyberpunk magazines, and its lifestyle from promotional spots for Budweiser, Carlsberg, Levi's, Pepsi, Nike, Sony, and Volkswagen.

Of course, the more this global pop culture draws from the hedonistic materialism of the transnational elite, the more the latter is driven toward innovative distinctiveness in its attitudes and material ensembles. The more self-consciously stylish the transnational bourgeoisie, the more tongue-in-*chic* the wannabes and the cyberpunks. As this dialectic has unfolded (via global networks in television and advertising), more people have come to see their lives through the prisms of others' lives, as presented by mass media. Consequently, *fantasy* has become a social practice characteristic of global metropolitanism. But these fantasies, too, become caught in the sociospatial dialectic of the fast world, the result being the further confusion of spatial and temporal boundaries and the collapse of many of the conventions that formerly distinguished fantasy from reality. The cognitive space of world cities, emptied of traditional referential signifiers, thus comes to be filled with *simulations*: iconographies borrowed from other times, other peoples, and other places (Olalquiaga 1992). Nowhere is this more apparent than in the large, set-piece developments that have come to characterize the built environment of world cities (and aspiring world cities): the "variations on a theme park" (Sorkin 1992) that constitute the landscapes of transnational power.

These, however, are but one dimension of the reorganization of space within and between world cities. The "space of flows" of the "informational economy" is made manifest through a series of sociocultural flows that both reflect and reproduce global metropolitanism. Though they are by no means isomorphic in structure, these flows may be conceptualized in categorical terms. Appadurai (1997) suggests that there are five principal categories:

- *Technoscapes*, produced by flows of technology, software, and machinery disseminated by transnational corporations, supranational organizations, and government agencies.
- *Finanscapes*, produced by rapid flows of capital, currency, and securities, and made visible not only through teleports and concentrations of financial service workers but also through the rapidly changing geography of investment and disinvestment.
- *Ethnoscapes*, produced by flows of business personnel, guestworkers, tourists, immigrants, refugees, etc.
- *Mediascapes*, produced by flows of images and information through print media, television, and film.
- *Ideoscapes*, produced by the diffusion of ideological constructs, mostly derived from Western worldviews, e.g., democracy, sovereignty, citizenship, welfare rights.

To these I would add a sixth category – "Commodityscapes" in Appadurai's terminology – produced by flows of high-end consumer products and services: the ensembles of clothes, interior design, food, and personal and household objects that are the signifiers of taste and distinction within the culture-ideology of consumption propagated by the transnational producer–service class. Taken together, these flows are just as important to the organization of global space and to core–periphery patterns as were the flows of raw materials and manufactured products to earlier phases of capital accumulation.

World Cities as Places

The geographical overlay of the economic, built, and social environments that creates a sense of place for both residents and visitors is inevitably ruptured by the need to restructure both metropolitan form and metropolitan labor markets in response to the imperatives of global capital accumulation. Just as inevitably, the ubiquity of transnational architectural styles, retail chains, fast-food chains, clothing styles, and music, together with the ubiquitous presence of transcontinental immigrants, business visitors, and tourists, tends to propagate a sense of placelessness and dislocation (Augé 1995; Iyer 2000). Ethnographers have often stressed this in terms of *deterritorialization*: "As groups migrate, regroup in new locations, reconstruct their histories, and reconfigure their ethnic 'projects,' the *ethno* in ethnography takes on a slippery, non-localized quality. Groups are no longer tightly territorialized, spatially bounded, historically unselfconscious, or culturally homogeneous" (Appadurai 1991: 191).

Yet the common experiences engendered by globalization are still mediated by local reactions. The structures and flows of global metropolitanism are variously embraced, resisted, subverted, and exploited as they make contact with specific political economies and sociospatial settings. In the process, places are reconstructed rather than effaced. Often (and perhaps unexpectedly) it involves *reterritorialization* and a revaluation of place. Strassoldo (1992: 46–7) puts it this way:

> Postmodern man/woman, just because he/she is so deeply embedded in global information flows, may feel the need to revive small enclaves of familiarity, intimacy, security, intelligibility, organic–sensory interaction in which to mirror him/herself . . . The possibility of being exposed, through modern communication technology, to a near infinity of places, persons, things, ideas, makes it all the more necessary to have a center in which to cultivate one's self. The easy access of the whole world, with just a little time and money, now gives new meaning to a need for a subjective center – a home, a community, a locale – from which to move and to which to return and rest.

World cities have also come to be special kinds of cultural spaces, sites for the construction of new cultural and political identities, for new discourses, texts, and metaphors through which the struggle for place is enacted. One way of interpreting the outcomes of these processes is in terms of *loyalty shifts* (DiMuccio and Rosenau 1992). In terms of the organization of global space, the most significant of these are *outward* shifts, whereby loyalties are redirected toward entities (e.g., transnational employers), classes (e.g., Sklair's transnational producer–service class), supranational organizations (e.g., the European Union), or movements (e.g., global ecology, human rights).

But in terms of the organization of metropolitan space and the social construction of place the most significant loyalty shifts are *inward*. At one level, we can see these shifts in people's apparent need for stability, identity, and centeredness within the infinite relativism of postmodernity. This impulse has been articulated through housing markets (most strikingly through gentrification and through private master-planned communities) and commodified through neo-traditional urban design and the merchandising of local histories. Among the affluent within the fast world, reterritorialization thus results from colonizing or invading spaces that can be given both social meaning and spatial identity. But perhaps the most dramatic examples of reterritorialization are those deriving from the lived experience of low-income transnational migrants, exiles, and refugees. Within world cities such groups are able to establish new networks and new cultural practices that define new spaces for daily life. In addition to the transformation and adaptation of old neighborhoods and obsolete sociocultural spaces within consolidating ethnic enclaves, this involves the emergence of otherwise marginalized voices and alternate representations. These voices and representations can be carried over into the "host" society and carried back into the "homeland." The latter has the potential, at least, for a kind of "transnational grassroots politics" (Smith 1994). The former not only contributes to the cosmopolitanism of world cities but also has the potential (realized only in a few world cities – London, New York, San Francisco, Los Angeles, and Sydney) – to foster distinctive and innovative multicultural spaces – the latest phase in the sociospatial dialectic of global metropolitanism.

22 The Emerging Geographies of Cyberspace

Rob Kitchin and Martin Dodge

Since the invention of the telegraph in the early nineteenth century, information and communication technologies (ICTs) have become increasingly sophisticated and integral to social and economic life. The successive developments of the telephone, telex, fax, and mobile telephone have enabled significant space-time compression through instantaneous communication over great distances (Brunn and Leinbach 1991). Over the last thirty years, since the creation of the internet,[1] the role and importance of ICTs has grown significantly; an estimated 304 million people are connected to the internet, with this figure growing 30 percent annually (Nua 2000), and billions of dollars are being invested in infrastructure and content production by large corporations and venture capitalists. As discussed in chapters 2 and 21 of this volume, it is now well recognized that ICTs, and in particular the internet and intranets,[2] are instigating significant cultural, social, political, and economic effects at all geographic scales from the body to the global (Kitchin 1998). It has been argued recently that ICTs and the conceptual space they support – cyberspace – have a number of implications for how the following are constituted, conceived, and theorized: identity (Turkle 1995); the body (Haraway 1991); community (Smith and Kollock 1999); democracy (Loader 1997); employment (Castells 1996); urban and regional development (Graham and Marvin 1996); and accessibility to goods and services (Janelle and Hodge 2000).

In many of these accounts it is hypothesized that the changes are occurring because the role, importance, and nature of space is changing, with the relations between people and space being reconfigured in complex ways. It is contended that ICTs are important transformative agents which are helping to reconfigure the spatial logic of modern society (Mitchell 1996). ICTs are leading to massive time-space compression, with the instantaneous communications of the internet, intranets, and mobile telephony precipitating the large-scale reconfiguring of spatial and temporal boundaries. This reconfiguring, some speculate, will eventually lead to the eradication of geography (the death of distance) as a central organizing modality of society, in relation to both space and place (Cairncross 1997). Consequently, commentators such as Benedikt (1991: 10) have begun to question the "significance of geographical location at all scales," with ICTs seen as liberating and transcendent tools, freeing human life from the tyranny of material space.

Others have countered that while undoubtedly space-time relations are being reconfigured, the importance of space as an organizing principle and a constituent of social relations is not being eliminated and spatial differences and inequalities between places are, on many measures such as economic growth, becoming more pronounced (Dodge and Kitchin 2000). Moreover, there are still significant tensions and resistances between processes operating at different spatial scales, from the local to the global, so that while a complex global economic system is in place, significant variations in culture, social and political relations, and wealth remain across the globe. In this manner ICTs and cyberspace are inherently complex and often contradictory in their spatial outcomes.

Interestingly these debates over the changing nature of geographical relations and their role in understanding contemporary society have barely been extended to cyberspace itself. To date, cyberspace has been conceived and examined as largely aspatial and tellingly the *lack* of geography is considered one of the key features in the development and sustenance of online social relations. As such, many commentators have argued that cyberspace is essentially spaceless and placeless (Rheingold 1993). Indeed much of the populist rhetoric about cyberspace focuses on spacelessness as the key to its revolutionary potential (as in discussions of the marketing of online shopping and e-commerce). In this chapter we argue that this could not be further from the truth and that, to the contrary, cyberspace is ripe for geographic enquiry. We contend that the many domains of cyberspace possess both spatiality and geometry and illustrate this by examining the emerging geographies of cyberspace in relation to two themes: (1) community and (2) maps and spatializations. These are by no means the only geographies currently being examined by scholars, but they suffice to illustrate our arguments (Dodge and Kitchin 2000).

Communities in Cyberspace

> [Virtual communities are] social aggregations that emerge from the Net when enough people carry on those public discussions long enough, with sufficient human feeling, to form webs of personal relationships in cyberspace. (Rheingold 1993: 5)

Over the past decade it has frequently been argued that cyberspace allows the formation of communities that are free of the constraints of place. Instead of being founded on geographic propinquity, communities in cyberspace are sustained and grounded by communicative practice – a sense of community is based upon new modes of interaction (computer-mediated communication) and centered on common interests and affinity. As such, Rheingold (1993) suggests that personal intimacy, moral commitment, and social cohesion replace ties arising from shared location as the key constituents of maintaining a community identification and spirit. For him, cyberspace offers the unique opportunity of marrying *gemeinschaft* (where community relationships are tied to social status, public arenas, and bounded, local territory) and *gesellschaft* (where community relationships are individualistic, impersonal, private, and based on "like-minded" individuals), so that individualistic, like-minded people join forces to form public-based communities.

Online communities, according to Rheingold and others, are constructed around what their members think, say, believe, and are interested in, rather than on where they live or what the participants look like. These communities are facilitated by online media such as email, mailing lists, chat rooms, bulletin boards, and web pages within which it is thought that individual participants can circumvent the geographical constraints of the material world and take a more proactive role in shaping their own virtual community and their position within it. Jones (1995: 11) thus proclaims: "we will be able to forge our own places from among the many that exist, not by creating new places but by simply choosing from the menu of those available."

There is now little debate as to whether virtual communities exist: Anderson (1983) suggests that at a basic level all communities are imagined, and as long as members share a common imaginative structure, a community exists. Moreover, most commentators agree that many of these communities are self-sustaining and rich in diversity. Indeed Rafaeli and Sudweeks (1996) point out that people would not invest so much time and effort into online social interactions if they did not gain some sense of social cohesion or community from their virtual actions. They contend that the form and depth of interaction mean that many virtual communities are neither pseudo nor imagined, despite claims from critics (Robins 1995; Sardar 1995) because, for Rafaeli and Sudweeks, cyberspace possesses the qualities of what Castells (1996) terms "real virtuality," a reality that is entirely captured by the medium of communication and where experience is communication. Where there is significant divergence of opinion, however, is over the extent to which (1) these online communities provide an alternative to geographic communities and (2) they are really placeless. We deal with each of these debates in turn.

Cyberspace communities as an alternative to geographic communities

To Rheingold (1993), Mitchell (1996), and others communities in geographic space are fragmenting and losing cohesion due to cultural and economic globalization – a coalescing of cultural signs and symbols, increased geographic mobility, a designificance of the local, and changing social relations. In Relph's (1976: 90) terms, society is suffering increasingly from a condition of placelessness: "a weakening of the identity of places to the point where they not only look alike, but feel alike and offer the same bland possibilities for experience." For them, geographic communities no longer provide a coherent "sense of place"; instead, online communities (for the reasons discussed above) provide an alternative and an antidote to social alienation and placelessness experienced in geographic communities.

Robins (1995) has severely criticized this notion of online communities as alternatives to geographic communities. He argues that the former are at the very best self-selecting, pseudo-communities and that it is a serious misnomer to directly equate communication with communion and community, thereby questioning the quality of relationships forged and sustained through cyberspace (e.g., issues such as responsibility and respect), a sentiment echoed by Gray (1995):

> We are who we are because of the places in which we grow up, the accents and friends we acquire by chance, the burdens we have not chosen but somehow learn to cope with. *Real* communities are always local – places in which people have to put down some roots and are willing to put up with the burdens of living together. The *fantasy* of virtual communities is that we can enjoy the benefits of community without its burdens, without the daily effort to keep delicate human connections intact. Real communities can bear those burdens because they are embedded in particular places and evoke enduring loyalties. In cyberspace, however, there is nowhere that a sense of place can grow, and no way in which the solidarities that sustain human beings through difficult times can be forged. (Ibid: our emphases)

Wellman and Gulia (1999) critique the idea of cyberspace communities as alternatives to geographic communities using a different tack. They note that online and geographic communities are remarkably similar in some respects. For example, due to developments in long-distance transportation and telecommunication technologies throughout the modern period, it has long been the case that a person's community (their kith and kin) does not necessarily live within walking distance. Instead, geographic communities have been replaced by social networks spread out over a wide terrain, and sustained by letter writing, telephone conversations, and now various modes of computer-mediated communication. They observe that even when people share the same geographic space most social networks are actually sustained through telephone conversations and face-to-face contact. As such, they contend that the division between geographic and virtual is not helpful – one is simply an extension of the other. The relationship between people is what's important, not the medium of communication. Networks maintained exclusively in cyberspace are thus not pale imitations of "real" networks, or substitutions for these networks; they are merely another form of network, a subset of an individual's total network, much as pen-pals were in the era of letter writing.

Moreover, it should be noted that cyberspace is often used as a means to try to "reconnect" members of a community and foster a sense of place in a particular locale. Many Western cities now have websites and PENs (public electronic networks) devoted to community relations and development, many allowing citizens to discuss issues among themselves and with local statutory and voluntary agencies (Graham and Aurigi 1997). Further, many communities are using cyberspace to develop cross-community and cross-issue alliances to help fight particular concerns. Probably the most widely documented case of such use was by the Zapatistas of Chiapas (Mexico), who used the Web to garner international political support (Froehling 1997). In these cases, in contrast to Rheingold's replacement thesis, geographic communities are being augmented by online interactions.

Cyberspace communities as placeless communities

As noted in the introduction, cyberspace is commonly conceived as aspatial; it has no spatiality and thus no sense of place. This conception is now being challenged by a number of academics who argue that online interactions are often structured through a variety of geographic metaphors, employed to help create a "sense of place" and to provide a tangible spatiality. For example, cyberspace is replete with the vocabulary of place: nouns such as rooms, lobbies, highway, frontier, cafes; and verbs such as surf, inhabit, build, enter (Adams 1998). Couclelis (1998) describes the use of these geographic metaphors – the spatialization of cyberspace – as an attempt to translate information and communication media into domains familiar and comfortable to users. Cyberspace, these analysts contend, is literally built out of the ideas and language of place, and the employment of these metaphors to create sites of interaction engenders an online spatiality.[3] As a consequence, Taylor (1997: 190) states that "to be within a virtual world is to have an intrinsically geographic experience, as virtual worlds are experienced fundamentally as places." Indeed, if we take the definition of place provided by Jess and Massey (1995) – places are characterized as providing a setting for everyday activities, as having linkages to other locations, and providing a "sense of place" – then there can be little doubt that new places, and new spatialities, are being formed online. Batty (1997: 339) thus states that the many components that comprise cyberspace – web pages, mailing lists, chat rooms, bulletin boards, MUDs (multi-user domains), virtual reality environments, information databases, online stores, and game spaces – each have "their own sense of place and space, their own geography." As yet these spatialities have been little considered by geographers, but they are becoming increasingly prevalent in people's lives, particularly as businesses provide online services to reduce transaction costs (e.g., the promotion of online banking to facilitate the reduction of expensive physical premises). Here, we illustrate the extent to which online social relations are contextualized by spatiality, and the importance of understanding this spatiality in order to comprehend online communities, through the reporting of two case studies.

Correll's (1995) study of an online lesbian cafe describes how patrons constructed an elaborate cafe setting using textual descriptions and contextualized all their interactions within this setting (for example, patrons would "buy" drinks and hang out round the jukebox). She suggests that the construction (spatialization) of this shared setting created a common sense of reality which grounded communication. In essence, the locale needed for community in geographic space was simulated online, so that place and setting remained important. Indeed, for her the spatialization of the online meeting space was the secret to the community being a success, suggesting that without the shared "reality" of the bar the community might have dissolved. This bar, however, differed in significant ways from gay bars in geographic space, "where the games are for real" (ibid: 281). Here, patrons could explore their ideas and thoughts without fear of physical or mental retribution. As such, the bar served to augment offline lives by providing a surrogate community for a group who are often marginalized within geographic communities (Bell and Valentine 1995). In this case, the cafe was providing a relatively safe space, often denied to the women offline, in which they could express and explore their sexuality.

Smith's (n.d.) study charted the process of virtual place-making as performed in shared and immersive internet VR-type environments, such as AlphaWorld.[4] These virtual worlds are popular and AlphaWorld has been visited by over 800,000 unique users since its inception in the summer of 1995, many of whom have built homesteads (as of August 2000, 64.2 million objects had been placed by the inhabitants). In order to undertake his study, Smith created a new virtual world that any person could inhabit and build within. He then monitored in detail the building of urban structures and the social interaction of inhabitants over a 30-day period (starting November 30, 1998). The plot of land he used was 3 million square meters in size and capable of supporting 32 simultaneous users. No specific guidelines were provided, although inhabitants were encouraged to visit a website[5] which detailed the experiment, and a prize was offered for the best structure built during the 30 days. Inhabitants entered the world in a town square surrounded by message billboards. Nearby a builder's yard provided a wide range of generic building blocks from which users could build structures.

The experiment revealed a number of interesting results about the socio-spatial construction of virtual worlds. Most importantly, users built a diverse range of structures, and a strong core community, who met and interacted regularly, developed. The extent of the building is evident in figure 22.1, showing "satellite"-type land-use images of urban growth over the 30 days. The first 24-hour period in particular experienced considerable

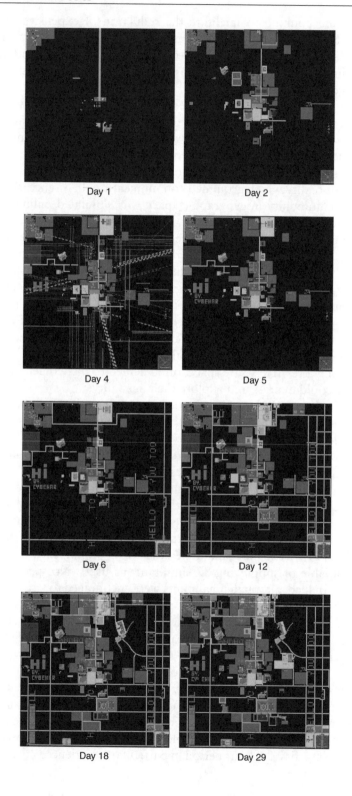

Figure 22.1   Thirty days in active worlds.

development, with 7,219 objects placed. In total, 27,699 objects were placed by 49 registered users and an unknown number of tourists, with 49 percent of all available land built on. Smith reports that a recognizable community of about ten users had already developed by the third day, appearing much sooner than he predicted. This group used the same nicknames and avatar appearances over the course of the 30 days.[6] The community developed throughout the experiment, and produced a number of communal structures (such as a temple) and undertook a number of communal events (such as all adopting Smith's avatar for a day).

In addition, the world experienced some of the more anti-social phenomena of virtual worlds like AlphaWorld. For example, on day 4 it was subjected to attack from what was self-described as the Activeworlds Terrorist Group. On this occasion over 85,000 objects were added to the world, as evidenced by the patterns of dashed lines in figure 22.1. Also some inhabitants took to "sky writing" – claiming sizeable tracts of land to spell out a message when the world is viewed from the air. The first of these appeared on day 5 ("Hi").

Using Smith's work it is possible to think of AlphaWorld as consisting of hybrid places lacking the materiality of geographic and architectural space, but yet having a powerful mimetic quality, containing enough geographical referents and structure to make them tangible. This, we suggest, engenders a level of spatiality beyond that found in other virtual media (such as email and web pages), with social interaction explicitly situated and grounded in a geographic context. As with textual MUDs, the place-like qualities of AlphaWorld provide a context in which specific forms of social interaction and experiments with identity are played out. In AlphaWorld the "sense of place" is centered around the activity of claiming land, designing and building homesteads, the means by which the space is transformed into meaningful *places*, and by social interaction between the inhabitants. Both lead to specific forms of sociospatial practice: the playing with identity, the creation of community, land disputes, virtual vandalism, and policing. These in turn are framed within a regulatory structure centered on citizenship. In essence what Smith's experiment reveals is that space, place, and sociospatial processes are central to online interactions within the Alphaworld environment, and by extension other social milieux (although the forms of spatialities might differ between domains: see Adams 1998).

The importance of spatiality in these communities is highlighted by Foster's (1997) analysis of an attempt to create a virtual community which he thinks failed because it did not achieve a "sense of place." In this case, the community was a PEN (public electronic network) seeking to revitalize a geographic locale. Instead of fostering integrated social interaction, however, the PEN disintegrated into monologues and separate spaces.

One of the principal reasons that so many analysts, particularly those of a utopian persuasion, have misunderstood cyberspace as placeless, spaceless media is because they have conceived cyberspace as a separate realm divorced from geographic space. This conception falls into the trap, as identified by Bingham (1999), of treating cyberspace as locations of the sublime (as powerful, dislocated, deterministic paraspaces).[7] We believe that cyberspace, rather than being a separate realm to geographic space, is merely an extension of it – as argued by Wellman and Gulia (1999). As such, we suggest that cyberspace is better conceived as embodied spaces (Dodge and Kitchin 2000).

Our reasoning for theorizing cyberspace as embodied spaces is because online and offline identities are not divorced. Donath (1999), in an application of Goffman's (1959) famous thesis, argues that online social interactions exhibit many of the same characteristics as those elsewhere, distinguished by "expressions given" (how one wishes to be perceived) and "expressions given off" (often unintentional messages that reveal aspects of character). In playing with identity in cyberspace many users are intentionally seeking to manipulate "expressions given" and limit those "given off." Messages "given off" almost inevitably translate disembodied spaces into embodied spaces. This is because we enter cyberspace from geographic space, and although we can play with our identity and seek to deny our geographic point of entry, our online personae are grounded in our experiences and memories of geographic space (which in turn adapt to accommodate online experiences); our online and offline identities are thus not divorced but are situated in relation to each other.

Mapping and Spatializing Cyberspace

In the previous section we discussed the extent to which cyberspace is placeless. In this section we continue that analysis to examine the extent to which it can be considered spaceless. Again, a number of analysts have speculated that cyberspace lacks space, that it is lacking geometrical (space-time) properties and is thus closed to cartographic visualization and geographic analyses. For example, Mitchell (1996: 8–9) describes cyberspace as

> profoundly *antispatial* . . . You cannot say where it is or describe its memorable shape and proportions or tell a stranger how to get there. But you can find things in it without knowing where they are. The Net is ambient – nowhere in particular but everywhere at once. You do not go *to* it; you log *in* from wherever you physically happen to be . . . the Net's despatialization of interaction destroys the geocode's key. (Original emphasis)

This, to a degree, is true. Many parts of cyberspace, due to their form (structured by the underlying network protocols and the end-user interface), lack a spatial quality (e.g., email or bulletin boards), and other spaces possess a very chaotic geometry that lacks Cartesian logic (e.g., websites). It is clear, however, from the wealth of research being conducted (Dodge and Kitchin

2000 provide an overview) that cyberspace is amenable to, and benefits from, geographic visualization and analysis. This is because cyberspace does possess space-time geometries; that in all cases there is a geography of sorts that bounds and helps define a domain and the interactions occurring within and between. For example, some domains clearly display recognizable spatial geometries, such as MUDs and virtual worlds; in other cases domains that lack a formal spatial quality have been (and can be) given one through processes of spatialization (a spatial structure is applied where no inherent or obvious one exists through the application of concepts such as hierarchy and proximity).

As such, cyberspace does have space-time geometries but they are highly complex and we are only just beginning to chart and understand them through techniques of mapping and spatialization. As we illustrate below, this project of mapping and spatializing cyberspace is important because (1) it has the potential to make cyberspace easier to search and navigate through and (2) it reveals more fully the complex relationships that exist between data and/or people online (relationships that are often hidden or difficult to determine when viewing text or hypermedia documents).

ET-map is a prototype spatialization application that provides a "big picture," an overview of the whole information space; it was developed by Hsinchun Chen and a research team in the University of Arizona's Artificial Intelligence (AI) Lab (Chen et al. 1998).[8] Its aim is to provide a navigable map of the Web, using the power of map categorization and visualization to make browsing for information easier. (It also reveals the wider, overall structure of a very complex site.) Essentially, ET-map constructs a hierarchical set of "category maps" which act as visual directories that can be interactively browsed to find particular web pages of interest (Chen, Schuffels, and Orwig 1996). Figure 22.2 displays the spatialization of over 110,000 entertainment-related web pages listed by the *Yahoo!* directory (Chen et al. 1998). The three images reveal how the maps are nested and can be browsed, in this case to locate websites related to jazz music. At each level the "category map" displays groupings of similar web pages as regularly shaped, homogeneous "subject regions," which can be thought of as virtual "fields" which all contain the same type of information "crop." The spatial extent of the subject regions is directly related to the number of web pages in that category. For example, the MUSIC subject area (figure 22.2a) contains over 11,000 pages and so has a much larger area than the neighboring area of LIVE, which only has some 4,300 pages. Clicking on a subject region with less than 200 pages takes one to a conventional text listing of the page titles. If a region has more than 200 pages, then a sub-map of greater resolution is created, with a finer degree of categorization (figure 22.2b and c). In addition, a concept of neighborhood proximity is applied so that subject regions that are closely related in content are plotted close to each other. For example, FILM and YEAR'S OSCARS, at the bottom left of figure 22.2a, are neighbors.

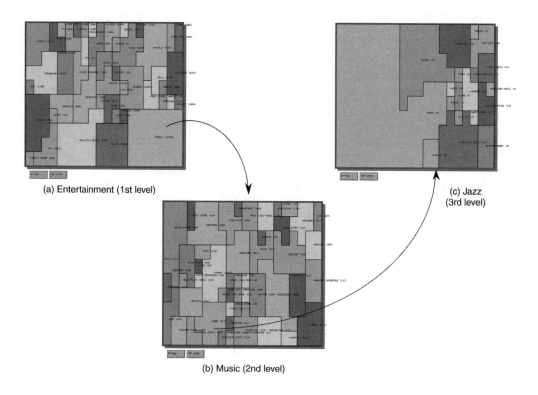

(a) Entertainment (1st level)

(c) Jazz
(3rd level)

(b) Music (2nd level)

Figure 22.2   ET-map: the hierarchical category map.

The maps are created using a sophisticated AI technique that automatically (i.e., no human supervision) analyses and classifies the semantic content of text documents like web pages (Chen et al. 1998). While quite successful, the technique is not without problems; for example, it is difficult to classify pages automatically from a very heterogeneous collection and it is not clear that the automatically derived categories necessarily match the conceptions of a typical user. From the limited usability studies conducted it appears they are good for conducting unstructured, "window shopping" browsing, but less useful for undertaking more directed searching.

There also have been attempts to spatialize the wider Web landscape with whole websites represented as singular, graphical objects. Figure 22.3 displays one such landscape created by Tim Bray (1996). In order to answer four questions (How big is it? How wide is it? Where is the center? How interconnected is it?) Bray used a large search engine index to calculate the key metrics on the structure of the known Web in 1995.[9] Examining the hyperlink structures of the Web, he found that interlinking between sites was surprisingly sparse. Most links were local, within a site, and a few key sites (e.g., *Yahoo!*) acted as super-connectors tying sites together. Bray

Figure 22.3   Webspace landscape.

derived two intuitive measures of website character based on hyperlinks: visibility and luminosity. Visibility is a measure of incoming hyperlinks, the number of external websites that have a link to a particular site. In 1995 the most visible website was that of the University of Illinois, Urbana-Champaign (UIUC), the home of the Mosaic browser. The vast majority of sites had very low visibility and nearly 5 percent had no incoming links. Measuring the reverse, the number of outgoing links, determines a site's luminosity. The most luminous sites carry a disproportional amount of navigational workload. *Yahoo!* was the most luminous site in 1995, and probably still is today.

Using these statistical characteristics Bray spatialized the key landmarks of the Web in 1995, highlighting the largest, most visible and connected sites. The resulting information landscapes, shown in figure 22.3, are dotted with 3D models which he termed ziggurats (ancient stepped pyramidal temples). Each ziggurat visualized the degree of luminosity and visibility of a single site, along with the size of the site and its primary domain (e.g., government, education, commercial, etc.). The basic graphic properties of the ziggurat – size, shape, and color – were used to encode these four dimensions. The overall height represented visibility; the width of the pole represented the size of the site, in terms of number of pages; the size of the globe atop the ziggurat indicated the site's luminosity; and color coding displayed the primary domain (green for university, blue for commercial, red for government agencies). The ziggurats were also labeled with the site's domain name for identification. The spatial layout of the ziggurats across the plane was based on the strength of the hyperlink ties between them. The model is three-dimensional and can be "flown through" and viewed from different positions. Figure 22.3 displays a field of ziggurats at the very core of the Web in 1995. Further from this core region there would be many thousands of other ziggurats, but most would be minuscule in relation to those at the heart.

These two examples demonstrate that cyberspace is not spaceless and reveal how spatializing cyberspace can aid navigation and provide a wider understanding of the Web. They are just two examples from a rapidly developing field being driven by strong commercial pressures to deliver better information interfaces and navigation tools. Interestingly, most research teams do not consist of geographers or cartographers, but information scientists (Dodge 2000 has a full catalogue of these efforts; Dodge and Kitchin 2000). At present, most maps and spatializations are experimental in nature, often with limited scope, and there is a long way to go before we really start to understand how cyberspace is organized spatially and how it might be more effectively reorganized. This has led some commentators to suggest that present maps and spatializations are little more than "eye candy" and are not effective, functional navigational aids (Nielsen 2000). While we concur with this sentiment, we are of the opinion that, over time, mappings and spatializations of cyberspace are going to become increasingly important tools for both navigating cyberspace and understanding relationships between people and data in cyberspace; in short, plotting the space-time geometries of cyberspace will be a significant area of study for geographers, cartographers, and others.

Conclusions

In this chapter we have countered the claims of some analysts that cyberspace has no geography and is essentially placeless and spaceless. We have done this through an examination of two key areas of study – community and mapping – detailing some of the emerging geographies of cyberspace with reference to both spatiality and geometry. Through these examples we have demonstrated the need for and utility of a geographic approach to cyberspace. At present, research with geographic perspectives is nascent, and while it would be unfair to say that cyberspace is a neglected area of research, more research is certainly needed before we understand more fully its spatialities and geometries.

In relation to the two areas of research, we have discussed briefly in this chapter how research needs to focus on exploring the relationship between space and online community, seeking to uncover the ways in which communities and spatialities are constructed, maintained, and disrupted, plus examining the nature of space in cyberspace and how to effectively measure and map its geometries. In the case of the former, it is important to remember that cyberspace is not a paraspace, a realm divorced from geographic space. Rather we need to consider it as a continuum, as the extension of geographic space into cyberspace, and how this affects social, cultural, political, and economic relations. We need to conceive of cyberspace as an embodied space, where online interactions are not divorced from those offline, but rather are contexted by them. This allows us to understand and embed the use of cyberspace into the context of other aspects of daily life, a practice that is lacking in much analysis, particularly that which is utopian in conception. Taking this approach, following Wellman and Gulia (1999), it is clear that cyberspace does not provide alternative communities to geographic ones, but rather supplements and augments social networks. A key research area is to investigate the processes of supplementation and augmentation.

In the case of the latter, we need to consider how cyberspace sits in relation to traditional conceptions of space and the practices of Western cartography. It is quite clear that cyberspace poses an ontological question of traditional understandings of space and those that seek to map and chart it (for a full discussion, see Dodge and Kitchin 2000). Composed of billions of lines of computer software, cyberspace is entirely a social production – it can be designed with various forms of spatial geometries that lack materiality and which are highly mutable. Charting these geometries is a difficult but an exciting challenge, which needs careful thought. As noted in relation to the traditional practice of mapmaking, this process and the products constructed also need to be scrutinized. As a consequence, we suggest five significant questions that need to be asked of those maps and spatializations that have so far been created: How "accurate" is the map? Is the map interpretable? What does the map *not* tell us? Why was the map drawn? Is the map ethical? We provide initial answers to these questions in Dodge and Kitchin (2000), but a more rigorous application is needed that extends our understanding of the spatial geometries of cyberspace and the means by which to measure and interpret them.

Notes

1   The internet consists of a global network of computers that are linked together by "wires": telecommunications technologies (cables of copper, coaxial, glass, as well as radio and microwaves). Each linked computer resides within a nested hierarchy of networks, from its local area, to its service provider, to regional, national, and international telecommunication networks. The links have all different speeds/capacities, and some are permanent, while many others are transient dial-up connections. While some networks are relatively autonomous, being self-contained spaces, almost all allow connections to other networks by employing common communication protocols (ways of exchanging information) to form a global system.

2　Intranets have the same functional forms as the internet, but are private, corporate networks linking the offices, production, and distribution sites of a company around the world. These are closed networks, using specific links leased from telecommunication providers, or employ new virtual private networking technologies with no, or very limited, public access to files (company employees with knowledge of the correct password might gain entry from a public network). For example, most banks and financial institutions have national, closed intranets connecting all its branches, offices, and ATMs (Automatic Teller Machines) to a central database facility which monitors transactions.

3　Note, there is a long history of using familiar metaphors and analogies to explain new, strange, and potentially hostile phenomena.

4　AlphaWorld is owned and managed by Activeworlds.com, Inc., a small firm based in Newburyport, MA, USA (http://www.activeworlds.com/).

5　http://www.casa.ucl.ac.uk/30days/

6　An avatar is a visual character that represents the user online.

7　Paraspace means "other space," a sublime space that has forms and practices alien to that in geographic space.

8　http://ai2.bpa.Arizona.edu/ent/

9　Comprising a mere 11 million pages from about 90,000 sites, compared to 800 million-plus in 1999 (Lawrence and Giles 1999).

Part V
Geoenvironmental Change

Introduction to Part V:
A Burden Too Far?

Very few human societies beyond the simplest have lived in harmony with their natural surroundings, so it has been rare for the environment not to be depleted as a consequence of human occupation. Some societies, such as those which practiced slash-and-burn agriculture, left the environment to recover from their depredations, but even then the prior state was rarely regained.

More complex societies have taken more from the environment than they have returned to it, and have stimulated very substantial change in their physical milieux (Simmons 1993b) – so much so that in some cases the land has never regained more than a small fraction of its former fertility (as in much of the Mediterranean littoral), whereas in many others its carrying capacity has been severely, and almost certainly permanently, depleted (Blaikie and Brookfield 1987). As a consequence, some societies collapsed because the satisfaction of their material wants was not sustainable. Others have maintained themselves by colonizing new areas, enabling them to replace those products which their home environments had formerly yielded: this colonization frequently involved marginalizing, if not destroying, pre-existing societies.

In many areas, of course, farmers learned how to restore the fertility of at least some lands – by the application of manure, for example – and through trial and error they developed methods which reduced the rate of depletion. This rate has been further slowed, though not arrested, in many parts of the world during the last five centuries or so by improvements in agricultural technology. The increased use of machinery in the past century has assisted in this reduction, as has contour ploughing; and the more recent rapid development of the chemical and biotechnology industries has aided the goal of increasing yields from the land, though in many cases with unforeseen consequences that have later created new environmental crises.

Changing Relationships with Nature

Despite this increased dominance of society over nature, and the rapid domestication of great tracts of the Earth's habitats, resource depletion has increased apace (Turner et al. 1990). By the late twentieth century many observers believed that, although much remains unknown about the physical, chemical, and biological processes involved, an environmental tragedy is looming. At some time in the relatively near future, it is argued, the Earth will be unable to meet the demands made upon it and the transformations wrought by human use of its renewable and non-renewable resources will become irreversible. Others claim that this doomsday scenario is overplayed, that we have managed to resolve such problems in the past and there is no reason to believe that human ingenuity will not cope again: in the United States, for example, food production has increased very substantially in recent decades, from a smaller proportion of the land; more, it seems, can be produced from less. The prospects of the impending biotechnology and GMO revolutions make this environmental and social problem even more charged.

Resolution of this debate may come too late, especially if the pessimists (who may well argue that "it is better to be a pessimist and be proved wrong than to be an optimist and be wrong") are correct. If the doomsday scenario is wrong – at least for the present – then we will lose little, except the time and human resources, which might be better expended elsewhere, in preparing to avoid it. Furthermore, our increased appreciation of the environmental constraints to action will assist the development of strategies to ensure sustainable development in the future (Adams 1990). Alternatively, if it is right, then if we don't seek to prevent its realization, our future is bleak indeed.

The pessimists' cause is partially supported by observation of current trends within society which are increasing the nature and pace with which we are ravaging the Earth. First, pressure on the Earth's resources is growing because of increasing population (though growth rates are declining virtually everywhere and the impact of AIDS changes the picture still further). Secondly, demands on those resources are increasing more rapidly than is the population because of greater per capita material expectations: average living standards are increasing (despite their decline in some regions), producing a variant on Malthus's still controversial predictions of nearly two centuries ago regarding the exponential growth of demand (Woods 1989). Thirdly, very few parts of the Earth's surface remain relatively untouched by these demands – the colonial "escape valves" of previous centuries have been removed – and conversion of increasing tracts to non-productive uses, through urbanization processes, reduces the amount of land available for exploitation. Fourthly, increasing production and consumption are both associated with increasing waste production and problems with its disposal – not only in the case of highly toxic wastes associated with many industrial processes and power production, but also simply with how to dispose of the great volume of wastes without despoiling vast tracts of land and/or water. Fifthly, technological advances have been such in recent decades that not only has our ability to ravage the earth been magnified many times over, but in addition we are ravaging it in new ways, through technologies that enter the core biological and chemical processes that maintain life forms.

These trends have been accentuated in recent years by the demands for "cheap food" in the world's more "developed" countries. Part of the response has been the "industrialization" of much agricultural and horticultural activity, with the main food providers (a small number of supermarket chains in most countries) dominating the production process in global chains linking farms to consumers, forcing down prices (and hence wages), while at the same time insisting on high quality. In this they are closely associated with major chemical industry firms, whose products are promoted – through technologies such as genetic modification of seeds making crops resistant to fungicides and herbicides – as effective means of increasing productivity. These new practices are stimulating unforeseen side-effects, however – as with the use of animal carcasses in feeds (in effect, animal cannibalism) which created BSE ("mad cow disease") and led to Creutzfeldt-Jakob disease in humans who ate contaminated beef; and with the rapid spread of foot-and-mouth disease among animals in the UK in 2001, which was exacerbated by a range of government policies, such as the closure of small abattoirs on health grounds, forcing long-distance movements of animals to slaughter, and the widespread movement of stock from farm to farm in order to claim subsidies. Some consumers have responded to these trends with protests against the new practices (as with the "trashing" of GM-modified crop trials in the UK) and switching to presumed healthier foods, both through vegetarianism and preferences for organically produced (and usually more expensive) foods, some of them sold by the producers at local farmers' markets rather than in supermarkets at the end of long product chains. Indeed the industrialization of agriculture by the new life-science industries – Monsanto, Novartis, and so on – has become one of the major planks of the anti-globalization movement. Large transnational entities with an eye on profit rather than health, supported by an armory of intellectual property rights and patents and organizations like the WTO, are, in this view, precisely why there has been much popular support for the famous protests and street battles in Seattle, Prague, and Genoa.

Alongside these concerns about our links with nature in the production of foodstuffs, there is also increasing awareness of potential environmental catastrophes. With these, as with the concerns over food quality, most people are unable to evaluate the arguments directly, since they rely on (at best) unproven scientific forecasts and scientific claims that are unobservable. (The hole in the ozone layer above parts of the Southern hemisphere is not directly apprehendable, for example, and all but a very small number of people have to rely on the claims of scientists – accompanied by images whose provenance is also indeterminable for most people – as to both the phenomenon's existence and its relationship with assumed consequences, such as the increase in skin cancers in Australia and New Zealand.) Whereas certain types of environmental problems and their consequences are readily understood – because they are local and clearly linked to particular natural events, such as volcanic eruptions and hurricanes – many are not, in part because trends are relatively long term, their causes are difficult to disentangle (climatic change linked to global warming rather than random disturbances and other oscillations), and the causal links are far from clear. This has been apparent in recent decades with regard to claims concerning global warming, its association with increased CO_2 production, and its links to short-term climatic changes (such as increased frequency of storms and flooding) and longer-term trends such as sea-level rise. As Trudgill (1990) has argued, there are many barriers to be faced and overcome before such problems are tackled; at the outset it is necessary to convince people that the problem is "real," that its cause is understood and its effects are appreciated, and that solutions are available; in many cases, too, this may involve convincing people of the need to take measures now because of the long-term potential harm that may follow if some of the worst-case scenarios turn out to be valid in the longer term, when a trend may prove irreversible. Taking action according to such a precautionary principle requires at least three barriers to be overcome: those involved with recognition that the problem exists, that the cause is understood, and that the proposed solutions are viable. Then it is necessary to get political agreement that action is necessary – increasingly an international barrier, given the nature of some of the most pressing contemporary issues, which will include both economic concerns (is the resolution affordable?) and social concerns (is the proposed resolution acceptable?). The process may be long and drawn-out and difficult to conclude, as the concerns over such issues as CFC contributions to the destruction of the ozone layer and CO_2 reductions to constrain global warming illustrate. President Bush's victory in the 2000 US elections has seemingly made both the political process – ratification of the Kyoto protocols – and the scientific question – is there evidence for global warming as a product of industrial emissions? – even more intractable. Post-Kyoto meetings in the Hague (2000) and Bonn (2001) have illustrated the problems of getting governments to agree on policies to counter global warming, with national interest being placed before the long-term common good.

Changing those Changing Relationships

Assuming some general validity to the pessimists' arguments and the need to apply the precautionary principle, what can be done to end the destruction of the resource base which sustains human life? Several major problems can be isolated.

First, there is a need to win acceptance that a major catastrophe could be on, or just beyond, the horizon, which calls for rapid immediate action if it is to be averted. This involves a major educational task, which is made difficult by the absence of conclusive evidence and is exacerbated by the lack of detailed knowledge of how sustainable development could be ensured given the existing pressures on the environment, let alone those likely to come in the next few decades as a consequence of growing demands for increased material standards. The science is not universally agreed upon: with climate change, for example, although most accept the evidence for recent global warming (compiled in large volume by an inter-governmental panel), many contest that it cannot be proven because of anthropogenic causes – and not all of those arguing to the contrary are financially supported by commercial interests which stand to lose from the proposed regulation. This first group of problems is associated with a second, which is largely political in its structure. Much of the "rape of the earth" to date has been undertaken either by or for the populations of "developed world" countries. With a changing world political order, the governments representing the peoples whose lands and livelihoods have been exploited in these unequal relationships argue that they should not pay the price of resource depletion, for which they have not been responsible and from which their populations have benefited very little. Suggestions that population pressure is a major cause of the oncoming problems, and that as a consequence high priority should be given to introducing and promoting birth control policies, often stimulate the response from "Third World" governments that for them to do so is to accept the basis of the current inequality within the world in an intrinsically unjust economic system (Harvey 1974). Against that, the US government in 2001 argued that the Kyoto protocols on the control of greenhouse gas production were unfair because only the "developed world" countries were committed to their reduction.

A third group of problems is linked to the common treatment of natural resources as private property. Much resource depletion and most environmental pollution comes about because of individual (including corporate) actions, each of which in itself is a very small, marginal contribution to the growing problems. Those actions take place within a mode of production whose political and other leaders increasingly promote: the private ownership of all means of production, including nature. Within capitalism, economic survival demands that resource exploiters continually increase their pressure on nature in order to sustain their competitive position in world markets (Johnston 1996).

How can this deleterious pressure be reduced? Education and persuasion are important, but unlikely to have major permanent impacts, because of the apparently conflicting goals of "saving the earth" (for posterity) and sustaining one's standard of living in the face of increased prices (now). Some individuals and groups may react positively to calls for changes in their behavior (especially in the more affluent countries), as might some relatively altruistic business leaders, but the dynamo of capitalism is against them (Pepper 1993): as a whole, if not as individuals, we are pressed to act against our long-term interests, and those of our successors, in order to maintain our short-term positions.

The necessary conclusion is that to ensure sustainable development in the future others must require individuals and firms to respond to the imperative to conserve rather than destroy: those others can only be governments, for no alternative institution to the state has the sovereign power with which to insist on actions to protect the environment (Johnston 1989a). Three main government strategies are available: public regulation of activities to ensure that pro-environment strategies are implemented; taxation of polluters, thereby making it more sensible and efficient for individuals to change their behavior; and state subsidies to assist change to a "greener" set of actions. All are being used. Some policies within individual countries – such as the reduction of air pollution from coal-burning in British cities since the 1960s (Brimblecombe 1987) – have been very successful, to a considerable extent because people perceive the benefits and are content to comply, including meeting their share of the costs, when they know that everybody else is doing so too; others, such as the international agreements (signed in Vienna and Montreal) to reduce and eventually end the production of CFCs, were agreed and implemented because the commercial interests involved in the developed world identified alternatives that they could manufacture and sell – though the US government challenged the details of the (for it, very small) additional aid budget needed to help "developing countries" change their technologies to the new products. But in many other cases change is very difficult to achieve, as with the use of cheap energy in the United States, where attempts to regulate fuel use are countered by the strong oil industry and other lobbies within and close to governments; shocks, such as the blackouts and price rises following the failures of post-privatization utility companies in California in 2001, may be the catalysts for change, but they may not.

Despite such successes, however, much environmental use remains either poorly controlled or entirely unregulated, and contributes to continued environmental degradation. In part this is because, in the pursuit of environmental policy goals, governments are constrained by the pressures of capitalism: if they make the costs of production more expensive within their territories, compared with other states where there is less environmental control, then they will potentially be making their local industries uncompetitive, so harming the employment chances and living standards of their populations relative to the residents of more "liberal" states. This was illustrated by President Bush's decision in early 2001 not to seek US ratification and observance of the Kyoto protocols with regard to CO_2 reductions because they threatened American jobs both directly and indirectly and because "Third World" countries were not included in the proposed reductions. (Earlier, in the negotiations over the protocols at Kyoto and then the Hague, the US offered to plant more trees as its contribution to the reduction of

CO_2 production impacts on global warming, in the as-yet unproven belief that trees absorb CO_2 in sufficient quantities to make a difference.) Furthermore, given the global interconnections within environmental systems, free-riding is not possible: if some states fail to implement environmental protection, and cannot be persuaded to do so, then eventually the land, water, and air of all states will be despoiled (Johnston 1989b).

Just as individuals within a state may be prepared to comply with environmental legislation when they know that everybody else is, so individual states may be prepared to act if all other states are. But there is no super-state which operates above individual states and to which they are prepared to cede sovereign power – with the partial, strongly contested, exception of the European Union. Thus international action will only come about if there is international agreement, and in many areas of environmental concern there is little evidence – despite a great deal of rhetoric – that such agreement is being reached and enacted. At the Rio Earth Summit in 1992, for example, the US president declined to sign the Biodiversity Convention because it "threatened American jobs": his successor did sign, but Bush's action, like others by a number of states (as with the Law of the Sea and the Ozone Treaty) and those of his son when he won the presidency eight years later, illustrates governments' unwillingness to sacrifice local economic advantage – especially in the developed countries which have most to lose (Johnston 1992).

In sum, we have an impasse. Many believe that environmental tragedies are rapidly approaching because we have so abused the Earth that it cannot continue to support human life, at least at its present magnitude and at the high material standards enjoyed by a small proportion of the total population (and coveted by many more). Others hope that this is not so: they await scientific evidence that the stories of the Earth's demise are much exaggerated, and press for continued scientific innovation which will reassert human hegemony over nature. Whether the optimists or the pessimists are right may be settled early in the twenty-first century, unless much more concerted action than is currently taking place is taken very soon to boost the optimists' cause.

23 The Earth Transformed: Trends, Trajectories, and Patterns

William B. Meyer and B. L. Turner II

The twentieth century, especially its second half, was one of unprecedented rapid change in the Earth's environment. Human impacts on the physical form and functioning of the Earth have reached levels that are global in character, and they have done so with escalating speed. Human-transformed landscapes on the Earth's surface now rival in extent the area of natural or lightly altered biomes; strictly speaking, indeed, there are no "natural" biomes that are not at least lightly altered. Human-induced changes, by and large, have escalated in recent decades and are likely to continue escalating through the twenty-first century.

If the degree of change is novel, so is the degree of concern – scientific and public – that it now raises. Throughout recorded history human societies have recognized their power to alter the Earth (Glacken 1967; Lowenthal 2000). Never before, though, has environmental change been a matter of more interest than it has become in the past few years, and never before has the interest been so tinged with pessimism. Humankind has become what many have long wished it to be. It is now a transforming force equal in its power and effects to natural forces. Yet that accomplishment is viewed with regret at least as much as with satisfaction. Humankind was expected to become more and more a force rationally and deliberately improving the environment while preserving and cultivating what is valuable in it. It appears more and more to have become, instead, a powerful agent that acts blindly and haphazardly; on a gloomy view it represents a novel force of nature wreaking havoc with the biosphere and hardly more conscious than the weather or plate tectonics of what it is doing. Its ordinary activities have profound and rapid, yet often unintended, unforeseen, and little-understood effects.

Views of Environmental Change

A topical world geography compiled for students at the beginning of the twentieth century would probably have devoted no separate chapter to the human alteration of the Earth's surface: not because there existed little notion that the Earth was being altered greatly, but because it seemed not to be a major problem nor likely to become one. At the most, such a volume would have dealt in passing with the need to conserve and use wisely the Earth's finite mineral and energy stocks. Forecasts at the end of the nineteenth century of what was in store for the twentieth by and large predicted further improvement of the environment by its human users: more deserts and near-deserts to be irrigated, more coal seams opened for use, more useless swamps and forests converted to productive cropland, the ocean's resources harvested ever more efficiently.

Most turn-of-the-century seers also forecast a rosier social future than has thus far arrived. Yet the change in environmental perceptions has been an even graver blow to faith in progress. The bettering of the environment had long been that faith's last resort, the trend to which it could appeal no matter how slowly or how little the behavior of human beings toward one another seemed to improve. The erosion of environmental confidence has been a fairly sudden one. Even at mid-century, a major scholarly assessment of *Man's Role in Changing the Face of the Earth* (Thomas 1956) documented in detail the human alteration of the globe but "displayed a remarkably low level of concern about human impacts" (Lowenthal 1990: p. 124); it did not convey "any feeling of urgency" (O'Riordan 1988b: 25). In contrast, the historian Paul Kennedy (1993) devoted a full chapter of his book *Preparing for the Twenty-First Century* to the harmful human impact on the globe. Even within a generally pessimistic volume, the tone of the chapter is not hopeful. The problems are grave, Kennedy concludes, and the decisions and reorganizations needed to address them more than superficially are unlikely to be taken. His is not an isolated view among those writing on the subject and may not even be a minority one. The belief that the Earth has been seriously damaged, and is today being damaged more severely and rapidly than ever, is at least a far more prevalent and respectable belief than it was only a few years, let alone a few decades, ago. Increasingly the assumption that the Earth is being improved requires a defense and an explanation, while the assumption it is being dangerously degraded requires none. Coping with global environmental change has become one of humankind's more pressing problems.

What is Global Change?

Not all environmental change is global, though how much is global depends on how that word is defined. In the strictest sense, no change is fully global, inasmuch as none occurs uniformly across the Earth. In the most expansive sense, all change is global, inasmuch as all changes are ultimately connected with one another through physical and social processes alike. Some intermediate definition is required if "global" is to remain a useful distinction.

One such definition is based on the physical character of the change. Human impacts on the environment can acquire a global character in one of two ways. In the first, "global refers to spatial scale or functioning of a system" (Turner et al. 1990: 15). Because of the planet-scale fluidity and connectedness of the system affected, changes in it have the potential to have effects around the world – albeit far from uniform ones. There are two systems that can operate in this way: the atmosphere and the oceans. Global warming induced by greenhouse gas emissions, sea-level rise as a result of such a warming, and the depletion of stratospheric ozone by chlorofluorocarbon releases are the classic examples of such globally systemic change. It is likewise through the global systems of the atmosphere and oceans that some trace-pollutant releases spread worldwide from their places of origin: fallout from atomic tests, DDT and its residues from farmland, lead from automobile exhausts.

A second kind of environmental change can also be considered global: "if it occurs on a worldwide scale, or represents a significant fraction of the total environmental phenomenon or global resource" (ibid: 15–16). Losses of forest, biodiversity, soil fertility, and wetlands widely repeated around the world subtract significant fractions of the net worldwide stocks of the resources affected. Some of these globally cumulative changes – deforestation, for example, which releases carbon dioxide – have direct connections to systemic changes as well. So, linked or not, they in themselves may pose threats to the resources on which the habitability of an ever more populous and more affluent globe will depend.

The systemic impacts of human activity – notably, climate warming and ozone depletion – were central to the emergence of the current scientific and popular interest in global change. It is, however, the cumulative global changes that are better documented, to date, more significant for human activities, and the subject of increasing assessments of their economic costs to society.

Documenting Global Change

Documentation of either kind of change, however, is no simple task. Some aspects of the environment fluctuate substantially and unpredictably even in the absence of significant human involvement. Hence the effects of such involvement are difficult to identify. Climate at the global and lower levels is one example. It remains open to debate to what extent global temperature changes in this century are natural fluctuations or the result of greenhouse gas emissions (Karl 1993). Marine fish populations (Hillborn 1990) experience significant variations that are not necessarily human-induced. The parts played by meteorological shifts and human land use in degrading the North African Sahel zone remain matters of scientific dispute (Tucker, Dregne, and Newcomb 1991).

Much is asserted about the degree of past, present, and future change, whether or not much is known. The world has never been so awash in statistics and assertions about the environment, including estimates of the economic value of the tropical forests and even the biosphere (Daily et al. 2000). Supply has responded to an explosion in demand: government agencies, scholars, journalists, and many others dealing with environmental matters require facts, preferably numbers, to illustrate their arguments and perhaps even to guide them. Yet these users, by and large, are ill-placed to ask how accurate, how representative, and how current those facts are. Because environmental change is very much a multidisciplinary and also an extra-academic affair of research, it loses one of the principal advantages possessed by established academic disciplines: that of quality control, imperfect though even that may be (Carpenter 1989). Bad numbers once in print are extremely hard to confine or root out, corrections hard to make stick.

Uncertainties about cumulative kinds of changes in the states or faces of the Earth arise from the poor reliability of even precise and officially guaranteed figures. In China, as Smil (1992: 434–6) observes, post-revolutionary accounts of "mass reforestation programs . . . appeared to convey a success story," duly recorded in annual United Nations FAO (Food and Agriculture Organization) land-use statistics as a net increase in China's forest area. "Realities, gradually disclosed since the 1970s, are quite different." Most new plantings did not survive, and existing forest has continued to decline both nationwide and regionally: "by 1988, the forest cover in Sichuan, China's most populous province and . . . one of the country's principal timber bases, was down to 12.6 percent from 19 percent in the early 1950s." One of the most widely discussed arenas of contemporary environmental change, that of tropical deforestation, "remains an area in which increasingly sound methodology is applied to very unsound data" (Brookfield 1992: 4). As for systemic change, evidence to test the theoretical prognostications of greenhouse warming, for example, will not be available until well into the twenty-first century. Climate systems are complex, involving many feedbacks and experiencing many other changes at the same time as they are being affected by greenhouse gas releases. Thus it cannot be certain that the observed and well-documented increases in greenhouse gases have yet led or will lead in the near future to a significant rise in the Earth's average temperature.

Some reasonably reliable generalizations can nonetheless be offered about the extent and trends of human-induced environmental change, in the global aggregate, drawing on a recent inventory covering the past three centuries (Turner et al. 1990). Indices of change compiled in that effort cover a representative array of major forms of environmental impact (figure 23.1; table 23.1). Taken together they offer a solid empirical base for several assertions. First, human activities are comparable to or greater than natural forces as drivers of many kinds of change. Secondly, most of them have only recently become so. Yet, thirdly, acceleration and intensification of human impact, though frequently and perhaps usually the case, has not always been so. Finally, human impacts have steadily expanded in variety and character, from involving mainly some of the landscape resources of the Earth – forests, soils, water, biota – to affecting the material and energy flows of the biosphere.

Human-induced environmental change is, of course, of long standing. Alterations amounting to transformations of local and regional environments go back to prehistory. Occasional environmental disasters may have occurred in premodern times. Sustained and knowledgeable use of the land was probably the norm, however, and clear cases of purely environmentally induced civilization collapse are rare or non-existent. To say so is not to endorse the "green legend" (Whitmore and Turner 1992), the myth of the environmentally nurturing premodern or precapitalist society that is so

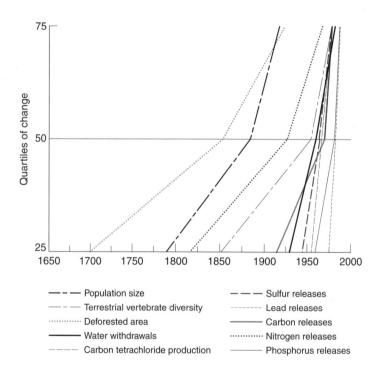

Figure 23.1 Trends in selected forms of human-induced transformation of environmental components

Table 23.1   The Great Transformation: selected forms of human-induced transformation of environmental components – chronologies of change.

Quartiles of change from 10,000 BC to mid-1980s

Form of transformation	Dates of quartiles		
	25%	50%	75%
Deforested area	1700	1850	1915
Terrestrial vertebrate diversity	1790	1880	1910
Water withdrawals	1925	1955	1975
Population size	1850	1950	1970
Carbon releases	1815	1920	1960
Sulfur releases	1940	1960	1970
Phosphorus releases	1955	1975	1980
Nitrogen releases	1970	1975	1980
Lead releases	1920	1950	1965
Carbon tetrachloride production	1950	1960	1970

prevalent in the literature today. Indeed, one of the authors has documented the kinds of environmental changes inflicted by such societies in the Western hemisphere (Turner and Butzer 1992; see also Denevan 1992). It is, rather, to contest another myth: that societies that have profoundly altered their environments in the pursuit of wealth and power have been punished by environmental catastrophe. Many, in fact, have transformed the lands that they occupied, and thrived. Claims of environment-related societal collapses in the distant past are contested and cannot be resolved by the state of the evidence at this time (e.g., Turner 1991).

The modern era of widespread concurrent change in the Earth's land cover – species transfer, the conversion of forests, grasslands, and wetlands to cropland along expanding frontiers of change – may be thought of as having been inaugurated by the Columbian Encounter in 1492 (Turner and Butzer 1992). Substantial change in the Earth's biogeochemical flows is a product of the Industrial Revolution, beginning in Western Europe in the late eighteenth century.

Set in motion by these processes, the major human-induced changes to date have begun to rival or surpass in their magnitude the rate of the operation of natural forces in the biosphere. World forest area has been reduced by some 20 percent and an area of land the size of South America converted from its original vegetation cover to cropping. Humankind may usurp as much as 40 percent of net primary productivity (Vitousek et al. 1997). Global fossil-fuel and industrial mobilization of sulfur, a major source of acid precipitation and other forms of pollution, now approximately equal the natural flow of the element, while human activity, primarily use of

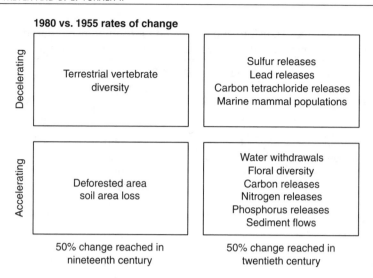

Figure 23.2   Recency and rate of change in human-induced transformation of environmental components.

fertilizer, exceeds the natural flow of nitrogen. The important atmospheric trace gases carbon dioxide and methane, both contributors to the greenhouse effect, have been increased by about 25 percent and 100 percent respectively over their preindustrial levels by human actions. Emissions of some trace pollutants released by human activities – a number of metals, for example – now greatly exceed natural flows, while releases of synthetic organic chemicals introduce novel substances that do not exist in nature.

The changes vary both in the lengths of time over which they have occurred and in their current trajectories (figure 23.2). Longstanding impacts on the Earth's land cover are represented by a decline in forest area and by terrestrial biodiversity; half of the total impact on each was registered as early as the nineteenth century. Not of such long standing, as a global problem, is human withdrawal of freshwater from the hydrologic cycle, though in some localities it was substantial, as a result of irrigation, millennia ago. It represents a rapidly accelerating process, per capita withdrawals having grown fourfold since 1950. Human releases of key elements of the biosphere – carbon, sulfur, nitrogen, and phosphorus – are all, except for carbon mobilization through land-cover change, relatively recent as global phenomena. Sulfur releases, at least, appear to have been leveling off by the mid-1980s compared with the previous decades, in consequence of fossil-fuel price increases and pollution regulations. Trace pollutant releases – for example, those of metals and synthetic organic chemicals – are quite recent in origin as significant phenomena. Both lead and CCl_4 releases, like those of sulfur, have shown some tendency to decelerate during the second half of the twentieth century, again in part as a result of the regulation to which they have been subjected in some countries; other trace pollutant emissions, though, and perhaps most, have continued to accelerate.

All of these changes are significant not only in themselves but for the further physical changes that they can drive: the alteration of the composition of the atmosphere, for example, may affect the climate, and climate change in turn reacts on forest area and composition. It is not only the magnitude of contemporary change, but its variety, that presents formidable challenges to societies' abilities to adapt to the consequences.

The Meaning of Global Change

Environmental change as defined in natural-science terms is, from the human point of view, neither good nor bad in itself. It is, as Zimmermann (1951) described the nature of "natural" resources, merely "neutral physical stuff." Environmental change is neither environmental improvement nor environmental degradation per se. The characteristics of the society interacting with physical phenomena are what make those phenomena either resources or hazards, or good and bad change (Meyer 2000). This does not mean that the physical characteristics of environmental change can safely be ignored or given only superficial attention, for these characteristics matter a great deal. A close acquaintance with them is necessary for an assessment of the meaning of change, but is not sufficient.

For most policy-makers, and probably for most of the public, the "bottom-line" concerns over global change are less those about its alteration of nature per se than about its economic and social consequences. How will the settings for economic activities be affected by substantial human alterations of the biosphere? How much would it cost to prevent those alterations? Deciding how desirable or undesirable an environmental change is, is no simple task. Interpreting the consequences of greenhouse gas-induced global warming for yields of current cropping systems requires combining a large variety of impacts: atmospheric CO_2 enrichment, changes in temperature, changes in water availability, changes in extreme weather events, changes in soil, and changes in pests and diseases (Rosenzweig and Hillel 1993). All are themselves uncertain, and results based on them must also be combined with speculations on future land uses, technologies, markets, and resource availability to produce meaningful projections or forecasts of the important human consequences.

In most environmental changes, even ones that overall are more harmful than not, there are individual winners as well as losers. Drought is a frequent form of natural climate fluctuation that would probably be increased in some areas by global warming. It often benefits farmers living in other areas, or farmers in the areas affected who have access to irrigation water, because reduced overall production raises agricultural prices. Intervening to prevent change would also produce winners and losers. Environmental-ists have proposed some courses of stringent carbon emissions regulations that economic analyses suggest would cost more for society overall, and much more for certain sectors of society, than would adapting to the changes as they occur or pursuing more moderate courses of regulation (e.g., Manne and Richels 1992; Nordhaus 1993). Some environmental-hazard studies suggest that the poor and unempowered are the most vulnerable to these impacts because they have less options. Yet it is not clear that the poor and unempowered would not also disproportionately bear the costs of action taken to prevent these impacts.

But selecting the "best" (by whatever criteria) responses to global environmental change is complicated not only by these distributive issues but also by the unprecedented scale, rates, and uncertainties of the physical changes in question, be they systemic or cumulative in kind. The cost of guessing wrong may be catastrophic. The size of human-induced change may be so great as to outstrip the ability of individuals and societies to adapt adequately to it. The whole course of environmental history, and particularly its modern course, suggests that "surprises" – unknown and unpredicted environmental impacts – will continue to occur. As human-induced change increases in scale and rate it seems probable that they will occur more frequently. The CFC–ozone relationship is a good example, one in which a new chemical compound thought to be environmentally benign rapidly accumulated in the atmosphere and began to erode the shield of ozone that protects the Earth's surface from ultraviolet radiation. We are ignorant of such potential threshold levels and non-linear relationships be-cause the conditions that produce them lie outside our experience. Even economists who doubt that stringent and immediate measures to curb car-bon emissions would, on the basis of current knowledge, be cost-effective, acknowledge that "the potential for catastrophic surprises" raises questions that cannot be conclusively answered by current models or by purely eco-nomic reasoning. Such efforts, as Nordhaus (1993: 23, 24) writes, can in-form but cannot dictate the political decision of "how to balance future perils against present costs."

Weighing heavily in that decision will be different assumptions about the nature of nature and about how it will react to human impacts: most broadly, the four "nature myths" discussed by Douglas (1992) (figure 23.3). Some see natural systems as resilient, little affected even by powerful shocks; some see them as resilient within limits but vulnerable when those limits are ex-ceeded; some see them as inherently fragile and likely to react sharply even to mild pressures; and some see them as capricious and largely unpredict-able. The first will see little need for restraint in exploiting and altering the environment, but the second and third will see a need for caution to avoid disaster; the fourth may display a fatalism that sees concern and regulation as futile. The four views, Douglas suggests, reflect different social bases, the view of nature as resilient, for example, reflecting a relatively unconstrained

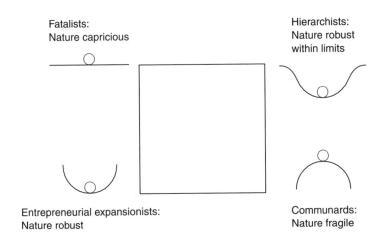

Fatalists:
Nature capricious

Hierarchists:
Nature robust
within limits

Entrepreneurial expansionists:
Nature robust

Communards:
Nature fragile

Figure 23.3 Risk and blame.

and entrepreneurial culture; that of nature as extremely fragile, an egalitarian and sectarian one. In any case, the same scientific data about what is occurring now will suggest different futures and different courses of action when interpreted through such different worldviews.

Global Change from a Regional Perspective

It is increasingly recognized that understanding the physical and social patterns and dynamics involved in global environmental change requires regional as well as global assessments (NRC 1999; Kates et al. 2001). The trends and patterns of global changes are aggregations of regional and local ones that display great variation. Releases of carbon dioxide, for example, are strongly concentrated in the industrialized world, including China, while rural areas with intensive wet-rice production, such as Southeast Asia, are major sources of methane, a less abundant but more potent greenhouse gas. The global pattern of increased deforestation and cropland expansion is not evident in Western Europe and the United States, areas in which much former agricultural and other cleared land has reverted to tree cover and where the land may be a net absorber of carbon from the atmosphere.

The human causes of changes in the biosphere also vary widely between regions and localities. At the global aggregate level, most changes correlate well with the variables in the "PAT" formula (Ehrlich and Holdren 1971), which ascribes human impact on the environment to the product of population, per capita resource use, and the technologies by which the resources are used. At the regional level, studies demonstrate the importance of other factors – institutions, policy and political structure, trade relations, beliefs and attitudes (Meyer and Turner 1992; Geist and Lambin 2001). At sub-global scales these forces often overshadow the PAT drivers of change.

The human consequences of global environmental change also tend to be different throughout the world. A doubling of atmospheric carbon dioxide may have a major fertilization effect on some crops, but little or none on some others in which the chemical process of photosynthesis follows a different pathway (Rosenzweig and Hillel 1993). Likewise, global warming, should it occur as forecast by current general circulation models (GCMs), would extend commercial crop production further northward in parts of Russia, Japan, and Canada, while increasing heat and water stress and substantially lowering yields on croplands nearer the equator. Putting these and other factors together, the food supply and economic impacts appear to be modest globally but severe for the tropical world (Rosenzweig and Parry 1994). That is the sector of the globe, moreover, where countries, organizations, and individuals are most constrained in their abilities to respond. Successful adaptations may require investment – in farming systems, in infrastructure, and in policy innovation – that can least be afforded by those lands that may most require it to deal with the challenges of global climatic change. It is also the developing (mainly low-latitude) world where agriculture and other climate-sensitive activities remain major elements of overall national economies, and where cumulative changes degrading land resource are most severe in any case.

Comparative case studies at finer scales can further help disentangle what is and is not shared between the global and sub-global scales and between different regions. These studies may range from comparative assessments of single facets of change (e.g., trajectories of deforestation) to more holistic assessments. An example of the latter is the Project on Critical Environmental Zones (Kasperson, Kasperson, and Turner 1993). It examined nine regions of the world where pressures of change and physical and socioeconomic vulnerabilities appear to have come together to produce environmental crises of unusual severity, jeopardizing continued use and occupancy at existing or projected levels of population and standards of living.

Only in one of these regions was the wealth or well-being of the region's population clearly declining as the result of adverse environmental changes. That region is the Aral Sea bordering Kazakhstan and Uzbekistan in former Soviet Central Asia. Water withdrawal from the tributary rivers for irrigated cotton production steadily reduced and continues to reduce the sea's water level and surface area (Micklin 1988), creating saline conditions that have destroyed the once lucrative fishing industry and rendering shipping facilities useless as the shoreline recedes (Kotlyakov 1991). The climate has become drier and harsher. Large dust storms, carrying salts from the exposed sea bed, have become more frequent and more damaging to agriculture. Water contamination by salts, fertilizers, and pesticides has seriously affected human health.

In each of the other regions, environmental change, though rapid and substantial, has not yet reduced the overall wealth and well-being of the population, severely though it may have affected some sub-groups. In many of them, it appears likely to do so in the near future, given the range of feasible responses. If only the Aral basin fitted the project's definition of environmental criticality; the Basin of Mexico, the Sundaland rain forest of Indonesia and Malaysia, the Ukambani region of southeastern Kenya, and the Ordos Plateau of northern China were classed in whole or part as environmentally endangered regions: ones in which trajectories of environmental change put the sustainability of current human uses and standards of living in question in the short- to medium-term future. Amazonia, the Llano Estacado, and the North Sea basin also incurred major environmental changes. Yet, overall, either the rates of environmental change were declining, the changes themselves, though substantial, did not seriously undermine the livelihoods or the health of the regional populations, or the wealth of the region provided a series of feasible options for its sustained use.

These nine regions are only a sample of an increasing list of areas where human use has so drastically altered the environment that alarm has been raised about their long-term occupancy and about the cumulative impacts at the global scale of a large number of regional environmental catastrophes. Such alarm cannot and should not be taken lightly; many people in many regions will suffer because of the environmental changes wrought now and their inadequate access to the means to alleviate the impacts of, or adapt to, these changes. Nevertheless, if confident predictions of the conquest of nature have not been borne out in the twentieth century, the literature of environmental history abounds in prophecies of doom that have also failed to come true.

The lessons from such comparative case studies are several. There exist large regions where environmental degradation is seriously jeopardizing the sustained livelihood and/or health of their inhabitants, where its rate is outstripping the ability to adapt, and where the medium-term costs either of permitting it to go on or of responding are likely to be enormous. Management response has typically not been impressive, even where the problems were clearly foreseen (as in the case of the Aral Sea, for example); a do-nothing response was often preferred, though it greatly increased the eventual costs. The fact that global environmental problems require negotiations and agreements among many nations would only complicate matters at the higher level and make response more difficult. Yet there are also areas where adaptation, often hard to foresee or predict, has deflected what would once have seemed an inexorable trajectory toward criticality.

Conclusions

Though the patterns, sources, and impacts of environmental change are not uniform across the globe, humankind has so altered the physical conditions of the Earth that we must recognize overall an "Earth transformed" and an Earth certain to be transformed further. These changes will find different expression and have different consequences in different regions, but they will be significant almost everywhere. The twenty-first century is likely to witness an Earth in which all lands are formally managed; few, if any, frontiers will remain. The demands for resources will increase, even should they change in character owing to technological changes, and they will mean increasing pressures on the earth to fulfill them. The question for humankind is whether it will be able to satisfy these demands and adapt to the foreseeable and unforeseeable consequences without destroying the only home it has.

24 The Earth as Input: Resources

Jody Emel, Gavin Bridge, and Rob Krueger

The plot of *Black Indies* (by Jules Verne) arises from one miner's refusal to believe that the Aberfoyle pit is exhausted. Early in the book, young Harry Ford exclaims to James Starr, "It's a pity that all the globe was not made of coal; then there would have been enough to last millions of years!" The older, wiser engineer responds that nature showed more forethought than human beings by forming the earth mainly of stones that cannot be burned. Otherwise, Starr comments, "the earth would have passed to the last bit into the furnaces of engines, machines, steamers, gas factories; certainly, that would have been the end of our world one fine day!" (R. Williams 1990: 197)

Introduction

As the mining slogan goes, "if it can't be grown, it has to be mined." Everything we use to communicate, move, stay warm, stay cool, sit, sleep, cook, and refrigerate comes from the 15 billion tons of raw material that humans extract from the earth each year. And although ideas have changed about the desirability of this massive appropriation of nature, mining, damming, timber cutting, and fishing proceed much as they have for the last couple of centuries. At the same time, a new order of natural resource dependency is emerging that consists of new plants, animals, and materials produced through biotechnology, together with human-engineered "natural" resources like fish farms, timber plantations, and the like. Cyborgs or "robo sapiens," chimeras, living patented objects, genetically modified organisms, and other new forms of "nature" inhabit this emergent order. Those who view this transition optimistically refer to it as the "next industrial revolution." Those who see traditional primary production accelerating with economic development, and who view the "new nature" more circumspectly, if not pessimistically, refer to this new era of development as the "coming age of scarcity" (Dobkowski and Wallimann 1998) or the "ephemeraculture" of "hyperecological" living (Lukes 1999). While, in chapter 4, Whatmore addresses some issues of biotechnology, this chapter analyzes the traditional basis of our modern society: earth as input.

Our purpose in this chapter is to examine trends in resource production and consumption, and to provide an overview of the many ways those trends are interpreted. In truth, a number of different conceptual frameworks exist for analyzing current resource trends. These differ in their level of theoretical sophistication and offer varying degrees of practical leverage for those interested in reshaping current patterns of consumption/production. All are agreed, however, that contemporary patterns of resource use pose sufficient political, economic, and/or socioecological impacts as to present a crisis of some sort, although they offer differing interpretations of precisely where that crisis lies.

Trends

Nearly ten years after the Earth Summit in Rio and almost thirty years after the environmental movement made significant global waves, we have gained only marginal ground in efforts to decrease overall dependency on non-renewable resources, and to reduce degradation of primary fisheries, forests, and water resources. The volumes of materials extracted and harvested have increased. Recent trajectories for selected major commodities show the continued expansion of resource extraction (see figure 24.1). Figure 24.2 (from Ayers 1992) illustrates the long-term trend in materials production.

In addition to those commodities in figure 24.1, other resources are also being produced and consumed at higher levels. For example, global water consumption rose sixfold between 1900 and 1995 – more than double the rate of population growth (World Resources Institute 1999). Global groundwater deficit (where pumping exceeds recharge) is conservatively estimated at 160 billion cubic meters per year – the amount of water used to produce one tenth of the world's grain supply (Brown et al. 2000). Water problems are most acute in Africa and West Asia, but lack of water is already a major constraint to industrial and socioeconomic growth in many other areas, including China, India, and Indonesia (UNEP 1999: 41). Projections of water stress vary, but if present consumption patterns continue, 2 out of every 3 persons will live in water-stressed conditions by the year 2025 (ibid).

Global petroleum consumption increased by 12 percent from 1989 to 1998, with OECD countries' consumption of petroleum expanding 15 percent (from 40,881 to 46,984 thousand barrels per day) between 1989 and 1998 (Energy Information Agency 2000). Between 1989 and 1998, non-OECD countries' petroleum consumption grew from 25,036 to 26,660 thousand barrels per day (ibid). The rise in total energy and oil consumption in

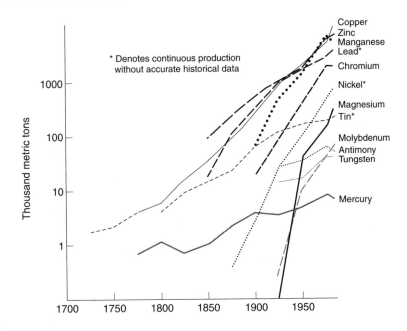

Figure 24.1   Annual worldwide production of selected metals from 1700 to 1983.

the newly industrializing countries and elsewhere in the Third World could be twice their 1990 levels by 2010 owing to industrialization, transportation needs, urbanization, and population growth (OECD 1993b).

Global wood consumption has increased 64 percent since 1961 to around 3.4 billion m³ per year, over half of which is burned for fuel. In Africa, where firewood (and other biomass sources) are a critical energy source for up to 90 percent of the population, the production and consumption of firewood and charcoal doubled between 1970 and 1994 and is expected to rise by another 5 percent by 2010 (UNEP 1999: 38; WRI 2000b). Commercial wood production is dominated by the developed world, though developing countries increased their share of industrial round wood output (i.e., all wood not used as fuel) from 17 percent in 1970 to 33 percent in 1997 (ibid). In Europe, wood comes mainly from managed forests and plantations, but logging from natural or virgin forests remains common in North America. Paper accounts for almost one-fifth of the world's wood harvest (Brown et al. 2000). Paper use is highly correlated with income levels: the biggest consumers are the US, Japan, and China which collectively represent 22 percent of the world's population, but 71 percent of paper use. In the future, the biggest demand for commercial wood will come from Asia, where demand is rising rapidly but reserves are already inadequate (UNEP 1999: 39).

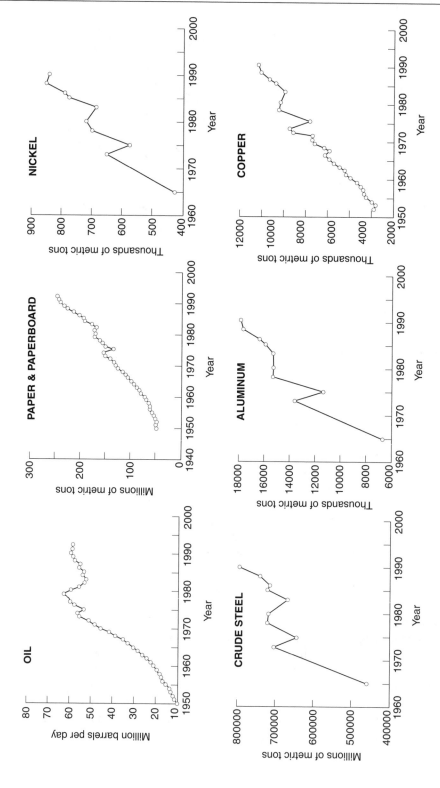

Figure 24.2   Annual worldwide production of selected commodities: recent trends.

Global marine fish catch rose from some 50 million tonnes in 1975 to more than 97 million tonnes in 1995 (ibid: 45). This increase comprises the harvesting of new species of fish and new fishing grounds, while older fisheries have been exploited and depleted. Aquaculture output, meanwhile, has grown dramatically, now accounting for almost 20 percent of all fish and shellfish production (ibid). The catch in Africa, Europe, and North America had begun to decline by 1990. Failures to control over-fishing, water pollution, and other factors resulted in the decline of approximately 60 percent of the world's ocean fisheries. Demand for fish is projected to increase from about 75 million tonnes in 1994–5 to 110–120 million tonnes in 2010, a quantity that can be met only through continued increases in aquaculture (ibid).

Nearly all of the resources examined are being consumed at larger volumes than in the past. There exists, however, a complex set of understandings about why this is occurring, whether it constitutes a crisis, and where the crisis is located. We now outline several of these interpretations.

Interpreting the Trends

Is there a resource crisis? The scarcity thesis

A traditional and intuitively appealing approach to thinking about the "Earth as input" has been to contend that, since supplies of non-renewable resources such as coal, copper, or oil are physically finite, current patterns of resource consumption must be unsustainable. At some point, this perspective argues, we will "run out" of key resources because consumption must necessarily reduce the supply available for future generations. To support this view of an emerging "scarcity crisis" proponents typically adopt a *physical* (as opposed to economic) definition of resources. This defines resource availability in terms of physical concepts (such as "crustal abundance" or groundwater in storage) that describe the total amount of a substance in the earth's crust independent of actual demand.

Increasing resource scarcity, the argument goes, will drive up primary commodity prices to the point that serious sociopolitical tensions emerge. These include slowing rates of economic growth (as feedstock resources become more costly), shortages of critical raw materials, and/or increasing conflict between those able to access dwindling supplies and those who cannot. The same argument is extended to renewable resources – such as some groundwater aquifers, soil, and fish – by claiming that current rates of use exceed rates of renewal so that these resources are effectively being "mined." As a result they will, like non-renewables, become increasingly scarce unless current rates of use are reduced to levels at which resource regeneration can occur.

This literal view of a crisis caused by physical resource scarcity has deep roots. It is most often traced to the late-eighteenth-century writings of

Thomas Malthus on the relationship between population growth and food production, but has appeared many times since in the debate over resource availability. Prominent recent proponents include Paul Ehrlich and Thomas Homer-Dixon (1999). The latter, for example, draws on recent conflicts such as those in Chiapas, Gaza, and Rwanda to argue that soil erosion and escalating water and fuelwood shortages can contribute to social unrest, generate mass refugee movements, and increase intraregional conflict. Protests in Europe and elsewhere over the high price of oil in the fall of 2000 are testimony to the acrimony and conflict that can develop over temporary shortages of an important resource.

No scarcity crisis, just a price crunch

The historical record, however, provides little support for the neo-Malthusian position, except in local or regional cases. Reserves of most major minerals are larger now than thirty to fifty years ago, despite decades of increased production and consumption. Reserves of crude oil in the US, for example, grew at the same rate as consumption for most of the twentieth century (McCabe 1998). Real prices of most resources have declined or maintained over time, suggesting that resources are less scarce now than in the past. These trends enable this second school of thought to be more sanguine about scarcity. For these "Cornucopians" resource scarcity is viewed in *economic* terms (rather than physical). While posing a temporary price crunch resulting from market adjustments to scarcity, it does not pose a hindrance to human development in the long term.

In this view, then, there is no long-term structural crisis in the relationship between resource consumption and the capacity of the Earth as a feedstock for the economy. Resource depletion can even be viewed as a positive occurrence. Why? Because during times of economic scarcity the prices for raw materials increase, which stimulates human ingenuity. From this perspective, human ingenuity overcomes scarcity in three ways: (1) through innovations in technology, (2) through the substitution of one material for another, and (3) by encouraging exploration in areas once considered marginal. Gains in the efficiency of use and recycling of many exhaustible resources provide such an example of innovation and substitution. In 1990, for example, recycled material provided nearly 75 percent of the lead and over 50 percent of the iron and steel consumed in the US; 30 percent of global aluminum demand was met with recycled material. In addition, substitutes such as plastics, ceramics, and other composites are competing with metals in cars, airplanes, building construction, communications, and pipelines. Development of oil reserves offshore and in remote cold climates are examples of exploration and production from marginal areas.

In this view the only crisis is that of the short term, and results from inefficiencies in existing markets which distort or delay adjustment mechanisms that would reduce prices over time by increasing the volume and range of resources that flow onto the market. Cases of physical resource depletion – such as that of fuelwood shortages around underground gold mining operations in California in the late nineteenth century – are interpreted as localized occurrences that provide evidence of the temporary and non-critical nature of resource shortages. In summary, if the market price for commodities is taken as the sole metric of resource availability – we will take issue with this assumption shortly – declining prices for most non-renewable resources suggest there is no scarcity crisis and that non-renewable resources are in fact "infinitely finite."

It's an "environmental" crisis, not a resource crisis

A third perspective shifts the resource question away from issues of resource *supply* towards the *impacts* of resource production and consumption on the environment. The concern here is not that we will "run out," but that the environmental costs of current rates of consumption will accumulate until they represent serious obstacles to sustainable forms of development. Rather than emphasizing the Earth as a "source," the focus is on the Earth as a "sink." For these "environmental economists," then, the debate over fossil fuel consumption shifts away from issues of resource depletion toward concern over the social costs of local and regional air quality (i.e., smog and acid precipitation) and the global carbon cycle (i.e., global warming and associated changes in sea surface temperatures and atmospheric circulation).

The field of environmental economics has grappled with these "sink" issues by framing them as *externalities*. These are defined as costs to the environment that, because they fall outside current accounting techniques, mean that current resources are under-priced. As a result, the price paid for raw materials (such as iron, oil, or cotton) does not reflect the true costs to society of producing those goods. Much current research within environmental economics, therefore, attempts to "internalize the externalities" by devising methodologies for valuing environmental impacts and adjusting the price of raw materials to incorporate these additional costs.

An incompatibility crisis: ecological economics and thermodynamic constraints

A fourth position – that of ecological economics – shares a focus on the environmental impacts of resource production, but is far less sanguine about the ability of modified accounting techniques to avoid environmental degradation. It argues that the problem is not simply one of market pricing, but of the fundamental biophysical laws of thermodynamics that govern the transformation of matter. Economic development requires access to cheap, plentiful energy in a highly intensive (rather than dissipative) form that is currently provided by fossil fuels. The ability of the green revolution technologies to improve agricultural yields, for example, depends on access to low-cost energy inputs in the form of petroleum-derived pesticides and fertilizers. It also relies on direct inputs of fossil fuel energy in the form of pumped irrigation water and mechanized farming equipment. These energy inputs are not fully recovered in the product since, following the second law of thermodynamics, there is always some leakage or wastage of energy. As a result, resource production and consumption are processes that involve an increase in entropy (that is, a decline in the energy that is available to human beings). To ecological economics, therefore, thermodynamics impose a biophysical limit on our ability to access usable forms of energy and raw materials indefinitely.

By way of illustration, ecological economists challenge the resource optimist position (described above) by arguing that technical improvements in the extractive sectors have made available previously uneconomic deposits only at the expense of more energy-intensive capital and labor inputs. They reframe the resource optimist story of increasing reserves over time as one of decreasing returns on energy investment: while resource production has kept pace with rising demand, it has been necessary to use more and more energy in order to extract lower grades of raw material. In the case of minerals such as copper, iron, and nickel, for example, the trend of increasing output illustrated by figures 24.1 and 24.2 has taken place in the context of declining ore grades and, therefore, has required ever larger amounts of energy to extract the same amount of usable material. In terms of pollution and externalities, ecological or biophysical economists would argue that "getting the price right" is meaningless in the context of the Earth's unyielding physical limits. Regardless of our willingness to pay to pollute, the Earth's absorptive capacity is finite, and if exceeded, poses ecological consequences.

Arguments for an ecologically based resource crisis are widely represented in the literature. Individual researchers, however, differ in the extent to which they embed this argument into a broader analysis of capitalist economies. Some researchers focus exclusively on thermodynamic constraints to resource production. Cleveland and Ruth (1996), for example, have developed biophysical models of resource depletion that show how the ability to pump oil from deeper and smaller fields, grow crops in less fertile soils, or extract copper metal from leaner and leaner copper ores is constrained by the availability of energy. Others pay less attention to the thermodynamic aspects of commodity production and emphasize how historic practices of externalizing environmental costs can prove contradictory for capital. For example, while factories or confined-animal feeding operations ("factory farms") can reduce costs in the short term by releasing waste products into the environment, such practices can undermine the conditions necessary for making a profit in the long term. As pollutants build up in the local environment, the environment's capacity for absorbing further wastes is diminished, eroding the value of such "free" environmental goods and services. Similarly, the emission of pollutants to air and water may reduce the costs of pollution control in the short term, but such actions can provoke extensive and sustained sociopolitical opposition from other interests in ways that drive up costs. The historic record of mining companies in sacrificing local environments in pursuit of ore deposits, for example, continues to haunt contemporary mining firms. In North America, for example, the ability of mining interests to avoid the costs of environmental restoration in the past by denuding landscapes, polluting groundwater, and generating unsustainable boom–bust economies contributed directly to the strength of contemporary popular opposition to mining.

There's no incompatibility crisis: self-regulation and ecological modernization

We have seen how the "cornucopian" perspective suggests that so-called "resource crises" are in fact only temporary price squeezes, and the environmental economics perspective contends we need only get everything priced right to resolve resource issues. A more sophisticated version of these arguments has emerged in the sociological perspective of "ecological modernization."

This broadly optimistic point of view contends that more ecologically friendly forms of production (and consumption) are emergent, and that core institutions need only be reformed to achieve desirable convergence between economic growth and ecological health. Capitalism, it is argued, has sufficient institutional flexibility to adapt to emergent environmental contradictions, and to incorporate them as a new source of profit. Proponents of this "eco-transition" point to gains made in the efficiency of use and recycling of many exhaustible resources. For example, since the energy crises of the 1970s, chemical producers within OECD countries have halved their energy consumption per unit of output (OECD 1993b). A similar pattern can be seen within mature industrial economies: the national output of the United States, for example, weighs little more than it did a century ago, notwithstanding massive increases in US GDP. This suggests a "decoupling" of economic growth from raw material use such that the generation of a unit of GDP now requires fewer resource inputs than historically. The chemical industry, for example, has seen raw-material requirements for a given unit of output fall an average of 1.25 percent per annum since the early part of the century (OECD 1993b).

Concepts of ecological modernization have been deployed at a range of analytical scales, from individual firms, to entire industrial sectors (where it is closely associated with industrial ecology), to full-scale societies. Some countries have been particularly active in trying to drive ecological modernization. Some local government bodies in Sweden, for example, are moving to reorganize business operations from linear to cyclical processes in a nationwide program called "The Natural Step." Lower Saxony (Germany) is promoting the installation of closed water cycles, while Kalundborg in Denmark is an oft-cited example of an industrial ecosystem in which the exchange of wastes, by-products, and energy between clustered firms minimizes releases to the environment (Ehrenfeld and Gertler 1997). Several governments have established agencies charged with driving forward developments in environmental technology and industrial ecology. The Japanese Ministry for International Trade and Industry (MITI), for example, founded the Research Institute for Innovative Technology on the Earth (RITE), whose primary concern is the development of environmentally friendly production processes and substitute substances.

While the origins of contemporary environmental problems may lie in the experience of industrialization and modernization, ecological modernization asserts that solutions to these problems lie in more – not less – modernization (Buttel 2000). Like the resource optimists, ecological modernization suggests that if crises occur as a result of waste emissions or a shortage of environmental goods and services, they are typically short-term and temporary since capitalism has the institutional flexibility to enable self-repair, especially through strong governmental–corporate partnerships.

Geography matters: boom and bust, social justice, and neoliberal development

A sixth perspective focuses on the geography of resource extraction and, in particular, the socioeconomic impacts of extraction in place and across space. The central argument here is that there is no crisis *inherent* to resource production, but that it is the characteristics of place and the relationships between places (of production and consumption) that determine whether resource production represents a crisis or not. In other words, geography makes a difference by mediating how potential resource challenges – such as those of economic growth, the socio-distribution of benefits, and environmental impacts – are worked out in practice. Geographers concerned with these contextual issues of resource production, trade, and consumption examine such topics as the impact of commodity price volatility on extractive communities, and the linkages between resource extraction, social injustice, and uneven development.

Commodity prices are notoriously cyclical, with prices swinging from periods of boom to bust. These price cycles are of acute significance to extractive communities because "boom and bust" translate into social dislocation and depression for many extractive areas. The initial extraction phase is associated with an influx of population, which can be both large and sudden, placing a strain on existing social facilities. The demographic characteristics of the immigrant population (typically young adult males, or young families) can make for significant personal and social problems, as both immigrants and host community struggle to adjust. Once the resources are gone, or are no longer economically viable, the population experiences unemployment, a decline in living standards, and the need for out-migration.

The influx of wealth from extractive development can be destabilizing for a locality that has not been involved extensively in economic activity and that has had a shallow tax base (Auty 1993). The wholesale dependence of many resource producers on a single commodity renders them vulnerable to the vagaries of the world market. In producer states (such as the oil economies of OPEC), resource revenues often accrue directly to state treasuries, giving the state responsibility for the realization of resource rents. Downward adjustments to the flow of resource revenues stress the state's ability to meet its domestic and international commitments, and can force the imposition of austerity measures on the population. The macroeconomic impacts of large mineral wealth have led some analysts to see a rich resource endowment as "curse" rather than benefit (Auty 1993).

Another way of looking at the relationship between resource extraction and human welfare is to focus on the sociospatial distribution of resource extraction activity. By comparing demographic and economic conditions between places that produce resources and those that consume them, one can make the argument that the geography of resource extraction is fundamentally inequitable (e.g., Geddicks 1993). While the benefits of resource production (in the form of revenues and the utility of finished materials) typically accrue to dominant social groups, the "costs" of resource extraction – in the form of economic volatility and the environmental impact of waste streams – are allocated to social groups who are politically (and often geographically) marginal. Rather than an economic or environmental crisis, then, current patterns of resource use represent a crisis of social justice as marginalized groups bear the costs of increasing raw material production.

From this perspective it is relations of wealth and power that structure the flow of resources. At the global scale, flows of natural resources from "developing" nations to "developed" nations are assured through an export-dependent model of development that has reached a zenith during the past fifteen years of neoliberal (or liberal-productivist) hegemony. Here's how it works. Most "developing" nations have accumulated large debts since the 1970s and now seek finance capital to build their economies. Structural adjustment policies implemented by the World Bank to ensure debt repayment require debtor countries to promote exports in order to generate foreign exchange earnings. They also require the adoption of economic policies to attract foreign investment, particularly investment in the export sector. To attract finance capital, a country must offer lucrative opportunities for investment. What many developing nations have to offer are minerals, trees, agricultural products, and inexpensive labor. For example, over 100 countries have "liberalized" their mining laws since 1980, shifting from a more protectionist orientation where the minerals were viewed as patrimony or common wealth, to a neoliberal orientation where foreign direct investment is invited and private property rights are assured. In its most extreme form this search for exports condemns regions (and countries) to the worst kind of productivism, with growth in living standards limited to a small section of society and little respect for the environment. Recent protests against the International Monetary Fund (IMF) and the World Bank recognize the waste of human life, animal life, and natural resources attendant to the massive debt that burdens many developing countries and the tremendous inequities wrought by neoliberal (free trade) markets.

Conclusion

A combination of population growth, an intensification and globalization of the culture of consumerism, and innovations in resource supply which have kept resource costs from rising have ensured that resource extraction trends continue their historical upward trajectory. Despite frequent assertions that we now live in a post-industrial society, the Earth remains an essential source of raw materials and a provider of environmental goods and services to the "new economy." While there is evidence to suggest that the rate of economic growth has become increasingly decoupled from the rate of raw material use, we have only to consider the popular icons of post-industrialism – the laptop, mobile phone, and double latté – to begin thinking about the complex flow of metals, petrochemicals, and food products that lie behind both corporate and personal economies in the contemporary period.

Indeed, the vision of a "new economy" that imposes few demands on the local resource base can be considered a geographical illusion made possible by the increased spatial separation of production and consumption. The restructuring of regional economies in developed countries away from heavy industry (and, in some cases, away from productivist agriculture) may have decreased traditional demands on regional ecosystems as "inputs" to these economies. Yet the experience of post-industrialism is critically dependent on extra-regional, long-distance flows, as resources such as minerals (oil, steel), fibers (cotton), and foodstuffs (grains, beverages, meats) are sourced externally to the post-industrial spaces of the global economy. Thus a typical resident of a developed country may develop attachments to place that are intensely *local* and be a citizen of a *national* territory, yet their experience of everyday life draws directly on a flow of resources that is *biospheric* in scale. The socioecological conditions under which these resources are produced and exchanged are largely obscured from view by the length and complexity of the commodity chain. When fueling the car, for example, we are materially connected to – yet largely ignorant of – the political–economic processes that precede the supply of gasoline onto the market. These can include the displacement of indigenous peoples from the site of extraction, the abrogation of traditional land use rights, and negative impacts to the quality of air, water, and land resources.

Unearthing the "hidden" geographies of commodities – the sociospatial patterns of commodity investment, production, trade, and consumption – provides a means of reembedding discussion of political–economic change in the materiality of daily life. Far from quaint anachronisms, practices of forestry, fishing, agriculture, and mining remain key components of the contemporary global economy. In many developing economies these sectors have long been significant in terms of their role in personal livelihoods, national income, and export earnings. Neoliberal reforms to encourage foreign investment and promote exports have further increased the significance of these sectors in countries with comparatively large resource endowments. In developed countries the emergence of the "new economy" has decreased the relative significance of primary industries, yet the products of these industries remain central to the experience of post-industrialism and are increasingly sourced from the peripheries of the global economy.

Any spatial accounting of material resource flows must also recognize that the way we think, speak, and write about the relationship between nature and society – and resources in particular – is one of the processes through which these flows are produced. That is, discourses about nature, commodities, and development are part of a set of relations that produce the material flows themselves. To speak of the Earth as "an input," for example, is to accept this formulation of the Earth as a feedstock, a set of discrete resources that can be separated, extracted, and developed at a rate determined by the needs of the economy. Whether codified into laws governing resource access, internalized as a personal worldview as part of the education process, or used as rhetoric to gain popular support for political projects, such discourses are complicit in – and fundamental to – the conversion of ancestral homelands into oil fields, tropical forests into bio-mines for pharmaceutical prospecting, and sentient animals into laboratory resources. One of the most notable features of the last few decades is the increasing struggle of many groups – both in the South and North – against this commodification of nature. Indigenous rights groups and eco-feminists have been among those promoting alternative discourses as a means of countering resource exploitation, whether it be the commodification of remote tropical habitat and the decollectivization of land ownership, or the increased commodification of the human body and the privatization of genetic code.

25 The Earth as Output: Pollution

David K. C. Jones

Introduction

Pollution is an important topic for geographers, for four main reasons. First, pollution (and individual pollutants) displays marked spatial variations in occurrence, intensity, and impact, thereby showing that geography really does matter. Second, it represents the adverse interaction between human society and the environment, a core aspect of the discipline Geography. Third, discussions of pollution range across the natural sciences, social sciences, and humanities, as does Geography. Fourth, as societies evolve so too do the causes, consequences, and preferred solutions to pollution, making the subject dynamic, contemporary, and relevant to notions of sustainability.

Developments in the media (television especially) have done much to reinforce the geographical characteristics of pollution, while at the same time emphasizing unsightliness and damage. The "Chernobyl Cloud," the "Exxon Slick," "Death on the Rhine," and the "Hole in the Ozone Shield" all invoke map images. However, despite the strength of these images, they fade quickly in the minds of most people. As a consequence, environmentalists often blame the "issue-attention cycle" (Downs 1972) of diverting attention towards the conspicuous crisis of the pollution *incident* and away from the *chronic* and *continuous* disposal of wastes from our (otherwise) sophisticated society into environmental media (land, air, water) whose capacities to absorb such increasingly complex wastes remain little known. Deeper analysis of media coverage of environmental issues (e.g., Hansen 1993) reveals much about the cultural and political aspects of the current crisis; geographers are able to balance these aspects with the physical science dimensions of pollution.

The first purpose of this chapter is to develop a meaningful definition of pollution and to consider its nature. Then the intention is to transcend the issue-attention cycle by briefly reviewing the evolution of pollution to its contemporary significance. It is then essential to ask whether the sociopolitical forces highlighted by a space-time framework point to future styles and outcomes of environmental decision-making on the control of pollution problems (at a cascade of scales from global to individual). Scale is a persuasive (but not a pervasive) concept in environmental policy-making. Haber (1993: 42) has stated that "The scale problem is also one of the main reasons for the failures or slow progress of international environmental politics . . . Successful and trustworthy environmental politicians distinguish themselves by an elevated scale-mindedness." Can we hope for such elevated scale-mindedness in the future?

The Nature of Pollution

Pollution, as anthropologists tell us (e.g., Douglas 1966), is culturally defined. It is a social construct: a label applied by individuals, groups, and societies to changes in the composition of air, water, land, and organisms that are perceived to detract from their quality or result in harm or loss. As a consequence, opinions differ as to what does or does not constitute pollution and there are no uncontested definitions.

Nevertheless, we need to look at two of the numerous available definitions in order to set the parameters (scientific, ethical, social, and political) for any judgment about the extent and seriousness of pollution and, thereby, the need for its control. Pollution has been defined by Murley and Stevens (1991) as:

> The introduction into the environment of substances that are potentially harmful to the health or well-being of human, animal or plant life, or to ecological systems.

This essentially biological definition is flawed because it focuses only on *substances* (not energy), assumes that anything harmful to organisms is pollution (e.g., pesticides? disinfectants?), does not specify how pollution differs from natural causes of environmental change (e.g., meteorites, variations in solar radiation, etc.), and fails to indicate the economic and aesthetic consequences (e.g., accelerated weathering of stone monuments). It is for these reasons that the somewhat older definition of Holdgate (1979) is still preferred:

> The introduction by man [sic] into the environment of substances or energy liable to cause hazards to human health, harm to living resources and ecological systems, damage to structures or amenity, or interference with legitimate uses of the environment.

The differences between these definitions are of fundamental importance. First and foremost, the term *pollution* is properly applied to substances (solid, liquid, gaseous) and energy (vibration, heat) introduced into the environment through *human activity*. Naturally carbonated and mineralized groundwaters, the variations in salinity of sea-water, hot springs and ash falls from volcanic eruptions are not examples of pollution but *contamination*. Second, to qualify as pollution the change in composition must be sufficient to cause recognizable harm or loss *from a human perspective*. Thus we are concerned not merely with the physical effects but also the *human reactions* to the physical effects which are not confined to health considerations but can be expressions of distaste, unpleasantness, distress, concern, or anxiety, all of which can be grouped together as *loss of welfare* (Pearce and Turner 1990). Thus applying the label "pollution" is dependent on the identification of harm to *valued* environmental attributes/resources and a consequent *loss of utility*. If the organisms affected are not valued (e.g., pests) or the changes are considered beneficial, then the term "pollution" cannot be applied.

The emphasis on perceived harm or loss is critical to differentiate pollution from contamination. Changes from ideal conditions (e.g., pure water), average conditions (e.g., mean salinity of sea-water), or baseline/background conditions (e.g., pre-industrial average composition of the lower atmosphere) are termed contamination, whether due to natural or human agencies. Natural contamination can reach levels where harm is threatened or exists (e.g., toxic water, dust storms). By contrast, the term "human contamination," whether due to substances unknown to nature (e.g., DDT, plastic) or naturally occurring (e.g., smoke, CO_2), is restricted to changes which do not exceed the impact/damage threshold and, therefore, do not result in perceived costs. However, once costs or losses are recognized, human-produced contamination becomes pollution. As contamination is a normal and increasingly diverse consequence of human activity, environmental management is often more concerned with limiting the concentrations of polluting substances/energy (pollutants) to below damage threshold levels, than attempting to ban substances entirely. This requires knowledge of *threshold levels* and *environmental capacities*.

Since Holdgate (1979) it has become standard practice to consider pollution in terms of the environmental processes upon which we depend when we discard waste into any medium as receptor: dispersion of that waste and its removal by physical or biological agencies. His geography of pollution adopts the systems approach of sources, pathways, stores, and targets: "where emissions of a substance to the environment are tolerated, controls need to be adjusted so that targets are not unduly hazarded (just what constitutes undue hazard depends on the nature of the target and the value set upon it)."

The nature and sources of pollutants are exceptionally diverse and cover the waste produced by humans themselves (sewage), domestic life, domesticated animals, agriculture, industrial activity, commercial operations, transport, warfare, civil emergencies, and accidents. Sources are both mobile (transport) and stationary, interacting to produce a complex geography characterized by linear belts of supply (transport routes), aerial releases (agriculture), and punctiform patterns of point sources, the last consisting of scattered large-scale emitters (factories, power stations, sewage works) set against a backdrop of innumerable smaller point sources (dwellings, commercial establishments) to yield spatially variable patterns of concentration and complexity. These patterns evolve over time, changing in nature and volume as populations grow and technologically based societies evolve.

What subsequently happens to these waste products depends on the ability of the main receiving environmental media – land, air, and water – to cope with the supply. This ability focuses on environmental capacity as displayed by physically defined environmental units such as rivers, lakes, estuaries, aquifers, airsheds, and geological outcrops, and is determined by (1) the nature and volume of waste production/supply, (2) the existing contamination/pollution state of the medium concerned, and (3) the ability of the medium to disperse or convert the supplied substances or energy, thereby giving rise to the classic statement "dilution is the solution to pollution." Should supply exceed capacity, then inevitable contamination becomes avoidable pollution with adverse consequences to humans, human activity, and the things that humans value – artefacts, organisms, biodiversity, landscape, amenity, quality of life. Broadening the framework to include these aspects means modifying the term "environmental media" to "environmental receptors" so as to place emphasis on the organic components of environment. Thus, while some pollutants can become concentrated due to the operation of physical processes, such as accumulations in lakes, land-locked seas, and in the lower atmosphere under stable anticyclonic conditions, others can accumulate in organisms (bioaccumulation) with resulting adverse consequences higher up the food chain due to biomagnification. Pollution is, therefore, an ecosystem problem and can have widespread ramifications due to adverse effects on the crucial biogeochemical cycles, as is the case with the discharge of greenhouse gasses. The difficulty is that, while we have a good level of understanding about the processes of dilution, dispersion, containment, and conversion operating in water, air, and land, understanding of the processes operating in these more catholicly defined receptors still needs to be improved.

As a consequence, pollution can variously be a chronic problem which may be progressively changing over time, or a repetitious phenomenon occurring on an annual (winter smoke-generated "smogs") or diurnal basis (photochemical smog), or a random feature due to a particular combination of circumstances.

Contamination becomes pollution when the damage threshold is crossed, but defining such a threshold is problematic. Reductionist science has faced the challenge of assessing the capacity of environmental media in two main ways. First, it has examined the toxicity of wastes by laboratory testing using plants and animals, a practice which many people find repugnant.

But toxicity testing is an extremely crude area of experimental science, by the evidence of its own practitioners (see authors in Nriagu and Lakshminarayana 1989); even when techniques yield usable results for a single or group of substances, the rate of innovation in the chemical field is so great that testing-bottlenecks abound. Also there are problems of scaling-up from laboratory animals to humans, of substances acting in combination, of synergistic effects, of changes in chemistry after release into the environment, and of changing threshold levels due to the interaction of both chemicals and environmental factors.

Second, science has traditionally explored the patterns of pollution in small units of land, water, and air, mainly in response to point sources. As a consequence, it was shocked by incidents such as the Chernobyl nuclear disaster (1986), with their very extensive repercussions (even though Rachel Carson (1962) warned about pesticides in the Arctic and Scandinavians had successfully complained about "imports" of sulfur dioxide from the rest of Europe, both indicating the transnational and international character of some pollutants). It has also had to cope with the widespread impacts of pollution from diffuse sources such as agriculture and by longer-lasting and transgenerational impacts, such as those claimed for some nuclear power plants. In some cases, cumulative effects combined with the complexity of environmental conditions may result in progressively increasing impacts for which cause–effect relationships are difficult to establish (elusive hazards), while in other cases cause–effect links may only become established at a future date (long-tailed risks, e.g., asbestos, CFCs).

Clearly laboratory tests in vitro have a (rightly) restricted impact on human perceptions of harm and therefore of risk. The "Catch 22" is that, once in vivo evidence is available, serious harm may have been done, at least to a sample of the population, or to non-human organisms which have taken the traditional role of the miner's canary. It is easy to see why profound ethical debates focus on such issues as whether the concept of rights (Taylor 1986) should be extended to other species and even inorganic components of the environment and the degree to which the precautionary principle should be applied to human activities (O'Riordan and Cameron 1994). A dichotomy exists, therefore, between instrumental rationality (linked to toxicity tests) on the one hand, and value rationality (rights to avoid harm and have risks distributed "fairly") on the other.

The disposal of waste into the environment is the classic externality problem associated with common property (Hardin 1968). Thus a major underlying cause of pollution is that it is usually easier and cheaper to discard waste into an environmental medium than it is to process the material so as to minimize environmental impact. While such ignorance, laziness, and greed (profit maximization) are on the decline, such tendencies still exist in the dubious nature of some "accidental" discharges of toxic waste, in fly-tipping, and in litter production, and contribute to the existence of heavily polluting factories in less developed parts of the world where political and economic aspirations have priority over environmental quality. The attitudes are well described by the statements "somebody else's problem, someone else should pay" and "private affluence, public squalor." In addition to the above, other categories of pollution are planned pollution, where it is economically or technically impossible to remove all waste discharge, and fugite pollution due to leaks, etc. However, the potential for damage is often only brought sharply into focus when "normal life" is disturbed by natural hazard impact (flood, hurricane, earthquake), hostilities, civil emergencies (e.g., strikes, riots), and accidents. It is then that pollution emergencies arise: severe, conspicuous events which result in dramatic scenes that invoke considerable, albeit shortlived, media attention. For example, few will forget the burning oil-fields of Kuwait (1991) or the blackened coastlines and heaps of dead birds following major tanker accidents. While good practices can minimize the background levels of pollutants, emergencies will still occur if for no other reason than human behavior and error (one of the reasons for continuing international suspicion of nuclear power and its waste stream).

Pollution emergencies focus attention on both the ambient processes and capacities to contain, dilute, disperse, and neutralize pollution and the inability of human societies adequately to predict consequences and undertake realistic risk assessments. The accidental release of radioactivity from the Chernobyl nuclear reactor in 1986 resulted in an unexpected pattern of dispersion due to prevailing atmosphere conditions (figure 25.1; Gould 1990). In Britain the resulting precipitation-induced contamination was spatially variable, with the further unexpected consequences that it was the upland sheep farmers of North Wales and the English Lake District that were the most affected by the accumulation of radionucleides (Wynne 1990). The "surprise" in this case was that the UK system for tracking radioactivity through farming systems had previously focused only on lowland soils and farming systems.

In the cases of the Seveso (1976) and Bhopal (1984) incidents, explosions resulted in the release of toxic substances over adjacent human settlements. At Seveso it was dioxin, although prompt evacuation of 1,000 people resulted in no impact even though sheep died in a field adjacent to the factory. Eventually it was shown that the sheep had died of bloat due to what they had eaten and their deaths were in no way connected to the accident, thereby showing the necessity of clearly establishing cause–effect relationships before drawing conclusions. In the case of Bhopal, however, 40 tons of methyl isocyanate were released from the Union Carbide plant killing over 5,000 people, injuring at least 20,000, and affecting over 300,000, for the most part inhabitants of the shantytowns around the plant (Weir 1987; Morehouse 1994). The proximity of so many people to a potentially dangerous plant working under pressure were the ideal ingredients for a disaster "that was waiting to happen."

Finally, attention must be paid to assimilative capacity. The volume and/ or toxicity of released substances/energy are not the only indicators of

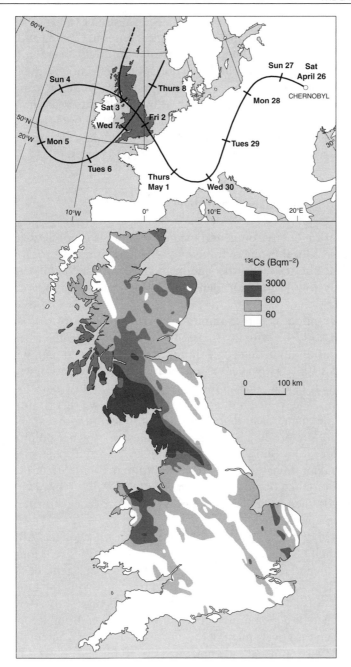

Figure 25.1   The temporal and spatial variability of pollution: Chernobyl cloud (1986).

potential impact: vulnerability as determined by resistance and resilience (ability to recover) are also important. The ecological crises in Canada and Scandinavia arising from the windblown import of sulfur acids were not primarily because of the amount of acid deposition, but a reflection of the inability of particular ecosystems to cope with the supply (low buffering capacity). The same arguments apply in the context of media debates regarding the long-term impacts of maritime and coastal oil spills. While optimists point to the "recovery" of affected bird populations, pessimists take a gloomier view based on the observed resilience of oiled mangroves and corals.

Pollution in Evolutionary Perspective

Contamination of environmental media by humans is not new. It probably first became a local problem following sedenterization due to the Neolithic Revolution, but attained little significance for thousands of years because of relatively low population densities. Urbanization was the catalyst for change.

Taking London as the example, historical archives record smoke nuisance from AD 1247, with laws passed in the late thirteenth century against smoke pollution involving some quite severe penalties (a man was hung to death in AD 1306!). However, it was the Industrial Revolution which stimulated the great change due to the combination of industrialization and urbanization. Waste became synonymous with wealth as mining and engineering expanded, so that environmental degradation came to be seen as the natural by-product of progress. In the case of water pollution, it was not until Dr. John Snow traced the source of an outbreak of cholera in London to contaminated water from a pump at Broad Street in 1854 and the severe befouling of the River Thames in 1858 ("The year of Great Stink" when 20,000 died of cholera in London) that the costs of water pollution began to be appreciated.

In the case of air pollution the process took rather longer. Although the gloomy, smelly, smoke-laden atmospheres above cities had long been remarked upon and the term "smog" applied to the persistent, dense, yellowish, sulfur-laden fogs that developed in winter (following the London event of 1905), it was not until the 1952 smog when records showed that 12,000 Londoners had died prematurely that action was taken (Brimblecombe 1987). But even then the government of the day was reluctant to act on the grounds that smogs were "normal" phenomena beyond human control. Eventually, the pioneering Clean Air Acts of 1956 and 1968 were passed and the cleanliness of urban atmospheres in Britain greatly improved, although questions remain as to the extent to which this was due to cultural changes rather than legislation (Jones 1991).

Smoke and sewage are traditional, local pollutants. The extent to which they exist today depends on political and economic considerations rather than technical ones. However, they have been replaced on the agenda of concern by new second- and third-generation pollutants with very different characteristics and environmental chemistry. First, there was acidification (often oversimplified as "acid rain") due to the distant down-wind transport of sulfur and nitrogen compounds (SO_x and NO_x respectively), resulting in the highlighting of pollution as a transnational problem. Then came the discovery that the thinning of the stratospheric ozone shield in high latitudes is due to various chemicals (most especially chlorine) released into the atmosphere, leading to the identification of problems at the regional scale due to global causes. Finally, came the recognition of global warming due to the release of greenhouse gases (CO_2, CFCs, methane, NO_x), a classic "elusive hazard" at the global scale with such a multitude of possible adverse consequences and implications for society that it is often referred to as a problematic, syndrome, or concatenation. At the same time, there have developed other more localized problems associated with the over-use and misuse of pesticides and the dramatic rise in use of road vehicles which has resulted in the widespread reproduction of the so-called Los Angeles smog problem (photochemical smog and low-level ozone).

These pollutants are all products of the changing relationship between human society and the natural environment ushered in by the Industrial Revolution – the so-called "Great Climacteric" of Burton and Kates (1986) – and to varying degrees share the common feature of sensory amodality, i.e., the agents cannot be detected by human sensory organs. This has had serious repercussions, for since the recognition of nuclear fall-out, which is also undetectable by humans, the natural environment has become a focus of concern and seen to be the medium by which various undetectable harms are transmitted. As a consequence, the terms "nature" and "natural environment" have come to be progressively replaced by "environment" as everexpanding human influences have confined "natural" ecosystems and their essential gene pools into ever-shrinking wilderness areas (table 25.1). While "natural environment" was seen to be pristine, reliable, and ambivalent, the simple term "environment" is increasingly used to mean human-modified, questionable, and potentially dangerous. The present situation, therefore, is one in which benign surroundings which were seen to be almost limitless in their ability to absorb disruption and consume waste, are now seen as finite and risk-laden, containing invisible threats that can only be determined by scientists. Thus at the beginning of the new millennium the physical environment has achieved a greater than ever significance. Therefore, at the same time as scientific and technological developments have enabled human activity to become increasingly divorced from the constraints imposed by the physical surroundings, human society has become increasingly concerned about the adverse consequences of its activities and troubled by its environment. This tendency is well displayed in the media, where the views of pessimistic Cassandras dominate over those of optimistic Pollyannas, as was evident in the UK *Sunday Times* color supplement of

Table 25.1   The shrinking terrestrial wilderness habitats.

Vegetation type	km²	%	No. of blocks
Tundra	20,047,533	41.7	100
Warm desert/semi-desert	9,329,531	19.4	389
Temperate needle-leaf forests	8,799,312	18.3	120
Tropical humid forests	3,006,855	6.3	77
Mixed mountain systems	1,973,391	4.1	76
Cold-water deserts	1,478,494	3.1	51
Tropical dry forests	1,424,099	3.0	120
Tropical grassland/savannahs	735,331	1.5	33
Temperate rainforests	450,215	0.9	15
Temperate broadleafed forests	290,646	0.6	20
Temperate grasslands	272,016	0.6	24
Evergreen sclerophyllous forests	170,885	0.4	7
Mixed island systems	910,649	0.2	7
Total	48,069,951		1039

Continent	km²	% of total	% of continent	No. of blocks
Antarctica	13,208,983	27.5	100	1
North America	9,077,418	18.9	37.5	85
Africa	8,232,382	17.1	27.5	434
Soviet Union	7,520,219	15.6	33.6	182
Asia	3,775,858	7.9	13.6	144
South America	3,745,971	7.8	20.8	90
Australasia	2,370,567	4.9	27.9	91
Europe	138,553	0.3	2.8	11
Total	48,069,951			

World habitat	162,052,691
% wilderness	29.7

Source: after McCloskey and Spalding (1989)

February 26, 1989 entitled "The World is Dying: What Are You Going To Do About It?" The extent to which such views are based on changing perceptions and increasing levels of expectation regarding health and safety is open to debate. As Douglas and Wildavsky (1983) remarked, "Is the world becoming more dangerous or are we becoming more afraid?"

It is these notions that underpin the *Risk Society* paradigm advanced by the sociologist Ulrich Beck (1992). He sees contemporary environmental threats as both greater and more problematic than at any stage in human history, due to the fact that they are typically invisible, affect society in exceedingly complex ways, and have consequences that are increasingly incalculable. His view of modernity is essentially a catastrophic one, as is that of Giddens (1991) who uses the term "late modernity" and considers pollution to be an element of "manufactured risk" as distinct from "external risk" due to geophysical processes and extraterrestrial phenomena (Giddens 1999). Both emphasize the development of new risks arising from technology, the universality of society due to the fact that developments at any one place can have immediate global consequences, the growing significance of the individual, and the recognition that risk colonizes the future, which thereby determines the present, whereas previously it was the past that was seen to determine the present. Thus environmental problems such as pollution are not so much problems of the environment but problems of people, or as Beck (1992: 81) states, "At the end of the twentieth century nature is society and society is also *nature*."

Think Globally, Act Locally

The change in scale and the scale of change of pollution, especially with respect to the variety, complexity, and catastrophe potential of modern releases, brings sharply into focus the issue of spatial scale and how to interrelate the global and the local, or "glocal" as it is sometimes called.

For the throw-away society of the twenty-first century, characterized by rapid innovation, materialism, and designed obsolescence, there turns out to be no such place as "away." Decisions about where and when pollution can occur remain concentrated at national to local levels, where gradients of inequity (in terms of outcome) are easiest to judge. The impact on communities or regions, in the face of universal NIMBY ("Not In My Back Yard") attitudes, has been to create rapid learning curves in the principles of sustainable development! The question remains, therefore, as to whether the "big issue" of pollution and its control is scientific and *global* or political and *local* – with the latter operating at the scale of problems as perceived by non-technical (often commercial) actors. Furthermore, the perception of the problem will depend heavily on those factors which determine people's appreciation of risk; there is no international culture of risk to rival that possessed by individuals and communities, as is well known to geographers studying environmental hazards (see Cutter 1993).

There is also a link on another scale – the pervasiveness of financial pressure to innovate and to develop resources under the capitalist system; failure to recognize this partly explains the negative aspects of the United Nations Conference on Environment and Development (the "Earth Summit") held in Rio in 1992. It is to the credit of those who drafted "Agenda 21" for the Conference follow-up that they homed in on local actors and communities and showed sensitivity to inequity, risk, and economic systems in this plan for sustainable futures. However, relating individual actions to global-scale problems remains a major challenge, especially where freedom of choice is curtailed or costs are incurred.

The Global Dilemmas

At the global scale, industrial development patterns based on fossil fuel have established a North–South contrast in the acidification of the atmosphere (figure 25.2). Cross-equatorial dilution and dispersion is not available and as a result the capacity of sensitive parts of the northern hemisphere has been exceeded, producing harm to ecosystems and cultural heritage to such an extent that geographically based controls on acid emissions have now been introduced (Elsom 1992). This "unseen plague of the industrial age" has been linked particularly with the widespread phenomenon of "forest death" or "Waldsterben," the analysis of which has revealed the dangers of rushing to simple cause–effect conclusions when dealing with complex environmental problems. It is now realized that the synchronized death of trees in many parts of the northern hemisphere is due to no one single mechanism (table 25.2), but varies spatially due to combinations of multiple stresses (including disease and drought), air pollution (top-down hypotheses), and soil acidification

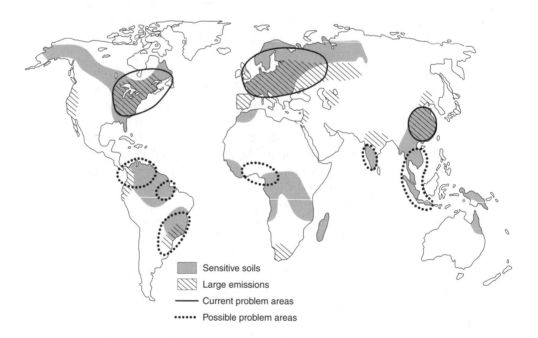

Figure 25.2   The progress of acidification.

Table 25.2 Primary causes of forest decline in the 1980s, ranked in order of importance by region.

	Western Europe	Eastern Europe	North America
Ozone	1	2	1
Acid deposition	2	3	5
Other gaseous pollutants (e.g., SO_2 and NO_x)	3	1	3
Excess nitrogen deposition	4	–	2
Growth-altering organic chemicals	5	–	6
Heavy metals	–	4	4

(bottom-up hypotheses). The actual role of acid deposition varies geographically, as does the significance of "dry" versus "wet" deposition and the proportion of sulfur-based acids (from power stations, etc.) to nitrogen-based acids (from motor exhausts).

The contrast in terms of spatial and temporal impacts between the output of acid pollutants and that of carbon dioxide, also largely from fossil-fuel combustion, illustrates a key component of the geography of pollution: acidification is an international (transboundary) problem but not yet a global one; it can be technically mitigated (emission controls) and by land and water management. Harm is reversible (over time) if the emissions of the pollutants are reduced or buffering capacity increased. By contrast, outputs of carbon dioxide, together with other "greenhouse" gases, will have a variable impact over the whole globe by changing climatic patterns, sea-level, ecosystems, patterns of disease and human activities (Houghton, Jenkins, and Ephramus 1990; Houghton 1997). It is not possible to "back-track" the CO_2 as it is for acidity, and extreme uncertainty exists about the pattern of possible effects, particularly on climate, despite the fact that global warming is now widely recognized to be a reality (figures 25.3 and 25.4). Sea-level rise is more easily predictable – it has already begun (figure 25.5)! Furthermore, any assessment of current or anticipated harm from greenhouse emissions must admit that there is drastic global inequality between those who have utilized the carrying capacity for carbon dioxide and those who will bear the brunt of the impacts (and whose future development, including that of human rights, will be compromised by forgoing further use of that capacity: see Grubb 1990).

One perceptual difficulty with the "greenhouse effect" is that it is not a problem involving toxic harm. Where the latter is at issue, "life and death" becomes the perceptual avenue for risk assessment, even when the assessment is made for non-human biota. This has been the case for acidification and was also the imperative when the true extent of stratospheric ozone

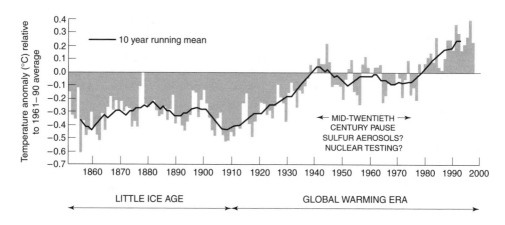

Figure 25.3   The progress of global warming.

depletion became apparent in the late 1980s (figure 25.6). Here a powerful combination of scientific explanation, visually stunning computer graphics, and the heightening of "dread" among politicians and the public by doom-laden predictions of increased numbers of cancers and regional ecocatastrophe, together with the powerful image of a life-threatening "hole in the ozone shield," resulted in an amazingly rapid international political response, as testified by the Montreal Protocol (1987) and its subsequent amendments and modifications (Smith and Warr 1991; Benedick 1998). However, in the case of the "greenhouse effect" the issues are less clear cut. Sources are diverse, changes in physical conditions continue to be difficult to predict because of the lack of scientific knowledge and the scale and complexity of the systems concerned, and the consequences for individual nations could vary from catastrophic (island states such as the Maldives) via generally neutral to potentially beneficial. Nevertheless, the postulated adverse consequences for ecosystems (accelerated extinction rate) and for human populations (environmental refugees) indicate that there will be in-creasing pressures for international cooperation to mitigate the worst ef-fects, despite the disappointing outcomes of the Kyoto Conference on Climate Change (1997) and the subsequent "summit" at the Hague (2000), and President Bush's rejection of the Kyoto Protocol (2001). It is now clear that *adaptation to* rather than *prevention of* climate change is likely to be the long-term outcome.

Global Harm – Relax: Gaia is Pro-Life

Science is both part of the problem – when it is reductionist – and part of the answer – when it is holistic – in the current environmental crisis. Reductionist

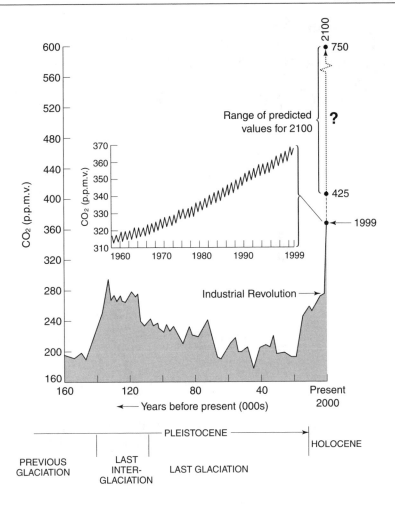

Figure 25.4   Changing levels of atmospheric CO_2 (160,000 BP–AD 2100).

science, in the guise of technology in the capitalist world economy, has led to linear concepts of "progress" based on production with inevitable waste production and pollution. Of the environmental sciences, ecology has experienced more holistic phases than most. The concept of ecosystems stresses that the impacts of pollution are distributed in space and time, not only by environmental media, but also by the connectivity of systems, e.g., through food chains as part of food webs. Ecology has also offered a defense of the ethical extension approach to harm because it stresses the supportive role of biodiversity in repairing and restoring mechanisms of global equilibrium (Reid and Miller 1989). Yet the political extension of ecological principles (including hierarchies and predation) through ecologism has been a feature of rightist rather than leftist tendencies in the twentieth century (Bramwell 1989).

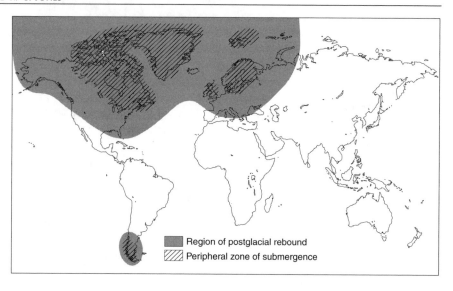

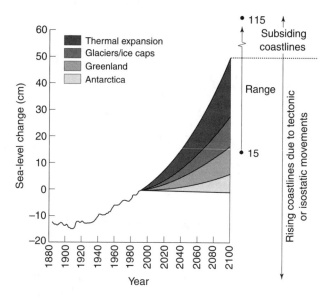

Figure 25.5   Variations in sea-level across the globe and with time (1880–2100).

The most serious recent theoretical contribution by holistic science to the global assessment of pollution impacts has been the Gaia Hypothesis (Lovelock 1979; 1988). Lovelock, who is feted by some greens for the transcendental qualities of his vision, is in fact optimistic about technology in the future. The global homeostasis achieved by the multiple feedbacks of Gaian cybernetics will, in Lovelock's view, always maintain the Earth's environment as a fit one

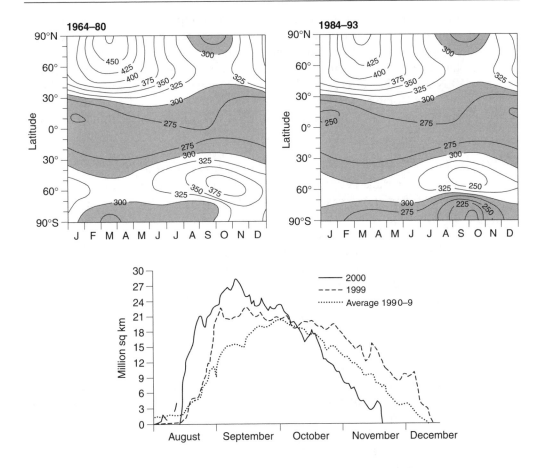

Figure 25.6 Variations in total ozone with latitude and season in Dobson units (1964–80 and 1984–93) and mean size of Antarctic ozone "hole" (1990–9 and 1999–2000).

for life. He does not take a local or regional ecosystems approach, but rather one which stresses the huge capacities of the global spheres – biosphere, atmosphere, hydrosphere – their storages and cycles and their teeming macroscopic organisms. The forms of life which co-evolve with the chemistry of the global spheres may not necessarily include humans in a prominent role, but he has optimism even for our species. His grounds are that even contemporary pollution-control problems show no signs of swamping the absolute capacity of atmospheric systems to repair and restore. Rather, it is the danger of unstable behavior (crossing the fine threshold of system state) as capacities are *approached* which Gaia warns about.

The Gaia Hypothesis has proved its scientific worth by generating debate about global-scale pollution-control priorities and the importance of atmospheric and ocean capacities and coupling. Global scientific research programs are now widespread and the resulting increases in knowledge about the geosystem have been impressive, not the least of which has been the clear identification of the El Niño-southern oscillation (ENSO) and its effects on global weather patterns (teleconnections). Gaia has also focused attention on the behavior of natural systems close to their homeostatic limits: the potential for abrupt, threshold changes is now a central concern of research on the enhanced "greenhouse effect." The rapidity with which, for example, protocols for ozone control were adopted internationally is not merely a lesson on the processes of diplomacy (Benedick 1998), but is also a manifestation of a new global concern. However, the case of ozone controls is unusual in that science spoke directly to power, with great clarity and certainty. Developed-world capitalists seemed keen, as the developing ozone "crisis" made headlines, to make use of and to develop substitutes (witness the great spray-can switch!), and consumers were concerned enough to refuse the old damaging products that were needlessly threatening the integrity of the geosphere.

The upsurge of interest in global environmental issues has focused attention on two problems. First, the recognition of a changed relationship between science and politics or research and policy. Whereas, traditionally, science was expected to provide "hard" objective facts upon which decision-makers could base their "soft" value-ridden policies, in the postmodern era of the risk society, scientists are increasingly forced to deliver "soft" uncertain "facts" for policy-makers to attempt to make difficult or "hard" decisions. But ignorance and uncertainty are well established concepts in science. That they have been lost sight of during the processes of risk communication in recent decades is unfortunate and it is important to note that the Intergovernmental Panel on Climate Change (IPCC) takes great pains clearly to indicate the uncertainties. Second, international transboundary environmental problems are only slowly being confronted through the development of legal principles as a basis for international environmental law due to the ad hoc, incremental growth of legal obligations in the form of treaties, etc. and non-binding but nevertheless influential declarations and action plans. Here, the United Nations Environment Program (UNEP) and NGOs have been important catalysts for progress.

Thus, the main signal at the global scale is that of the uncertainties and risk of approaching threshold conditions in relation to capacities to absorb pollutants. Because there is not yet a global polity able to assess these risks with a common perception and purpose, it remains necessary to look to more local perceptions and polities and their relationship to the one human global process – that of international capitalism and its production processes.

What are the Limits? Rates of Growth, Limits to Capitalism

In the 1970s the power of computer models was harnessed to create predictions about future resource use and pollution, given the headlong economic growth of the era. Doom-laden predictions emerged, soberly summarized as "limits to growth" (Meadows et al. 1972). These limits were quickly challenged by technocentric optimists. The limits were cast more in terms of resource depletion than pollution and a subsequent re-run of models (Meadows, Meadows, and Randers 1992) maintained the relatively optimistic perception by stressing the part being played by the slowly and variably reducing emissions of traditional pollutants and hence ambient concentrations/loads in the developed world. Nevertheless, if growth means growth in the number and variety of pollution sources and resulting pollutants, then it remains a danger to carrying capacities.

Even optimistic Gaian scientists question whether Gaia can cope with the present *rates* of environmental change. There are concerns regarding the extent to which biological homeostasis has been validated, and also that many of the crucial feedback processes identified may only operate over geological timescales. To quote Schneider and Boston (1991):

> Therefore, whether to risk radically modifying the Earth in pursuit of human goals is not a scientific question per se; rather, it is a fundamental *political value choice* that weighs the immediate benefits of population or economic growth versus the potential environmental or societal risk of a rapidly altered Earth. (Original emphasis)

Here is a dimension which reconciles the global with the local, but what values are involved and whose politics? It is also a fundamental *ethical* value choice, although ethics and politics do not necessarily correlate in an era of soundbite, spin-doctored opportunism. The biodiversity of world politics has itself declined markedly; the domination of the capitalist system makes it extremely difficult to examine the potential of a genuinely green political dimension. However, as Weale (1992) stresses, neither the ethical stance struck by an extension of rights to non-human biota, nor a rival political dimension in favor of precautionary "speed limits," is likely to break through when both approaches to pollution control breach the conventional structure of politics in terms of *interests*. Curtailing rates of growth to be compatible with long-term carrying capacities is against the growth imperative of national governments (see, for example, Pezzey 1991). Another infringement noticed by our political referees is that the green dimension will increasingly make it extremely difficult to separate domestic and foreign policy.

However, Weale has introduced the term "ecological modernization," a reconceptualization of the relationship between the economy and the environment in which environmental protection becomes a growth promoter, not a burden. This view tends to play into the hands of the science–technology–capitalism–innovation system of the developed world. Many industrial and commercial leaders in the developed world are alerted to pollution only because it is represented as industrial bad practice; they can now qualify for certificates of good environmental practice, carry out audits, and have special logos (doves and leaves are common!) on their goods, but they continue to rely upon the Earth's carrying capacities for much of their waste.

International capitalism has considerable impact on modern geographical patterns (Storper and Walker 1989); can this impact be reconciled with the spatial patterns necessary for sustainable environmental management, one component of which is pollution control? Slowly, "ecological modernization" is coming to mean the privatization of environmental management within a suitably rigged market whose ethical pillars are "stewardship" and "sustainability."

Certain characteristics of ecological modernization can form the basis of a pollution-control policy within world capitalism if the scale units are chosen with respect to the problems and the problems prioritized. River basins have shown this form of environmental politics in many parts of the world (Newson 1992).

Pollution: Just Act Locally

As Johnston (1989b) stated, it is impossible to undertake any practical policy-making on environmental issues without reference to the nation-state as a unit and the world capitalist system as a process. This neglects the fact that cities and city-regions are often the major causes of the pollution problems which policy seeks to cure or correct. Environmentalism may well require its own spatial organization, and in this it will battle further with capitalism, especially in the increasingly numerous and large cities.

Is it possible to develop environmental ethics which base the choice of right or wrong actions upon their community impacts? Those seeking a rights extension, such as Taylor (1986), are skeptical about such ecological attitudes to harm. The argument of ecosystem health can be just as anthropocentric and hierarchical, through notions of "stewardship," as that based directly on human self-interest. If, by contrast, the concern for human rights is extended to one of biotic rights, then there will exist a much more concrete, though more difficult, basis for judgment on the necessary policies for management. This also provides the opportunity to reintroduce the dimensions of culture and spatial units which geographers find so fascinating.

In the early days of the New Environmental Age, philosophers and pragmatists of green approaches settled on the need to set new scales and networks for living (e.g., Schumacher 1973); this obsession failed (Pepper 1991) largely because it was a romanticized rural obsession. Of what value is the commune vision to a "risk society" (Nohrstedt 1993) when most maps of pollution risk show urban peaks? Murray Bookchin has always stressed (e.g., Bookchin 1986) the human-rights aspects of pollution and other forms of environmental degradation, seeking a "red/green" communalistic solution around reconceptualized city polities.

The uncertainty of environmental science also has a spatial relevance. As many analysts now conclude, the postmodern element of the environmental crisis is the inability of science to provide traditional mechanistic explanations and therefore remedies based on technological solutions alone. As was discussed earlier, caveats and expressions of uncertainty tend to be lost sight of during the process of risk communication, with unfortunate consequences when revealed outcomes do not conform to predictions. As a consequence, trust in certain branches of science has diminished over recent years, often through no fault of the scientist concerned, which is of considerable significance in this era of greater inclusivity in risk management decision-making. In the words of O'Riordan and Rayner (1991: 98): "Under . . . conditions of high uncertainty and high societal stakes, risk management decisions cannot be justified according to purely technical criteria or even clinical judgment."

Decisions therefore need to seek acceptance and be interactive with a concerned population in a way that conventional budgetary and foreign-affairs political decisions cannot be. Since pollution is culturally defined, remedies need to be culturally adapted. Such adaptation is said to be the secret of the supposedly harmonious environmental relations enjoyed by indigenous peoples. In seeking sustainable pollution-control policies the polity of our decision-making must be structured to make most use of citizen roles (Portney 1991). Principle 10 in the Rio Declaration on Environment and Development states that "Environmental issues are best handled with the participation of all concerned at the relevant level."

Politicians are rushing to sign up for environmental management units such as bioregions (Gore 1992) or regional community-based power structures for environmental management (Taylor 1992). If capitalism is forced into similar units, as a basis for pollution control via the use of economic instruments (in which industry is allowed to trade permits to pollute into regional "pollution bubbles"), a heady mixture of sustainability and subsidiarity may gain credence.

We are not yet a global "EMU" (environmental management unit); within the Gaian milieu of the spheres there are a myriad local and regional cultural traits that are of relevance to the risk society and its long-term success or failure in controlling pollution. Here is the proper focus for the phrase, "It's all geographical."

26 Sustainable Development?

W. M. Adams

Introduction

Sustainable development has proved a spectacularly popular phrase in both general and academic writing about environment and development. It has roots and a large following in many disciplines, but probably none is so relevant to it as geography. Where else can the science of the environment be married with an understanding of the economic, political, and cultural change that we call development? What other discipline offers insights into both environmental change and environmental management, and who but geographers can cope with the diversity of environments and countries, and the sheer range of spatial scales, at which it is necessary to work to understand processes of human use of nature and the dynamics of the environment?

Kates (1987) identifies the human environment as one of the four key traditions of academic geography. David Lowenthal described *Man and Nature*, written in 1864 by George Perkins Marsh, as "the most important and original American geographical work of the nineteenth century" because of the revolution it brought about in understanding the ways that people transformed their surroundings (Lowenthal 1958: 246). Marsh's ideas were recalled and claimed for geography at the symposium held by the Wenner-Gren Foundation at Princeton, New Jersey, in 1955, *Man's Role in Changing the Face of the Earth* (Thomas 1956), chaired by Carl Sauer, Marston Bates, and Lewis Mumford. That meeting, in its turn, was recalled forty years later in a volume that both renewed the commitment of geographers to the study of human impacts upon the earth, and noted the scale and extent of that impact (Turner et al. 1990).

Sustainable development draws on both the technocentrist and ecocentrist axes of environmentalism (O'Riordan 1981). On the one hand, many people writing about sustainable development propose rational, technical solutions to environmental problems (better environmental planning, and clean technologies, for example). Other proponents believe that sustainable development must involve more radical changes to economy and society (for example, zero growth or local self-sufficiency), or new attitudes to nature (for example, a concern for the rights of other species). Geographers have tended to emphasize the technocentric end of this spectrum. Resource management has long been an important theme in geography (e.g., Burton and Kates 1965; O'Riordan 1971; Mitchell 1979; c.f., Johnston 1983), as has the application of these ideas to the Third World, both in general terms (e.g., Omara-Ojungu 1992) and in more specific studies (arid lands, e.g., Heathcote 1983, or water resources, e.g., Agnew and Anderson 1992). These ideas have been particularly strong in North America, and have been extensively influenced by ideas about "rational utilization" or "wise use" of natural resources in the utilitarian conservation philosophy developed in the USA in the first decades of the twentieth century (Hays 1959). The notion of efficiency in resource use played an important role in the agencies created to implement Roosevelt's New Deal in the 1930s (Wallach 1991), when organizations like the Bureau of Reclamation and the Tennessee Valley Authority inscribed the principles of rational resource management on American landscapes.

Technocentric ideas about environmental management were strengthened in European geography by the growth of empires, particularly in Africa. In England the Royal Geographical Society moved from exploration to a concern for geography in universities in the nineteenth century (Stoddart 1986), while in France geographers were closely involved in French imperial ambitions (Heffernan 1990). In the twentieth century geographers mapped the new colonial territories, sanctioning by their skills the partitioning and expropriation of land. Geography was well fitted to colonial service: climate, hydrology, soils, land use, settlements, markets, and transportation were all grist to the geographer's mill as the discipline became established in European universities, and developed its own institutional life.

Nonetheless, despite its apparent natural advantages, geography failed to provide the intellectual leadership for the environmental revolution of the 1960s and 1970s (Kates 1987). Notwithstanding experience in areas like natural hazards, neither the quantitative revolution nor the explosion of radical thought provided a strong foundation for understanding the key issues of environment and development that began to emerge.

However, if geographers are interested in following the "road still beckoning" (Kates 1987) to play a major role in debates about people and environment, there is now a far richer range of debates within the discipline that are relevant. Developments in physical geography, and techniques such as modeling, offer greater sophistication in scientific understanding of environmental change (e.g., Roberts 1994; Huggett 1993); radical scholarship has placed the rather static and complacent debates of some resources geography into political economic context (e.g., Watts 1983; Hewitt 1983; Blaikie 1985; Peet and Watts 1996; Rocheleau, Thomas-Slayter, and Wangari 1996; Bryant and Bailey 1997); and cultural geographers have looked anew at nature and at the discourses of science about it, to challenge the conventional ideas that underpin the "environment–development" debate (e.g., Fitzsimmons 1989; Evernden 1992; Mcnaughten and Urry 1998).

Sustainable development has many definitions (e.g., Pearce, Markandya, and Barbier 1989). The most widely quoted is still that of the Brundtland Commission in *Our Common Future*: "development which meets the needs of the present without compromising the ability of future generations to meet their own needs" (Brundtland 1987: 43). Sustainable development fulfilled LéLé's (1991) prediction and became the development paradigm of the 1990s. Indeed, it became a vital buzzword in the speeches of politicians, company chairs, and NGO activists alike. It has come to carry on its deceptively simple back a vast range of hopes and ideals, from near-revolutionary environmentalism to the blandest hint of green in corporate capitalism. Academically, sustainable development seems to contain the promise of new joined-up research, unlocking the doors separating disciplines, and to break down the barriers between academic knowledge and action. In politics it promises to tackle both poverty and environmental degradation without requiring that anybody gets any less well off, or affecting corporate profits or shareholder value. Sustainable development implies win-win-win solutions (across environment, society, and economy), perfect for the decade of the televised soundbite and the spin doctor.

Despite its simplicity, the phrase "sustainable development" is capable of carrying a wide range of meanings, sometimes allowing quite different ideas about what should be done to exist side by side without the conflict between them being clear. Both radical environmentalists and conventional development-policy pragmatists have seized the phrase and used it to express and explain their ideas about development and environment. In the process they have created a theoretical maze of great complexity (LéLé 1991; Redclift 1996; Redclift and Benton 1994; Adams 2001). The superficial conformity of writing and thinking about sustainable development hides very real and fierce battles behind the scenes over meanings.

Mainstream Sustainable Development

The principles of sustainable development were first developed in a coherent way in 1980, in *The World Conservation Strategy* (WCS) prepared by IUCN (the International Union for the Conservation of Nature and Natural Resources) with finance provided by the UN Environment Program and the World Wildlife Fund (IUCN 1980). It was further developed in the report of the World Commission on Environment and Development, *Our Common Future* (Brundtland 1987), and the follow-up to the WCS, *Caring for the Earth* (IUCN 1991), before its appearance in *Agenda 21* at the Rio Conference in 1992. These documents differ, but have a remarkably consistent core of ideas, a "mainstream" that has persisted through the two decades between Stockholm and Rio (Adams 2001). At the heart of these documents is a vision of sustainable development strongly influenced by science, by wildlife conservation, by concerns for multilateral global economic relations, and by an emphasis on the rational management of resources to maximize human welfare.

Concern about the environment as a central issue in development began to grow in the 1960s and 1970s. The idea was slowly incorporated into developmental and environmental debates by international organizations. Throughout this process the idea of sustainability was pushed most strongly and effectively by scientists (like Sir Julian Huxley, first Director of UNESCO), particularly by ecologists engaged in international scientific collaboration (Boardman 1981; Worthington 1983). Scientific conservationists began to engage with the impacts of development on ecosystems, publishing books such as *The Careless Technology* (Farvar and Milton 1973) and the handbook for development planners, *Ecological Principles for Economic Development* (Dasmann, Milton, and Freeman 1973).

The United Nations Conference on the Human Environment in Stockholm in 1972 is usually identified as a watershed in the emergence of sustainable development, although it was only partly, and belatedly, concerned with the environmental and developmental problems of the emerging Third World. The growing influence of concern for the environmental aspects of economic development was by no means accepted unopposed. "Developing" countries saw discussions of global resource management as an attempt by industrialized countries to take away control of their resources (Biswas and Biswas 1984), and feared that the environment was being given too high a priority compared with development; far from suffering the impacts of pollution caused by industrialization, poor countries suffered the "pollution of poverty." Faced with this concern, the phrase "sustainable development" was used explicitly to smooth over the apparent dichotomy between economic growth based on industrialization, and associated adverse environmental impacts. The phrase captured perfectly the idea (or the belief) that it was possible to have development without adverse environmental side-effects (Clarke and Timberlake 1982). The essence of this idea, and the power of its political appeal, changed very little in the two decades that elapsed before the Stockholm Conference's successor at Rio de Janeiro in 1992.

The most conspicuous result of the Stockholm Conference was probably the creation of the United Nations Environment Program (UNEP), which played an important role in the preparation of *The World Conservation Strategy*, published in 1980. This identified three objectives for conservation: first, the maintenance of essential ecological processes; second, the preservation of genetic diversity; and third, the sustainable development of species and ecosystems (IUCN 1980). *Our Common Future*, published seven years later, had a different emphasis, blending environmental concerns (e.g., the need to achieve a sustainable level of population, and the desirability of conserving the resource base and reorientating technology) with development concerns (e.g., the fundamental goal of meeting basic needs and the need to build environmental factors into economic decision-making). Economic growth was seen as the only way to tackle poverty, and hence to achieve environment–development objectives. It must, however, be a new form of growth: sustainable, environmentally aware, egalitarian, integrating economic and social development: "material- and energy-intensive and more equitable in its impact" (Brundtland 1987: 52). The Brundtland Report's vision of sustainable development rested firmly on the need to maintain and revitalize the world economy.

Caring for the Earth (IUCN 1991) selected elements from both its predecessors. It opened with a chapter setting out nine "principles to guide the way towards sustainable societies," and these nine principles offer a structure for the rest of the report. They blend both the ethical ("respect and care for the community of life"), the humanitarian ("improve the quality of human life"), the scientific/environmentalist ("keep within the earth's carrying capacity"), and the pragmatic ("provide a national framework for integrating development and conservation"). Its central argument was much the same as those of its predecessors, although more carefully and fully expressed: "we need development that is both people-centered, concentrating on improving the human condition, and conservation-based, maintaining the variety and productivity of nature. We have to stop talking about conservation and development as if they were in opposition, and recognize that they are essential parts of one indispensable process" (ibid: 8).

Sustainable Development at Rio, 1992

The United Nations Conference on Environment and Development (UNCED) in Rio in 1992 (the "Earth Summit") was attended by 128 heads of state and in total by the representatives of some 178 governments. Debate at the conference drew very directly on the mainstream ideas about the environment and development that had evolved during the 1980s. The Rio Declaration and the much larger Agenda 21 were the fruit of endless negotiation at a series of Preparatory Commission meetings and at Rio itself, between teams of diplomats determined to surrender as little as possible of their national interest. A key feature of these debates was the gap between countries in the industrialized "North" and the underdeveloped "South," which became steadily more glaring in the run-up to, and during, the conference. The issues of climatic change and biodiversity that dominated the Rio Conference are vitally important to certain countries (especially those vulnerable to sea-level rise), but they are not the principal environmental problems faced by most countries of the South (Redclift 1997).

At Rio there were differences between Northern and Southern countries over the key problems (for the industrialized countries, global atmospheric change and tropical deforestation; for unindustrialized countries, poverty and the problems that flow from it), over who should bear the burden of responsibility for finding (and funding) solutions. As at Stockholm in 1972, there was fear on the part of Third World countries that their attempts to industrialize would be stifled by restrictive international agreements on atmospheric emissions, and that their freedom to use natural resources within their boundaries would be constrained by industrialized countries responding to their domestic conservationist concerns (e.g., on tropical forest clearance).

Non-governmental organizations (NGOs) were excluded from the official negotiating sessions of the Rio Conference, which were intergovernmental. They held a parallel Global Forum, physically separated from the main conference, and a specially convened "Earth Parliament." NGOs fought hard to win a place at the various preparatory meetings that prepared the conference documents, and a few (mostly the dozen or so largest and mostly corporately organized American-based conservation organizations) were able to maintain close links with government delegations through the conference itself (Chatterjee and Finger 1994). NGOs did, however, have a significant influence on the way issues like "empowerment" were approached in *Agenda 21*. Chapter 27 of *Agenda 21* concerns NGOs, and recognized their important role in achieving sustainable development. The importance of grassroots groups to future debates about environment and development became widely recognized (e.g., Ekins 1992; Ghai and Vivian 1992), although the conference also revealed the distance between the powerful, wealthy, and influential NGOs of industrialized countries and grassroots organizations formed among the poor of the urban and rural Third World.

The Rio documents fix in time what has been an ongoing debate between First and Third World governments, between governments and NGOs, between developers and environmentalists, between those who wish to exploit natural resources and those who wish to conserve them, between conservationists and the rural poor. They highlight the increasingly important difference in views between poor and rich (globally, nationally, and sometimes locally) about why the environment is important: as a source of wealth alone, as a source of non-material values alone, or some kind of mixture of the two. Here Rio continued the debate begun at Stockholm two decades earlier.

The tensions between different interests are clear in the UNCED documents. Although intended as a short, clear, and inspiring statement of intent, the Rio Declaration ended up as a bland list of 27 Principles (Holmberg, Thornson, and Timberlake 1993). It was long-winded, and in places self-contradictory: even after long debate, the US delegation released an "interpretative statement" that effectively dissociated it from a number of the principles agreed. It dissented from Principle 3, that there was a right to development, the Americans arguing that development was not a right but simply a shared goal, and from any interpretation of Principle 7 that suggested that there was an international liability to make development sustainable (i.e., for rich countries to pay for it). For many countries, choices about development and environment were matters for individual countries to make their minds up about, not an issue over which they should be subject to international opinion.

Agenda 21 was even more burdened by divergent opinion, becoming a bloated volume of more than 600 pages in 40 separate chapters. Its scope was enormous, covering issues from biodiversity and water quality to the role of women, children, and organized labor in delivering sustainable development. It reflected previous mainstream thinking about sustainable development in several ways. First, it made the need for economic growth a central theme, as in the Brundtland Report. In the sustainable development mainstream, everything is predicated on economic growth, both globally and nationally. Second, *Agenda 21* emphasized the familiar straightforward issues of environmental management: all the familiar environmental issues from the *World Conservation Strategy* appear, developed but unmistakable. Third, *Agenda 21* was technocentrist. The first six key themes make this quite clear: growth will power and technology will direct the evolution of policy towards more efficient use of the environment and hence towards a more sustainable world economy. The "essential means" to achieve sustainability also reflect this technocentrism, building on information, science, and environmentally sound technology (United Nations 1993). Fourth, *Agenda 21* presumed a multilateral benefit to be derived from a sustainable development strategy, as did the Brundtland Report. It suggested that change would arise from the mutual interest of industrialized and non-industrialized countries, and from the concern of present generations about the future. It suggested that this shared interest would cause international financial resources and technology to flow, directed and promoted by international agencies and structured and regulated by international legal instruments. Fifth, like its predecessors, *Agenda 21* called for sustainable development through participation. As in *Caring for the Earth*, women, children, young people, indigenous people, trade unionists, businesses, industry farmers, local authorities, and scientists are all summoned to play a role, a rainbow coalition to put flesh on the endless skeleton of the text of *Agenda 21*.

Forests and Biodiversity

There was a great deal of debate at the Rio Conference about forests. It had been hoped that a Convention on Forests could be agreed and signed. This did not happen, and instead a much shorter and lesser document of forest management was agreed. Pressure for specific action on forests came from First World governments, driven by public opinion and the lobbying of Northern environmental organizations concerned at the rate of clearance of tropical moist forests (rainforests). This pressure was resisted by Third World rainforest countries, who argued that their forests were a vital economic re-source that they should be allowed to exploit in the optimal way – just as industrialized countries had done (losing up to 90 percent of their forest cover and the vast majority of their Old Growth Forest, along the way). There was also the awkward fact that logging practices in those parts of the North which still had Old Growth Forests (the USA and Scandinavia, for example) were patently unsustainable. Third World countries dug their heels in over forests, hoping to gain concessions on aid. The hopes of Northern NGOs that Rio might generate significant constraints on rates of tropical deforestation were not realized. The "Forest Principles" negotiated were a poor substitute: their title reveals their character, and the problems that beset the Convention for which they substituted: a "Non-legally binding authoritative statement of principles for a global consensus on the management, conservation and sus-tainable development of all types of forests" (Sullivan 1993).

The other great hope of the Northern environmental movement was for a global convention on conservation, and here there was more success, for the Convention on Biological Diversity was signed at Rio. There had been pressure for such an international agreement for some years, and a draft Convention had been prepared by IUCN in the mid-1980s. Negotiations were initiated by UNEP in 1990, in good time for Rio. The initial focus was on biodiversity loss, especially in tropical rainforests. However, at the sec-ond Geneva PrepCom meeting, the G77 countries (in fact the group of 128 less-developed and less-industrialized countries set up as a counter lobby to the developed G7 countries) demanded inclusion of the issue of bio-prospecting and biotechnology, and the sharing of wealth generated by the exploitation of biodiversity in the South by Northern biotech companies. In this odd hybrid form the Convention was agreed at Rio and signed by 156 countries (Chatterjee and Finger 1994). The USA refused to sign at that time, although it did so subsequently. The Convention on Biological Diver-sity came into force on December 29, 1993, and by 1997 it had been rati-fied by 162 countries.

The aim of the Convention on Biological Diversity is to conserve biologi-cal diversity, promote the sustainable use of species and ecosystems, and ensure the equitable sharing of the economic benefits of genetic resources. Signatory nations commit themselves to the development of strategies for conserving biological diversity, and for making its use sustainable. Bio-diversity conservation can be achieved *in situ* (i.e., through conventional methods such as the designation of systems of protected areas) or *ex situ* (e.g., through captive breeding), but the Convention also requires cross-cutting measures (for example, affecting forestry or fishing). While all these elements of the Convention are qualified by a get-out clause (all is to be done "as far as possible and appropriate"), this program is a logical devel-opment of the traditional conservationist concern for sustainable ecosys-tems use, and draws directly on the thinking in the *World Conservation Strategy* and *Caring for the Earth*.

The novel dimension of the Convention is its provisions for the exploitation of genetic resources through biotechnology. In principle, these are no different from the provisions of the earlier mainstream documents that species and ecosystems should provide resources for human benefit, if used in such a way that their availability was sustained. By 1992, developments in genetic science had opened up vast new areas of potential exploitation of genetic diversity, including the creation of novel organisms (which might perhaps be patented by the corporation that created them) and products (drugs for example) derived from wild species. It was perceived that this technology had the potential to generate vast wealth; however, the biotechnological capacity was almost uniquely held by industrialized countries (because of the high costs of research laboratories, research infrastructure, and training), and moreover was increasingly held by private corporations within those countries and not by states themselves. Third World countries feared stripping of their genetic resources by bioprospectors, and loss of access to economic benefits derived by First World corporations (Shiva 1997). First World countries (particularly the USA, which dominated in this area of science) feared restriction of economic opportunity if trade in biotechnology were restricted by a benefit-sharing agreement. The Convention reflects the balance of these opposite fears, and does contain provision for sharing benefits from commercial exploitation of genetic resources (Article 15). The Convention on Biological Diversity is therefore much more than a conservation agreement.

Climate Change

The Rio Conference also saw agreement on the Framework Convention on Climate Change. The debate about what emerged as the Framework Convention reflected the divergent reactions to the report of the Intergovernmental Panel on Climate Change (IPCC) in 1990 (Houghton 1997; Jäger and O'Riordan 1996). The IPCC's consensus view of the importance of fossil-fuel consumption and carbon dioxide (CO_2) output cut directly at the heart of the interests of the industrialized Northern countries, while also having significant implications for rapidly industrializing countries in the South such as India and China. There was a broad divergence between industrialized and non-industrialized countries, with the North urging the priority of environmental protection and that any measures agreed should be cost-effective, while the South pushed the need for development and industrialization, and the principle of historical responsibility (Rowbotham 1996). Most industrialized countries were unwilling to countenance a significant reduction in CO_2 output. Oil producing states were also opposed to this, while small island states vulnerable to sea-level rise wanted urgent action. The EU favored agreement on targets and a timetable for implementation, the USA was reluctant (the USA even refusing, in the run-up to Rio, to agree to cut back emissions in the year 2000 to 1990 levels). In April 1992, at the last Intergovernmental Negotiating Committee meeting, the compromise was agreed on a non-binding call for an attempt to return to 1990 emissions of CO_2 and other greenhouse gases not controlled by the Montreal Protocol.

The Convention was a delicate balance between divergent political and economic interests, and full of rather pious intentions (Rowbotham 1996). The Framework Convention was signed by over 150 states and the EU at Rio, and came into force in March 1994. The technical demands of the evolving regime on climate change placed a considerable and growing burden on the technical and policy capacities of Southern countries, and (like the rest of the Rio process) themselves demand considerable financial support from industrialized countries over and above that required to address wider issues of environment and poverty.

Rio in Perspective

Many commentators have been critical of the achievements of Rio. Many of the problems on which *Agenda 21* focused continued to get worse (Brown 1997). Poverty has deepened and widened, and the gap between rich and poor countries has grown. The enhanced resource flows needed to implement *Agenda 21* were not forthcoming, and indeed countries such as the UK have reduced their aid budget. The activities of transnational companies have not been significantly influenced. Moreover, many countries failed to respond to the Framework Convention on Climate Change and curb CO_2 emissions (Parikh, Babu, and Kavi Kumar 1997), and there are still intractable issues of sovereignty over genetic resources. The attempt to re-energize the response of governments through a special session of the UN General Assembly to review *Agenda 21* (the so-called "Earth Summit II") in 1997 was not hugely successful.

Arguably, the Rio Conference should not be interpreted as a single event that could be assessed in terms of success or failure, but as part of a framework for tackling global environmental problems, including "soft law," voluntary reporting and action by communities and other non-state actors at local, national, and international levels (Roddick 1997). Therefore, while perhaps Rio did little to promote sustainable development as such, it clearly did at least open the debate about choices in development, about the ways in which the biosphere is restructured in pursuit of profit, of the costs of technology, of the inequalities in wealth, technologies, environmental hazard, and life chances (Brown 1997). Rio led to some solid achievements, in new international conventions, in new international institutions (e.g., the Commission on Sustainable Development: Robinson 1993), in the actions of national and local governments (e.g., Local Agenda 21).

One clear failure of the Rio Conference, however, was that it did not secure the scale of financial support necessary to implement *Agenda 21*. Before the Fourth UNCED PrepCom started in New York, Maurice Strong estimated the cost of implementing the Rio agreements at US$125 billion in new aid every year between 1992 and 2000 (Chatterjee and Finger 1994). At the conference itself, the Secretariat estimated that implementation would cost something like US$600 billion per year, of which US$125 billion would have to be in the form of gifts and loans. This was more than twice the current total disbursement of official development assistance to Third World countries, and close to the official UN target of 0.7 percent of GNP from First World countries (Grubb et al. 1993). The money available has been a tiny fraction of that. Of the US$125 billion per year needed to take necessary actions, only about $2.5 billion was pledged (Holmberg, Thornson, and Timberlake 1993). The chosen channel for new funds was the Global Environment Facility (GEF). This had been set up in 1990 by the World Bank, UNDP, and UNEP. By September 1998 (seven years after establishment), the GEF had allocated $US2 billion (World Bank 2000). While the GEF has brought much needed funds to some sectors (National Parks for example), the overall influence of the GEF's funds, in the context of debt and total resource flows between First and Third World, has been very limited (Brown, Adger, and Turner 1993).

One piece of unfinished business at Rio was a Convention on Desertification. This was proposed as part of the Rio process, but negotiation fell behind in the run-up to the Conference. In the event, a formal commitment was made at Rio to negotiate a convention on desertification after the Conference was over. This was duly done, and the Convention to Combat Desertification was open for signature by June 1994, coming into force in December 1996.

The Economics of Sustainable Development

The tension between radicalism and conformism within sustainable-development thinking is slowly but surely being resolved in favor of a pragmatic reform of existing systems of decision-making, particularly through the incorporation of environmental assessment and environmental economics into the design of development projects and programs (e.g., Adams 2001). The broad consensus of mainstream sustainable development is based on a set of ideas that Low and Gleeson (1998) label "Market environmentalism," the utilitarian, individualistic, and anthropocentric idea that the market is the most important mechanism for mediating between people and regulating their interaction with the environment. Market environmentalism argues that the further market exchange penetrates into the environment, the greater the efficiency of environmental management. Policy proposals therefore pursue the setting of prices for environmental "goods" and "services."

Market environmentalism is the result of a growing engagement in the sustainable development mainstream by economists. Partly because of the involvement of the World Bank in the development of practical policy responses to the Rio Conference, it is the "dismal science" of economics that has led the way in the greening of governments and development bureaucracies, in some ways a remarkable case of the poacher turned gamekeeper. The discipline of environmental economics has undergone rapid theoretical development and expansion through the 1980s in universities and government agencies. In particular, there has been reform of the way economies are measured to take account of the consumption or degradation of natural resources (for example, in environmentally adjusted national accounts), and reform of the economic methods that support development decision-making to take account of environmental impacts (for example, in the development of cost–benefit analysis).

In practice, development and the environment have a two-way relationship: economic growth can have both good and bad impacts on the environment. It creates the possibility of "win-win" opportunities through poverty reduction and improved environmental stewardship in both high- and low-income countries. However, growth also brings risks of serious side-effects if markets fail to capture environmental values and deal with externalities, and if the public regime of regulation at local, national, and international scales is inadequate. Environmental economics values environmental resources more precisely, and allows those values to be taken into account in economic decision-making (Munasinghe 1993; Barbier, Burgess, and Folke 1994; Barbier 1998).

Economic approaches to sustainability build on long-established ideas of maximizing flows of income while maintaining the stock of assets (or capital) from which they come (Munasinghe 1993). What is new is the distinction drawn between natural and human-made capital (Berkes and Folke 1994). Economists have traditionally defined capital as things people have built that have value, such as roads or factories – which environmental economists define as human-made capital. Natural capital is created by biogeophysical processes rather than human action, and represents the environment's ability to meet human needs, whether through providing raw materials (fish or timber), or through "services" such as the maintainance of the ecological conditions suitable for human life through global biogeochemical cycles, or the moderation of floods or adsorption of pollutants by wetlands (Barbier 1998).

Using these economic concepts, sustainability can be defined in various ways. The first, and most obvious, is as constant stocks of both human-made and natural capital maintained over time. Economic development has always implied increasing capital stocks, but the insights of environmental economics suggests the need to account for stocks of both human-made and natural capital separately. Thus development that increases stocks of human-made capital only by the depletion of natural capital of an equivalent value is not sustainable. This view of sustainability is commonly referred to as "strong" sustainability (Beckerman 1994). However, its requirement for zero natural capital depreciation, if imposed at project level, is likely to stultify development, since it effectively makes it impossible to do anything that damages the environment at all. An alternative approach is the notion of "weak sustainability," which allows the possibility of trade-offs between losses to natural capital in one project and gains elsewhere and the substitution of either human-made capital or human-induced "natural capital" for lost natural capital (Pearce, Markandya, and Barbier 1989; Barbier, Burgess, and Folke 1994; Jacobs 1995).

Developments that reduce natural capital can be balanced by other action specifically designed to compensate by creating new natural capital elsewhere, "shadow projects" within a development program to compensate for degradation other projects cause (Pearce, Markandya, and Barbier 1989). Thus, in industrialized economies such as the UK, developers proposing a project that has adverse environmental impacts are now routinely expected to offer as part of the development "ecosystem restoration" schemes such as new areas of woodland or wetlands to replace those lost (Owens and Cowell 1994). However, conservationists argue that natural capital is not like human-made capital. Unlike bricks and mortar, it can be argued that some species or ecosystems (e.g., the Giant Panda or rainforests) are irreplaceable and should not be conceived of as "capital" to be replaced or exchanged for some other form of natural capital. Those parts of natural capital that cannot be replaced if lost (or, at least, not within feasible time frames), are referred to as "critical natural capital."

Trade-Offs and Sustainability

The complex trade-offs involved in the search for sustainable development have important geographical dimensions. Trade-offs can take place in both space (between losses in natural capital or some other measure here and gains somewhere else) or in time (between losses or gains now, and those coming in the future). Principles of inter- and intragenerational equity demand that a balance be struck between the needs of present and future generations. However, this can be problematic because of uncertainty about future technologies and future resources, the capacity of future generations to provide different environmental goods and the values they will attach to the environment, and hence how they will view decisions made now on their behalf (Dovers and Handmer 1992). There is also scientific uncertainty about the ways in which ecosystems will behave in the future, so it is hard to predict the value of flows of benefits from existing and future natural capital. Resource use that is judged sustainable in the short term may well not be sustainable over longer periods (Dixon and Fallon 1989). The impacts of development can be delayed far into the future, either because the impact itself is delayed (for example, the impacts of poorly stored toxic waste that starts to leak after a few decades), or because ecological impacts take time to work their way through the ecosystem (for example, the responses of floodplain ecosystems to dam construction: see Thomas and Adams 1997). The true severity of environmental impacts may only be revealed when new economic opportunities come to be developed (for example, an attempt to irrigate using water from an aquifer contaminated with heavy metals).

Trade-offs in space are integral to discussions about sustainability. Governments routinely trade-off sustainability at one location to meet national goals. The benefits of a development project (for example a mine) might be at the cost of reduced sustainability elsewhere (for example, pollution downstream: see Low and Gleeson 1998). Internationally, industrialized countries seeking to make their policies sustainable may do so at the expense of other places by importing resources or exporting wastes. Both ecological and political boundaries (local, regional, national, or international) are relevant to the assessment of sustainability. Debates about the sustainability of particular developments might very easily descend into arguments about boundaries, and different actors (for example, governments, non-governmental organizations, and transnational corporations) may base conflicting assessments of the sustainability of controversial projects on different choices of boundaries for analysis.

Natural systems may provide appropriate boundaries for sustainability analysis; however, these are not always easy to define. In particular, the ecosystem is effectively an arbitrary analytical category, not a natural entity whose characteristics are endogenously determined. Furthermore, in practice development planning usually takes place within political jurisdictions, not within natural boundaries. Human-made boundaries rarely fit the spatial patterns of natural systems, and ecosystems (and impacts such as pollution) often straddle political boundaries.

Conclusion

The pursuit of sustainability demands choices about the distribution of costs and benefits in space and time. One way of understanding this is in terms of political ecology. Mainstream sustainable development skates over political and economic issues (Adams 2001). Mainstream sustainable-development thinking is built upon the conventional vision of a managed Keynesian world economy, mutual trading to mutual advantage, and the environmentalism of both the 1960s and 1970s. The documents of mainstream sustainable development identify the problems of poverty and inequality, but they have little to say in the way of explanation of their causes, and hence perhaps hardly begin to formulate ways to overcome them. Mainstream sustainable development is a synthesis of conventional ideas about development.

It takes no position on contemporary debates in development (for example, the role of the state, or of aid, or the structure of the world economy), let alone developing new theoretical ideas of its own. Radical thoughts about the world economy or about the nature of this "development" (be they Marxist, anarcho-communist, or eco-feminist) are simply excluded. The sustainable development debate has had remarkably little to contribute to the "impasse" of development theory in the 1990s (Schuurman 1993).

The lack of radical edge in sustainable development is not surprising. The mainstream texts are all the product of the political process of negotiation in search of widespread international support. They avoid shaking the status quo, and all sides of debates about environment and development are able to find substantial numbers of things on which they agree in these documents, precisely because they are not analytically very precise (Redclift 1987; O'Riordan 1988a; LéLé 1991).

At the same time, the issue of environment and development offers a fierce challenge to complacency. The reciprocal and synergistic links between poverty and environmental degradation forces what Blaikie described as the "desperate ecocide" of the poor (Blaikie 1985: 138). Access to and control over cultivable land, fuelwood, or other usable attributes of nature is unequal. Blaikie emphasized the political dimensions of rights over resources, stressing the need for those seeking to understand environment–development problems to explore the links between environment, economy, and society. Blaikie and Brookfield argued that "land degradation can undermine and frustrate economic development, while low levels of economic development can in turn have a strong causal impact on the incidence of land degradation" (Blaikie and Brookfield 1987: 13). Poverty and environmental degradation, driven by the development process, interact to form a grim world of risk and hazard within which urban and rural people are trapped. Understanding the reality of this world, and the environmental, economic, and political factors that create it, is a widespread contemporary concern for sustainable development. The burgeoning field of political ecology is addressing the political dimensions of environment–society relations (Blaikie 1985; Peet and Watts 1996; Rocheleau, Thomas-Slayter, and Wangari 1996; Bryant and Bailey 1997). To geographers, the exercise of power (both over people and non-human nature) at the discursive as well as the material level offers a considerable research challenge (Bryant and Bailey 1997).

Today, geographers have at hand the empirical and theoretical tools with which to enter and lead pragmatic debates about sustainable development, and with which to explore and understand their implications. Both tasks are important, for the discipline and for their potential contribution to the growing international debate.

27 Environmental Governance

Simon Dalby

Environmental Geopolitics

In the aftermath of the Cold War we are apparently living in a world of interconnections, a globalized world, one where humanity supposedly shares common identities and dangers. One of the prominent dangers to our well-being is supposedly a combination of matters of concern that are broadly designated "environmental." In the face of dangers governance is apparently needed; administrative arrangements and technical practices are invoked to manage planet Earth. Environmental treaties and international regimes are negotiated, production of harmful chemicals is regulated, trade in endangered species is banned, and Northern consumers rest assured that the experts will take care of things as long as they do their bit by recycling newspapers, pop cans, beer bottles, and plastic containers.

But if one probes a little deeper into these matters and asks very simple geographical questions about specifically what environments are endangered where, and by whom, matters quickly become much more complex. If remote valleys are being flooded to build dams to provide water supplies and electricity for distant cities then the water in our suburban bathtub, and the electricity to warm it, have ecological consequences for the residents, human and non-human, of that valley far away from our immediate desires for relaxation. Who lives in the neighborhood near the sewage treatment plant, through which our bathwater subsequently flows, also matters, because they too bear some of the consequences of our aqueous pleasures. But these dams and sewage treatment plants are often in some other jurisdiction, across some political boundary line, in another municipality, or county, or country. On an even bigger scale the herbs in our scented bubblebath may come from some tropical land, the plastic bottle containing the fragrant bubbles from a factory somewhere else.

Important though the recognition of such simple connections between routine life in one place and consequences elsewhere is, it underlies only some of the discussion of global environmental concern. Frequently assumptions of common dangers and imprecise designations of the threat to environment obscure the importance of these interconnections (Dalby 2000a). The assumption that environment is something outside us, a separate threatened place that is nonetheless somehow important for our health and well-being, is part of the problem in how the geography of these matters is usually presented. The international dimensions all too frequently get lost in the finer points of diplomacy and technical negotiations, but the assumptions that nation-states are the appropriate units for administering agreements and making environmental decisions nearly always shape these discussions.

When environment is confronted with conventional modern assumptions of how to administer matters for the collective good, how to make political decisions, and how to think about both these things in light of some claim to morality, cultural appropriateness, or people's rights, the spatial assumptions of such thinking become unavoidable, but at the same time also seem increasingly inadequate to deal with contemporary problems. But this is the problematic of environmental governance and it raises questions both of how to think about what precisely is being administered as well as the appropriate geographical concepts for collective decision-making. Faced with the very real and accelerating changes that human activity is causing to "natural" systems at the largest of scales, these matters are now especially pressing.

They are so urgent because it is now becoming obvious that the powerful modern assumption that nature is somehow separate from humanity is itself now no longer tenable, because human actions are re-engineering, or "terraforming," the whole biosphere (Braun and Castree 1998). With genetically modified crops now loose in the environment, with land-use changes related to commercial farming rapidly changing many ecosystems, and atmospheric emissions of greenhouse gases literally changing the air that we all breathe, the distinction between natural and artificial is even less convincing now than it used to be. Put this way the question of environmental governance is thus an impossibly grandiose one: "What kind of planet are we making for the future of what kind of humanity?"

And yet somehow scholars and citizens have to think about these things intelligently and reflect on the commonsense assumptions that have facilitated the rapid expansion of the global impact of at least the richest parts of humanity. To tackle these questions all too briefly this chapter first looks to questions of governance, then to matters of the relationship of the global North and South. Later sections emphasize the importance of understanding the interconnections across such boundaries and point out that the environmental crisis is very much a crisis of modern modes of living and the assumptions about who we are who live this way (Taylor 1999).

Global Governance

In the words of the Commission on Global Governance (1995: 2):

> Governance is the sum of the many ways individuals and institutions, public and private, manage their common affairs. It is a continuing process through which conflicting or diverse interests may be accommodated and cooperative action may be taken. It includes formal institutions and regimes empowered to enforce compliance, as well as informal arrangements that people and institutions either have agreed to or perceive to be in their interest.

Thus governance relates to both formal government institutions as well as the less formal arrangements of social movements and non-governmental organizations now frequently called civil society (Lipschutz 1996). But the understandings of the nature of environmental problems and the appropriate measures needed to deal with these issues are frequently very different. Where international institutions have been established to deal with many problems that have an obviously environmental component, in many cases these effectively work to regulate business activities and coordinate trading arrangements rather than to build the new institutions that might make sustainable societies possible.

In part this is because the existing understanding is based on the dominant political assumption that liberal governments' role is to legislate and regulate to assure public safety in the face of various obvious threats, but that business and the "private sector" ought to make things and provide services. Economic growth supposedly provides the largest benefits for the largest number, and while its obvious excesses need curbing, its overall function is socially beneficial. But now the sheer scale of changes and the complexity of issues of climate change, genetically modified organisms, ozone depletion, and numerous other matters demands a response that looks forward and produces goods and services that do not endanger the long-term future of environments that humanity needs.

Such was the logic that led to the formulation of ideas on "sustainable development" in the 1980s (World Commission on Environment and Development 1987; see chapter 26). In turn, these gave rise to global discussions about climate conventions, forestry agreements, and related matters in the lead up to the United Nations Conference on Environment and Development (UNCED) held amid massive publicity in Rio de Janeiro in 1992. Subsequent international meetings in Kyoto, Berlin, and other places have gradually produced some agreements on curtailing the emissions of carbon dioxide and other greenhouse gases (Sooros 1997). But apart from some notable reductions of emissions from states in the former Soviet Union and Eastern Europe, due to the collapse of their economies, and some more modest reductions in Western Europe, the practical effect of these initiatives has apparently been minimal. Sales of automobiles show no signs of slowing as globalization spreads the American suburban way of life as the universal human aspiration.

As environmentalists often put it, development is made sustainable but the key theme of developing sustainability has frequently been lost in the corporate specifications of ecological modernization. Campaigns by many environmental activists remain caught in the logic of local actions to protect particular places, or in hopelessly general campaigns to change everything. They are also fighting an uphill battle to get even simple information about such things as climate change and the continued dangers of ozone depletion due to the illegal use of chlorofluorocarbons, into the mainstream media (Nissani 1999). The logic of ecological modernization, where management and technological innovation is assumed to have all the answers to "environmental" problems, remains the dominant discourse of official environmental discussion (Hajer 1995).

When coupled to the priority given to trade agreements and economic arrangements in international affairs, even the possibility of seriously engaging the major issues of environmental governance repeatedly gets shunted aside in official discussions. Air pollution, water quality, land-use zoning, and environmental regulation of numerous matters are rendered as technical issues, matters for government experts and consulting engineers, not philosophical inquiry, democratic debate, and public discussion beyond the battle between protesters and public relations spokespeople about what is safe or technically accurate. And yet public attention to the bigger issues, in part driven by continued scientific attention to matters of climate change, repeatedly raises these questions. Specific campaigns have had successes in curbing the worst excesses of environmental destruction, but the larger trends continue in many worrisome directions (Brown et al. 2000). These trends, it now seems, also demand that people rethink their own identities as consumers, and consider their responsibilities to the unseen producers of the commodities they consume wherever they may be in whatever distant part of the planet.

North and South

The tentative admission by most governments in the early 1990s that global climate change is a potentially serious concern was immediately followed by acrimonious discussions as to how responsibility for dealing with the issue ought to be allocated. Given the huge injustices in the world and the obvious wealth of the industrialized states in the North, many argued that the rich ought to pay for changing technologies and helping poor states onto non-fossil fuel based development paths. This logic was worked into the agreements to phase out production of chlorofluorocarbons (CFCs), even if there have been considerable difficulties with the finer points of financial arrangements and compliance with the agreements (Litfin 1994).

In the case of climate change matters are much more difficult. Even the numbers concerning where the sources of greenhouse gases were coming from were contentious. Were methane emissions from the subsistence plots of peasant farmers in the South to be equated with the luxury emissions of Northern consumption? Some estimates that did not distinguish between subsistence agriculture activities and the CO_2 emissions from such activities as North American recreational boating implied just this. The critics were quick to spot these assumptions and begin questioning how the apparently objective figures of gas emissions hid all sorts of political assumptions (Lutes 1998).

Among other things they argued that the industrial North had apparently got rich by using fossil fuels for generations. Why should those in the South forgo the same possibilities just because they came on the development scene a little later? The insistence by many Southern politicians that their CO_2 emissions should not be curtailed has repeatedly been used by those in the North to argue that climate change agreements are unfair in that not everyone is equally constrained. In the process the historical inequities and the fact that only some states are, at least as yet, serious contributors to climate change emissions gets lost. Southern advocates then argue that the claim to equality is in fact a device for ensuring that the South remains poor and undeveloped. Fears about the future are used to try to extend Northern control over Southern development (Shiva 1994).

Equity arguments also extended to the question of how to count the "sinks" of carbon, those natural systems that remove carbon from the atmosphere. If forests, which work to remove carbon from the air, are counted then obviously Russia and Canada, with huge forests, can make the case that they are not net emitters of CO_2 because their forests absorb the carbon that they put into the atmosphere. The argument is then made that these sinks might be traded and in the process provide incentives to reduce fuel consumption and so increase efficiencies and sell the savings to further increase profits. Trading emissions quotas may however work in ways that will enhance long-term Northern control over resources. If Northern states buy Southern forests to count against Northern emissions, Northern control over Southern resources thus expands. Does this then constrain Southern development unfairly?

The geography of these matters is also more complicated than the simple division of emissions and sinks according to national boundaries suggests. Should the natural gas flared off oil fields in Nigeria be counted as an emission by Nigeria, or rather counted as an emission caused indirectly, but caused nonetheless, by the drivers in European states whose cars consume the petroleum products that come from oil wells in Nigeria? This question shifts the matter of emissions from national figures to the more difficult matter of who actually uses the resource. Northern lifestyles thus become the focus of attention rather than Southern production, a shift of focus that more directly raises questions of responsibility. But if Northern consumers are the problem then the geography of these matters becomes even more complicated, as it is not difficult to argue that North American landscapes in particular have been constructed to make car ownership and use practically unavoidable.

If consumers are partly responsible what then about corporate roles? Should Shell Corporation, a major investor in oil production in Nigeria, be in some way responsible for these emissions, and for the political violence against those that protest the destruction of cultures and environments as a result of the oil industry? The corporation has gone to considerable lengths to disclaim any responsibility for the death of Ken Saro-Wiwa and other activists there, an effort that itself suggests that the issue is not a simple one of non-political business operation on the one hand, and a matter of politics as only what governments do on the other (De Larrinaga 2000).

Global Political Ecology

Discussing matters in these terms suggests that environmental governance is about much more than either territorial states or conventional practices of administration (Bryant and Bailey 1997). While obviously most international agreements on curtailing waste trade, protecting endangered species by preventing the sale of ivory and other products, or banning the production and importation of CFCs are a matter of the international coordination of national legislation (Vogler 2000), the larger politics of these matters stretches across boundaries and raises important questions of the responsibility for distant consequences of modern ways of life. If car driving in Europe is causing violence and environmental destruction in Nigeria then the politics of this are not easily solved by thinking in terms of national governments having sole responsibility for what happens within their borders. It is precisely such considerations that drive various contemporary international consumer boycotts (De Larrinaga 2000).

But such considerations also lead to questions of the role of international organizations and in particular those institutions that provide aid for development in the South. Much critical attention has focused on the World Bank and its role in financing dams, pipelines, highways, and other infrastructure. In particular the Bank has long been criticized for supporting projects undertaken by undemocratic governments which have severe human rights consequences for local populations who are in the way of "developments" (Rich 1994). While the campaigns in the 1980s and 1990s to stop the World Bank funding dam building in the Amazon and in the Narmada valley in India are perhaps most famous, the general pattern of concern about who is deciding what kind of development happens in the South is of continuing importance precisely because, in many ways, such development projects are more about integrating the South into the global economy than in practically and directly changing the lives of local inhabitants for the better. Indeed in most cases large infrastructure developments displace people, uproot traditional ways of life, and accelerate the commercialization of agriculture and subsistence. They frequently enrich local urban elites at the expense of the rural poor.

More so than this the current pattern of development has involved large pharmaceutical and agricultural corporations attempting to patent seed varieties and develop chemicals based on traditional knowledge of medicines. The incorporation of indigenous knowledge into patents which the companies then use to try to stop traditional uses of seeds and remedies further incorporates ecology into corporate practice and constrains agricultural innovation, while usually refusing to pay royalties to anyone who might have had something to do with the generations of experimentation and cross-breeding that produced numerous seed varieties. Huge protests over such things in India have focused attention on the further enclosures of nature and the transformation of ways of life into commodities and on the politics of who decides these matters – international corporations or local farmers?; national governments or international trade organizations?; local knowledge passed on through the generations or scientific innovation controlled by gene companies?

As the new scholarly field of environmental history is making increasingly clear, these patterns are not very new, although the scope of their operation is now worldwide. They follow on from centuries of European colonization. Rubber grown in Malaysia is derived from trees taken from Brazil in the nineteenth century. Concerns about population growth and political stability in the South have their precursors in colonial anxieties about the possibilities of revolts against colonial rule. Climate change, too, is a concern derived in part from colonial fears of land degradation and deforestation. But all these are related directly to the administration of a system that served to provide commodities to the growing urban economies of the European metropoles (Grove 1997). The expansion of international trade since the decolonization process of the mid-twentieth century has accelerated the earlier patterns of resource expropriation and consequently spread the environmental impact of modernity yet further.

But if the modern modes of living, dependent as they are on fossil fuel consumption which is changing the atmosphere, and the importation of resources from all over the globe which frequently displace rural peoples and disrupt traditional ecological patterns, are unsustainable in the long run, then the question for environmentalists is both how to reduce the disruptions and to reduce the pressure that drives the disruptions. Can the affluent learn to live lightly on the Earth? Many in the South obviously think that is the most pressing priority, but it is one that most North Americans and Europeans are not likely to be favorably disposed towards. George Bush, President of the United States during the Earth Summit in Rio de Janeiro in 1992, went on record at the time suggesting explicitly that the North American lifestyle was not for negotiation at the summit. Just as

Southern elites often object to Americans lecturing them on human rights, so too North Americans may object to Southern environmentalists trying to point out that the majority of the world's consumption of resources and production of greenhouse gases occurs in the North, and that Northern environmental ideas fail to deal with the practical conditions of people in the South, and in particular some of the gendered assumptions that Northerners frequently bring to their analyses (Agarwal 1998).

The Geography of Conservation

Much of the discussion of environmental matters is tied into discussions of conservation and the preservation of habitats and landscapes. The images of environment in nature magazines and in television documentaries are repeatedly of exotic places with wildlife of wondrous variety. The formulation of such places, in strict contrast to the predominantly urban locales of those who read such magazines and watch the television programs, is significant – not least because in a continuation of the colonial mode of thinking these exotic places are rendered as endangered, in need of management, preservation, and above all threatened by the usually vaguely defined problem of human encroachment. Parks and nature reserves are to be managed and controlled, their "Edenic" existences unpolluted by the depredations of humanity gaining subsistence within their borders.

But this series of aesthetic and managerial assumptions is increasingly untenable as a mode of thinking and dealing with environmental change. In part this is because ecology doesn't recognize artificial boundaries; park fences may keep some animals in and most humans out, but they do not stop many species spreading and often cause quite drastic ecological change precisely because of their attempts to artificially maintain a changing ecosystem in a stable state. Over the last few decades the park model of environmental preservation has come in for increasing criticism because of its failures to understand ecosystems as dynamic and open (Botkin 1990). Likewise the presence of humans in ecosystems is increasingly being understood as a necessary component of many environments; the assumption of humans as an external threat is gradually being replaced by a recognition of many human activities as essential to ecosystem function precisely because indigenous inhabitants have long been part of these ecosystems.

The latter understanding suggests quite clearly that in many cases the survival of endangered animals and the simultaneous prevention of conflict over environmental resources require the involvement of local people in establishing and maintaining institutions and rules for environmental governance. Indeed the local populations often have the power to render conservation arrangements unworkable if they are dispossessed in the process of establishing game reserves or similar arrangements. Understanding indigenous historical practice as part of environmental management, not as an obstacle to conservation, suggests the need to rethink the basic urban assumption of environment as outside and separate from people. Parks in cities are not a useful model for thinking about conservation; the aesthetic appeal of "natural" spaces in these terms overlooks the complexity of ecologies. Game reserves, breeding animals for tourists to shoot either with rifles or cameras, are frequently more farms than wilderness (Grove 1997).

At the largest scale this change from environment as separate, as a space out there to be managed and controlled, to one that is to be lived in, is related to the contested appropriation of the symbol of the planet as a single globe. Both environmentalists and corporate advertisers use the planet as a symbol, at once a place of danger and a dangerous place, an entity to be lived in or a place to be mastered by corporate reach (Dalby 2000b). Whether we understand ourselves primarily as inhabitants of a small planet, or as consumers served by the global capabilities of a corporation, matters in terms of how environmental governance is conceived. If we understand ourselves primarily as consumers then the assumptions of ecological modernization, of environment as a technical problem of governance to be solved by corporations and government regulators, seem likely to continue to dominate discussions. But if we understand ourselves as inhabitants of a small planet where the rich and powerful frequently are either ignorant of the effects of their actions on ecological systems, or little concerned about the consequences of their actions for either poor people or their places, then a different series of governance priorities emerges (Athanasiou 1996).

Scenarios of Doom

The concern expressed by many in the environmental justice movement in the United States, and in other environmental movements elsewhere (Harvey 1998), that these connections are not seen and that governance, such as it is, is devoid of moral solicitude for either other peoples or other species (Wolch and Emel 1998), has frequently encountered arguments that nothing will change until some large disaster strikes and brings people to their senses. Then action will be taken and the false sense of security provided by sophisticated urban technological systems will finally be overcome. Disaster scenarios are, in other words, part of the environmental discourse that is invoked when the need for governance is discussed (Lichterman 1999). Security is frequently the overarching rationale for government action.

The strategy of invoking imminent disaster is of course itself always in danger, not least because the credibility of those who raise the alarm is entirely dependent on either the eventual appearance of a disaster or, if it does not appear, on a very convincing explanation of the efficacy of their preventive measures. But the invocation of threat also empowers agencies and modes of governance that might not be appropriate for dealing with environmental difficulties. This has been a primary concern of the critics of the environmental security discourse in the United States for the last decade (Deudney 1999). The military and state powers of coercion and surveillance among the powerful countries of this world are not the agencies best equipped to deal with the complicated interconnected matters of environmental governance.

The important question here is why the poor are presented as a threat to global order at the present time. Why is this threat, in part a recycled concern over revolt in the colonies, gathering so much attention when the environmental damage done by an increasingly global consumer society is, literally, driving the processes of change (Barnett 2000)? Related to this is the geopolitical premise that environmental change is a matter that is indigenous to the states in the South, a matter of endogenous causes, not something that is at least partly related to the impact of the global economy. Competition over environmental resources by global corporations may explain much more about contemporary insecurities than neo-Malthusian explanations that blame poor peoples for environmental dangers (Redclift 1999).

This is not to suggest that the environmental situation is not serious either in many places, or most importantly, in global terms as a sum of numerous specific disruptions of natural processes. If the critics are even partly correct about such things as the limits of the authoritative Intergovernmental Panel on Climate Change projections on climate change, then matters are very serious indeed, at least for the more vulnerable poor majority of humanity who live in dangerous places and have few economic alternatives. The projections of gradual climate change that underlie the Framework Convention on Climate Change, the Kyoto Protocol, and related international agreements, ignore such important factors as the impact of terrestrial land-use changes and deforestation and may also seriously underestimate the interconnections between different phenomena set in motion by atmospheric temperature changes (Bunyard 1999). The possibilities of dramatic atmospheric disturbances due to the rapid release of frozen methane hydrate deposits on the ocean floor have become the basis of a scenario for at least one science-fiction novel that links global change directly to environmental security (Barnes 1994).

Sustainable Institutions

Such thinking only emphasizes the fact that conventional political reasoning and the assumptions that underlie formal international agreements of environmental governance are seriously out of line with the scientific thinking about ecology, and the political thinking that raises questions of the possibilities of alternatives to yet more "development," sustainable or otherwise, as the answer to all problems (Rahnema and Bawtree 1997). Environmental governance is thus very much about the tacit assumptions we make about the relationship of humanity to the non-human world. It is about the assumption that the planet is large and nature infinitely malleable to the manipulations of our corporate scientists, who can both make virtually anything the advertising agencies can convince us to buy, and promise to provide a technical means to clean up the mess made in the process. It is about the silence over the origin of the consumer items that feed, shelter, and entertain us while providing the cultural matrix within which the destruction of ecosystems and their peoples continues in so many places.

The appropriate institutions for tackling these issues often seem to be missing. The state system itself has long actively promoted life based on the private automobile, building the roads and infrastructure to facilitate private transportation (Paterson 2000). In the latter years of the twentieth century most states, whatever their public rhetoric, were obviously much more concerned with development, economic growth, and business than with matters of global environmental concern. The United Nations has never given much priority to its Environmental Program, although it has been a valuable conduit for information and international scientific collaboration. Despite much hype about the event, little of substance was accomplished at the huge United Nations conference in Rio in 1992 on environment and development, although the process for establishing the Framework Convention on Climate Change was set in motion there (Sooros, 1997).

Nonetheless there have been many efforts to establish international regimes to regulate aspects of environmental matters by getting governments to agree to cooperate and administer matters that inevitably cross state borders (Vogler 2000). Some of these have directly confronted environmental degradation by focusing on the cross-boundary trading dimensions of environmental issues. The most obvious, and perhaps the most successful, regimes of this sort include the Convention on the Trade of Endangered Species (CITES), which has long restricted the international trade of animals, or animal parts, in an attempt to stop the hunting of species at risk; the Basel convention banning the trade of hazardous wastes; and more specifically the agreements on the export of toxic wastes from the North to the South, as well as the Montreal Protocol on protecting the ozone layer which also restricts trade of CFCs (Bryner 1997).

While such agreements are clearly important innovations and have had some beneficial effects, in the 1990s various environmental phenomena have been once again threatened by the new international trading rules of the World Trade Organization. The power of such international organizations, and in particular the environmental consequences of decisions about trade rules, should not be underestimated. Here, in organizations which frequently explicitly claim to be dealing with things that are not environmental, decisions are made about how commodities are produced and traded that have profound effects on peoples and environments in many places. In particular the policies of free trade, encapsulated in the principles of non-discrimination and equal treatment, often work to undermine national standards by allowing governments and corporations to claim that these discriminate against companies based in other states (Conca 2000). The lessons for environmental governance are, first, that environment is frequently not the highest priority in either state policy-making or international politics and, second, that institutions, rules, and laws change in ways that may undermine earlier environmental gains. Perhaps more significant in the very long run may be the activities of the non-governmental conference at Rio in 1992, the networking of citizen groups and social movements dedicated to establishing principles for living sustainably (McCormick 1995). The implicit cultural shift away from assuming that more is necessarily better in economic terms shows few major successes as yet, but clearly the colonial assumptions of the planet as a resource to be exploited by the rich and powerful is no longer quite so unchallenged (Leslie 2000). Mainstream publications are increasingly recognizing that discussions of global warming are deadlocked on the crucial issues of greenhouse gas emissions and some suggest that the focus on vulnerabilities and adaptation instead offers more useful policy agendas (Sarewitz and Pielke 2000). But despite these recognitions, the larger problem of turning profound political questions, raised by global environmental concern, into technical discussions of the details of such things as the state-administered rules for trading emission quotas, continues to act to marginalize big questions of environmental governance (Hajer 1995).

The reluctance of the American political establishment to move seriously on tackling greenhouse gas emissions is noteworthy (Adams 2000). The ability that American power has to dominate international arrangements is due in part to its large financial input to, and hence large voting power in, such organizations as the International Monetary Fund. But, as in the case of numerous aspects of the Law of the Sea negotiations, it is also clear that international arrangements are unlikely to be effective if the most powerful state is not part of the agreement and does not recognize its constraints. This is not to suggest that other states cannot, should not, or don't take the lead on formulating policy and negotiating regimes. But to be effective the largest military, industrial, and financial power has to agree to its terms and support its implementation (Vogler 2000). So long as governance is a matter primarily for states and international regimes these structural difficulties remain a major obstacle to rethinking how to live on a small planet and how humanity might remake its circumstances.

The sheer speed of technological innovation, however, ironically allows for the future to be different. Precisely the possibilities opened up by innovation and research in the twentieth century allow societies to make very different things and do so in different ways. The environmental crisis is not a technological one, it is very much a political crisis of governance, institutions, and the limits of the modern identity (Sachs, Loske, and Linz 1998). The question now, and for the coming decades, is whether humans, caught in the contemporary structures of power at a global scale, can be ingenious enough to invent new institutions that work on the assumptions that we are part of a biosphere, rather than "on" Earth, connected in numerous ways, rather than living in separate spaces, and responsible for the consequences of our modes of life, rather than autonomous consumers.

Part VI
Conclusion

28 Remapping the World: What Sort of Map? What Sort of World?

Peter J. Taylor, Michael J. Watts, and R. J. Johnston

Introduction

We organized our contributors around five overarching themes each containing a set of chapters. The authors, on the other hand, had other ideas; they have broken free of the strictures imposed by our neat organization, and that is how it should be. The division into parts with "geo" prefixes was simply a pedagogic stance, since global change is intrinsically multidisciplinary in nature. Hence we can find discussion on postmodernism and the environmental crisis in Part I, "Geoeconomic Change"; on cultural relativities and social movements in Part II, "Geopolitical Change"; on political breakdowns, economic restructuring, and environmental problems in Part III, "Geosocial Change"; on new divisions of labor, and the economic and political consequences of communication in Part IV, "Geocultural Change"; and on economic growth and the limitations of politics in Part V, "Geoenvironmental Change." All of this is very worthy of geography in its best holistic tradition, as the liberal arts degree that is most relevant for the world in which students are going to spend the rest of their lives. This interweaving of a rich variety of global themes cannot, should not, be too neatly summarized therefore, but should be left to stand in its own connected diversity.

In this concluding editorial chapter we highlight one basic similarity and one key difference that provide a dual focus on the inherent complexities of contemporary global change. Both draw upon our discussion of the space-time concepts introduced in the opening chapter, but here we emphasize the various ways in which the authors have informed our original discussion. First, all the chapters show an acute awareness of the dangers of globalization as a superficial slogan. Global change does not in any sense make other geographical scales disappear, quite the reverse in fact: the rise of "globalization" coincides with a simultaneous affirmation of "localization" as places both of control (e.g., world cities) and of resistance (e.g., new nationalisms, anti-globalization movements). There is much debate about the form that the global–local nexus takes at the beginning of the twenty-first century, but there is no doubt as to its importance. Second, the chapters reveal a diversity of terms to describe the rapid changes they analyze: "restructuring" is the customary way of describing geoeconomic change; geopolitical change is about "new order"; in geosocial change "transitions" are commonplace; geocultural change is about "new identities," and it is in discussions of geoenvironmental change that the harbingers of "crisis" and "catastrophe" are most likely to be met. All of these terms have different meanings for the nature of change, not just in terms of substantive content but also in terms of degree of change. We investigate the latter below.

Global–Local Nexus

The treatment of geographical scale in the chapters above has ranged across the whole gamut of possible positions. Some contributors have argued very strongly that what is global is local and vice versa; others, that scale is a crucial dimension for understanding social practices, as, for instance, with representation in democratic theory and as the necessary ordering framework for understanding society–nature relations. Can such obverse positions be equally true beyond their specific subject matters? This is a difficult question and our answer hinges on how we deal with the perennial problem of the relation between theory and practice in social science. Clearly, in conceptualizing our world we must not fall into the trap of thinking that geographical scales exist separately from the social practices that create and continue to modify them. Logically, it is not possible to carry out a social activity at one particular scale but not at other scales. But this does not mean scale is neutral or inert in social activities. It is equally clear that some practices are facilitated by focusing at particular scales because effects of practices are typically concentrated at particular scales (Taylor 1993b: 40–7). But since scales are not simply given but are themselves products of social activities, interpretations must be measured, reflecting the messy world we live in. Slogans such as "act locally, think globally" have important propaganda value in some circumstances, but they are inadequate for a reasoned theoretical, or indeed political, understanding of global change. For instance, the implication of by-passing the state in the above slogan is especially problematic for many social practices, as is recognized above, and not just by the authors writing about geopolitical change.

Recent concern for geographical scale has concentrated on the global and the local as opposite ends of the range of social-space possibilities. But even these two seemingly straightforward concepts are by no means unproblematic (Smith 1993). The global implies a worldwide universalism, whereas the reality is that the processes of globalization are quite uneven. In the communications revolution, for example, the majority of humanity is "out of the loop" and there is little prospect of large swaths of the "South" being hooked into the system in the foreseeable future (Castells 1993). Similarly we can ask how local is local? If local implies community then only small neighborhoods and villages can be "true" communities based on face-to-face interactions. Most studies treat localities as the local scale, which is defined in terms of a town or city and its dependent region. This involves many definitional problems, but there is an underlying notion of local economy and society in a symbiotic relationship facing the outside world (Cooke 1989). However, despite these problems the chapters have been able to uncover a global–local nexus bridging a wide range of social practices. By "nexus" we mean that a complex connectivity exists between the two limiting geographical-scale possibilities. Five such examples can be easily identified from the chapters above.

Localities are often portrayed as "economic victims" of global forces, where investment decisions made thousands of miles away can make or break communities. This is sometimes known as the regional dilemma in a new market-led world. But life is never that simple. Localities are not inert population aggregates; they are constituted of people and their social networks that can, and do, devise practices to attract, retain, boost, and otherwise ameliorate forces that seem to be beyond control – and can compete with similar alliances elsewhere for various forms of investment. This global–local nexus is, of course, enormously complex both economically and politically. This complexity is often related to a second nexus concerning the recent rise of multiple ethnic rebellions, religious revivals, and nationalisms. Each can be interpreted in part as local resistances to the homogenizing global political forces that favor larger and larger political spaces to counter economic globalization. These movements attempt to generate a politics sensitive to local needs, in reaction to the destructive and destabilizing impact of economic restructuring. A third nexus is a broader formulation to what lies behind ethnic revivals. The postmodern celebration of diversity in all its forms – gender, race, sexuality, physical ability, as well as religion and ethnicity – derives from a critique of the metanarratives of modernity as a sort of intellectual globalization. The global implied by modernized space is countered by the local identified through diversity in places. But diversity is not universally accepted and intolerance has created a fourth "new world disorder" nexus wherein global changes are translated into numerous local conflicts. Sometimes surrogate wars for outside powers, the contemporary world is ablaze with political "flash points," civil wars that destroy localities, creating millions of refugees – the places we gasp at in horror on our television screens in the comfortable world. The fifth nexus treats the broader notion of destruction, the overloading of ecosystems locally to the point of producing uninhabitable localities, which can culminate in the destruction of the Earth as a living system.

The key point about these interlocking global–local nexuses is that they each represent real tensions between activities and consequences that are separated by geographical scale. The conundrum is that there is no easy way to overcome such problems of remoteness. No sooner is a solution found to one particularly onerous and dangerous situation than the world has moved on, spawning another series of related problems. For instance, many people looked forward to the end of the Cold War as a means of solving some very crucial problems, notably the threat of nuclear war. The Cold War is over but the problems have not disappeared; worries about nuclear proliferation, and many more problems, have arisen. All we can say is that the world is now different, not necessarily better. And so we must return to the one unchanging fact of our world – that it is forever changing.

Remapping

Our restless world continually moves on; as soon as we write something about the contemporary situation it is likely to become obsolescent (if not entirely obsolete), and new stories emerge which illustrate, and in some cases take forward, the processes we have identified. Africa has come to the fore with stories of massive contrasting fortunes in recent years, for example. In South Africa a successful all-race election took place in 1994 with little violence and a fair ballot in most provinces, and a second election five years later was also seen as an exemplar of democracy and the rule of law coming to a formerly divided society. But the 1994 election of Nelson Mandela coincided with an unprecedented slaughter of people in Rwanda, where a civil war produced hundreds of thousands of corpses in just a few weeks. In neighboring Uganda, Lake Victoria was declared a disaster zone due to the pollution of rotting bodies floating down the rivers from Rwanda. Rwanda immediately became a watchword for brutality and descent into an earthly hell. Five years later, when Thabo Mbeki was elected as Mandela's successor, other neighboring countries were the focus of conflict and disaster: in Mozambique, heavy rainfall stimulated widespread flooding and the loss of foodstuffs; in Zaire, a corrupt president was overthrown and a new regime sought to "modernize" and "join the international community" while a civil war (in which neighboring states were participating) continued to rage within large parts of its territory; and in Zimbabwe an increasingly beleaguered government sought to deflect local economic concerns by pressing for state takeover of white-owned farms, and encouraging occupations, some involving violence. The situation in central Africa as a whole remains unstable and there is good reason to think that the lines on the map are still being redrawn. The apocalypse that was Sierra Leone and Somalia have both stabilized but warlords, child combatants, and internecine struggle are seemingly intractable parts of the political landscape.

This mixture of hope and despair extends far beyond one continent. In Gaza and the West Bank the fraught peace process involving Palestinians taking political control of their lands as a first step toward peace in this part of the Middle East has encountered Israeli intransigence on key issues (such as the permanency of some settlements and control of Jerusalem) -- with the election in 2001 of a "hardline" Israeli prime minister being linked to increased violence on both sides. And in the United States, the election of President Bush (albeit in chaotic circumstances which led to much scorn being poured on this bastion of democracy) saw major changes in that country's policies on a range of international issues – and an early stand-off with China over an accident involving a US spy plane in international air space close to the Chinese coast when the US was considering whether to conclude a major sale of arms to Taiwan. Meanwhile, the inter-ethnic strife following an armed coup in Fiji remains unresolved, the Northern Ireland peace talks are stalled, there are increased tensions over asylum seekers ("illegal immigrants") being smuggled into the UK, and war looks likely to break out again in the Balkans – the push for a "Greater Serbia" having been repulsed, at least temporarily, a parallel push for a "Greater Albania" appears to have been launched, which may involve NATO states such as Greece and Turkey as well as Bulgaria. By the time these words are being read the world will have moved on again, creating a different geography of hope and despair.

Geopolitical changes tend to hog the newspaper headlines but they are not necessarily the most important changes that are happening. For instance, the 1994 signing of the new GATT treaty will have profound effects on the livelihoods of millions of people in the coming years – though not without considerable public protest at the impact of the World Trade Organization's policies for global inequalities and the environment, as at major meetings in Seattle and Prague in 2000 and Gothenburg and Genoa in 2001: and while the powerful nations were planning their vision of the world's economic future, the leaders of the 41 poorest countries were meeting in Zanzibar to decide how to counter it, and determined to withhold their crucial voting support for G8 plans at the next WTO conference if their own demands were not met. The impact on race relations in Europe of the legitimation of neo-fascists by their inclusion in Austria's government for the first time since the defeat of Nazism following a general election in 1999 remains unclear. In contrast, one of the most conservative of churches, the Anglican community, has accepted the ordination of women in its heartland, England, providing an important ideological boost for the cause of equality between the sexes – although it still refuses to create female bishops. The question is, how can we interpret all this change? Our contributors have identified a range of ideas about the nature of this change. Social change may be speeding up, it may exhibit regular rhythms, or perhaps these arguments for continuity have got it wrong and we are experiencing a discontinuity, a rupture with what has gone before. We have at our disposal quite a large lexicon for describing this change – restructuring, transitions, cycles, trends, and numerous concepts that have the prefix "post" – but as yet no clear articulation of terminology to encompass the wide range of global changes covered in this book.

For the title of both this chapter and the book we have described change using the cartographic metaphor of remapping. Given its spatial connotations this may be deemed particularly appropriate for a geography text, but our motives go beyond such niceties. We treat the notion of remapping as a means for thinking about open-ended processes of social change in order to dispel some pernicious myths about what is happening. We identify seven remappings that we believe to be fundamental for a basic understanding of the current trajectory of our world.

First, there is a reconfiguration of capitalism occurring with new divisions of labor, an enhanced role for finance, and new possibilities of control through communication and computing-technology innovations. This package of changes is disrupting traditional capital–labor relations within countries, but there is no simple trend to economic globalization. Related to this is the second remapping, which involves the promotion of markets as resource allocators in an increasingly deregulated world. But markets do not exist in a social vacuum. Evangelical marketeers in both the old communist world and the Third World have confused market principles with market practices. All social activity is premised on the existence of rules: deregulation presupposes an alternative reregulation – and many argue that the state in many of those countries is relatively impotent in the face of criminal and other organizations, not least those involved in the production and distribution of drugs whose sale and use is criminalized in many "developed world" countries. Which brings us to the third remapping, the challenge to state sovereignty. Clearly the functions of the state have changed appreciably, but this does not mean the demise of the state. The legitimacy it gives to markets and other social activities continues to be a necessary prop to cope with social change. Without the state, capitalism's creative destruction, which is continually reproducing the system, degenerates into private mafias such as the market warlords in the former USSR, and short-term economic destruction. What is required is the fourth remapping, the construction of new civil societies in which peoples can engage with social change through their states. Through much of the world, social institutions have been eliminated in massive reform programs but without, as yet, new democratic institutions to take their place – as the American political scientist Robert Putnam (2000) has argued in his seminal book *Bowling Alone* on the decline of social capital and trust. Increasingly, governments are seeking a "Third Way" between free-market liberalism, on the one hand, and socialism, on the other, but their attraction to the rhetoric of the options on offer ("communitarianism," "direct democracy," etc.) has produced little in the way of "revolutionary" action. This challenges the multi-ethnic nature of most states, since democracy presupposes a single "demos," or people, through which to operate.

The fifth remapping genuinely transcends the state in the form of environmental change. The environment is no respecter of political boundaries, and when "world leaders" such as the Presidents George Bush promise to put America's national interests first, we can appreciate the necessity for a new international politics that environmentalists are struggling to create. The sixth remapping encompasses much that is implied in the previous remappings. Modernity is being debated, which means that our centuries-old faith in automatic social progress is being reevaluated. The rise of the so-called Post-Development School of "anti-development theory" which sees postwar assistance to the Third World as a colossal failed modernity (Rahnema 1999; Escobar 1995) is simply one case it point. But does this latest critique of modernity represent something quite different from the long history of challenges to the notion of Progress, or Science, or Enlightenment? The answer to this question turns on whether one views postmodernism as taken from the same cloth as those earlier movements that bemoaned the loss of custom, tradition, and some mystic past.

And so we come to the seventh remapping, which some might call a "demapping." The rapid expansion of information technologies over the last decade has led some commentators to identify an "end of geography" paralleling the "end of history" that Fukuyama associated with the triumph of liberal democracy in the late 1980s. Not only do we now have millions of computers connected with each other through networks that enable virtually immediate contact between machines anywhere on the globe (enabling rapid searches for information and the ability to download masses of data – perhaps to be used in writing essays and project papers associated with courses using this book as a text!), but increasingly we can do this with hand-held machines: wherever and whenever we are, we can be in contact with everywhere else – as long as the technology is available there too. But does that mean the end of geography, the creation of a "single global village" of "community without propinquity" that forecasters were prophesying a few decades ago? Or is geography still important? These questions are addressed by several authors (such as Kitchin and Dodge in this book) and the answers suggest that, at the very least, claims of geography's death are somewhat premature.

All such mappings and remappings, because they are the products of complex local–global articulations, also involve something else: a radical intermixing of the old and the new, custom and modernity, left and right, culture and nature. It is for this reason that the words "hybridity" and "cosmopolitanism" have acquired such weight in contemporary debates over millennial capitalism (Comaroff and Comaroff 2001; Gaonkar 2001; Mitchell 1999). Geographical analysis must address these hybridities if it is to meet the challenges of a global world that cannot be encapsulated by the old homilies of progress, linear history, or economic reductionism.

All of these remappings imply profound social change, although whether this is in the form of continuity – more of the same – or discontinuity – a turning point to something different – remains open for debate. As editors we disagree among ourselves on this. But our purpose here is not to polish the crystal ball and predict the unpredictable; rather, we conclude by considering the remappings as geographies of global change. And at one level that geography is very simple. Contemporary trends are producing a more polarized world: according to the United Nations Development Program, between 1960 and 1990 differentials between the richest and poorest countries increased by 20 percent (UNDP 1993) and at a greater rate in the following decade. At the personal level this is good news to most readers of this book since you are likely to be part of the comfortable world of relative affluence. With average luck, each of you should be able to find a pleasant niche in this world – which is, of course, one of the reasons you are reading this book, as part of reading for a degree. This is not a callous view, it is a realistic one defined by three authors who are also fully integrated into the comfortable world. But what of the rest of humanity, in their world of struggle?

While the comfortable world debates who will win the head-to-head economic competition between America, Europe, and Japan (Thurow 1992), those who are not combatants in this great contest already know who the real losers are. While the "Big Three" have hammered out a trade agreement through GATT and have got the rest of the world to go along with them, it should come as no surprise that sub-Saharan Africa is predicted to be the region to suffer most from the treaty's trade provisions (hence the Zanzibar July 2001 meeting of the leaders of the world's poorest countries). And this is to follow years of net outflows of billions of dollars from the continent, to a degree that it almost disappeared from the world economic map altogether. (In the late 1990s public pressure in the "developed world" led governments there to announce that they would write-off the international debts of many of the poorest countries – but this seemed to be more of a rhetorical than an actual commitment over the next few years, in part because of the terms that some of those governments insisted upon. But even President Bush, faced with such inequities, has suggested that the World Bank might consider its activities as grants rather than loans.) No wonder the Mexican Zapatist guerrilla leader, Subcomandante Marcos, has described NAFTA and the general opening up of the international economy as a "death sentence" for the poor. For many millions that is what it will be. And these will include people in the former "Second World." So-called "shock therapy" in Russia and Eastern Europe is not only taking a massive economic toll in unemployment; precipitous collapses in entitlements and welfare are causing a rise in death rates among vulnerable groups – in Russia the death rate as a whole went up 22 percent in 1992 alone (*New York*

Times, March 1, 1994). Overall, the prospects for vast tracts of the former Second World and Third World are increasingly bleak. And this relates, of course, to the political turmoil in these regions. For all the democratic hype at the end of the Cold War, the political victory of liberal democratic states over communist states has no direct bearing on the transferability of liberal democracy across the world, beyond an initial "honeymoon period" of winner's euphoria. As we have emphasized before, social institutions have to be constructed, and that requires a suitable context. Increasing economic polarization both between and within countries is a most unsuitable context for creating institutions that define a political equality. Of course the idea of diffusing liberal-democratic practices is not new. It was tried after decolonization and failed almost everywhere. As before, instead of creating new liberal-democratic states, the likely outcome is liberal-democratic interludes where the military or authoritarian movements take advantage of the political opportunities that will inevitably arise from the continuing economic difficulties of the poorer countries (Taylor 1993c: 270–6). Kaplan's (1994: 48–60) image of a bifurcated world emerges to haunt us: "Part of the globe is inhabited by Hegel's . . . Last Man, healthy and well-fed . . . The other part by Hobbes' First Man, condemned to a life that is 'poor, nasty, brutish and short'."

Whatever happened to "one world" in all of this? However unevenly, our world is interconnected through the vortex of globalization. Polarization could work as a sustainable system if the world were populated by rather dim economic men and women. But it is not; it is full of human beings with hopes and dreams and expectations. These will have to be accommodated across the world to prevent a politics of polarization destroying all. An interconnected world is an easily sabotaged world. Terrorism, the politics of the weak, has followed a definite upward trend. It used to be confused with Cold War conflict but, with the latter out of the way and terrorism showing no signs of abating, we can expect more and greater upheaval of our comfortable lives. Globally this throws up some nice ironies: in 1993 US tourists were avoiding Europe in case they got bombed, while European tourists were canceling bookings in the USA in case they got mugged – and then again in 2001 US tourists were avoiding the UK because of "scare stories" about murder rates in its cities, the "closing of its countryside" because of foot-and-mouth disease, and the lack of safety of many foods, and most media attention regarding the Tamil Tiger bombing of Colombo airport in July 2001 focused on its impact for tourists from rich countries. Furthermore, new forms of terrorism have introduced new politics of the weak – however weakness is to be defined. As the examples of computer hackers and computer-virus-spreaders show, it is possible for both commercial and government operations to be "invaded" and (at least temporarily) "disabled" by their challengers without resorting to physical violence, from remote (perhaps unidentifiable and unassailable) sites.

The world of struggle interpenetrates the world of comfort, and increasingly so with the growth of polarization within countries. When will the protection costs of comfort outweigh the economic benefits of living in a comfortable world? Let us hope our world leaders realize that increasing polarization is not a sustainable condition for humanity, before the lessons of terrorism or of simply falling off the world map, become clear for all to see.

Bibliography

ACSH (American Council on Science and Health) (1997) *Global climate change and human health*. New York: ACSH.

Adams, P. (1998) Network topologies and virtual place. *Annals of the Association of American Geographers* 88, 1, 88–106.

Adams, P. C. (1996) Protest and the scale politics of telecommunications. *Political Geography* 15, 5, 419–41.

Adams, P. G. (ed.) (2000) *Climate Change and American Foreign Policy*. New York: St. Martin's Press

Adams, W. M. (2001) *Green Development: Environment and Sustainability in the Third World*, 2nd edn. London: Routledge.

Agarwal, B. (1998) The gender and environment debate. In R. Keil, D. V. J. Bell, P. Penz, and L. Fawcett (eds.) *Political Ecology: Global and Local*. London: Routledge.

Agnew, C. and Anderson, E. (1992) *Water Resources in the Arid Realm*. London: Routledge.

Agnew, J. (1987) *Place and Politics: The Geographical Mediation of State and Society*. London: Allen Unwin.

Agnew, J. and Knox, P. L. (1989) *The Geography of the World Economy*. London: Edward Arnold.

Agulhon, M. (1981) *Marianne in Battle: Republican Imagery and Symbolism in France, 1798–1880*. Cambridge: Cambridge University Press.

Alber, J. (1988) Is there a crisis of the welfare state? Cross-national evidence from Europe, North America and Japan. *European Sociological Review* 4, 3: 181–207.

Alexander, M. J. (1994) Not just (any) body can be a citizen: the politics of law, sexuality and postcoloniality in Trinidad & Tobago and the Bahamas. *Feminist Review*, 48: 5–24.

Alibek, K. (1999) *Biohazard*. New York: Random House.

Allen, J. and Hamnett, C. (1995) *A Shrinking World?* Oxford: Oxford University Press.

Allison, G., Cote, O., Falkenrath, R., and Miller, S. (1996) *Avoiding Nuclear Anarchy: Containing the Threat of Loose Russian Nuclear Weapons and Fissile Material*. Cambridge, MA: MIT Press.

Al-Mazrou, Y. et al. (1997) A vital opportunity for global health. *Lancet* 350, 750–1.

Alvares, C. (1992) *Science, Development and Violence*. Delhi: Oxford University Press.

Alvarez, S., Dagnino, E., and Escobar, A. (eds.) (1998) *Cultures of Politics, Politics of Cultures*. Boulder, CO: Westview Press.

Amin, A. and Thrift, N. (1992) Neo-Marshallian nodes in global networks. *International Journal of Urban and Regional Research* 16, 571–87.

Amin, A. and Thrift, N. (2000) What kind of economic theory for what kind of economic geography? *Antipode* 32, 4–9.

Amin, A., Cameron, A., and Hudson, R. (1999) Welfare as work? The potential of the UK social economy. *Environment and Planning*, A, 31, 2033–51.

Amin, S. (1976) *Unequal Development: An Essay on the Social Formations of Peripheral Capitalism*. New York: Monthly Review Press.

Ampel, N. M. (1991) Plagues – what's past is present: thoughts on the origin and history of new infectious diseases. *Reviews of Infectious Diseases* 13, 658–65.

Anderson, B. (1983) *Imagined Communities*. London: Verso; 2nd edn. 1991.

Anderson, B. (1992) The new world disorder. *New Left Review* 193, 3–14.

Anderson, J. (1988) Nationalist ideology and territory. In R. J. Johnston, D. B. Knight, and E. Kofman (eds.) *Nationalism, Self-Determination and Political Geography*. London: Croom Helm.

Anderson, K. (1992) *Vancouver's Chinatown: Racial Discourse in Canada, 1875–1980*. Montreal: McGill-Queen's University Press.

Annas, G. (1998) Human rights and health. *New England Journal of Medicine* 339, 1778–80.

Anon (1995) Heat-related mortality – Chicago July 1995. *Mortality and Morbidity Weekly Report* 44, 31, 577–9.

Anzaldua, G. (1987) *Borderlands/La Frontera: The New Mestiza*. San Francisco: Spinster/Aunt Lute.

Appadurai, A. (ed.) (1986) *The Social Life of Things: Commodities in Cultural Perspective*. Cambridge: Cambridge University Press.

Appadurai, A. (1990) Disjuncture and difference in the global cultural economy. In M. Featherstone (ed.) *Global Culture: Nationalism, Globalization and Modernity*. Newbury Park, CA: Sage.

Appadurai, A. (1991) Global ethnoscapes: notes and queries for a transnational anthropology. In R. G. Fox (ed.) *Recapturing Anthropology: Working in the Present*. Santa Fe: School of American Research Press.

Appadurai, A. (1997) *Modernity at Large: Cultural Dimensions of Globalization*. Minneapolis: University of Minnesota Press.

Appadurai, A. and Breckenridge, C. A. (1988) Why public culture? *Public Culture* 1, 5–9.

Ariàs, P. (1962) *Centuries of Childhood*, trans. R. Baldick. London: Jonathan Cape.

Arnold, G. (1993) *The End of the Third World*. London: Macmillan.

Ashley, R. (1987) The geopolitics of geopolitical space: towards a critical social theory of international politics. *Alternatives* 12, 403–34.

Athanasiou, T. (1996) *Divided Planet: The Ecology of Rich and Poor*. Athens, GA: University of Georgia Press.

Augé, M. (1995) *Non-Places: Introduction to an Anthropology of Supermodernity*, trans. J. Howe. London: Verso.

Austin, J. and Coventry, G. (2001) *Emerging Issues on Privatized Prisons*. WDC: National Council on Crime and Delinquency/Bureau of Justice Assistance Monograph (NCJ 181249). February.

Auty, R. (1993) *Sustaining Development of Mineral Economies*. London: Routledge.

Avery, D. (1995) The world's rising food productivity In J. L. Simon (ed.) *The State of Humanity*. Oxford: Blackwell.

Ayers, R. U. (1992) Toxic heavy metals: materials cycle optimization. *Proceedings National Academy of Sciences* 89, 815–20.

Baker, J. N. L. (1931) *A History of Geographical Discovery and Exploration*. London: Harrap.

Ball (1990) The Process of International Contract Labour Migration from the Philippines: The Case of Filipino Nurses. Unpublished Ph.D. dissertation, University of Sydney, Australia.

Bandyopadhyay, J. and Shiva, V. (1988) Political economy of ecology movements. *Economic and Political Weekly*, June, 1223–32.

Barbier, E. B. (1998) *The Economics of Environment and Development: Selected Essays*. Cheltenham: Edward Elgar.

Barbier, E. B., Burgess, J. C., and Folke, C. (1994) *Paradise Lost? The Ecological Economics of Biodiversity*. London: Earthscan.

Barff, R. and Austen, J. (1993) "It's Gotta be da Shoes": domestic manufacturing, international subcontracting, and the production of athletic footwear. *Environment and Planning* A, 25, 8, 1103–14.

Barnes, J. (1994) *Mother of Storms*. New York: Tor.

Barnes, T. (1992) Reading the texts of theoretical economic geography. In T. Barnes and J. Duncan (eds.) *Writing Worlds: Discourse, Text and Metaphor in the Representation of Landscape*. London: Routledge.

Barnes, T. (1995) Political economy 1: "the culture, stupid." *Progress in Human Geography* 19, 423–31.

Barnet, R. J. and Muller, R. E. (1974) *Global Reach: The Power of the Multinational Corporations*. New York: Simon and Schuster.

Barnett, H. and Morse, C. (1963) *Scarcity and Growth: The Economics of Natural Resource Availability*. Baltimore, MD: Johns Hopkins University Press for Resources for the Future.

Barnett, J. (2000) Destabilizing the environment–conflict thesis. *Review of International Studies* 26 2, 271–88.

Barrett, M. (1991) *The Politics of Truth: From Marx to Foucault*. Cambridge: Polity Press.

Bartov, O. (1996) *Murder in our Midst: The Holocaust, Industrial Killing, and Representation*. Oxford: Oxford University Press.

Batty, M. (1997) Virtual geography. *Futures* 29, 4/5, 337–52.

Baudrillard, J. (1988) *Jean Baudrillard: Selected Writings*. Cambridge: Polity Press.

Bauer, P. T. (1966) *Economic Analysis and Policy in Underdeveloped Countries*. London: Routledge and Kegan Paul.

Bauman, Z. (1998) *Globalization: The Human Consequences*. New York: Columbia University Press.

Bayley, D. (1985) *Patterns in Policing*. New Brunswick, NJ: Rutgers University Press.

Beaverstock, J. V., Smith, R. G., and Taylor, P. J. (1999) The long arm of the law: London's law firms in a globalizing world economy. *Environment & Planning* A, 31, 10, 1857–76.

Beaverstock, J. V., Taylor, P. J., and Smith, R. G. (1999) A Roster of World Cities.

Cities 16, 6, 445–58.

Beck, U. (1992) *Risk Society: Towards a New Modernity.* London: Sage.

Beck, U. (1999) *World Risk Society.* Cambridge: Polity Press.

Beckerman, W. (1994) Sustainable development: is it a useful concept? *Environmental Values* 3, 191–209.

Bell, D. (1994) Erotic topographies. *Antipode* 26, 96–100.

Bell, D. and Valentine, G. (1995) *Mapping Desire.* London: Routledge.

Bell, D. and Valentine, G. (1997) *Consuming geographies.* London: Routledge.

Bell, D., Binney, J., Cream, J., and Valentine, G. (1994) *Landscapes of Desire.* London: Routledge.

Bell, D. A. (1999) Which rights are universal? *Political Theory* 27, 849–56.

Bello, W. (1994) *Dark Victory.* Oakland, CA: Food First.

Bendiner, B. (1987) *International Labour Affairs: The World Trade Unions and the Multinational Companies.* Oxford: Clarendon Press.

Benedick, R. E. (1998) *Ozone Diplomacy: New Directions in Safeguarding the Planet.* Cambridge, MA: Harvard University Press.

Benedikt, M. (ed.) (1991) *Cyberspace: First Steps.* Cambridge, MA: MIT Press.

Benenson, A. S. (1990) *Control of Communicable Diseases in Man*, 15th edn. Washington, DC: American Public Health Association.

Beneria, L. (1992) Accounting for women's work: the progress of two decades. *World Development* 20, 11, 1547–60.

Beneria, L. and Feldman, S. (1992) *Unequal Burden, Economic Crises, Persistent Poverty, and Women's Work.* Boulder, CO: Westview Press.

Benhabib, S. (ed.) (1996) *Democracy and Difference: Contesting the Boundaries of the Political.* Princeton, NJ: Princeton University Press.

Berger, J. (1980) *About Looking.* New York: Pantheon.

Berger, M. T. (1993) Civilizing the South: the US rise to hegemony in the Americas and the roots of "Latin American Studies" 1898–1945. *Bulletin of Latin American Research* 12, 1, 1–48.

Berkes, F. and Folke, C. (1994) Investing in cultural capital for the sustainable use of natural capital. In A. Jansson, M. Hammer, C. Folke, and R. Costanza (eds.) *Investing in Natural Capital: The Ecological Economics Approach to Sustainability.* Washington, DC: Island Press.

Berlin, I. (1958) *Two Concepts of Liberty.* Oxford: Clarendon Press.

Bhabha, H. K. (ed.) (1990) *Nation and Narration.* London: Routledge.

Bhashkar, R. (1989) *Reclaiming Reality.* Oxford: Verso.

Bingham, N. (1999) Unthinkable complexity? Cyberspace otherwise. In M. Crang, P. Crang, and J. May, (eds.) *Virtual Geographies.* London: Routledge.

Biswas, M. R. and Biswas, A. K. (1984) Complementarity between environment and development processes. *Environmental Conservation* 11, 35–43.

Blaikie, P. (1985) *The Political Economy of Soil Erosion.* London: Longman.

Blaikie, P. and Brookfield, H. (1987) *Land Degradation and Society.* London: Routledge.

Blair, B. (1993) *The Logic of Accidental Nuclear War.* Washington, DC: Brookings Institution.

Blaut, J. (1994) *The Colonizer's Model of the World: Geographical Diffusionism and Eurocentric History.* New York: Guilford.

Blonsky, M. (1992) *American Mythologies.* New York: Oxford University Press.

Bloom, D. E. and Canning, D. (2000) The health and wealth of nations. *Science*

287, 1207–9.

Bluestone, B. and Harrison, B. (1982) *The Deindustrialization of America.* New York: Basic Books.

Bluestone, B. (1990) The impact of schooling and industrial restructuring on recent trends in wage inequality in the United States. *AEA Papers and Proceedings* 80, 2, 303–7.

Blumstein, A. and Beck, A. (1999) Population Growth in U.S. Prisons 1980–1996. in M. Tonry and J. Petersilia (eds.), *Prisons.* Chicago: University of Chicago.

Boardman, R. (1981) *International Organizations and the Conservation of Nature.* Bloomington: Indiana University Press.

Bobbio, N. (1996) *Left and Right: The Significance of a Political Distinction.* Cambridge: Polity Press.

Bongaarts, J. (1985) The fertility inhibiting effects of the intermediate fertility variables. In F. Shorter and H. Zurayk (eds.), *Population Factors in Development Planning in the Middle East.* Cairo: Population Council.

Bonifice, P. and Fowler, J. (1993) *Heritage and Tourism in "The Global Village."* London: Routledge.

Bookchin, M. (1986) *The Modern Crisis.* Philadelphia: New Society Publishers.

Bordo, S. (1990) Feminism, postmodernism and gender-scepticism. In L. Nicholson (ed.) *Feminism/Postmodernism.* London: Routledge.

Borzaga, C. and Maiello, M. (1998) The development of social enterprises. In C. Borzaga and A. Santuari (eds.), *Social Enterprises and New Employment in Europe.* Trento: Regione Autonoma Trentino-Alto Adige and European Commission DGV.

Botchwey K. et al. (1998). *External Evaluation of the ESAF.* Washington, DC: IMF.

Botkin, D. (1990) *Dischordant Harmonies: A New Ecology of the Twenty First Century.* New York: Oxford University Press.

Bouvier, M., Pittet, D., Loutan, L., and Starobinski, M. (1990) Airport malaria: a mini-epidemic in Switzerland. *Schweizerin Medizin Wochenschreiber* 120, 1217–22.

Bowman, I. (1921) *The New World.* New York: World Books

Boyer, R. (1990) *The Regulation School: A Critical Introduction.* New York: Columbia University Press.

Bradshaw, W., Noonan, R., Gash, L., and Sershen, C. B. (1993) Borrowing against the future: children and Third World indebtedness. *Social Forces* 71, 3, 629–56.

Braithwaite, J. and Drahos, P. (2000) *Global Business Regulation.* Cambridge: Cambridge University Press.

Bramwell, A. (1989) *Ecology in the 20th Century: A History.* New Haven, CT: Yale University Press.

Brandt Commission Report (1980) *North–South: A Program for Survival.* Cambridge, MA: MIT Press.

Braudel, F. (1980) *On History.* London: Weidenfeld and Nicolson.

Braudel, F. (1984) *Civilization and Capitalism, 15th–18th Century. Vol. 3: The Perspective of the World.* New York: HarperCollins.

Braun, B. and Castree, N. (eds.) (1998) *Remaking Reality: Nature at the Millennium.* London: Routledge.

Bray, T. (1996) Measuring the Web. Fifth International Conference World Wide Web, 6–10th May 1996, Paris. http://www5conf.inria.fr/fich_html/papers/P9/

Overview.html

Brecher, J. and Costello, T. (1994) *Global Village or Global Pillage*. Boston: South End Press.

Brecher, J., Brown Childs, J., and Cutler, J. (eds.) (1993) *Global Visions: Beyond the New World Order*. Boston: South End Press.

Brenner, N. (1998) Global cities, glocal states: global city formation and state territorial restructuring in contemporary Europe. *Review of International Political Economy 5*, 1–37.

Brenner, R. (2001) *Turbulence in the World Economy*. New York: Verso.

Brimblecombe, P. (1987) *The Big Smoke*. London: Routledge.

Brohman, J. (1996) *Popular Development: Rethinking the Theory and Practice of Development*. Oxford: Blackwell.

Brookfield, H. C. (1992) "Environmental colonialism," tropical deforestation, and concerns other than global warming. *Global Environmental Change 2*, 93–6.

Brown, K. (1997) The road from Rio. *Journal of International Development 9*, 383–9.

Brown, K., Adger, N. W., and Turner, R. K. (1993) Global environmental change and mechanisms for North–South resource transfers. *Journal of International Development 5*, 571–89.

Brown, L., Starke, L., and Halweil, B. (eds.) (2000) *Vital Signs 2000: The Environmental Trends that are Shaping Our Future*. New York: W. W. Norton.

Brown, L. R., Renner, M., Halweil, B., and Starke, L. (2000) *Vital Signs 1999–2000: The Environmental Trends that are Shaping Our Future*. London: Earthscan.

Brown, L. et al. (2000) *State of the World 2000*. New York: W. W. Norton.

Brown, L. R. (2000) The world's growing water deficit threatens its food supply. *International Herald Tribune* Thursday June 29, 12

Brundtland, G. H. (1987) *Our Common Future*. Oxford: Oxford University Press for the World Commission on Environment and Development

Brunn, S. D. and Leinbach, T. R. (1991) *Collapsing Space and Time: Geographic Aspects of Communication and Information*. London: HarperCollins Academic.

Bryant, R. and Bailey, S. (1997) *Third World Political Ecology*. London: Routledge.

Brydon, L. and Chant, S. (1984) *Women in the Third World: Gender Issues in Rural and Urban Areas*. London: Edward Elgar.

Bryner, G. C. (1997) *From Promises to Performance: Achieving Global Environmental Goals*. New York: W. W. Norton.

Budge, I. and Newton, K. (1997) *The Politics of the New Europe*. Harlow: Longman.

Bulatao, R., Bos, E., Stephens, P., and Vu, M. (1990) *World Population Projections 1992–3 Edition*. Baltimore, MD: World Bank/Johns Hopkins University Press.

Bunker, S. (1985) *Underdeveloping the Amazon: Extraction, Unequal Exchange, and the Failure of the Modern State*. Urbana: University of Illinois Press.

Bunker, S. (1991) *The Political Economy and Ecology of Raw Materials Extraction and Trade*. Unpublished ms. Madison: University of Wisconsin.

Bunyard, P. (1999) How climate change could spin out of control. *The Ecologist 29*, 2, 68–75.

Burger, J. (1990) *Gaia Atlas of First Peoples: A Future for the Indigenous World*. New York: Anchor Books.

Burke, T. (1996) *Lifebuoy Men, Lux Women: Commodification, Consumption and Cleanliness in Modern Zimbabwe*. Durham, Duke University Press.

Burton, I. and Kates, R. W. (eds.) (1965) *Readings in Resource Management and*

Conservation. Chicago: University of Chicago Press.

Burton, I. and Kates, R. W. (1986) The great climacteric, 1748–2048: the transition to a just and sustainable environment. In R. W. Kates and I. Burton (eds.) *Themes from the Work of Gilbert F. White*. Chicago: University of Chicago Press.

Busch, G. K. (1983). *The Political Role of International Trades Unions*. New York: St. Martin's Press.

Busch, L., Bonnano, A., and Lacy, W. (1989) Science, technology and the restructuring of agriculture. *Sociologia Ruralis* 29, 2, 118–30.

Bush, G. (2000) Remarks during the South Carolina Republican debate, 15 February, 2000. Transcript available from http://www.cnn.com

Bushrui, S., Ayman, I., and Lazlo, E. (eds.) (1993) *Transition to a Global Society*. Oxford: One World Publications.

Butler, J. (1998) Merely cultural. *New Left Review* 27, 33–44.

Butler, J. (1999) *Subjects of Desire: Hegelian Reflections in 20th Century France*. New York: Columbia University Press.

Butler, R. and Parr, H. (eds.) (1999) *Mind and Body Spaces: Geographies of Disability and Impairment*. London: Routledge.

Buttel, F. (2000) Ecological modernization as social theory. *Geoforum* 31, 1, 57–65.

Cairncross, F. (1997) *The Death of Distance: How the Communications Revolution Will Change Our Lives*. Boston: Harvard Business School Press.

Cannan, E. (1895) The probability of a cessation of the growth of population in England and Wales during the next century. *Economic Journal* 5, 505–15

Cardoso, F. H. (1972) Dependency and development in Latin America. *New Left Review* 74 (July–August), 83–95.

Cardoso, F. H. and Faletto, E. (1979) *Dependency and Development*. Berkeley: University of California Press.

Carpenter, R. A. (1989) Do we know what we are talking about? *Land Degradation and Rehabilitation* 1, 1–3.

Carson, R. (1962) *Silent Spring*. Boston: Houghton-Mifflin.

Carter, F. W. (1993) Ethnicity as a cause of migration in Eastern Europe. *Geojournal* 30, 241–8.

Castells, M. (1983) *The City and the Grassroots*. Berkeley: University of California Press.

Castells, M. (1989) *The Informational City: Information Technology, Economic Restructuring and the Urban–Regional Process*. Oxford: Blackwell.

Castells, M. (1993) The informational economy and the new international division of labour. In M. Carnoy et al. (eds.) *The New Global Economy in the Information Age*. University Park: Pennsylvania State University Press.

Castells, M. (1996) *The Rise of Network Society*. Oxford: Blackwell.

Castells, M. (1999) Grassrooting the space of flows. *Urban Geography* 20, 294–302.

Center for Contemporary Cultural Studies (CCCS) (1982) *The Empire Strikes Back*. London: Hutchinson.

Center for Defense Information (2000) CDI website: http://www.cdi.org

Chandra, R. (1992) *Industrialization and Development in the Third World*. London: Routledge.

Charney, E. (1999) Cultural interpretation and universal human rights: a response to Daniel A. Bell. *Political Theory* 27, 840–8.

Chase-Dunn, C. (1985) The systems of world cities 800 AD–1975. In M. Timberlake (ed.) *Urbanization in the World-Economy*. New York: Academic Press.

Chase-Dunn, C. (1989) *Global Formation*. Oxford: Blackwell.

Chatterjee, P. and Finger, M. (1994) *The Earth Brokers: Power, Politics and World Development*. London: Routledge.

Chatterjee, P. et al. (1999) Back with a vengeance. www.humanscepeindia.org/hs05999/hs59910t.htm

Chen, H., Schuffels, C., and Orwig, R. (1996) Internet categorization and search: a machine learning approach. *Journal of Visual Communications and Image Representation* 7, 1, 88–102.

Chen, H., Houston, A. L., Sewell, R. R., and Schatz, B. R. (1998) Internet browsing and searching: user evaluations of category map and concept space techniques. *Journal of the American Society for Information Science* 49, 7, 582–603.

Chevigny, P. (1995) *Edge of the Knife*. New York: New Press.

Chinn, S. (2000) *Technology and the Logic of American Racism*. London: Continuum.

Chisholm, M. (1982) *Modern World Development*. Totowa, NJ: Barnes and Noble.

Christopherson, S. (1991) *The Service Sector: A Labour Market for Women?* OECD/GD (91) 212. Paris: Organization for Economic Cooperation and Development.

Cirincione, J. (2000) *Repairing the Regime: Preventing the Spread of Weapons of Mass Destruction*. New York: Routledge.

Citizens Budget Commission (1991) *Managing the Department of Parks and Recreation in a Period of Fiscal Stress*. New York: Citizens Budget Commission.

Clark, G. L. (ed.) (1992) Special issue on "real" regulation. *Environment and Planning* A, 24, 615.

Clark, G. L. (1993) Global interdependence and regional development: business linkages and corporate government in a world of financial risk. *Transactions, Institute of British Geographers* N.S. 18, 309–25.

Clarke, R. and Timberlake, L. (1982) *Stockholm Plus Ten: Promises Promises? The Decade since the 1972 UN Environment Conference*. London: Earthscan.

Cleaver, H. (1999) Computer-linked social movements and the global threat to capitalism. polnet.html @www.eco.utexas.edu

Cleveland, C. and Ruth, M. (1996) Interconnections between the depletion of minerals and fuels: the case of copper production in the United States. *Energy Sources* 18, 355–73.

Cleveland, C. J., Constanza, R., Hall, C. A. S., and Kaufmann, R. (1984) Energy and the US economy: a biophysical perspective. *Science* 225, 890–7.

Cliff, A. D. and Haggett, P. (1985) *The Spread of Measles in Fiji and the Pacific: Spatial Components in the Transmission of Epidemic Waves through Island Communities*. Canberra: Australian National University.

Cliff, A. D. and Haggett, P. (1988) *Atlas of Disease Distributions: Analytical Approaches to Epidemiological Data*. Oxford: Blackwell.

Coale, A. (1991) Excess female mortality and the balance of the sexes: an estimate of the number of missing females. *Population and Development Review* 17, 3, 517–23.

Coffey, W. J. (1996) The "newer" international division of labor. In P. W. Daniels and W. F. Lever (eds.) *The Global Economy in Transition*. London: Longman.

Cohen, J. E. (1995) *How Many People Can the Earth Support?* New York: W. W.

Norton.

Cohen, J. L. (1985) Strategy and identity: new theoretical paradigms and contemporary social movements. *Social Research* 52, 4, 663–716.

Cohen, R. and Kennedy, P. (2000) *Global Sociology*. London: Macmillan.

Cohen, S. (2000) *Failed Crusade: America and the Tragedy of Post-Communist Russia*. New York: W. W. Norton.

Collier, P. and Gunning, J. W. (1999) *The IMF's Role in Structural Adjustment WPS 99–18*. Washington: World Bank.

Comaroff, J. and Comaroff, J. (2001) Millennial capitalism. *Public Culture* 3, 291–343.

Commission of the European Community (1988) *The Future of Rural Areas*. Brussels.

Commission on Global Governance (1995) *Our Global Neighbourhood*. Oxford: Oxford University Press.

Conca, K. (2000) The WTO and the undermining of global environmental governance. *Review of International Political Economy* 7, 3, 484–94.

Conway, D., Keshav, B., and Shrestha, N. R. (2000) Population–environment relations at the forested frontier of Nepal: Tharu and Pahari survival strategies in Bardiya. *Applied Geography* 20, 221–42.

Cook, I. *Journal of Geography in Higher Education*.

Cook, I. and Crang, P. (1996) The world on a plate: culinary culture, displacement and geographical knowledge. *Journal of Material Culture* 1, 131–53.

Cooke, P. (1989) *Localities*. London: Unwin Hyman.

Cooke, P., Moulaert, F., Swyngedouw, E., Weinstein, O., and Wells, P. (1992) *Towards Global Localization*. London: University College London Press.

Cooper, F. (2001) The problem with globalization. *Critique Internationale* 4, 1.

Corbridge, S. (1986) *Capitalist World Development*. London: Macmillan.

Corbridge, S. (1993a) *Debt and Development: IBG Studies in Geography*. Oxford: Blackwell.

Corbridge, S. (1993b) Colonialism, post-colonialism and the political geography of the Third World. In P. J. Taylor (ed.) *Political Geography of the Twentieth Century*. London: Belhaven.

Corbridge, S. E. (1991) Third world development. *Progress in Human Geography* 15, 3, 311–21.

Coronil, F. (2000) Towards a critique of globalcentrism: speculations on capitalism's nature. *Public Culture* 12, 2, 351–74.

Correll, S. (1995) The ethnography of an electronic bar: the lesbian cafe. *Journal of Contemporary Ethnography* 24, 270–98.

Cosgrove, D. and Daniels, S. (eds.) (1988) *The Iconography of Landscape*. Cambridge: Cambridge University Press.

Couclelis, H. (1998) Worlds of information: the geographic metaphor in the visualization of complex information. *Cartography and Geographic Information Systems* 25, 209–20.

Crush, J. (ed.) (1995) *The Power of Development*. London: Routledge.

Cutter, S. (1993) *Living with Risk*. London: Edward Arnold.

Daily, G. C. and Ehrlich, P. R. (1995) Development, global change and the epidemiological environment. http://dieoff.com/page106.htm

Daily, G. C., Söderqvist, T., Aniyar, S., Arrow, K., Dasgupta, P., Ehrlich, P. R., Folke, C., Hansson, A., Jansson, B.-O., Kautsky, N., Levin, S., Lubchenco, J.,

Mäler, K.-G., Simpson, D., Starrett, D., Tilman, D., and Walker, B. (2000). The value of nature and nature of value. *Science* 289, 395–6.

Dalby, S. (1990) *Creating the Second Cold War*. London: Pinter.

Dalby, S. (2000a) Geopolitics and ecology: rethinking the contexts of environmental security. In M. Lowi and B. Shaw (eds.) *Environment and Security: Discourses and Practices*. London: Macmillan.

Dalby, S. (2000b) Globalizing environment: culture, ontology and critique. In H. Koechler (ed.) *Democracy versus Globality: The Changing Nature of International Relations in the Era of Globalization*. Vienna: International Progress Organization and Jamahir Society for Culture and Philosophy.

Danaher, K. and Burbach, R. (eds.) (2000) *Globalize This! The Battle Against the World Trade Organization*. Monroe, ME: Common Courage Press.

Daniels, S. (1993) *Fields of Vision: Landscape Imagery and National Identity in England and the United States*. Cambridge: Polity Press.

Daroesman, R. (1979) An economic survey of East Kalimantan. *Bulletin of Indonesian Studies* 15, 3, 43–82.

Darrow, W. W. et al (1986) The social origins of AIDS. In D. A. Feldman and T. M. Johnson (eds.) *The Social Dimensions of AIDS*. New York: Praeger.

Dasmann, R. F., Milton, J. P., and Freeman, P. H. (1973) *Ecological Principles for Economic Development*. Chichester: Wiley.

DaVanzo, J. and Adamson, D. (1997) Russia's demographic crisis: how real is it? Center for Russian and Eurasian Studies, Issue Paper IP-162. Santa Monica: Rand Corporation.

Davidson, B. (1992) *The Black Man's Burden: Africa and the Curse of the Nation-State*. New York: Times Books.

Davis, A. Y. and Gordon, A. (1998) Globalism and the prison industrial complex: an interview. *Race and Class* 40 (2/3): 1.

Davis, H. (1998) *A Million a Minute: Inside the World of Traders*. London: Nicholas Brealey.

Davis, J. and Goldberg, R. (1957) *A Concept of Agribusiness*. Boston: Harvard Business School.

Davis, M. (1992) Beyond Blade Runner: Urban Control. The Ecology of Fear. *Open Magazine* Pamphlet 23. New Jersey: Westfield.

De Larrinaga, M. (2000) (Re)Politicizing the discourse: globalization is a s(h)ell game." *Alternatives* 25, 2, 145–82.

Dematteis, G. (2001) Shifting cities. In C. Minca (ed.) *Postmodern Geography: Theory and Practice*. Oxford: Blackwell.

Denevan, W. M. (1992) The pristine myth: the landscape of the Americas in 1492. *Annals of the Association of American Geographers* 82, 369–85.

Denitch, B. (1992) *After the Flood: World Politics and Democracy in the Wake of Communism*. Hanover, NH: Wesleyan University Press.

Descombes, V. (1993) *The Barometer of Modern Reason: On the Philosophies of Current Events*. Oxford: Oxford University Press.

Deudney, D. (1999) "Environmental security: a critique. In D. Deudney and R. Matthew (eds.) *Contested Grounds: Security and Conflict in the New Environmental Politics*. Albany: State University of New York Press.

Devall, B. and Sessions, G. (1985) *Deep Ecology: Living as if Nature Mattered*. Salt Lake City, UT: Peregrine Smith.

Diamond, I. and Quinby, L. (eds.) (1988) *Feminism and Foucault: Reflections on*

Resistance. Boston: Northeastern University Press.

Diamond, L. (ed.) (1992) *The Democratic Revolution: Struggles for Freedom and Pluralism in the Developing World*. New York: Freedom House.

Dicken, P. (1992) *Global Shift: The Internationalization of Economic Activity*, 2nd edn. New York: Guilford.

Dicken, P. (1998) *Global Shift*, 3rd edn. London: Chapman.

Dickens, E. (1996) The Federal Reserve's low interest rate policy in 1970–1972: determinants and constraints. *Review of Radical Political Economics* 28, 3, 115–25.

DiMuccio, R. B. A. and Rosenau, J. N. (1992) Turbulence and sovereignty in world politics. In Z. Mlinar (ed.) *Globalization and Territorial Identities*. Aldershot: Avebury Press.

Dixon, J. A. and Fallon, L. A. (1989) The concept of sustainability: origins, extensions and usefulness for policy. *Society and Natural Resources* 2, 73–84.

Dobkowski and Wallimann (1998) *The Coming Age of Scarcity: Preventing Mass Death and Genocide in the Twenty-First Century*. Syracuse, NY: Syracuse University Press.

Dodge, M. (2000) Atlas of Cyberspaces. Centre for Advanced Spatial Analysis, University College London. http://www.cybergeography.org/atlas/

Dodge, M. and Kitchin, R. (2000) *Mapping Cyberspace*. London: Routledge.

Dollar, D. and Wolff, E. (1993) *Competitiveness, Convergence, and International Specialization*. Cambridge, MA: MIT Press.

Domash, M. (1991) Towards a feminist historiography of geography. *Transactions, Institute of British Geographers* N.S. 16, 95–104.

Donaghu, M. T. and Barff, R. (1990) Nike just did it: international subcontracting, flexibility, and athletic footwear production. *Regional Studies* 24, 537–52.

Donald, J. and Hall, S. (1986) *Politics and Ideology*. Milton Keynes: Open University Press.

Donath, J. S. (1999) Identity and deception in the virtual community. In P. Kollock and M. Smith (eds.) *Communities in Cyberspace*. London: Routledge.

Dos Santos, T. (1970) The structure of dependence. *American Economic Review* 60 (May), 231–6.

Douglas, M. (1966) *Purity and Danger: An Analysis of Concepts of Pollution and Taboo*. London: Penguin Books.

Douglas, M. (1992) A credible biosphere. In M. Douglas (ed.) *Risk and Blame: Essays in Cultural Theory*. London: Routledge.

Douglas, M. and Wildavsky, A. (1983) *Risk and Culture: An Essay on the Selection of Technological and Environmental Dangers*. Berkeley: University of California Press.

Dovers, S. R. and Handmer, J. W. (1992) Uncertainty, sustainability and change. *Global Environmental Change* December, 262–76.

Downs, A. (1972) Up and down with ecology: the issue attention cycle. *The Public Interest* 10, 38–50.

Dreze, J. and Sen, A. K. (1989) *Hunger and Public Action*. Oxford: Clarendon Press.

Driver, F. (1992) Geography's empire: histories of geographical knowledge. *Environment and Planning* D: *Society and Space* 10, 23–40.

Dryzek, J. S. (1996) Political inclusion and the dynamics of democratization. *American Political Science Review* 90, 475–87.

Du Bois, W. E. B. (1989) *The Souls of Black Folk.* New York: Bantam.

Dunbar, M. J. (1974) Arctic ecosystems and arctic resources. In E. Bylund (ed.) *Ecological Problems of the Circumpolar Area.* Lulea: Nordbottens Museum.

Duncan, J. and Duncan, N. (1988) (Re)-reading the landscape. *Environment and Planning* D: *Society and Space* 6, 17–26

Duncan, J. and Gregory, D. (eds.) (1998) *Writes of Passage.* London: Routledge.

Duncan, J. and Ley, D. (eds) (1994) *Place/Culture/Representation.* London: Routledge.

Duncan, J. S. (1989) The power of place in Kandy, Sri Lanka: 1780–1980. In J. A. Agnew and J. S. Duncan (eds.) *The Power of Place: Bringing Together Geographical and Sociological Imaginations.* London: Unwin Hyman.

Duncan, J. S. (1990) *The City as Text: The Politics of Landscape Interpretation in the Kandyan Kingdom.* Cambridge: Cambridge University Press.

Dunning, J. H. (1981) *International Production and the Multinational Enterprise.* London: Allen and Unwin.

Dunning, J. H. (1983) Changes in the level and structure of international production: the last one hundred years. In M. Casson (ed.) *The Growth of International Business.* London: Allen and Unwin.

Dunning, J. H. and Cantwell, J. A. (1987) *IRM Directory of Statistics of International Investment and Production.* Basingstoke: Macmillan.

Dunning, J. H., Cantwell, J. A., and Corley, T. A. B. (1986) The theory of international production: some historical antecedents. In P. Hertner and G. Jones (eds.) *Multinationals: Theory and History.* Brookfield, VT: Gower.

Durning, A. (1992) *How Much is Enough?* New York: W. W. Norton.

Dyck, I. (1995) Hidden geographies: the changing life worlds of women with disabilities. *Social Science in Medicine* 40, 307–20.

Dymski, G. and Veitch, J. (1992) *Race and the Financial Dynamics of Urban Growth.* Working Paper 92–91, University of California, Riverside, Department of Economics.

Dymski, G. and Veitch, J. M. (1996) Financial transformation and the metropolis. *Environment and Planning* A, 28, 1233–60.

Dyson, T. (1996) *Population and Food: Global Trends and Future Prospects.* London: Routledge.

Eatwell, J. and Taylor, L. (2000) *Global Finance at Risk: The Case for International Regulation.* New York: New Press.

Economist, The (1993) *Survey of the Food Industry,* December 4, 1993.

Edid, M. (1994) *Farm Labor Organizing.* Ithaca, NY: Cornell/ILR.

Egerton, J. (1995) *Speak Now Against the Day.* New York: Knopf.

Ehrenfeld, J. and Gertler, N. (1997) Industrial ecology in practice: the evolution of interdependence at Kalundborg. *Industrial Ecology* 1, 1, 67–80.

Ehrlich, P. and Holdren, J. (1971) Impact of population growth. *Science* 171, 1212–17.

Ehrlich, P. R. and Ehrlich, A. H. (1990) *The Population Explosion.* New York: Simon and Schuster.

Ekins, P. (1992) *A New World Order: Grassroots Movements for Global Change.* London: Routledge.

Ekland-Olsen, S. (1992) Crime and incarceration: some comparative findings from the 1980s. *Crime and Delinquency* 38, 3, 392–416.

Elsom, D. (1992) *Atmospheric Pollution.* Oxford: Blackwell.

Emmanuel, A. (1972) *Unequal Exchange: A Study of the Imperialism of Trade*. New York: Monthly Review Press.

Energy Information Agency (2000) World consumption of primary energy by selected country groups, 1989–1998. Available on the web at eia.doe.gov/emeu/iea/table11.html

Enloe, C. (1989) *Bananas, Beaches and Bases: Making Feminist Sense of International Politics*. London: Pandora Press.

Environmental Project of Central America (1987) Militarization – the environmental impact in Central America. *Cultural Survival Quarterly* 11, 3, 38–45.

Escobar, A. (1984) Discourse and power in development: Michel Foucault and the relevance of his work to the Third World. *Alternatives* 10, 377–400.

Escobar, A. (1992a) Culture, economics, and politics in Latin American social movements: theory and research. In A. Escobar and S. E. Alvarez (eds.) *The Making of Social Movements in Latin America*. Boulder, CO: Westview Press.

Escobar, A. (1992b) Imaging a post development era. *Social Text* 31, 20–54.

Escobar, A. (1995) *Encountering Development*. Princeton, NJ: Princeton University Press.

Esping-Andersen, G. (1990) *The Three Worlds of Welfare Capitalism*. Cambridge: Polity Press.

Esteva, G. (1992) Development. In W. Sachs (ed.) *The Development Dictionary*. London: Zed Books.

Eudey, A. A. (1988) Another defeat for the Nam Choan Dam, Thailand. *Cultural Survival Quarterly* 12, 2, 13–16.

Evernden, N. (1992) i. Baltimore: Johns Hopkins University Press.

Falk, P. and Campbell, C. (eds.) (1997) *The Shopping Experience*. London: Sage.

Falk, R. (2000) The quest for humane governance in an era of globalization. In D. Kalb et al. (eds.) *The Ends of Globalization*. Lanham, MD: Rowman and Littlefield.

Fanon, F. (1961) *The Wretched of the Earth*. New York: Grove.

Farvar, M. T. and Milton, J. P. (eds.) (1973) *The Careless Technology: Ecology and International Development*. London: Stacey.

Feldman, A. (1991) *Formations of Violence*. Chicago: University of Chicago Press.

Fellner, J. and Mauer, M. (1998) *Losing the Vote: The Impact of Felony Disenfranchisement Laws in the United States*. NY: Human Rights Watch/The Sentencing Project (G1003).

Ferguson, J. (1994) *The Anti-Politics Machine*. Minneapolis: University of Minnesota Press.

Fernandes, W. and Thukral, E. (eds.) (1989) *Development, Displacement and Rehabilitation*. New Delhi: ISI.

Findji, M. T. (1992) From resistance to social movement: the indigenous authorities movement in Colombia. In A. Escobar and S. E. Alvarez (eds.) (1992) *The Making of Social Movements in Latin America*. Boulder, CO: Westview Press.

Findlay, A. M. and Hoy, C. (2000) Global population issues: towards a geographical research agenda. *Applied Geography* 20, 207–19.

Fine, B. (1999) The developmental state is dead – long live social capital? *Development and Change* 30, 1, 1–19.

Fine, B. and Leopold, E. (1993) *The World of Consumption*. London: Routledge.

Fisher, W.F. (ed.) 1997 *Toward Sustainable Development: Struggling Over India's Narmada River*. Jaipur: Rawat Publications.

Fishman, J. (1972) *Language and Nationalism*. Rowley, MA: Newbury House.

Fitzsimmons, M. (1989) The matter of nature. *Antipode* 21, 2, 106–120.

FitzSimmons, M. (1990) The social and environmental relations of the US agricultural regions. In P. Lowe, T. Marsden, and S. Whatmore (eds.) *Technological Change and the Rural Environment*. London: David Fulton.

Flax, J. (1990) *Thinking Fragments: Psychoanalysis, Feminism and Postmodernism in the Contemporary West*. Berkeley: University of California Press.

Flora, P. (1983) *State, Economy and Society in Western Europe 1815–1975: A Data Handbook, Vol. I*. London: Macmillan.

Food and Agriculture Organization (FAO)(1999) *The State of Food Insecurity in the World 1999*. Rome.

Food and Agriculture Organization (FAO) (2000) *The State of Food Insecurity in the World 2000*. Rome.

Foster, D. (1997) Community and identity in the electronic village. In D. Porter (ed.) *Internet Culture*. London: Routledge.

Foster, H. (ed.) (1985) *Postmodern Culture*. London: Pluto Press.

Foucault, M. (1970) *The Order of Things: An Archeology of the Human Sciences*. London: Tavistock.

Foucault, M. (1977a) *Discipline and Punish: The Birth of the Prison*. London: Penguin Books.

Foucault, M. (1977b) *The Archeology of Knowledge*. London: Tavistock.

Foucault, M. (1978) *The History of Sexuality, Part I*. London: Penguin Books.

Foucault, M. (1980) *Power/Knowledge*. Brighton: Harvester Press.

Frank, A. G. (1967) *Capitalism and Underdevelopment in Latin America*. New York: Monthly Review Press.

Franke, R. and Chasin, B. (1992) Kerala: development without growth. *Earth Island Journal* spring, 25–6.

Fraser, N. (1995) From redistribution to recognition? Dilemmas of justice in a post-socialist age. *New Left Review* 212, 68–93.

Fraser, N. (1997) *Justice Interruptus: Critical Reflections on the Post-socialist Condition*. London: Routledge.

Fraser, N. (1998) Heterosexism, misrecognition and capitalism: a response to Judith Butler. *New Left Review* 228, 140–9.

Fraser, N. (2000) Rethinking recognition. *New Left Review* 3, 2nd series, 107–20.

Freeman, C. (1973) Malthus with a computer. In H. Cole et al. (eds.) *Thinking About the Future: A Critique of the Limits to Growth*. London: Chatto and Windus.

Fridenson, P. (1986) The growth of multinational activities in the French motor industry, 1890–1979. In P. Hertner and G. Jones (eds.) *Multinationals: Theory and History*. Brookfield, VT: Gower.

Friedland, W. (1991) Introduction. In W. Friedland, L. Busch, F. Buttel, and A. Rudy (eds.) *Towards a New Political Economy of Agriculture*. Boulder, CO: Westview Press.

Friedland, W., Banton, A., and Thomas, R. (1981) *Manufacturing Green Gold*. Cambridge: Cambridge University Press.

Friedmann, H. (1982) The political economy of food: rise and fall of the postwar international food order. *American Journal of Sociology* 88, supplement 246–86.

Friedmann, H. and McMichael, P. (1989) Agriculture and the state system: the rise and decline of national agricultures, 1870 to the present. *Sociologia Ruralis* 29, 93–117.

Friedmann, J. (1986) The world city hypothesis. *Development and Change* 17, 69–83.

Friedmann, J. (1992) *Empowerment*. Oxford: Blackwell.

Friedmann, J. (1994) Where we stand: a decade of World City research. In P. Knox and P. J. Taylor (eds.) *World Cities in a World-System*. Cambridge: Cambridge University Press.

Frobel, F., Heinrich, J., and Kreye, O. (1980) *The New International Division of Labor*. Cambridge: Cambridge University Press.

Froehling, O. (1997) The Cyberspace "war of ink and Internet" in Chiapas, Mexico. *The Geographical Review* 87, 291–307.

Frognier, A., Quevit, M., and Stenbock, M. (1982) Regional imbalances and centre–periphery relationships in Belgium. In S. Rokkan and D. Unwin (eds.) *The Politics of Territorial Identity*. London: Sage.

Frye, A. (1999) *Toward an International Criminal Court?* New York: Council on Foreign Relations.

Fujita, K. (1991) A world city and flexible specialization: restructuring the Tokyo metropolis. *International Journal of Urban and Regional Research* 15, 269–84.

Fukuyama, F. (1992) *The End of History and the Last Man*. New York: Free Press.

Gadgil, M. and Guha, R. (1995) *Ecology and Equity*. London: Routledge.

Gagen, E. A. (2000) Playing the part: performing gender in America's playgrounds. In S. L. Holloway and G. Valentine (eds.) *Children's Geographies: Playing, Living, Learning*. London: Routledge.

Gainsborough, J. and Mauer, M. (2000) *Diminishing Returns: Crime and Incarceration in the 1990s*. Washington, DC: The Sentencing Project.

Gallie, W. B. (1978) *Philosophers of Peace and War*. Cambridge: Cambridge University Press.

Gaonkar, D. (2001) *Alternative Modernities*. Durham, NC: Duke University Press.

Garrett, L. (2000) *The Betrayal of Trust*. New York: Hiperion.

Geddicks, A. (1993) *The New Resource Wars: Native and Environmental Struggles Against Multinational Corporations*. Boston: South End Press.

Geist, H. and Lambin, E. (2001) *What Drives Tropical Deforestation? A Meta-Analysis of Proximate and Underlying Causes of Deforestation Based on Subnational Case Study Evidence*. LUCC Report Series No. 4. Luovain-la-Neuve, Belgium: LUCC International Project Office.

Gellner, E. (1992) *Postmodernism, Reason and Religion*. London: Routledge.

Ghai, D. and Vivian, J. M. (1992) (eds.) i. Routledge, London.

Gibson, D. V., Kozmetsky, G., and Smilor, R. W. (1992) *The Technopolis Phenomenon: Smart Cities, Fast Systems, Global Networks*. Lanham, MD: Rowman and Littlefield.

Giddens, A. (1985) *The Nation-state and Violence*. Cambridge: Polity Press.

Giddens, A. (1991) *Modernity and Self-Identity: Self and Society in the Late Modern Age*. Cambridge: Polity Press.

Giddens, A. (1999) *Runaway World: How Globalization is Reshaping our Lives*. London: Profile.

Gillespie, M. (1995) *Television, Ethnicity and Cultural Change*. London: Routledge.

Gillis, M. (1987) Multinational enterprises and environmental and resource management issues in the Indonesian tropical forest sector. In C. Pearson (ed.) *MNCs,*

Environment, and the Third World. Durham, NC: Duke University Press.

Gillis, M. (1988) Indonesia: public policies, resource management, and the tropical forest. In R. Repetto and M. Gillis (eds.) *Public Policies and the Misuse of Forest Resources*. Cambridge: Cambridge University Press.

Gilmore, J. S. (1975) *Boom Town Growth Management: A Case Study of Rock Springs – Green River, Wyoming*. Boulder, CO: Westview Press.

Gilmore, R. W. (1991) Decorative beasts. *California Sociologist* 14, 1/2, 113–35.

Gilmore, R. W. (1998) Globalization and US prison growth. *Race and Class* 40, 2 and 3: 171–88.

Gilmore, R. W. (1999) You have dislodged a boulder. *Transforming Anthropology* 8, 1 and 2, 12–38.

Gilmore, R. W. (2002) Fatal Couplings. *The Professional Geographer* forthcoming, spring.

Gilmore, R. W. (forthcoming) *Golden Gulag*. Berkeley: University of California Press.

Gilroy, P. (1993) *The Black Atlantic: Modernity and Double Consciousness*. London: Verso.

Gilroy, P. (2000) *Against Race*. Cambridge, MA: Harvard University Press.

Ginzburg, R. (1962) [1988] *100 Years of Lynching*. Baltimore, MD: Black Classic Press.

Githeko, A. K. et al. (2000) Climate change and vector-borne diseases. *Bulletin of the World Health Organization* 78, 1136–47.

Glacken, C. (1967) *Traces on the Rhodian Shore*. Berkeley: University of California Press.

Gleb, A. (1988) *Oil Windfalls: Blessing or Curse?* Washington, DC: Oxford University Press for the World Bank.

Glennie, P. D. and Thrift, N. J. (1993) Modern consumption: theorizing commodities and consumers. *Environment and Planning* D: *Society and Space* 11, 603–6.

Global Policy Forum (2000) Pace of globalization quickening. www.globalpolicy.org/globaliz/define/sotw.htm

Godfrey, B. J. and Zhou, Y. (1999) Ranking cities: multinational corporations and the global urban hierarchy. *Urban Geography* 20, 268–81.

Godlewska, A. and Smith, N. (eds.) (1994) *Geography and Empire*. Oxford: Blackwell.

Goffman, I. (1959) *The Presentation of Self in Everyday Life*. New York: Doubleday.

Goldman, R. and Papson, S. (1998) *Nike Culture: The Sign of the Swoosh*. London: Sage.

Goodin, R. E. and Le Grand, J. (1987) *Not Only The Poor: The Middle Classes and the Welfare State*. London: Allen and Unwin.

Goodman, C. (1979) *Choosing Sides: Playground and Street Life on the Lower East Side*. New York: Schocken Books.

Goodman, D. (1999) Agro-food studies in the age of ecology: nature, corporeality, biopolitics. *Sociologia Ruralis* 39/1, 17–40.

Goodman, D. and Redclift, M. (1989) *The International Farm Crisis*. London: Macmillan.

Goodman, D. and Redclift, M. (1991) *Refashioning Nature*. London: Routledge.

Goodman D. and Watts, M. (eds.) (1997) *Globalizing Food: Agrarian Questions and Global Restructuring*. London: Routledge.

Goodman, D., Sorj, A., and Wilkinson, J. (1987) *From Farming to Biotechnology*.

Oxford: Blackwell.

Gordon, A. (1988) Globalism and the prison industrial complex: an interview with Angela Y. Davis. *Race and Class* 40, 2/3.

Gordon, L. (1994) *Pitied But Not Entitled*. New York: Free Press.

Gore, A. (1992) *Earth in the Balance*: *Forging a New Common Purpose*. London: Earthscan; Boston: Mifflin.

Goss, J. (1993) The "magic of the mall." *Annals, Association of American Geographers* 83, 18–47.

Goulborne, H. (1991) *Ethnicity and Nationalism in Post-imperial Britain*. Cambridge: Cambridge University Press.

Gould, P. (1990) *Fire in the Rain: The Democratic Consequences of Chernobyl*. Cambridge: Polity Press.

Graham, S. and Aurigi, A. (1997) Virtual cities, social polarization, and the crisis in urban public space. *Journal of Urban Technology* 4, 1, 19–52.

Graham, S. and Marvin, S. (1996) *Telecommunications and the City: Electronic Spaces, Urban Places*. London: Routledge.

Gramsci, A. (1971) *Selections from the Prison Notebooks*, trans. Q. Haore and G. N. Smith. New York: International Publishers.

Gray, J. (1995) The sad side of cyberspace. *Guardian* April 10.

Greene, J. (2001) The Rise and Fall – and Rise Again – of the Private Prison Industry. *American Prospect*, 12 (16): 23–7.

Gregory, D. (1994) *Geographical Imaginations*. Oxford: Blackwell.

Gregory, D. and Urry, J. (eds.) (1985) *Social Relations and Spatial Structures*. London: Macmillan.

Gregson, N. and Crewe, L. (1997) Performance and possession: rethinking the act of purchase in the light of the car boot sale. *Journal of Material Culture* 2, 241–63.

Grigg, D. (1993) *The World Food Problem*. Oxford: Blackwell.

Grove, R. (1997) *Ecology, Climate and Empire: Colonialism and Global Environmental History, 1400–1940*. Cambridge: White Horse.

Grubb, M. (1990) *Energy Policies and the "Greenhouse" Effect*. Aldershot: Dartmouth.

Grubb, M., Koch, M., Thompson, K., Munson, A., and Sullivan, F. (eds.) (1993) *The 'Earth Summit' Agreements: A Guide and Assessment*. London: Earthscan (for the Royal Institute of International Affairs, London).

Guha, R. (1989) The Problem. *Seminar* (March), 12–15.

Haber, W. (1993) Environmental attitudes in Germany: the transfer of scientific information into political action. In R. J. Berry (ed.) *Environmental Dilemmas: Ethics and Decisions*. London: Chapman & Hall.

Habermas, J. (1976) *Legitimation Crisis*. London: Heinemann.

Haggett, P. (1992) Sauer's "Origins and dispersals": its implications for the geography of disease. *Transactions, Institute of British Geographers* N.S. 17, 387–98.

Hajer, M. A. (1995) *The Politics of Environmental Discourse: Ecological Modernization and the Policy Process*. Oxford: Oxford University Press.

Hall, P. (1996) The global city. *International Social Science Journal* 147, 15–24.

Hall, S. (1980) *Race, Articulation, and Societies Structured in Dominance*. UNESCO.

Hall, S. (1984). The state in question. In G. McLennan, D. Held, and S. E. Hall (eds.) *The Idea of the Modern State*. Milton Keynes: Open University Press.

Hall, S. (1991a) The local and the global. In A. King (ed.) *Culture, Globalization and the World System*. Binghampton: Department of Art and Art History, State University of New York.

Hall, S. (1991b) Ethnicity: identity and difference. *Radical America* 23, 9–20.

Hall, S. (1992) Old and new identities, old and new ethnicities. in A. D. King (ed.) *Culture, Globalization and the World-System*. London, Macmillan.

Hall, S. et al. (1978) *Policing the Crisis: Mugging, the State, and Law and Order*. New York: Holmes and Meier.

Handelman, S. (1995) *Comrade Criminal: Russia's New Mafia*. New Haven, CT: Yale University Press.

Hannertz, H. (1991) *Cultural Complexity*. New York: Columbia University Press.

Hannerz, U. (1992) The cultural role of world cities. In A. Cohen and K. Fukui (eds.) *The Age of the City*. Edinburgh: Edinburgh University Press.

Hansen, A. (ed.) (1993) *The Mass Media and Environmental Issues*. Leicester: Leicester University Press.

Haraway, D. (1991a) *Simians, Cyborgs and Women*. London: Free Association Press.

Haraway, D. (1991b) Situated knowledge: the science question in feminism and the privilege of partial perspective. In *Simians, Cyborgs and Women: The Reinvention of Nature*. London: Routledge.

Hardin, G. (1968) The tragedy of the commons. *Science* 162, 1243–8.

Harrison, P. E. and Lederberg, J. (1998) *Antimicrobial Resistance*. Washington, DC: National Academic Press.

Hart, R., Katz, C., Iltus, S., and Mora, M. R. (1992) International student design competition of two community elementary schoolyards. *Children's Environments* 9, 2, 65–82.

Hartwick, E. (1998) Geographies of consumption: a commodity-chain approach. *Environment and Planning* D: *Society and Space* 16, 423–37.

Harvey, D. (1969) *Explanation in Geography*. London: Edward Arnold.

Harvey, D. (1989) *The Condition of Postmodernity: An Inquiry into the Origins of Cultural Change*. Oxford: Blackwell.

Harvey, D. (1996) *Justice, Nature and the Geography of Difference*. Oxford: Blackwell.

Harvey, D. (1998) What's green and makes the environment go round? In F. Jameson and M. Miyoshi (eds.) *The Cultures of Globalization*. Durham, NC: Duke University Press.

Harvey D. (2000) *Spaces of Hope*. Edinburgh: Edinburgh University Press.

Harvey, D. W. (1974) Population, resources and the ideology of science. *Economic Geography* 50, 256–77.

Harvey. D. W. (1990) Between space and time: reflections on the geographical imagination. *Annals of the Association of American Geographers* 80, 418–34.

Harvey, D. W. (1993) From space to place and back again. In J. Bird et al. (eds.) *Mapping the Futures*. London: Routledge.

Haubrich, J. G. and Wachtel, P. (1993) Capital requirements and shifts in commercial bank portfolios. *Economic Review* (Federal Reserve Bank of Cleveland) 29, 3, 2–15.

Hays, S. P. (1959) *Conservation and the Gospel of Efficiency: The Progressive Conservation Movement 1890–1920*. Cambridge, MA: Harvard University Press.

Health Canada (2000) West Nile virus. www.hc-sc.gc.ca/hpb/lcdc/publicat/info/wnv_e.html

Heathcote, R. L. (1983) *The Arid Lands: Their Use and Abuse*. London: Longman.

Hecht, S. and Cockburn, A. (1990) *The Fate of the Forest*. New York: HarperCollins.

Heffernan, M. J. (1990) Bringing the desert to bloom. French ambitions in the Sahara Desert during the late nineteenth century – the strange case of la mer intérieure. In D. Cosgrove and G. Petts (eds.) *Water, Engineering and Landscape: Water Control and Landscape Transformation in the Modern Period*. London: Belhaven.

Held, D. (1987) *Models of Democracy*. Cambridge: Polity Press.

Held, D. (1991) Democracy, the nation-state, and the global system. In D. Held (ed.) *Political Theory Today*, Stanford, CA: Stanford University Press.

Held, D. (1995) *Democracy and Global Order: From the Modern State to Cosmopolitan Governance*. Stanford, CA: Stanford University Press.

Held, D., McGrew, A., Goldblatt, D., and Perraton, J. (1999) *Global Transformations: Politics, Economics and Culture*. Cambridge: Polity Press.

Herbertson, A. J. (1910) Geography and some of its present needs. *Geographical Journal* 36, 468–79.

Herder, J. G. von (1968) [1783] *Reflections on the Philosophy of the History of Mankind*, abridged and trans. T. Manuel. Chicago: University of Chicago Press.

Herod, A. (1995) The practice of international labor solidarity and the geography of the global economy. *Economic Geography* 71, 4, 341–63.

Herod, A. (1997a) Back to the future in labor relations. In L. Staeheli et al. (eds.) *State Devolution in America*. Thousand Oaks, CA: Sage.

Herod, A. (1997b) Labor as an agent of globalization and as a global agent. In K. Cox (ed.) *Spaces of Globalization: Reasserting the Power of the Local*. New York: Guilford.

Herod, A. (1998a) Of blocs, flows and networks: the end of the Cold War, cyberspace, and the geo-economics of organized labor at the fin de millénaire. In A. Herod, G. Ó Tuathail, and S. Roberts (eds.) *An Unruly World? Globalization, Governance and Geography*. London: Routledge.

Herod, A. (1998b) Theorizing unions in transition. In J. Pickles and A. Smith (eds.) *Theorizing Transition: The Political Economy of Change in Central and Eastern Europe*. London: Routledge.

Herod, A. (2000) Workers and workplaces in a neoliberal global economy. *Environment and Planning* A, 32, 1781–90.

Herod, A. (2001) *Labor Geography: Workers and the Landscapes of Capitalism*. New York: Guilford.

Herod, A., Ó Tuathail, G., and Roberts, S. (eds.) (1998) *An Unruly World? Globalization, Governance and Geography*. London: Routledge.

Hershkovitz, L. (1993) Tiananmen Square and the politics of place. *Political Geography* 12, 395–420.

Hertner, P. (1986) German multinational enterprise before 1914: some case studies. In P. Hertner and G. Jones (eds) *Multinationals: Theory and History*. Brookfield, VT: Gower.

Hevly, B. and Findlay, J. (1998) *The Atomic West*. Seattle: University of Washington Press.

Hewison, R. (1987) *The Heritage Industry*. London: Methuen.

Hewitt, K. (1983) *Interpretations of Calamity: From the Viewpoint of Human Ecology*. Hemel Hempstead: Allen and Unwin.

Hildyard, N. and Wilks, A. (1998) An effective state? but effective for whom? *IDS*

Bulletin 29, April 2, 49–55.

Hillborn, R. (1990) Marine biota. In B. L. Turner II et al. (eds.) *The Earth as Transformed by Human Action*. Cambridge: Cambridge University Press.

Hirschmann, A. O. (1958) *The Strategy of Economic Development*. New Haven, CT: Yale University Press.

Hobsbawm, E. (1990) *Nations and Nationalism since 1780*. Cambridge: Cambridge University Press.

Hobsbawm, E. (1994) *The Age of Extremes: The Short Twentieth Century*. London: Weidenfeld and Nicolson.

Hobsbawm, E. and Ranger, T. (eds.) (1986) *The Invention of Tradition*. Cambridge: Cambridge University Press.

Hoffman, D. (1999a) Old satellites give Russia dangerous blind spots. *International Herald Tribune* February 11, 1, 10.

Hoffman, D. (1999b) When the nuclear alarms went off, he guessed right. *International Herald Tribune* February 11, 2.

Holdgate, M. W. (1979) *A Perspective of Environmental Pollution*. Cambridge: Cambridge University Press.

Holm, H.-H. and Sorensen, G. (eds.) (1995) *Whose World Order?* Boulder, CO: Westview Press.

Holmberg, J., Thornson, K., and Timberlake, L. (1993) *Facing the Future: Beyond the Earth Summit*. London: Earthscan/International Institute for Environment and Development.

Homer-Dixon, T. (1999) *Environment, Scarcity, and Violence*. Princeton, NJ: Princeton University Press.

Hönkopp, E. (1993) East–West migration: recent developments concerning Germany and some future prospects. In *The Changing Course of International Migration*, OECD. Paris: OECD.

Hooks, G. (1991) *Forging the Military Industrial Complex*. Ithaca, NY: Cornell University Press.

Hooks, G. et al. (forthcoming) *The Prison Industry: Carceral Expansion in Employment in US Counties, 1969–1994*.

Hornung, M. (1993) *Critical Loads: Concept and Applications*. London: HMSO.

Horsman, R. (1981) *Race and Manifest Destiny: The Origins of American Racial Anglo-Saxonism*. Cambridge, MA: Harvard University Press.

Houghton, J. (1997) *Global Warming: The Definitive Guide*, 2nd edn. Cambridge: Cambridge University Press.

Houghton, J., Jenkins, G. J., and Ephramus, J. J. (1990) *Climate Change: The IPCC Scientific Assessment*. Cambridge: Cambridge University Press.

Howland, M. (1993) Technological change and the spatial restructuring of data entry and processing services. *Technological Forecasting and Social Change* 43, 185–96.

Huggett, R. J. (1993) *Modelling the Human Impact on Nature: Systems Analysis of Environmental Problems*. Oxford: Oxford University Press.

Huggins, M. K. (1998) *Political Policing*. Durham, NC: Duke University Press.

Hughes, R. (1993) *Culture of Complaint: The Fraying of America*. New York: Oxford University Press.

Hugill, P. J. (1999) *Global Communications Since 1844*. Baltimore, MD: Johns Hopkins University Press.

Huling, T. L. (1999) *Yes in My Backyard*. Documentary. Galloping Girls/WSKG

Production.

Huling, T. L. (2000a) Prisoners of the census. *MoJo Wire*. www.motherjones.com

Huling, T. L. (2000b) *Prisons as a Growth Industry in Rural America: An Exploratory Discussion of the Effects on Young African-American Males in the Inner-Cities*. Washington, DC: US Commission on Civil Rights.

Hung, W. (1991) Tiananmen Square: a political history of monuments. *Representations* 35, 84–117.

Huntington, E. (1924) *The Character of Races*. New York: Scribner's.

Huntington, S. (1998) *The Clash of Civilizations*. New York: Touchstone.

Hyams, M. (2000) "Pay attention in class . . . [and] don't get pregnant": a discourse of academic success among adolescent Latinas. *Environment and Planning* A, 32, 635–54.

Hymer, S. (1972) The multinational corporation and the law of uneven development. In J. Bhagwati (ed.) *Economics and World Order*. New York: Free Press.

Hymer, S. (1976) *The International Operations of National Firms: A Study of Direct Foreign Investment*. Cambridge, MA: MIT Press.

Hymer, S. (1979) *The Multinational Corporation: A Radical Approach*. New York: Cambridge University Press.

Ignatieff, M. (1993) The Balkan tragedy. *New York Review of Books* May 13, 3.

Innis, H. A. (1967) The importance of staple products. In W. T. Easterbrook and M. H. Watkins (eds.) *Approaches to Canadian Economic History*. Toronto: McClelland and Stewart.

Institute for Economics and Industrial Engineering (1993) *Report delivered by the IEIE, Siberian Branch of the Russian Academy of Sciences*, Novosibirsk, at Clark University, Worcester, MA, on December 16, 1993.

Institute of Medicine (1992) *Emerging Infections*. Washington, DC: National Academy Press.

Intergovernmental Panel on Climate Change (IPCC) (1996) *Climate change 1995*, vol. 2. Cambridge: Cambridge University Press.

Intergovernmental Panel on Climate Change (IPCC) (2001) Climate change 2001: the scientific bans. www.meto.govt.uk/sec5/CR_div/ipcc/wgl/WGI-SPM.pdf

International Institute of Strategic Studies (2000) *The Military Balance 2000/2001*. Oxford: Oxford University Press.

Isard, W. (1956) *Location and Space-Economy*. Cambridge, MA: MIT/Wiley.

Itoh, M. (1992) The Japanese model of post-Fordism. In M. Storper and A. J. Scott (eds.) *Pathways to Industrialization and Regional Development*. London: Routledge.

IUCN (1980) *The World Conservation Strategy*. Geneva: International Union for Conservation of Nature and Natural Resources, United Nations Environment Program, World Wildlife Fund.

IUCN (1991) *Caring for the Earth: A Strategy for Sustainable Living*. Gland: International Union for Conversation of Nature and Natural Resources.

Iyer, P. (2000) *The Global Soul: Jet Lag, Shopping Malls, and the Search for Home*. New York: Knopf.

Jackson, P. (1988) Street life: the politics of Carnival. *Society and Space* 6, 213–27.

Jackson, P. (1993) Towards a cultural politics of consumption. In J. Bird, B. Curtis, T. Putnam, G. Roberston, and L. Tickner (eds.) *Mapping the Futures*. London: Routledge.

Jackson, P. (1999) Commodity cultures: the traffic in things. *Transactions, Insti-*

tute of British Geographers 24, 95–108.

Jackson, P. and Penrose, J. (eds.) (1993) *Constructions of "Race," Place, and Nation*. Minneapolis: University of Minnesota Press.

Jackson, P. and Taylor, J. (1996) Geography and the cultural politics of advertising. *Progress in Human Geography* 20, 356–71.

Jackson, P. and Thrift, N. (1995) Geographies of consumption. In D. Miller (ed.) *Acknowledging Consumption*. London: Routledge.

Jacobs, M. (1995) Sustainable development, capital substitution and economic humility: a response to Beckerman. *Environmental Values* 4, 57–68.

Jäger, J. and O'Riordan, T. (1996) The history of climate change science and politics. In T. O'Riordan, and J. Jäger (eds.) *Politics of Climate Change: A European Perspective*. London: Routledge.

James, C. L. R. (1980) *Fighting Racism in World War II*. New York: Pathfinder.

Jameson, F. (1991) *Postmodernism, or, the Cultural Logic of Late Capitalism*. London: Verso.

Janelle, D. and Hodge, D. (2000) *Information, Place, and Cyberspace*. Berlin: Springer-Verlag.

Jarman, N. (1992) Troubled images. *Critique of Anthropology* 12, 179–91.

Jess, P. and Massey, D. (1995) The conceptualization of place. In D. Massey and P. Jess (eds.) *A Place in the World? Places, Cultures and Globalization*. Oxford: Oxford University Press.

Jessop, B. (1989) Conservative regimes and the transition to post-Fordism: the cases of Great Britain and West Germany. In M. Gottdiener and N. Komninos (eds.) *Capitalist Development and Crisis Theory: Accumulation, Regulation and Restructuring*. New York: St. Martin's Press.

Jessop, B. (1990) Regulation theories in retrospect and prospect. *Economy and Society* 19, 2, 153–216.

Jessop, B. (1992) Fordism and post-Fordism: a critical reformulation. In M. Storper and A. J. Scott (eds.) *Pathways to Industrialization and Regional Development*. London: Routledge.

Joekes, S. (1987) *Women in the World Economy: An INSTRAW Study*. New York: Oxford University Press.

Johns, R. (1998) Bridging the gap between class and space: US worker solidarity with Guatemala. *Economic Geography* 74, 3, 252–71.

Johnson, N. C. (1992) Nation-building, language and education: the geography of teacher recruitment in Ireland, 1925–55. *Political Geography Quarterly* 11, 170–89.

Johnson, N. C. (1994) Sculpting heroic histories: celebrating the centenary of the 1798 rebellion in Ireland. *Transactions of the Institute of British Geographers* N.S. 19, 78–93.

Johnston, R. J. (1983) Resource analysis, resource management and the integration of physical and human geography. *Progress in Physical Geography* 7, 127–46.

Johnston, R. J. (1989a) The individual and the world economy. In R. J. Johnston and P. J. Taylor (eds.) *The World in Crisis*. Oxford: Blackwell.

Johnston, R. J. (1989b) *Environmental Problems: Nature, Society and the State*. London: Belhaven.

Johnston, R. J. (1992) Laws, states and superstates. *Applied Geography* 12, 211–20.

Johnston, R. J. (1993) The rise and decline of the corporate welfare state: a comparative analysis in global context. In P. Taylor (ed.) *Political Geography of the Twentieth Century*. London: Belhaven.

Johnston, R. J. (1996) *Nature, State and Economy: A Political Economy of the Environment*. Chichester: Wiley.

Johnston, R. J. (1999a) Political spaces and representation within the state. In M. Pacione (ed.) *Applied Geography: Principles and Practice*. London: Routledge.

Johnston, R. J. (1999b) The United States, the triumph of democracy, and the end of history. In D. Slater and P. Taylor (eds.) *The American Century*. Oxford: Blackwell.

Johnston, R. J. and Taylor, P. J. (eds) (1986) *World in Crisis?* Oxford: Blackwell.

Johnston, R. J., Taylor, P. J., and O'Loughlin, J. (1987) The geography of violence and premature death: a world-systems approach. In R. Vayrynen (ed.) *The Quest for Peace*. London: Sage.

Johnston, R. J., Taylor, P. J., and Watts, M. J. (eds.) (1995) *Geographies of Global Change: Remapping the World in the Late Twentieth Century*. Oxford: Blackwell.

Jones, D. K. C. (1991) Human occupancy and the physical environment. In R. J. Johnston and V. Gardiner (eds.) *The Changing Geography of the United Kingdom*. London: Routledge.

Jones, J. (1992) *The Dispossessed*. New York: Basic Books.

Jones, J. P., III, Nast, H., and Roberts, S. (eds.) (1997) *Thresholds in Feminist Geography*. New York: Rowman and Littlefield.

Jones, S. G. (1995) *CyberSociety: Computer-Mediated Communication and Community*. Thousand Oaks, CA: Sage.

Kahn, H. and Cooper, C. L. (1993) *Stress in the Dealing Room: High Performers Under Pressure*. London: Routledge.

Kain, K. C. et al. (2001) Malaria deaths in visitors to Canada and in Canadian travellers. *Canadian Medical Association Journal* 164, 654–9.

Kandiyoti, D. (1991) Identity and its discontents: women and nation. *Millennium: Journal of International Studies* 20, 429–43.

Kane, E. J. (1990) Incentive conflict in the international risk-based capital agreement. *Economic Perspectives* (Federal Reserve Bank of Chicago) 14, 3, 33–6.

Kaplan, D., Schiraldi, V., and Ziedenberg, J. (2000) *Texas Tough: An Analysis of Incarceration and Crime Trends in the Lone Star State*. Washington, DC: Justice Policy Institute.

Kaplan, R. (1994) The coming anarchy. *Atlantic Monthly* February, 44–86.

Kapstein, E. B. (1992) Between power and purpose: central bankers and the politics of regulatory convergence. *International Organization* 46, 1, 265–87.

Kapur, J. (1999) Out of control: television and the transformation of childhood in late capitalism. In M. Kinder (ed.) *Kids' Media Culture*. Durham, NC: Duke University Press.

Karl, T. R. (1993) Missing pieces of the puzzle. *Research and Exploration* 9, 234–49.

Kasperson, J. X., Kasperson, R. E., and Turner, B. L., II (eds.) (1993) *Regions at Risk: International Comparisons of Threatened Environments*. Tokyo: United Nations University.

Kates, R. W. (1987) The human environment: the road not taken and the road still beckoning. *Annals of the Association of American Geographers* 77, 4, 525–34.

Kates, R. W., Turner, B. L., II, and Clark, W. C. (1990) The great transformation. In B. L. Turner II et al. (eds.) *The Earth as Transformed by Human Action*.

Cambridge: Cambridge University Press.

Kates, R. W., Clark, W. C., Corell, R., Hall, J. M., Jaeger, C. C., Lowe, I., McCarthy, J. J., Schellenhuber, H. J., Bolin, B., Dickson, N. M., Faucheaux, S., Gallopin, G. C., Grübler, A., Huntley, B., Jäger, J., Jodga, N. S., Kasperson, R. E., Mabogunje, A., Matson, P., Mooney, H., Moore III, B., O'Riordan, T., and Svedin, U. (2001) Sustainability science. *Science* 292, 641–2.

Katz, C. (1991) Sow what you know: the struggle for social reproduction in rural Sudan. *Annals of the Association of American Geographers* 81, 3, 488–514.

Katz, C. (1998a) Disintegrating developments: global economic restructuring and the eroding ecologies of youth. In T. Skelton and G. Valentine (eds.) *Cool Places: Geographies of Youth Cultures.* London: Routledge.

Katz, C. (1998b) Whose nature, whose culture? Private productions of space and the preservation of nature. In B. Braun and N. Castree (eds.) *Remaking Reality: Nature at the End of the Millennium.* London: Routledge.

Katz, C. (2001) Vagabond capitalism and the necessity of social reproduction. *Antipode* 33, 708–27.

Katz, C. (forthcoming) *Disintegrating Developments: Global Economic Restructuring and Children's Everyday Lives.*

Katz, C. and Kirby, A. (1991) In the nature of the thing: the environment and everyday life. *Transactions, Institute of British Geographers* N.S. 16, 259–71.

Kay, C. (1989) *Latin American Theories of Development and Underdevelopment.* London: Routledge.

Keck, M. E. and Sikkink, K. (1998) *Activists Beyond Borders.* Ithaca, NY: Cornell University Press.

Kennedy, P. (1993) *Preparing for the Twenty-First Century.* New York: Random House.

Keyfitz, N. (1991) Population and development within the ecosphere: one view of the literature. *Population Index* 57, 5–22.

Keysshar, A. (2000) *The Right to Vote: The Contested History of Democracy in the United States.* New York: Basic Books.

Kidron, M. and Segal, R. (1984) *The New State of the World Atlas.* New York: Simon and Schuster.

King, A. D. (1990) *Global Cities: Post-Imperialism and the Internationalization of London.* London: Routledge.

King, A. D. (ed.) (1997) *Culture, Globalization and the World-System.* Basingstoke: Macmillan.

Kitchin, R. (1998) *Cyberspace: The World in the Wires.* Chichester: John Wiley and Sons.

Kitson, M. and Michie, J. (1995) Trade and growth: a historical perspective. In J. Michie and J. Grieve Smith (eds.) *Managing the Global Economy.* Oxford: Oxford University Press.

Klak, T. (ed.) (1998) *Globalization and Neoliberalism: The Caribbean Context.* Lanham, MD: Rowman and Littlefield.

Knorr-Cetina, K. (2001) Financial markets, screens and object relations. *Theory, Culture and Society* 18 (forthcoming).

Kobayashi, A. and Peake, L. (1994) Unnatural discourse, "Race" and gender in geography. *Gender, Place, and Culture* 1 (2): 225–43.

Konadu-Agyemang, K. (2000) The best of times and the worst of times: structural adjustment programs and uneven development in Africa: the case of Ghana. *Pro-*

fessional Geographer 52, 469–83.

Konrad Adenauer Foundation (ICAF) (2000) HIV/AIDS: a threat to the African renaissance? Bonn: KAF.

Kotlyakov, V. M. (1991) The Aral Sea Basin: a critical environmental zone. *Environment* 33, 1, 4–9, 36–8.

Kuhn, Thomas (1996) *The Structure of Scientific Revolutions*. Chicago: University of Chicago Press.

Kuletz, V. (1998) *The Tainted Desert: Environment and Social Ruin in the American West*. New York: Routledge.

Laclau, E. (1985) New social movements and the plurality of the social. In D. Slater (ed.) *New Social Movements and the State in Latin America*. Amsterdam: CEDLA.

Lake, A. (2000) *Six Nightmares: Real Threats in a Dangerous World and How America Can Meet Them*. Boston: Little Brown.

Lam, Lawrence (1993) Focus on Southeast Asian refugees. *Refuge* 13, 5, 1.

Landry, D. and Maclean, G. (1993) *Materialist Feminisms*. Oxford: Blackwell.

Laponce, J. A. (1984) The French language in Canada: tensions between geography and politics. *Political Geography Quarterly* 3, 91–105.

Lappé, F. M., Collins, J., and Rosset, P. (1998) *World Hunger: 12 Myths*. London: Earthscan.

Last, J. (1998) *Implications of Climate Change for Human Health in Canada*. Ottawa.

Latouche, S. (1993) *In the Wake of the Affluent Society: An Exploration of Post-Development*. London: Zed Books.

Latour, B. (1993) *We Have Never Been Modern*. Brighton: Harvester Wheatsheaf.

Laub, J. (1983) Patterns of offending in urban and rural areas. *Journal of Criminal Justice* 11: 129–42.

Law, J. (1994) *Organizing Modernity*. Oxford: Blackwell.

Lawrence, S. and Giles, C. L. (1999) Accessibility of information on the web. *Nature* 400, 107–9.

Lawson, V. (1999) Questions of migration and belonging: understandings of migration under neoliberalism. *Ecuador International Journal of Population Geography* 5, 1–16.

Lawson, V. and Klak, T. (1993) An argument for critical and comparative research on the urban geography of the Americas. *Environment and Planning* A, 25, 8, 1071–84.

Le Heron (1988) Food and fibre production under capitalism: a conceptual agenda. *Progress in Human Geography* 12, 3, 409–30.

Leach, M. and Fairhead, J. (2000) Challenging neo-Malthusian deforestation analyses in West Africa's dynamic forest landscapes. *Population and Development Review* 26, 17–43.

Leduc, J. W. et al. (1993) Hantaan (Korean hemorhagic fever) and related rodent zoonoses. In S. S. Morse (ed.) *Emerging Viruses*. New York: Oxford University Press.

Lee, E. (1997) *The Labour Movement and the Internet: The New Internationalism*. London: Pluto Press.

Lee, K. and Walt, G. (1995) Linking national and global population agendas: case studies from eight developing countries. i 16, 257–72

Lefebvre, H. (1991) *The Production of Space*. Oxford: Blackwell.

LéLé, S. M. (1991) Sustainable development: a critical review. *World Development*

19, 607–21.

Leslie, D. and Reimer, S. (1999) Spatializing commodity chains. *Progress in Human Geography* 23, 401–20.

Leslie, D. A. (1995) Global scan: The globalization of advertising agencies. *Economic Geography* 71, 402–26.

Leslie, J. (2000) Running dry: what happens when the world no longer has enough freshwater? *Harpers Magazine* 301(1802), July, 37–52.

Levins, R. et al. (1994) The emergence of new diseases. *American Scientist* 82, 1, 52–60.

Lewis, B. (1982) *The Muslim Discovery of Europe*. New York: W. W. Norton.

Lewis, M. (1992) *Green Delusions: An Environmentalist Critique of Radical Environmentalism*. Durham, NC: Duke University Press.

Lewontin, R. C., Rose, S., and Kamin, L. J. (1984) *Not in Our Genes*. New York: Pantheon.

Ley, D. and Olds, K. (1988) Landscape as spectacle: world's fairs and the culture of heroic consumption. *Environment and Planning* D: *Society and Space* 6, 191–212.

Lichterman, J. D. (1999) Disasters to Come. *Futures* 31, 593–607.

Linebaugh, P. and Rediker M. (2000) *The Many-Headed Hydra*. Boston: Beacon Press.

Lipietz, A. (1986) New tendencies in the international division of labor: regimes of accumulation and modes of accumulation. In A. J. Scott and M. Storper (eds.) *Production, Work, Territory: The Geographical Anatomy of Industrial Capitalism*. Boston: Allen and Unwin.

Lipietz, A. (1987) *Mirages and Miracles: The Crises of Global Fordism*. London: Verso.

Lipietz, A. (1992a) The regulation approach and capitalist crisis: an alternative compromise for the 1990s. In M. Dunford and G. Kafkalas (eds.) *Cities and Regions in the New Europe*. London: Belhaven.

Lipietz, A. (1992b) *Towards a New Economic Order: Postfordism, Ecology and Democracy*. Cambridge: Polity Press.

Lipietz, A. (1996) Social Europe: the post-Maastricht challenge. *Review of International Political Economy* 3, 369–80.

Lipschutz, R. D. (1996) *Global Civil Society and Global Environmental Governance: The Politics of Nature from Place to Planet*. Albany: State University of New York Press.

Litfin, K. (1994) *Ozone Discourses: Science and Politics in Global Environmental Cooperation*. New York: Columbia University Press.

Liu, L. Y. (2000) The place of immigration in studies of geography and race. *Social and Cultural Geography* 1, 2, 169–82.

Liverman, D. M. (1990) Vulnerability to global environmental change. In R. E. Kasperson, K. Dow, D. Golding, and J. X. Kasperson (eds.) *Understanding Global Environmental Change: The Contributions of Risk Analysis and Management*. Worcester, MA: ET Program.

Livingstone, D. N. (1992) *The Geographical Tradition*. Oxford: Blackwell.

Loader, B. (ed.) (1997) *The Governance of Cyberspace*. London: Routledge.

Loftus, B. (1990) *Mirrors: William III and Mother Ireland*. Dundrum: Picture Press.

Love, J. L. (1980) Raul Prebisch and the origins of the doctrine of unequal exchange. *Latin American Research Review* 15, 3, 45–72.

Lovelock, J. E. (1979) *Gaia: A New Look at Life on Earth*. Oxford: Oxford University Press.

Lovelock, J. E. (1988) *The Ages of Gaia*. Oxford: Oxford University Press.

Low, N. and Gleeson, B. (1998) *Justice, Society and Nature: An Exploration of Political Ecology*. London: Routledge.

Lowenthal, D. (1958) *George Perkins Marsh: Versatile Vermonter*. New York: Columbia University Press.

Lowenthal, D. (1990) Awareness of human impacts: changing attitudes and emphases. In B. L. Turner II et al. (eds.) *The Earth as Transformed by Human Action*. Cambridge: Cambridge University Press.

Lowenthal, D. (1991) British national identity and the English landscape. *Rural History* 2, 205–30.

Lowenthal, D. (2000) *George Perkins Marsh: Prophet of Conversation*. Seattle: University of Washington Press.

Lubeck, P. (1999) The antinomies of the Islamic revival. In R. Cohen and S. Rai (eds.) *Global Social Movements*. London: Routledge.

Lubeck, P. and Britts, L. (2001) Muslim civil society in urban public spaces. In J. Eade and C. Mele (eds.) *Urban Studies*. Oxford: Blackwell.

Lukes, T. (1999) *Capitalism, Democracy, and Ecology: Departing from Marx*. Urbana: University of Illinois Press.

Lutes, Mark W. (1998) Global climatic change. In R. Keil, D. V. J. Bell, P. Penz, and L. Fawcett (eds.) *Political Ecology: Global and Local*. London: Routledge.

Lyotard, J.-F. (1986) *The Postmodern Condition*. Manchester: University of Manchester Press.

Mabogunje, A. L., Matson, P., Mooney, H., Moore III, B., O'Riordan, T., and Svedin, U. (2001) Sustainability science. *Science* 292, 641–2.

McCabe, P. (1998) Energy resources: cornucopia or empty barrel? *AAPG Bulletin* 5, 82, 2110–34.

McCally, M. and Cassel, C. K. (1990) Medical responsibility and global environmental change. *Annals of Internal Medicine* 113, 467–73.

McClintock, A. (1995) *Imperial Leather: Race, Gender and Sexuality in the Colonial Contest*. London, Routledge.

McCloskey, J. M. and Spalding, H. (1989) A reconnaissance-level inventory of the amount of wilderness remaining in the world. *Ambio* 18, 4, 221–7.

McCormick, J. S. (1989) *Reclaiming Paradise: The Global Environmental Movement*. Bloomington: Indiana University Press.

McCormick, J. S. (1995) *The Global Environmental Movement*. Chichester: John Wiley.

McCrone, D. (1992) *Understanding Scotland: The Sociology of a Stateless Nation*. London: Routledge.

McDowell, L. (1991) The baby and the bathwater: deconstruction, diversity and feminist theory in geography. *Geoforum* 22, 123–34.

McDowell, L. (1994) Polyphony or commodified cacophony: making sense of other worlds and pedagogic authority. *Area* 26, 241–8.

McDowell, L. (1999) *Gender, Identity and Place: Understanding Feminist Geographies*. Cambridge: Polity Press.

McDowell, L. (2000) Acts of memory and millennial hopes and anxieties: the awkward relationship between the economic and the cultural. *Society and Culture* 1.

Mace, R. (1976) *Trafalgar Square: Emblem of Empire*. London: Lawrence and

Wishart.

McGirr, L. (2001) *Suburban Warriors*. Princeton, NJ: Princeton University Press.

McGranaham, G. (1993) Household environmental problems in low-income cities. *Habitat International* 17, 105–21.

McGrew, A. (1992) A global society? In S. Hall, D. Held, and T. McGrew (eds.) *Modernity and its Futures*. Cambridge: Polity Press.

Machimura, T. (1992) The urban restructuring process in Tokyo in the 1980s: transforming Tokyo into a world city. *International Journal of Urban and Regional Research* 16, 114–28.

Mack, P. (1990) *Viewing the Earth: The Social Construction of the Landsat Satellite*. Cambridge, MA: MIT Press.

Mackinder, H. J. (1904) The geographical pivot of history. *Geographical Journal* 23, 421–42.

Mackinder, H. J. (1919) *Democratic Ideals and Reality*. London: Constable.

McLuhan, M. (1962) *The Gutenberg Galaxy*. Toronto: University of Toronto Press.

McMichael, P. and Myhre, D. (1991) Global regulation versus the nation-state: agro-food systems and the new politics of capital. *Capital and Class* 43, 83–105.

Mcnaughten, P. and Urry, J. (1998) *Contested Natures*. London: Sage.

McNeill, W. W. (1993) Patterns of disease emergence in history. In S. S. Morse (ed.) *Emerging Viruses*. New York: Oxford University Press.

MacShane, D. (1996) *Global Business: Global Rights*. London: Fabian Society (Pamphlet 575).

McWilliams, C. (1939) [1999] *Factories in the Field*. Berkeley: University of California Press.

Maddison, A. (1995) *Monitoring the World Economy, 1820–1992*. Paris: OECD.

Mann, M. (1988) *States, War, and Capitalism*. Oxford: Blackwell.

Manne, A. S. and Richels, R. G. (1992) *Buying Greenhouse Insurance: The Economic Costs of CO_2 Emission Limits*. Cambridge, MA: MIT Press.

Mansell, R. (1993) *The New Telecommunications: A Political Economy of Network Evolution*. London: Sage.

Marchand, M. H. and Parpart, J. L. (eds.) (1995) *Feminism/Postmodernism/Development*. London: Routledge.

Marglin, F. A. and Marglin, S. A. (eds.) (1990) *Dominating Knowledge: Development, Culture and Resistance*. Oxford: Clarendon Press.

Markusen, A. et al. (1991) *The Rise of the Gunbelt*. New York: Oxford University Press.

Marsden T., Munton, R., Whatmore, S., and Little, J. (1986) Towards a political economy of capitalist agriculture. *International Journal of Urban and Regional Research* 4, 498–521.

Marshall, B. (ed.) (1991) *The Real World*. Boston: Houghton Mifflin.

Martin, B. (1988) Feminism, criticism and Foucault. In I. Diamond and L. Quinby (eds.) *Feminism and Foucault: Reflections on Resistance*. Boston: Northeastern University Press.

Martin, R. and Sunley, P. (2000) Rethinking the economic in economic geography: broadening our vision or losing our focus? *Antipode* 32.

Masquelier, A. (1992) Encounter with a road siren. *Visual Anthropology* 8, 1, 56–69.

Massey, D. (1984) *Spatial Divisions of Labour*. London: Macmillan; New York: Methuen.

Massey, D. (1991) A global sense of place. *Marxism Today* June, 24–9.

Massey, D. (1994) *Space, Place and Gender.* Cambridge: Polity Press.

Massey, D. S. and Denton, N. A. (1993) *American Apartheid.* Cambridge, MA: Harvard University Press.

Mattelart, A. (1979) *Multinational Corporations and the Control of Culture.* Brighton: Harvester.

Meadows, D. H., Meadows, D. L., and Randers, J. (1992) *Beyond the Limits: Global Collapse or Sustainable Future?* London: Earthscan.

Meadows, D. H., Meadows, D. L., Randers, J., and Behrens, W. (1972) *Limits to Growth.* New York: Universe Books.

Melman, S. (1974) *The Permanent War Economy.* New York: Simon and Schuster.

Melucci, A. (1989) *Nomads of the Present.* London: Radius.

Melucci, A. (1996) *Challenging Codes.* Cambridge: Cambridge University Press.

Meyer, W. B. (2000) *Americans and Their Weather.* Oxford: Oxford University Press.

Meyer, W. B. and Turner, B. L., II (1992) Human population growth and global land-use/land-cover change. *Annual Review of Ecology and Systematics* 23, 39–61.

Meyers, N. (1994) Pre-debate statement. In N. Meyers and J. L. Simon (eds.) *Scarcity or Abundance? A Debate on the Environment.* New York: W. W. Norton.

Micklin, P. (1988) Desiccation of the Aral Sea: a water management disaster in the Soviet Union. *Science* 241, 1170–6.

Miller, D. (1987) *Material Culture and Mass Consumption.* Oxford: Blackwell.

Miller, D., Jackson, P., Thrift, N., Holbrook, B., and Rowlands, M. (1998) *Shopping, Place and Identity.* London: Routledge.

Mills, G. (2000) *AIDS and the South African Military.* In KAF op-cit.

Mingione, E. (ed.) (1993) The new urban poverty and the underclass. Special issue of *International Journal of Urban and Regional Research* 17, 3.

Mink, G. (1995) *The Wages of Motherhood.* Ithaca, NY: Cornell University Press.

Mintz, S. W. (1960) *Worker in the Cane: A Puerto Rican Life History.* New Haven, Yale University Press.

Mintz, S. W. (1985) *Sweetness and Power: The Place of Sugar in Modern History.* New York: Penguin Books.

Mirza, H. S. (ed.) (1997) *Black British Feminism: A Reader.* London: Routledge.

Mitchell, B. (1979) *Geography and Resource Analysis.* London: Longman.

Mitchell, B. R. (1975) *European Historical Statistics 1750–1970.* London: Macmillan.

Mitchell, B. R. (1983) *International Historical Statistics: The Americas and Australasia.* London: Macmillan.

Mitchell, B. R. (1998a) *International Historical Statistics 1750–1993: Europe.* London: Macmillan.

Mitchell, B. R. (1998b) *International Historical Statistics 1750–1993: The Americas.* London: Macmillan.

Mitchell, D. (2000) *Cultural Geography.* Oxford: Blackwell.

Mitchell, K. (1997) Transnational discourse: bringing geography back. *Antipode* 29, 101–14.

Mitchell, T. (2001) *Out of Egypt.* Berkeley: University of California Press.

Mitchell, W. J. (1996) *City of Bits: Space, Place and the Infobahn.* Cambridge, MA: MIT Press.

Momsen, J. and Kinnaird, V. (eds.) (1992) *Different Places, Different Voices*. London: Routledge.

Monro, J. M. (1993) World population forecasts. *Nature* 363, 215–16.

Moody, R. (ed.) (1988) *The Indigenous Voice*, 2 vols. London: Zed Books.

Moore, W. D. (1996) *Constitutional Rights and Powers of the People*. Princeton, NJ: Princeton University Press.

Morehouse, W. (1994) Unfinished business: Bhopal ten years after. *The Ecologist* 24, 5, 164–8.

Morley, D. (1992) *Television, Audiences and Cultural Studies*. London: Routledge.

Morris, M. (1988) The pirate's fiancee: feminists and philosophers, or maybe tonight it'll happen. In I. Diamond and L. Quinby (eds.) *Feminism and Foucault: Reflections on Resistance*. Boston: Northeastern University Press.

Morris, M. (1992) The man in the mirror: David Harvey's "condition" of postmodernity. *Theory Culture and Society* 10, 253–79.

Morse S. S. (1995) Factors in the emergence of infectious diseases. *Emerging Infectious Diseases* 1, 7–15.

Mosse, G. L. (1975) *The Nationalization of the Masses*. New York: Howard Fertig.

Moulaert, F. and Djellal, F. (1995) Information technology consultancy firms: economies of agglomeration from a wide-area perspective. *Urban Studies* 32, 105–22.

Mountjoy, A. B. (1963) *Industrialization and Underdeveloped Countries*. London: Hutchinson.

Mudimbe, V. Y. (1988) *The Invention of Africa*. Bloomington: Indiana University Press.

Mulgan, G. J. (1991) *Communication and Control: Networks and the New Economies of Communication*. Cambridge: Polity Press.

Munasinghe, M. (1993) Environmental issues and economic decisions in developing countries. *World Development* 21, 1729–48.

Munton, R. (1992) The uneven development of capitalist agriculture: the repositioning of agriculture within the food system. In K. Hoggart (ed.) *Agricultural Change, Environment and Economy: Essays in Honour of W. B. Morgan*. London: Mansell.

Murdoch, J., Marsden, T., and Banks, J. (2000) Quality, nature, embeddedness: some theoretical considerations in the context of the food sector. *Economic Geography* 76, 107–25.

Murley, L. and Stevens, M. (eds.) (1991) *NSCA Pollution Glossary*. Brighton: National Society for Clean Air and Environmental Protection.

Murray, C. J. L. and Lopez, A. D. (1996) *The Global Burden of Disease*. Cambridge, MA: Harvard School of Public Health.

Naiman, R. and Watkins, N. (1999) *A Survey of the Impacts of IMF Structural Adjustment in Africa*. Washington, DC: Center of Economic and Policy Research.

Nairn, T. (1993) All Bosnians now? *Dissent* fall, 403–10.

Nandy, A. (1984) Culture, state and rediscovery of Indian politics. *Economic and Political Weekly* 19, 49, 2078–83.

Nasaw, D. (1985) *Children of the City*. Garden City, NY: Anchor Press/Doubleday.

Nash, C. (2000) Historical geographies of modernity. In B. Graham and C. Nash (eds.) *Modern Historical Geographies*. Harlow: Longman.

National Research Council (NRC) (1999) *Our Common Journey: A Transition toward Sustainability*. Washington, DC: National Academy Press.

Nederveen Pieterse, J. (2000) After post-development. *Third World Quarterly* 21,

2, 175–91.

Negri, T. (1988) *Revolution Retrieved*. London: Red Notes.

Newson, M. D. (1992) *Land, Water and Development: River Basin Systems and their Sustainable Management*. London: Routledge.

Newson, M. D. (1994) *Scales and their Appropriateness for Integrating Land-Use Management: Science, Subsidiarity and Sustainability*. Aberdeen: Macauley Land Use Research Institute.

Nielsen, J. (2000) *Designing Web Usability*. Indianapolis: New Riders Publishing.

Niess, F. (1990) *A Hemisphere to Itself: A History of US–Latin American Relations*. London: Zed Books.

Nissani, M. (1999) Media coverage of the greenhouse effect. *Population and Environment* 21, 1, 27–43.

Noble, D. (1999) *The Religion of Technology*. New York: Penguin Books.

Nohrstedt, A. (1993) Communication in the risk-society: public relations strategies, the media and nuclear power. In A. Hansen (ed.) *The Mass Media and Environmental Issues*. Leicester: Leicester University Press.

Nordhaus, W. D. (1993) Reflections on the economics of climate change. *Journal of Economic Perspectives* 7, 11–25.

Nriagu, J. O. and Lakshminarayana, J. S. S. (1989) *Aquatic Toxicology and Water Quality Management*. New York: John Wiley and Sons.

Nua (2000) How many online? http://www.nua.ie/surveys/how_many_online/index.html

Nunn, S. and Stulberg, A. (2000) The many faces of modern Russia. *Foreign Affairs* 79, 2, 45–62.

Ó Tuathail, G. (1996) *Critical Geopolitics*. Minneapolis: University of Minnesota Press.

Ó Tuathail, G., Dalby, S., and Routledge, P. (1998) *The Geopolitics Reader*. London: Routledge.

O'Brien T. F. (1999) *The Revolutionary Mission: American Enterprise in Latin America, 1900–1945*. Cambridge: Cambridge University Press.

O'Connor, J. (1973) *The Fiscal Crisis of the State*. New York: St. Martin's Press.

Odom, W. (1998) *The Collapse of the Soviet Military*. New Haven, CT: Yale University Press.

OECD (1985) *The State of the Environment*. Paris: OECD.

OECD (1992a) *Trends in International Migration*. Paris: OECD.

OECD (1992b) *Development Cooperation: 1992 Report*. Paris: OECD.

OECD (1993a) *The Changing Course of International Migration*. Paris: OECD.

OECD (1993b) *World Energy Outlook to the Year 2010*. Paris: IEA/OECD.

Offe, C. (1985) New social movements: challenging the boundaries of institutional politics. *Social Research* 52, 4, 817–68.

Ohmae, K. (1990) *The Borderless World: Power and Strategy in the Interlinked Economy*. London: Collins.

Olalquiaga, C. (1992) *Megalopolis: Contemporary Cultural Sensibilities*. Minneapolis: University of Minnesota Press.

Oliver, M. L. and Shapiro, T. M. (1995) *Black Wealth/White Wealth*. New York: Routledge.

Omara-Ojungu, P. H. (1992) *Resource Management in Developing Countries*. London: Longman.

Omi, M. and Winant, H. (1986) *Racial Formation in the United States*. New York:

Routledge.

Oncu, A. and Weyland, P. (1997) *Space, Culture, and Power: New Identities in Globalizing Cities*. Atlantic Heights, NJ: Zed Books.

Ong, A. (1987) *Spirits of Resistance and Capitalist Development: Factory Women in Malaysia*. Binghamton: State University of New York Press.

O'Riordan, T. (1971) *Perspectives on Resource Management*. London: Pion.

O'Riordan, T. (1981) *Environmentalism*, 2nd edn. London: Pion.

O'Riordan, T. (1988a) The politics of sustainability. In R. K. Turner (ed.) *Sustainable Environmental Management: Principles and Practice*. Boulder, CO: Westview Press.

O'Riordan, T. (1988b) The earth as transformed by human action: an international symposium. *Environment* 30, 1, 25–8.

O'Riordan, T. and Cameron, J. (eds.) (1994) *Interpreting the Precautionary Principle*. London: Earthscan.

O'Riordan, T. and Rayner, S. (1991) Risk management for global environmental change. *Global Environmental Change* 1, 2, 91–108.

O'Riordan, T. and Turner, R. K. (1983) *An Annotated Reader in Environmental Planning and Management*. Oxford: Pergamon Press.

Owens, S. E. and Cowell, R. (1994) Lost land and limits to growth: conceptual problems for sustainable land use change. *Land Use Policy* 11, 168–80.

Palafox, J. (2000) Opening up borderland studies. *Social Justice* 27, 3, 56–72.

Pangle, T. L. (1992) *The Ennobling of Democracy: The Challenge of the Postmodern Age*. Baltimore, MD: Johns Hopkins University Press.

Parikh, J., Babu, P. G., and Kavi Kumar, K. S. (1997) Climate change, North–South cooperation and collective decision-making post-Rio. *Journal of International Development* 9, 403–13.

Parnell, P. C. (1992) Time and irony in Manila squatter movements. In C. Nordstrom and J. Martin (eds.) *The Paths to Domination, Resistance and Terror*. Berkeley: University of California Press.

Pateman, C. (1989) *The Disorder of Women: Democracy, Feminism and Political Theory*. Stanford, CA: Stanford University Press.

Pateman, C. and Gross, E. (eds.) (1986) *Feminist Challenges*. Boston: Northeastern University Press.

Paterson, M. (2000) Car culture and global environmental politics. *Review of International Studies* 26, 2, 253–70.

Patz, J. (1998) Climate change and health. *Health and Environment Digest* 12, 7, 49–53.

Patz, J. et al. (2000) The potential health impacts of climate variability and change for the US. *Environmental Health Perspectives* 108, 367–76.

Paz, O. (1992) *In Search of the Present*. New York: Harper.

Pearce, D., Markandya, A., and Barbier, E. (1989) *Blurprint for a Green Economy*. London: Earthscan.

Pearce, D. W. and Turner, R. K. (1990) *Economics of Natural Resources and the Environment*. London: Harvester Wheatsheaf.

Pearl, R. and Gould, S. (1936) World population growth. *Human Biology* 8, 399–419.

Peck, J. (2001) *Workfare States*. New York: Guilford.

Peet, R. (1991) *Global Capitalism*. London: Routledge.

Peet, R. and Watts, M. (1993a) Development theory and environment in an age of market triumphalism. *Economic Geography* 69, 3, 227–53.

Peet, R. and Watts, M. (eds.) (1993b) Environment and development, Parts I and II. *Economic Geography* July and October, 69, 3 and 4.

Peet, R. and Watts, M. (eds.) (1996) *Liberation Ecologies*. London: Routledge.

Pepper, D. (1984) *The Roots of Modern Environmentalism*. London: Croom Helm.

Pepper, D. (1991) *Communes and the Green Vision*. London: Green Print.

Pepper, D. (1993) *Eco-Socialism*. London: Routledge.

Petrella, R. (1991) World city-states of the future. *New Perspectives Quarterly* 8, 59–64.

Pezzey, J. (1991) *Impacts of "Greenhouse" Gas Control Strategies on UK Competitiveness*. London: HMSO.

Phillips, A. (1997) From inequality to difference: a severe case of displacement. *New Left Review* 224, 143–53.

Phillips, A. (1999) *Which equalities matter?* Cambridge: Polity Press.

Philo, C. (1989) "Enough to drive one mad": the organization of space in nineteenth century lunatic asylums. In J. Wolch and M. Dear (eds.) *The Power of Geography*. London: Macmillan.

Pickles, J. and Smith, A. (eds.) (1998) *Theorizing Transition: The Political Economy of Post-Communist Transformations*. London: Routledge.

Pierson, C. (1991) *Beyond the Welfare State? A New Political Economy of Welfare*. University Park: Pennsylvania State University Press.

Pinteiro, F. P. and Corber, S. J. (1997) Global situation of dengue and dengue hemorrhagic fever and its emergence in the Americas. *World Health Statistical Quarterly* 50, 161–8.

Platanov, A. E. et al. (2001) Outbreak of west Nile virus infection, Volgograd region, Russia, 1999. *Emerging Infectious Diseases* 7, 128–32.

Polanyi, K. (1944) *The Great Transformation*. Boston: Beacon Press.

Pollard, J. (1996) Banking at the margins: a geography of financial exclusion in Los Angeles. *Environment and Planning* A, 28, 1209–32

Pollock, G. (1996) *Generations and Geographies in the Visual Arts: Feminist Readings*. London: Routledge.

Popke, E. J. (1994) Recasting geopolitics: the discursive scripting of the International Monetary Fund. *Political Geography* 13, 3, 255–69.

Population Reference Bureau (1994) *World Population Data Sheet*. Washington, DC: PRB.

Porter, R. (1993) Baudrillard: history, hysteria and consumption. In C. G. Rojek and B. S. Turner (eds.) *Forget Baudrillard?* London: Routledge.

Portney, K. E. (1991) Public environmental decision making: citizen roles. In R. A. Chechile and S. Carlisle (eds.) *Environmental Decision Making: A Multidisciplinary Perspective*. New York: Van Nostrand Reinhold.

Power, M. (1998) The dissemination of development. *Environment and Planning* D: *Society and Space*, 16, 577–98.

Pravda, A. and Ruble, B. A. (eds.) (1986) *Trade Unions in Communist States*. Boston: Allen and Unwin.

Pred, A. (2000) *Even in Sweden*. Berkeley: University of California Press.

Pred, A. and Watts, M. (1993) *Reworking Modernity*. New Brunswick, NJ: Rutgers University Press.

Prendergast, C. (1992) *Paris in the Nineteenth Century*. Oxford: Blackwell.

Pretes, M. (1997) Development and infinity. *World Development* 25, 9, 1421–30.

Price, J. (1945) *The International Labour Movement*. London: Oxford University Press.

Prison Focus (2001) *Turkey Prisoners Protest SHUs*. CPF 14.

Pritchard, B. (2000) The transnational corporate networks of breakfast cereals in Asia. *Environment and Planning* A 32, 789–804.

Pritchett, L. (1997) Divergence, big time. *Journal of Economic Perspectives* 11, 3, 3–17.

Przeworski, A. (1991) *Democracy and the Market: Political and Economic Reforms in Eastern Europe and Latin America*. Cambridge: Cambridge University Press.

Pulido, L. (2000) Rethinking environmental racism. *Annals of the AAG* 90, 1, 12–40.

Putnam, R. D. (1993a) *Making Democracy Work: Civic Traditions in Modern Italy*. Princeton, NJ: Princeton University Press.

Putnam, R. D. (1993b) The prosperous community: social capital and public life. *The American Prospect* 13, 35–42.

Putnam, R. D. (2000) *Bowling Alone*. New York: Simon and Schuster.

Rabinow, P. (ed.) (1984) *The Foucault Reader: An Introduction to Foucault's Thought*. London: Penguin Books.

Radcliffe, S. and Westwood, S. (1996) *Remaking the Nation: Place, Identity and Politics in Latin America*. London: Routledge.

Radcliffe, S. A. and Westwood, S. (eds.) (1993) *"Viva" Women and Popular Protest in Latin America*. London: Routledge.

Rafaeli, S. and Sudweeks, F. (1996) Networked interactivity. *Journal of Computer Mediated Communications* 2, 4. http://www.ascusc.org/jcmc/vol2/issue4/rafaeli.sudweeks.html

Rahnema, M. (ed.) (1999) *The Post Development Reader*. London: Zed Books.

Rahnema, M. and Bawtree, V. (eds.) (1997) *The Post-Development Reader*. London: Zed Books.

Rainwater, C. (1922) *The Play Movement in the United States*. Chicago: University of Chicago Press.

Redclift, M. (1987) *Sustainable Development: Exploring the Contradictions*. London: Methuen.

Redclift, M. (1996) *Wasted: Counting the Costs of Global Consumption*. London: Earthscan.

Redclift, M. (1997) Development and global environmental change. *Journal of International Development* 9, 391–401.

Redclift, M. (1999) Environmental security and competition for the environment. In S. C. Lonergan (ed.) *Environmental Change, Adaptation and Security*. Dordrecht: Kluwer.

Redclift, M. R. and Benton, T. (eds.) (1994) *Social Theory and the Global Environment*. London: Routledge.

Rees, J. (1990) *Natural Resources: Allocation, Economics, and Policy*. London: Routledge.

Reich, R. B. (1991) *The Work of Nations: Preparing Ourselves for 21st-Century Capitalism*. New York: Knopf.

Reid, W. and Miller, R. R. (1989) *Keeping Options Alive. The Scientific Basis for Conserving Biodiversity*. Washington, DC: World Resources Institute.

Relph, E. (1976) *Place and Placelessness*. London: Pion.

Rheingold, H. (1993) *The Virtual Community: Homesteading on the Electronic Frontier*. New York: Addison-Wesley.

Rich, B. (1994) *Mortgaging the Earth: The World Bank, Environmental Impoverishment and the Crisis of Development*. Boston: Beacon Press.

Rimmer, P. (1993) Reshaping western Pacific rim cities. In K. Fujita and R. C. Hill (eds.) *Japanese Cities in the World Economy*. Philadelphia: Temple University Press.

Robbins, P. (1999) Meat matters: cultural politics along the commodity chain in India. *Ecumene* 6, 399–423.

Roberts, N. (ed.) (1994) *The Changing Global Environment*. Oxford: Blackwell.

Roberts, S. (1997) Instrumental Globality. Paper presented at the Inaugural International Critical Conference on Geography, Vancouver, BC, August 9–13.

Robertson, R. (1990) Mapping the global condition. In M. Featherstone (ed.) *Global Culture: Nationalism, Globalization and Modernity*. Newbury Park, CA: Sage.

Robertson, R. (1991) Social theory, cultural relativity, and the problem of globality. In A. D. King (ed.) *Culture, Globalization and the World-System*. Basingstoke: Macmillan.

Robertson, R. (1992) *Globalization: Social Theory and Global Culture*. London: Sage.

Robins, K. (1989) Reimagined communities? European image spaces, after Fordism. *Cultural Studies* 3, 2, 145–65.

Robins, K. (1995) Cyberspace and the world we live in. In M. Featherstone and R. Burrows (eds.) *Cyberspace, Cyberbodies and Cyberpunk: Cultures of Technological Embodiment*. London: Sage.

Robinson, M. (1993) Governance, democracy and conditionality: NGOs and the new policy agenda. In A. Clayton (ed.) *Governance, Democracy and Conditionality: What Role for NGOs?* Oxford: INTRAC.

Robinson, W. I. (1996) *Promoting Polyarchy: Globalization, US Intervention and Hegemony*. Cambridge: Cambridge University Press.

Rocheleau, D., Thomas-Slayter, B., and Wangari, E. (eds.) (1996) *Feminist Political Ecology: Global Issues and Local Experiences*. London: Routledge.

Roddick, J. (1997) Earth summit north and south: building a safe house in the winds of change. *Global Environmental Change* 2, 147–65.

Rodhe, H. and Herrera, R. (1988) Acidification in tropical countries. *SCOPE Report 26*. New York: Wiley.

Rodriguez, N. P. and Feagin, J. R. (1986) Urban specialization in the world-system. *Urban Affairs Quarterly* 22, 187–219.

Rogerson, P. A. (1997) The future of global population modelling. *Futures* 29, 381–92.

Rolston, B. (1991) *Politics and Painting: Murals and Conflict in Northern Ireland*. Toronto: Fairleigh Dickinson University Press.

Rorty, R. (1989) *Contingency, Irony and Solidarity*. Cambridge: Cambridge University Press.

Rose, G. (1993) *Feminism and Geography: The Limits of Geographical Knowledge*. Cambridge: Polity Press.

Rosenau, P. M. (1992) *Postmodernism and the Social Sciences: Insights, Inroads and Intrusions*. Princeton, NJ: Princeton University Press.

Rosenzweig, C. and Hillel, D. (1993) Agriculture in a greenhouse world. *Research and Exploration* 9, 208–21.

Rosenzweig, C. and Parry, M. L. (1994) Potential impact of climate change on world food supply. *Nature* 367, 133–8.

Rostow, W. (1960) *The Stages of Economic Growth: A Non-Communist Manifesto*. Cambridge: Cambridge University Press.

Routledge, P. (1993) *Terrains of Resistance: Nonviolent Social Movements and the Contestation of Place in India*. Westport, CT: Praeger.

Routledge, P. (1998) Going globile: spatiality, embodiment and mediation in the Zapatista insurgency. In S. Dalby and G. ÓTuathal (eds.) *Rethinking Geopolitics*. London. Routledge.

Rowbotham, E. J. (1996) Legal obligations and uncertainties: the climate change convention In T. O'Riordan and J. Jäger (eds.) *Politics of Climate Change: A European Perspective*. London: Routledge.

Roy, O. (1994) *The Failure of Political Islam*. Cambridge, MA: Harvard University Press.

Royal Society of London (1994) *Population: The Complex Reality*. London: Royal Society.

Ruddick, S. (1998) *Youth and Globalization: Rethinking the Politics of Social Reproduction*. Paper presented at NSF Workshop on Geographies of Youth and Children, San Diego, California, November 12–17.

Sachs, W. (ed.) (1992) *The Development Dictionary*. London: Zed Books.

Sachs, W., Loske, R., and Linz, M. (1998) *Greening the North: A Post-Industrial Blueprint for Ecology and Equity*. London: Zed Books.

Sadler, D. P. (1998) The globalization of public health. www.law.indiana.edu/glsj/vol5/nol/fidler.html

Sahabat Alam Malaysia (1987) Sarawak – Orang Ulu fight logging. *Cultural Survival Quarterly* 11, 4, 20–3.

Said, E. (1978) *Orientalism*. London: Routledge; New York: Pantheon.

Said, E. (1993) *Culture and Imperialism*. New York: Knopf.

Salvatore, R. D. (1998) The enterprise of knowledge: representational machines of informal empire. In G. M. Joseph, C. C. Legrand, and R. D. Salvatore (eds.) *Close Encounters of Empire*. Durham, NC: Duke University Press.

Samuel, R. (ed.) (1989) *Patriotism: The Making and Unmaking of British National Identity*. London: Routledge.

Sanders, R. (1978) *Lost Tribes and Promised Lands*. New York: HarperPerennial.

Sandoval, C. (1991) US, Third World feminism: the theory and method of oppositional consciousness in the postmodern world. *Genders* 10, 1–24.

Sardar, Z. (1995) alt.civilisations.faq: cyberspace as the darker side of the West. *Futures* 27, 777–94.

Sarewitz, D. and Pielke, R. (2000) Breaking the global warming gridlock. *Atlantic Monthly* 286, 1, July, 55–64.

Sassen, S. (1991) *Global City*. Princeton, NJ: Princeton University Press.

Sassen, S. (1997) *Losing Control? Sovereignty in an Age of Globalization*. Chichester: Wiley.

Sassen, S. (1999) Global financial centres. *Foreign Affairs* 78, 1, 75–87.

Satcher, D. (1998) Testimony on global health. www.hhs.gov/progorg/asl/testify/t980303d.html

Sayer, A. (1994) Cultural studies and "the economy, stupid." *Environment and Planning* D: *Society and Space* 12, 635–7.

Sayer, A. (1995) *Radical Political Economy: A Critique*. Oxford: Blackwell.

Sayer, A. and Walker, R. (1992) *The New Social Economy: Reworking the Division of Labor.* Cambridge, MA: Blackwell.

Scharf, M. P. (2000) The politics behind the US opposition to the International Criminal Court. *New England International and Comparative Law Annual* vol. 5.

Schell, J. (1998) *The Gift of Time: The Case for Abolishing Nuclear Weapons Now.* New York: Henry Holt.

Schivelsbuch, W. (1986) *The Railway Journey: The Industrialization of Time and Space in the Nineteenth Century.* Berkeley: University of California Press.

Schneider, S. H. and Boston, P. J. (1991) *Scientists on Gaia.* Cambridge, MA: MIT Press.

Schoenberger, E. (1988) Multinational corporations and the new international division of labor: a critical appraisal. *International Regional Science Review* 11, 2, 105–20.

Schulman, B. (1994) *From Cotton Belt to Sunbelt.* Durham, NC: Duke University Press.

Schumacher, E. F. (1973) *Small is Beautiful.* London: Sphere.

Schumpeter, J. (1952) *Socialism, Capitalism and Democracy.* London: Thames.

Schuurman, F. J. (ed.) (1993) *Beyond the Impasse: New Directions in Development Theory.* London: Zed Books.

Schuurman, F. J. (2000) Paradigms lost, paradigms regained? Development studies in the twenty-first century. *Third World Quarterly* 21, 1, 7–20.

Scott, A. and Storper, M. (eds.) (1993) *Pathways to Industrialization and Regional Development.* London: Routledge.

Segal, L. (1999) *Why Feminism? Gender, Psychology and Politics.* Cambridge: Polity Press.

Segal, M. J. (1953) The international trade secretariats. *Monthly Labor Review* April, 372–80.

Sen, A. (1981) *Poverty and Famines.* Oxford: Clarendon Press.

Sen, A. (1990a) More than 100 million women are missing. *New York Review of Books* December 20.

Sen, A. (1990b) Food, economics and entitlements. In J. Dreze and A. Sen (eds.) *The Political Economy of Hunger.* Oxford: Clarendon Press.

Senelle, R. (1989) Constitutional reform in Belgium: from unitarism towards federalism. In M. Forsyth (ed.) *Federalism and Nationalism.* Leicester: Leicester University Press.

Seton-Watson, H. (1977) *Nations and States: An Enquiry into the Origins of Nations and the Politics of Nationalism.* London: Methuen.

Seydlitz, R., Laska, S., Spain, D., Triche, E. W., and Bishop, K. L. (1993) Development and social problems: the impact of the offshore oil industry on suicide and homicide rates. *Rural Sociology* 58, 1, 93–110.

Shackley, S. (1997) The Intergovernmental Panel on Climate Change: consensual knowledge and global politics. *Global Environmental Change* 7, 77–9.

Shaikh, A. and Ahmet Tonak, E. (1994) *Measuring the Wealth of Nations.* New York: Cambridge University Press.

Sharp, J. (2000) *Condensing Communism.* Minneapolis: University of Minnesota Press.

Shell, R. (2000) Halfway to the holocaust, in KAF op cit.

Sheridan, A. (1980) *Michel Foucault: The Will to Truth.* New York: Tavistock.

Shiva, V. (1989) *Staying Alive: Women, Ecology and Survival in India*. London: Zed Books.

Shiva, V. (1994) Conflicts of global ecology: environmental activism in a period of global reach. *Alternatives* 19, 2, 195–207.

Shiva, V. (1997) *Biopiracy: The Plunder of Nature and Knowledge*. Boston: South End Press.

Short, J. and Kim, Y.-H. (1999) *Globalization and the City*. Harlow: Longman.

Short, J., Kim, Y.-H., Kuus, M., and Wells, H. (1996) The dirty little secret of world city research-data problems in comparative analysis. *International Journal of Urban and Regional Research* 20, 4, 697–719.

Shrestha, N. and Patterson, J. (1990) Population and poverty in dependent states. *Antipode* 22, 93–120.

Shurmer-Smith, P. (2000) *India: Globalization and Change*. London: Arnold.

Sibanda, A. (2000) A nation in pain. *International Journal of Health Services* 30, 717–38.

Simmons, I. G. (1989) *Changing the Face of the Earth*. Oxford: Blackwell.

Simmons, I. G. (1993a) *Interpreting Nature: Cultural Constructions of the Environment*. London: Routledge.

Simmons, I. G. (1993b) *Environmental History: A Concise Introduction*. Oxford: Blackwell.

Simon, D. (1998) Rethinking (post)modernism, postcolonialism and posttraditionalism: South–North perspectives. *Environment and Planning* D: *Society and Space* 16, 219–45.

Simon, J. L. (1995) *The State of Humanity*. Oxford: Blackwell.

Skelton, T. and Valentine, G. (1998) *Cool Places: Geographies of Youth Cultures*. London: Routledge.

Sklair, L. (1995) *Sociology of the Global System*. London: Prentice-Hall.

Sklair, L. (1999) Competing conceptions of globalization. *Journal of World-Systems Research* 5, 143–64.

Slater, D. (1992) On the borders of social theory: learning from other regions. *Society and Space* 10, 307–27.

Slater, D. (1993a) The political meanings of development: in search of new horizons. In F. J. Schuurman (ed.) *Beyond the Impasse: New Directions in Development Theory*. London: Zed Books.

Slater, D. (1993b) The geopolitical imagination and the enframing of development theory. *Transactions: An International Journal of Geographic Research* December, 18, 419–37.

Slater, D. (1995) Challenging western visions of the global: the geopolitics of theory and north–south relations. *The European Journal of Development Research* 7, 2, December, 366–88.

Slater, D. and Taylor, P. J. (eds) (1999) *The American Century*. Oxford: Blackwell.

Smil, V. (1992) China's environment in the 1980s: some critical changes. *Ambio* 21, 431–6.

Smil, V. (1994) How many people can the earth feed? *Population and Development Review* 20, 255–92.

Smith, A. (n.d.) Social and spatial analysis of virtual space: 30 days in Activeworlds. Unpublished manuscript.

Smith, A. D. (1986) *The Ethnic Origin of Nations*. Oxford: Blackwell.

Smith, F. B. and Clark, M. J. (1989) Airborne debris from the Chernobyl incident.

Met. Sci. Paper 2. London: HMSO.

Smith, M. and Kollock, P. (1999) *Communities in Cyberspace.* London: Routledge.

Smith, M. P. (1994) Can you imagine? Transnational migration and the globalization of grassroots politics. In P. Knox and P. J. Taylor (eds.) *World Cities in a World-System.* Cambridge: Cambridge University Press.

Smith, N. (1990) *Uneven Development.* Oxford: Blackwell.

Smith, N. (1992) Contours of a spatialized politics. *Social Text* 33, 54–81.

Smith, N. (1993) Homeless/global: scaling places. In J. Bird et al. (eds.) *Mapping the Futures.* London: Routledge.

Smith, N. (1996) *New Urban Frontier: Gentrification and the Revanchist City.* London: Routledge.

Smith, N. (2000) What happened to class? *Environment and Planning* A, 32, 1011–32.

Smith, P. M. and Warr, K. (eds.) (1991) *Global Environmental Issues.* Oxford: Oxford University Press.

Smith, T. F., Srinivasan, A., Schochetman, G., Marcus, M., and Myers, G. (1988) The phylogenetic history of immunodeficiency viruses. *Nature* 333, 573–5.

Sooros, M. (1997) *The Endangered Atmosphere: Preserving a Global Commons.* Columbia: University of South Carolina Press.

Soper, K. (1990) *Troubled Pleasures: Writings on Politics, Gender and Hedonism.* London: Verso.

Soper, K. (1993) Postmodernism, subjectivity and the question of value. In J. Squires (ed.) *Principled Positions: Postmodernism and the Rediscovery of Value.* London: Lawrence and Wishart.

Sorkin, M. (1992) *Variations on a Theme Park: The New American City and the End of Public Space.* New York: Hill and Wang.

Southall, H. (1989) British artisan unions in the New World. *Journal of Historical Geography* 15, 2, 163–82.

Spivak, G. C. (1988) Can the subaltern speak? In C. Nelson and L. Grossberg (eds.) *Marxism and the Interpretation of Culture.* Urbana: University of Illinois Press.

Spivak, G. C. (1991) Identity and alterity: an interview. *Arena* 97, 65–76.

Spivak, G. C. (2000) *Globalizing globalization.* Paper presented at Rethinking Marxism Conference, Amherst, Massachusetts, September 21–24.

Springborg, P. (1992) *Western Republicanism and the Oriental Prince.* Cambridge: Polity Press.

Squires, J. (ed.) (1993) *Principled Positions: Postmodernism and the Rediscovery of Value.* London: Lawrence and Wishart.

Stannard, D. (1992) *American Holocaust.* Oxford: Oxford University Press.

Stark, D. (1996) Recombinant property in East European capitalism. *American Journal of Sociology* 104, 4, 993–1027.

Stephanson, A. (1995) *Manifest Destiny.* New York: Hill and Wang.

Stoddart, D. R. (1982) Geography – a European science. *Geography* 67, 289–96.

Stoddart, D. R. (1986) *On Geography and its History.* Oxford: Blackwell.

Storper, M. (1997) Territories, flows and hierarchies in the global economy. In K. Cox (ed.) *Spaces of Globalization.* New York: Guilford

Storper, M. and Walker, R. (1989) *The Capitalist Imperative: Territory, Technology and Industrial Growth.* Oxford: Blackwell.

Strange, S. (1982) Cave! Hic Dragones: a critique of regime analysis. *International Organization* 36, 2, 479–96.

Strangeland, P. (1984) Getting rich slowly – the social impact of oil activities. *Acta Sociologica* 27, 3, 215–37.

Strassoldo, R. (1992) Globalism and localism: theoretical reflections and some evidence. In Z. Mlinar (ed.) *Globalization and Territorial Identities*. Aldershot: Avebury Press.

Strohmeyer, J. (1993) *Extreme Conditions: Big Oil and the Transformation of Alaska*. New York: Simon and Schuster.

Strukova, E. B. et al. (2000) *Macroassessment of environment-related human health damage cost for Russia*. Washington, DC: World Bank.

Sturken, M. (1991) The wall, the screen, and the image: the Vietnam Veterans Memorial. *Representations* 35, 118–42.

Sturmthal, A. (1950) The International Confederation of Free Trade Unions. *Industrial and Labor Relations Review* 3, 3, 375–82.

Sudbury, J. (2000) Transatlantic visions: resisting the globalization of mass incarceration. *Social Justice* 27, 3, 133–49.

Sullivan, F. (1993) Forest principles. In M. Grubb, M. Koch, K. Thompson, A. Munson, and F. Sullivan (eds.) *The 'Earth Summit' Agreements: A Guide and Assessment*. London: Earthscan (for the Royal Institute of International Affairs).

Susman, P. (1989) Exporting the crisis: US agriculture and the third world. *Economic Geography* 65, 4, 293–313.

Swift, J. (1989) Why are rural people vulnerable to famine? *IDS Bulletin* 20, 8–15.

Swyngedouw, E. (1997) Neither global nor local: "glocalization" and the politics of scale. In K. Cox (ed.) *Spaces of Globalization*. New York: Guilford.

Taussig, M. (1993) *Mimesis and Alterity*. London: Routledge.

Taylor, A. (1992) *Choosing our Future: A Practical Politics of the Environment*. London: Routledge.

Taylor, J. (1997) The emerging geographies of virtual worlds. *The Geographical Review* 87, 172–92.

Taylor, M. J. and Thrift, N. (eds.) (1982) *Multinationals and the Restructuring of the World Economy*. London: Croom Helm.

Taylor, P. J. (1985) *Political Geography: World-Economy, Nation-State, Locality*. London: Longman.

Taylor, P. J. (1993a) Full circle, or new meaning for the global. In R. J. Johnston (ed.) *The Challenge for Geography*. Oxford: Blackwell.

Taylor, P. J. (1993b) Geopolitical world orders. In P. J. Taylor (ed.) *Political Geography of the Twentieth Century*. London: Belhaven.

Taylor, P. J. (1993c) *Political Geography: World-Economy, Nation-State and Locality*, 3rd edn. Harlow: Longman.

Taylor, P. J. (1994) The state as container: territoriality in the modern world-system. *Progress in Human Geography* 18, 151–62.

Taylor, P. J. (1996a) Embedded statism and the social sciences. *Environment and Planning* A, 28, 1917–28.

Taylor, P. J. (1996b) *The Way the Modern World Works: World Hegemony to World Impasse*. Chichester: Wiley.

Taylor, P. J. (1999) *Modernities*. Minneapolis: University of Minnesota Press.

Taylor, P. W. (1986) *Respect for Nature*. Princeton, NJ: Princeton University Press.

Temin, P. (1999) Globalization. *Oxford Review of Economic Policy* 15, 4, 76–89.

Thomas, D. H. L. and Adams, W. M. (1997) Space, time and sustainability in the Hadejia-Jama'are wetlands and the Komodugu Yobe basin, Nigeria. *Transac-*

tions of the Institute of British Geographers N. S. 22, 430–49.

Thomas, W. L., Jr. (ed.) (1956) *Man's Role in Changing the Face of the Earth.* Chicago: University of Chicago Press.

Thompson, E. P. (1987) The rituals of enmity. In D. Smith and E. P. Thompson (eds.) *Prospectus for a Habitable Planet.* London: Penguin Books.

Thrift, N. J. (1990) Transport and communication 1730–1914. In R. L. Dodgshon and R. Butlin (eds) *A New Historical Geography of England and Wales*, 2nd edn. London: Academic Press.

Thrift, N. J. (1994) On the social and cultural determinants of international financial centres. In S. Corbridge, N. J. Thrift and R. L. Martin (eds.) *Money, Power and Space.* Oxford: Blackwell.

Thrift, N. J. (2000) Commodities. In R. J. Johnston, D. Gregory, G. Pratt and M. Watts (eds.) *The Dictionary of Human Geography*, 4th edn. Oxford Blackwell.

Thrift, N. J. and Leyshon, A. (1994) A phantom state? The de-traditionalization of money, the international financial system and international financial centres. *Political Geography* 13, 299–327.

Thurow, L. (1992) *Head to Head: The Coming Economic Battle among Japan, Europe and America.* New York: Warner.

Tickell, A. and Peck, J. (1992) Accumulation, regulation, and the geographies of post-Fordism. *Progress in Human Geography* 16, 2, 190–218.

Todd, E. (1987) *The Causes of Progress: Culture, Authority and Change.* Oxford: Blackwell.

Touraine, A. (1985) An introduction to the study of social movements. *Social Research* 52, 4, 749–87.

Toye, J. (1993) *Dilemmas of Development*, 2nd edn. Oxford: Blackwell.

Trudgill, S. T. (1990) *Barriers to a Better Environment.* London, Belhaven.

Trustees for Alaska (1993) *Trustees Battles Lease Sale Adjacent to Arctic Refuge.* Summer.

Tucker, C. J., Dregne, H. E., and Newcomb, W. W. (1991) Expansion and contraction of the Sahara Desert from 1980 to 1990. *Science* 253, 299–301.

Tucker, V. (ed.) (1997) *Cultural Perspectives on Development.* London: Frank Cass.

Tucker, V. (1999) The myth of development: a critique of a eurocentric discourse. In R. Munck and D. O'Hearn (eds.) *Critical Development Theory.* London: Zed Books.

Turkle, S. (1995) *Life on the Screen: Identity in the Age of the Internet.* New York: Simon and Schuster.

Turner, B. (ed.) (2001) *Stateman's Yearbook 2001.* London: Palgrave.

Turner, B. L., II (1991) Comment on Ponting's view of environment and prehistory. *Environment* 23, 4, 2–3, 45.

Turner, B. L., II and Butzer, K. W. (1992) The Columbian encounter and land-use change. *Environment* 43, 16–20.

Turner, B. L., II et al. (eds.) (1990) *The Earth as Transformed by Human Action: Global and Regional Changes in the Biosphere over the Past 300 Years.* Cambridge: Cambridge University Press.

Udall, L. (1997) The international Narmada campaign: a case of sustained advocacy. In W. F. Fisher (ed.) *Toward Sustainable Development: Struggling Over India's Narmada River.* Jaipur: Rawat Publications.

UNAIDS (2000) *Report on the Global HIV/AIDS Epidemic.* Geneva: UNAIDS.

UNAIDS and WHO (1999) *AIDS Epidemic Update: December 1999.* Geneva: Joint

United Nations Program on HIV/AIDS and World Health Organization.

UNCTC (1983/1988) *Transnational Corporations in World Development: Third Survey.* New York: United Nations.

UNDP (1992, 1993) *Human Development Report.* New York: Oxford University Press for the UNDP.

UNDP (1996) *The Human Development Report.* New York: Oxford University Press.

UNDP (2000) *Human Development Report.* New York: Oxford University Press.

UNEP (1991) *Environmental Data Report 1991.* Oxford: UNEP/Blackwell.

UNEP (1999) *Global Environmental Outlook 2000.* London: UNEP and Earthscan Publications.

United Nations (1977) *Desertification: Its Causes and Consequences.* Oxford: Pergamon Press.

United Nations (1993) *World Investment Report.* New York: United Nations.

United Nations (1999) *Transnational Investment Report.* New York: United Nations.

United States Bureau of Land Management (1991) *Alaska Minerals: Facts, Figures and Trivia.* Anchorage: Bureau of Land Management, Alaska State Office, Division of Mineral Resources.

United States Council on Environmental Quality (1982) *Environmental Quality 1981: 12th Annual Report of the Council on Environmental Quality.* Washington, DC: USCEQ.

Urry, J. (1990) *The Tourist Gaze.* London: Sage.

Valentine, G. (1993) Negotiating and managing multiple sexual identities: lesbian time-space strategies. *Transactions, Institute of British Geographers,* N.S. 18, 237–48.

Valentine, G. (1999) A corporeal geography of consumption. *Environment and Planning* D: *Society and Space* 17, 329–51.

Van der Pijl, K. (1998) *Transnational Classes and International Relations.* London: Routledge.

Van Holthoon, F. and Van der Linden, M. (eds.) (1988) *Internationalism in the Labour Movement 1830–1940, Vols. 1 and 2.* London: E. J. Brill.

Vernon, R. (1992) Transnational corporations: where are they coming from, where are they headed? *Transnational Corporations* 1, 2, 7–35.

Visvanathan, S. (1985) From the annals of the laboratory state. *Lokayan Bulletin* 3, 4, 23–47.

Vitet, C. R. and Wharton, M. (1998) Diphtheria in the former Soviet Union. www.cdc.gov/ncidod/eid/vol4no4/vitek.htm

Vitousek, P. M., Mooney, H. A., Lubchenco, J., and Melillo, J. M. (1997) Human domination of earth's ecosystems. *Science* 277, 494–500.

Vogler, J. (2000) *The Global Commons,* 2nd edn. Chichester: John Wiley.

Wade, R. and Veneroso, F. (1998) The Asian crisis. *New Left Review* 228, 3–24.

Wagner-Pacifini, R. and Schwartz, B. (1991) The Vietnam Veterans Memorial: commemorating a difficult past. *American Journal of Sociology* 97, 376–420.

Waldinger, R. and Bozorgmehr, M. (ed.) (1996) *Ethnic Los Angeles.* New York: Russell Sage Foundation.

Walker, R. B. J. (1993) *Inside/Outside: International Relations as Political Theory.* Cambridge: Cambridge University Press.

Wallace, I. (1985) Towards a geography of agribusiness. *Progress in Human Geog-*

raphy 9, 491–514.

Wallach, B. (1991) *At Odds with Progress: Americans and Conservation*. Tucson: University of Arizona Press.

Wallach, L., Sforza, M., and Nader, R. (2000) *The WTO: Five Years of Reasons to Resist Corporate Globalization. Open Media Pamphlet*. New York: Seven Stories Press.

Wallerstein, I. (1974) *The Modern World-System: Capitalist Agriculture and the Origins of the European World-Economy in the Sixteenth Century*. New York: Academic Press.

Wallerstein, I. (1984) *Politics of the World-Economy*. Cambridge: Cambridge University Press.

Wallerstein, I. (1989) *Unthinking Social Science*. Cambridge: Polity Press.

Wallgren, T. (1998) Political semantics of "globalization": A brief note. *Development* 41, 2, 30–2.

Walton, J. and Seddon, D. (1994) *Free Markets and Food Riots: The Politics of Global Adjustment*. Oxford: Blackwell.

Ward, M. (1983) *Unmanageable Revolutionaries: Women in Irish Nationalism*. Dingle: Brandon Press.

Ware, V. (1992) *Beyond the Pale*. London: Verso.

Warf, B. (1996) International engineering services. *Environment and Planning* A, 28, 667–86.

Waring, M. (1988) *If Women Counted*. New York: Harper and Row.

Warner, M. (1985) *Monuments and Maidens: The Allegory of the Female Form*. London: Picador.

Wasserman, J. (2001) Parks go to seed for lack of green. *Daily News* August 13, 4–5.

Waterman, P. (1993) Internationalism is dead! Long live global solidarity? In J. Brecher, J. Brown Childs, and J. Cutler (eds.) *Global Visions: Beyond the New World Order*. Boston: South End Press.

Waters, M. (1995) *Globalization*. London: Routledge.

Watson, R., Baldwin, T., and Webster, P. (2001) Blair vows to help those who help themselves. *Guardian* April 27.

Watts, M. J. (1983) *Silent Violence: Food, Famine and Peasantry in Northern Nigeria*. Berkeley: University of California Press.

Watts, M. J. (1984) State, oil, and accumulation: from boom to crisis. *Environment and Planning* D: *Society and Space* 2, 403–28.

Watts, M. J. (1991) Visions of excess. *Transition* 51, 124–41.

Watts, M. J. (1992a) Capitalisms, crises, and cultures, I: notes toward a totality of fragments. In A. Pred and M. J. Watts, *Reworking Modernity*. New Brunswick, NJ: Rutgers University Press.

Watts, M. J. (1992b) Space for everything (a commentary). *Cultural Anthropology* 7, 115–29.

Watts, M. J. (1996) Mapping identities: place, space, and community in an African city. In P. Yaeger (ed.) i. Ann Arbor: University of Michigan Press.

Watts, M. J. and Bohle, H. G. (1993) The space of vulnerability: the causal structure of hunger and famine. *Progress in Human Geography* 17, 43–67.

Weale, A. (1992) *The New Politics of Pollution*. Manchester: Manchester University Press.

Weaver, M. A. (2000) The real Bin Laden. *The New Yorker* January 24, 32–8.

Webber, M. J. and Rigby, D. L. (1996) *The Golden Age Illusion: Rethinking Post-war Capitalism*. New York: Guilford.

Weihe, W. H. (1979) *Climate, Health and Disease: Proceedings of the First World Climate Conference*. Geneva: World Meteorological Organization, Paper No. 537.

Weinberg, A. K. (1963) *Manifest Destiny: A Study of Nationalist Expansion in American History*. Chicago: Quadrangle Books.

Weir, D. (1987) *The Bhopal Syndrome: Pesticides, Environment and Health*. London: Earthscan.

Weiss, B. (1996) Coffee breaks and coffee connections. In D. Howes (ed.) *Cross-Cultural Consumption: Global Markets, Local Realities*. London: Routledge.

Wellman, B. and Gulia, M. (1999) Virtual communities as communities: net surfers don't ride along. In M. A. Smith and P. Kollock (eds.) *Communities in Cyberspace*. London: Routledge.

Welsch, W. (1999) Transculturality: the puzzling form of cultures today. In M. Featherstone and S. Lash (eds.) *Spaces of Culture*. London: Sage.

Whatmore, S. (1991) *Farming Women: Gender, Work and Family Enterprise*. London: Macmillan.

Whatmore S. and Thorne, J. (1997) Nourishing networks: alternative geographies of food. In D. Goodman and M. Watts (eds.) *Globalizing Food*. London: Routledge.

Whitmore, T. M. and Turner, B. L., II (1992) Landscapes of cultivation in Mesoamerica on the eve of the conquest. *Annals of the Association of American Geographers* 82, 402–25.

WHO (World Health Organization) (1996) *Climate Change and Human Health*. Geneva: WHO.

WHO (1997) *Health and Environment in Sustainable Development*. Geneva: WHO.

WHO (1998a) *Life in the Twenty-First Century*. Geneva: WHO.

WHO (1998b) Dengue and dengue hemorrhagic fever. www.who.int/info-fs/en/fact117.html

WHO (1998c) Malaria. www.who.int/info-fs/en/fact094.html

WHO (1999a) *Making a Difference*. Geneva: WHO.

WHO (1999b) *Report on Infectious Diseases*. Geneva: WHO.

WHO (2000a) Tuberculosis. www.who.int/info-fs/en/fact104.html

WHO (2000b) *Climate Change and Human Health*. Geneva: WHO.

WHO Commission on Health and the Environment (1992) *Report of the Panel on Food and Agriculture*. Geneva: WHO.

Williams, B. (1992) Contact cultures. *Panel, Annual Meeting of the Association of American Anthropologists*, San Francisco, December.

Williams, C. H. and Smith, A. D. (1983) The national construction of social space. *Progress in Human Geography* 7, 502–18.

Williams, F. (1989) *Social Policy: A Critical Introduction*. Cambridge: Polity Press.

Williams, M. (1990) Forests. In B. L. Turner et al. *The Earth as Transformed by Human Action*. Cambridge: Cambridge University Press.

Williams, R. (1976) *Keywords*. Oxford: Oxford University Press. Also (1983) London: Fontana.

Williams, R. (1985) *The Country and the City*. London: Chatto and Windus.

Williams, R. (1990) *Notes on the Underground*. Cambridge, MA: MIT Press.

Willis, S. (1991) *A Primer for Daily Life*. London: Routledge.

Windmuller, J. P. (1980) *The International Trade Union Movement*. Deventer, Netherlands: Kluwer.

Wines, M. (2000) An ailing Russia lives a tough life that's getting shorter. *New York Times* December 3.

Wolch, J. and Emel, J. (eds.) (1998) *Animal Geographies: Place, Politics, and Identity in the Nature–Culture Borderlands*. London: Verso.

Wolch, J. R. (1990) *The Shadow State: Government and Voluntary Sector in Transition*. New York: The Foundation Centre.

Wolfenson, J. (2000) Speech to UN Security Council meeting on HIV/AIDS in Africa. New York: United Nations.

Wolin, S. (1960) *Politics and Vision: Continuity and Innovation in Western Political Thought*. Boston: Little, Brown.

Wolin, S. (1989) *The Presence of the Past: Essays on the State and the Constitution*. Baltimore, MD: Johns Hopkins University Press.

Wolosky, L. (2000) Putin's plutocrat problem. *Foreign Affairs* 79, 2, 18–31.

Wood, D. and Beck, R. J. (1994) *Home Rules*. Baltimore, MD: Johns Hopkins University Press.

Woodehouse, T. (1990) *Replenishing the Earth: The Right Livelihood Awards, 1986–1989*. Bideford: Green Books.

World Bank (1986) *World Development Report*. Oxford: Oxford University Press.

World Bank (1992) *World Development Report 1992: Development and the Environment*. New York: Oxford University Press for the World Bank.

World Bank (1993) *World Development Report 1993*. Oxford: Oxford University Press.

World Bank (1997) *The World Development Report 1997*. New York: Oxford University Press.

World Bank (1999) *World Development Report 1999–2000*. Oxford: Oxford University Press.

World Bank (2000) *World Development Report 2000/2001: Attacking Poverty*. Washington, DC: World Bank.

World Commission on Environment and Development (1987) *Our Common Future*. Oxford: Oxford University Press.

World Resources Institute (1991) *World Resources*. Oxford: Oxford University Press.

World Resources Institute (1996) *World Resources 1996–7: The Urban Environment*. New York: Oxford University Press.

World Resources Institute (1998) *World Resources 1998–9: Environmental Change and Human Health*. New York: Oxford University Press:

World Resources Institute (1999) *World Resources 1998–1999*. Washington, DC: World Resources Institute.

World Resources Institute (2000a) *Water: Critical Shortages Ahead?* Washington, DC: World Resources Institute.

World Resources Institute (2000b) *World Resources 2000–2001: People and Ecosystems, the Fraying Web of Life*. Washington, DC: World Resources Institute.

Worthington, E. B. (1983) *The Ecological Century: A Personal Appraisal*. Cambridge: Cambridge University Press.

Wright, P. (1985) *On Living In an Old Country*. London: Verso.

Wyatt, E. (2000) School laptops paid with ads called feasible. *New York Times*, September 20, B1, B8.

Wynne, B. (1990) Sheepfarming after Chernobyl: a case study in communicating scientific information. In H. Bradby (ed.) *Dirty Words: Writings on the History and Culture of Pollution*. London: Earthscan.

Wynne, B. (1992) Uncertainty and environmental learning. Reconceiving science and policy in the preventive paradigm. *Global Environmental Change* 2, 2, 111–27.

Young, I. M. (1990) *Justice and the Politics of Difference*. Princeton, NJ: Princeton University Press.

Young, J. E. (1992) Mining the Earth. *Worldwatch Paper* 109. Washington, DC: Worldwatch Institute.

Zimmermann, E. W. (1951) *World Resources and Industries*. New York: Harper.

Index

Page references to figures and tables are in *italic*.

accumulation
  flexible, 25
  regime of, 144
  see also capital accumulation
acid precipitation, 369
acidification 399, *402*, 402–3
  see also acid precipitation; carbon
    dioxide emissions
actor-network theory, 294
Afghanistan, 180, 214
Africa, 11, 239, 310, 378, 446
  state formation in, 132
  see also Algeria; Kenya, Green Belt
    Movement
Agenda 21, 401, 411, 416, 418, 421
  development through economic
    growth, 416, 418
  efficiency, 418
  participation, 418
  recognition of grassroots, 417
  and technocentrism, 418
  see also Convention on Desertification;
    Convention on Biological Diversity;
    Earth Parliament; Global Forum;
    Rio Earth Summit/UNCED
agri-business, 59
agricultural futures, trading, 64
agriculture
  capitalist restructuring of, 58, 60,
    62–3
  green revolution type, effect of, 64–5
  industrialized, 359
  market handicap, 60

new political economy of, 58–60
slash-and-burn, 357
  see also agri-food systems; food
agri-food chain, 59, 62
agri-food complex, 59
agri-food regimes, 59
agri-food systems, 23
  contemporary, 59–60, *61*, 62–4
  defined, 60
  global, 23
  industrial, 62–4
  off-farm sectors, 60, 62, 64
AIDS *see* human immunodeficiency virus,
    HIV/AIDS
Albania, 107, 447
Algeria, 213
analysis, discursive and textual, 305
Anderson, B., 342
Anglo-Saxon supremacy, 90
annihilation of space by time, 258
annual growth rate, 200
Anti-Ballistic Missile Defense Treaty, 188
anti-globalization movement, 359
  see also Seattle; Zapatista(s)
appropriationism, 62
Aral Sea, 374, *375*
armaments industry, 109–10
  biological weapons, 109–10
  black market, 110
Asian Tigers, 23
assimilative capacity, 396, *397*
Atlantic Charter, 146
atmospheric trace gases, 370
atomism of the city, 33
automatic teller machines (ATMs), 37,
    38

back office functions, 76
Balkans, 447
  rewriting of history, 280
barometers of modernity, 29–30, 33
  see also legibility/illegibility; space of
    flows; time–space compression
Barrett, M., 299, 300
Beck, U., 400
Belgium, linguistic tension, 133
Bhopal, India (Union Carbide plant
    disaster), 396
bilateral arms control agreements, 184
  START I (Strategic Arms Reduction
    Treaty), 185, 187
  START II, 183–4
  START III, 184
bilateral investment, 22
bioaccumulation, 394
Biodiversity Convention/Biological
    Diversity Convention/Treaty see
    Convention on Biological Diversity;
    Rio Earth Summit/UNCED
biosphere, human-induced change, 373
biotechnology
  and the agri-food system, 61, 62,
    revolution, 358
  at Rio Earth Summit, 419, 420
birth-control programs/policies see
    family-planning programs/policies
birth rate, crude, 200
Black Atlantic hybrid culture, 304
Black Power movement, 267
Blaikie, P., 426
  and Brookfield, H., 426
Bolshevik Revolution, 80
boom and bust cycles, 386–7
boomerang effect, 186
Bosnia-Herzegovina, conflict, 107, 108
Bowling Alone, 448
Brandt Commission, 210
Bray, T., 351, 352
Brazil, 314, 320, 323, 433
  Amazonian resources, conflicts, 320,
    432
  Movimento Sem Terra (MST), 323
Bretton Woods, 27, 71, 145, 154
  effects of Accords, 135
Britain/UK
  dual tracking, 245–6
  effects of Chernobyl radioactive
    release, 395, 396

foot and mouth disease, 359
  resurgence of nationalism, 135
  unemployment, 240
broad banding, effects of, 246
Brookfield, H. (and Blaikie, P.), 426
Brundtland Commission, 312
  definition of sustainable development,
    416
Brundtland Report, Our Common
    Future, 312, 414, 415
  vision of sustainable development, 416
buffering capacity, 397
Bulova Watch, 68, 75
Burkina Faso, NAAM Movement, 321
Bush, G. W., 112, 117, 362, 404, 423,
    447, 449

Canada
  female job loss, 241–2
  impact of women's status, 239–40, 241
capital
  fluidity/mobility, 251, 252, 253
  foreign direct investment (FDI), 21, 22,
    23
  hypermobility, 23
  internationalization of, 143
  mobility, 11
  transnationalization of, 21, 23
capital accumulation
  agri-food sector, 58, 59
capital–labor relations, 448
capitalism, 30, 39
  global, 30–1, 41
  international, 408, 410
  reconfiguration of, 448
  and the Regulation School, 144
  transnational, and global
    metropolitanism, 329, 335
capitalist order, global, 39, 40
capitalist production, 253–4
carbon dioxide emissions, 362, 373, 399,
    403, 405, 420, 429, 431
  regulation of, 372, 420
carbon sinks, 431
Careless Technology, The, 415
Caring for the Earth, 414, 418, 419
carrying-capacity, 205
  see also Malthusianism; neo-
    Malthusianism
cash crops, 314
Castells, M., 8, 30–1, 342

and network society, 8–9
Central Park, New York City, 255–6,
    257
centralization vs. decentralization, 336
centralized authority, reassertion of, 126
CFCs, 362, 395, 399, 430, 432, 437
    and ozone relationship, 360, 372
change
    human induced, 302, 366, 368, 372
    quantitative and qualitative, 335
    and states, 159
    see also climate change; environmental
        change; global change
Chernobyl nuclear disaster (radioactive
    release), 395, 396, 397
Chiapas, 382
    see also Zapatista(s)
children
    China, 106, 193, 214, 373, 378, 420,
        447
    and globalization, 248–60
    and household consumption, 249
    New York City, 253–7
    Single Child Law, 193
    Sudan, 257–60
Christian Labor Union International
    (CISC/World Confederation of
    Labor), 80
cities
    flight from, 32
    modern, 30, 32, 33
    see also world cities
citizenship vs. the state, 119–21
civil rights movement, 267
civil society, 250, 256, 429
    global, 15–17
    NGOs and, 429
    role of, 26
class, 165
    and sociopolitical change, 95
class analysis, Marxist, failures of, 95
climate change, 360, 366, 420, 429, 430,
    433
    and health/disease, 229, 230, 231, 232
    and lifestyle/consumption, 431–2
    and North–South relations, 430–2
Climate Change, Framework Convention
    on, 420–1, 422
    see also Rio Earth Summit/UNCED
Cold War, 103, 107, 110, 174, 178, 179,
    180, 181, 427, 446, 451

and democracy, 118–19
    see also détente
Colombia, Indigenous Authorities
    Movement, 318
colonial escape valves, 358
colonialism
    and labor integration, 313
    use of geography, 413
Columbian Encounter (1492), 369
commodities
    agricultural, 59, 63
    commodity chain(s), 291
    geography of, 389
    hidden history, 289
commodity circuitry, 292
    spatialization, 293
    systems of provision approach, 293
    unveiling/unmasking, 293
    vs. circuits and networks, 291–3, 294
commodity cultures, 286–7
    bananas, 286–7
    soap, 289–90
    sugar, 287–9
Commonwealth of Independent States
    (CIS), 130
communal/sector ties, 125
communism, collapse of, 121
Communist Party, and political rights,
    120
community
    organizations, 250
    without propinquity, 449
comparative advantage, 50, 52
Comprehensive Test Ban Treaty, 187,
    188
Conference on Environment and
    Development, UN (UNCED) see Rio
    Earth Summit/UNCED; Agenda 21
connectedness, global, 330
consciousness raising, 124
consumers
    and food, 57
    marginal, position of, 65
    resistance to industrialized food
        products, 66
consumption
    in everyday life, 285–6
    geography of, 283
    of marine fish, 381
    meanings, 284
    of petroleum, 378

consumption (*cont'd*)
  and social relations, 285
  theory (theorizing), 284–5
  of water, 378
  of wood, 379
  see also natural resources
contested institutions, 50
contingencies of place, 215
continuity/discontinuity, 450
Convention on Biological Diversity, 363,
  419–20
  see also Rio Earth Summit/UNCED
Convention on Desertification, 422
  see also Rio Earth Summit/UNCED
Convention on Trade of Endangered
  Species (CITES), 437
convergent space, 323–4
Converse, 74
core–periphery model, and dependency
  theory, 313
Cornucopians, 199, 207, 215, 382, 385
corporate capitalism, 414
corporate welfare state *see* welfare state
corporations and integration, 68–79
corporatism, 162
cosmopolitanism, cultural, 449
Critical Environmental Zones, Project
  on, 374–5
Cuba
  Americanization, 91
  missile crisis, 174
  revolution and beyond, 96–7
  US intervention, 90, 91–2
cultural legitimacy, 132
cultural unity, 131
culture
  assimilation, 280
  of consumerism, 388
  global, content of, 278
  globalization of, 12, 278
  indigenous, 303
  late capitalist, 281
  traditions, 281
cyber communities, 341–2
  Alpha World, 345, 347
  as alternative, 342–4
  critiques of, 343–4
  lesbian cyber café, 345
  as placeless communities, 344–8
  and social networks, 343
cyberspace, 340

  as aspatial, 341
  and communicative practice, 342
  components, 344
  embodied space, 348, 352
  geographic approach to, 352
  geography of, 340–53
  vs. geometric space, 348
  identity, 348
  mapping, 348–52, 353
  spatialization of, 344, 345, 348

dealers/dealing rooms, 34–6, 37
death rate, 200
decentralization
  vs. centralization, 336
  economic, 331
decolonization, 433
  and language choice, 134
deforestation, 366, 367, 369, 373
  and population growth, 211–12
  Sarawak, 320–1
demapping, 449
democracy, 117–19
  barrier to economic development, 127
  and the Cold War, 121–3
  and economic liberalization, 127
  post-Cold War, 128–9
  and the territorial state, 123–4
  theory and practice, 119–21
  threat of group rights, 125
  threats to, 126–8
  transferability, 450
  transition to, 122–3
democratization, 110–11
demographic indicators, *201*
dependency
  analysis of, 313–15
  and foreign aid, 314
  and labor, 314
  and multinational corporations, 314
  theory, 94, 313
desperate ecocide, 426
détente, 103
  see also Cold War
deterritorialization, 338
developed countries/world, 312
  advantage of, 313
developing countries/world, 63
  see also newly industrializing
    countries; Third World
development, 89

as dependence, 313–15
as discourse, 92, 97, 310–13
and environment, 415
gender dimension, 98
and geopolitics of knowledge, 97–9
independence and colonialism, 312
Manifest Destiny and, 90
Marxist/neo-Marxist vision, 89, 94–7
neo-liberal, 93–4
Occidental visions, 89, 90–4
radical views, 426
uneven, 10, 11, 12
US role, 90–4
Western project, 90–4
see also sustainable development
development discourse, 89
    disadvantaging countries of the South,
        313
    and discourses of underdevelopment,
        310–11
    diseases of transition, 92
    economic dependency, 313–14
    economic growth to alleviate poverty,
        311
    environmental effects, 310, 325
    green revolution, 314–15
    and indigenous/traditional systems,
        311, 316
    problems caused by, 310, 325
    purposes of, 312
    as Westernization, 90–4
    development theory, 88–98, 311
    see also development; modernization
        theory
development–environment debate,
    416–18
diarrheal diseases, 219
diaspora, 278
difference, 296, 297
    forms of, 307–9
    politics of, 125–6, 297
differential of contemporaneity, 336
differentiation
    in farming restructuring, 64–5
    vs. homogenization, 336
disability-adjusted life years (DALYs)
    217, 218
discourse, 302, 307, 311
    and Orient, 303
    and truth, 303
    see also development discourse

disease burden, global, 221
diversification, 66
diversity, 300
    theorists, 297–8
Dominican Republic, 91
Dominican Rural Women Study, 238
doomsday scenarios, 358, 435–6
double burden, 217
double consciousness theory, 307
doubling time, 201
dual tracking, 245–6
dualistic theory, critique, 76
Dumbarton Oaks, draft of UN Charter,
    146

Earth Parliament, 417
Earth Summit see Rio Earth Summit/
    UNCED; Agenda 21
Earth Summit II, 422
Earth, the
    changes in the environment, 364
    damage to, 388
    and environmental goods, 388
    land cover, 370
    see also climate change; environmental
        change; environmental issues
East Asia
    low-cost labor, 74–5
Eastern Europe, 57, 72
    and carbon dioxide emissions, 429
    communist collapse, 104
    labor, 85–7
    nationalism, 104
    rights of linguistic minorities,
        133–4
    shock therapy, 24, 450
eco-transition, 385
ecocatastrophe, 404
ecological economics, 383
ecological modernization, 385–6,
    409–10, 430, 435
Ecological Principles for Economic
    Development, 415
ecological processes, essential, 423
ecologism, 405
ecology, 405
economic activity, transnational, 328–9,
    330
economic alliances, regional, 151
economic commissions, regional 147
economic liberalism, 93–4

economic liberalization, and democracy, 127
economic restructuring, 24, 25
  costs of, 25
Economic and Social Council, UN, 147
economic stagnation, 122
economism, Marxist, 94
economy, global, transformative process, 239
educational attainment of women, 246
efficiency
  and democracy, 126–7
  in resource use, 413
Ehrlich, P., 382
Ehrlichs, P. and A., 199
El Niño Southern Oscillation (ENSO), 408
El Salvador, 320
electronic ghettos, 39, 41
emissions controls/regulation, 403
emissions trading, 431
employment, part-time, definitions, 238
end of history, 449
entitlements, 210–11
environment, human impacts, 357, 359, 412
  nature and pace, 358
  obstacles to response, 360
  and technology, 359
environment–development debate, 414, 415, 416–18
  and boundaries, 425
  see also development; Rio Earth Summit/UNCED; sustainable development
environment–poverty links, 423
environmental capacities, 393, 394
environmental change, 436, 449
  and disease, 224–5, 227, 229
  documenting/measuring, 366–70
  evaluation of, 370–1
  human-induced, 357, 358, 364, 366, 367–70, 372
  and policy decisions, 371
  scales of, 364–5
  societal collapse, 368, 369
  views of, 365
  winners and losers, 371
  see also global change
environmental contradictions, 385
environmental degradation

education, role of, 361–2
  effects on livelihood/health, 374
  obstacles to response, 361–3
  and poverty, 423
  pressures of capitalism, 362
environmental disasters/catastrophes, 359, 368
  see also environmental change, societal collapse
environmental discourse, 435
  and security discourse, 436
environmental economics, 423
environmental governance, 427–38
  problematic of, 428
environmental history, 433
environmental issues
  endangered regions, 375
  geopolitics, 98
  and policy-making, 410
environmental justice movement, 435
environmental management, technocentrism, 413, 418
environmental management units, 411
environmental preservation, parks and reserves model, 434–45
environmental protection and social movements, 319–22
environmental receptors, 394
environmental revolution and geography, 413
environmentalism
  and agriculture, 66
  ecocentrist, 413, 418
  of the poor, 319
  radical alternatives, 426
  technocentrist, 413, 418
epistemic violence, 304
ET-map, 349, 350
  category map, 349
  information crop, 349
  subject region, 349
  virtual field, 349
ethnic cleansing, 13, 107, 125–6
ethnic enclaves, 339
ethnic minorities and welfare benefits, 166
ethnic revival and rebellion, 445
ethnic/tribal conflict, 13
ethnicity, 165
  and welfare benefits, 166
ethno-nationalism, 278

ethnoscapes, 337
eugenics, 263
Europe
  rural farmers, 65
  as social, 169–71
  and World War II, 71
European Community (EC)
  agro-industrial hot-spots, 65
  farmers, marginal, 65
  industrial agri-food system, 64–5
European Economic Community (EEC),
  71
European Exchange Rate System (ERM),
  34–6
European Social Model, 169–71
European Union (EU), 26, 27, 34–6, 170,
  363, 420
expansionism, 121
export goods, 314
Export Processing Zones (EPZs), 55, 75,
  76
externality, 383, 384, 395
  see also pollution

factor endowments, 52
factory farms, 384
facts, hard vs. soft, 408
fair trade, 54
family-planning programs/policies, 193
FAO (Food and Agricultural
  Organization), 148, 150
farmers, marginal, 65
farming
  changing role, 64–6
  family based structure, 64
  and global warming, 374
  industrialized agro-food system, 64–5
  traditional, 65
fast world, 329, 335, 336
feminism, 300, 301–2
  critique, 303
  right to inclusion, 301
feminist theories, 97–8, 301
fertility rate, global trends, 201–3
Feyerabend, P., 296
finance and business services, women in,
  242–3
financial crises, 35
finanscapes, 337
flexible accumulation, 25
follow-the-leader strategy, 71

food, 57
  scarcity/uncertainty, 57
  supply forecasts, global, 203–5, 206
foot and mouth disease, 359
foreign aid, 314
foreign direct investment (FDI)/foreign
  investment, 21, 22–3, 69, 75, 388,
  389
  global geography, 70–1, 72
  trends, 22
foreign exchange dealing, 34–6
foreign investment, role in South, 314
forest death (Waldsterben), 402
Forest Peoples Alliance, 320
Foucault, M., 301, 302, 307
  redefinition of power, 302–3
fragmented globality, 249
Fraser, N., 307, 308
free-riding, 363
free trade, 54
free trade advocates, 50, 52
  and neoliberal ideology, 52
Friedmann, H., 59

G7, 15
G8, 273, 447
Gaia, 406, 407, 408, 409
  cybernetics, 406
  homeostasis, global, 406
Gaia Hypothesis, 406–7
  Lovelock, J., 406
gender, 165
  discrimination/inequality, 195
  and mortality, Asia, 195
  and nationalism, 139–41
  and welfare benefits, 165–6
  see also feminism; women
General Agreement of Tariffs and Trade
  (GATT), 22, 26, 46, 52, 71, 153,
  154, 149, 273, 447, 450
  agricultural trade negotiations, 64
  generalized system of preferences,
  53
  most favored nation principle, 53
  national treatment rule, 53
  rule-oriented approach, 53
  Special Agreements, 53
  Uruguay Round, 52–3
General Assembly, UN, 147, 151
  see also United Nations
General Motors, 74

genetically modified organisms (GMOs), 428, 429, 358
  GMO revolution, 358
geocultural, 6
geoeconomic change, 6
  see also economic restructuring
geoenvironmental, 6
geographers
  analyses of the other, 298, 301, 303
  ethnographic fieldwork, 304
  and people/environment issues, 413, 425, 426
  theoretical perspectives, 297
  theories of difference and diversity, 300
geographic language, 305
geographical explanation, 297–8
geography
  agricultural, 60
  alternative positions, 299–306
  and colonialism, 413
  of commodities, hidden, 389
  of conservation, 434–5
  end of, 449
  and environmental management, 413
  of everyday life, 285–6
  and globalization, 4–5
  holistic tradition, 443
  of investment, 69, 70
  of labor, transnational, 81–4
  multiple perspectives, 298
  new debates, 306
  and new histories, 307
  of pollution, 403
  production of race, 262
  and race, study of, 262–3
  and radical approach to environment, 413
  radical geographers in, 300
  of resource extraction, 386
  and resource management, 413
  and scale, 444–5
geopolitical
  change/transitions, 103–5, 447
  worlds, 178
geopolitics
  critical geopolitics, 177
  definition, 176–7
  discourse, 176, 178, 179, 180
  of knowledge and development, 97–9
  main concepts, 177–9

orthodox, 177, 188
  of techno-nationalism, 332
  US interventions, 90–2
geosocial change, 6
Ghana, 212–13
Giddens, A., 400
glasnost, 130
global, as scale of study, 4–5, 6
global change, 40, 436
  atmospheric, 417
  cumulative, 366, 367
  documentation of, 366–70
  economic and social consequences, 371
  meaning of, 370–3
  response, 372
  scale of, 365–6
  study of, multidisciplinary, 367, 443
  systemic, 366
  and uncertainty, scientific, 367
  and vulnerability, 372
  see also climate change; environmental change
global consumer society, 436
global corporations, 21
  see also transnational corporations
global dangers, 186, 187–8
global demographic regimes, 194
global environmental problems/issues
  interest in, 391, 408
  nation-state, the, 428, 437, 438
  North–South relations, 428, 430–2
Global Forum, Rio, 417
  see also Agenda 21; Earth Parliament; non-governmental organizations; Rio Earth Summit/UNCED
global governance, 429–30
global health and disease, 217–35
global media, 279
global metropolitanism, 335, 335–6, 337
  populist dimension, 336–7
  structures and flows, 338
  tension and opposition, 336
global regulation, 144–57
  case study (ICC), 151–3
  from above, 144
global village, 5, 7, 449
global warming, 360, 366, 367, 371, 399, 438
  and crop production, 374
  see also environmental change; global

change; greenhouse effect/warming
global–local nexus, 444–6
  nexus, definition of, 445
globalization, 278
  aspects of, 3
  as a contention, 1
  communication, role of, 7, 9
  definitions, 2–3
  health consequences, 219
  and infectious disease, 217, 218–19
  vs. parochialism, 14
  re-scaling, 6, 7
  technology, 7, 9
  uneven, 9, 10, 11, 15, 445
glocal, 400
glocalization, 48
Goodman, D., et al., 60
  and Redclift, M., 60
Gorbachev, M., 130
god-trick, 253
grassroots groups, 321
green legend, 368–9
green revolution, 384
  creating dependency, 314–15
greenhouse effect/warming, 403, 408
greenhouse gases, 361, 403, 421, 428,
  438
Gregory, D., 305, 309
group rights, 124, 125
Gulf War, 106, 111

Haraway, D., 309
Harvey, D., 30, 31, 32–3, 130, 305
health, *161*, 162
health services and skilled migrants,
  243–4
hegemonic state, 177
heritage industry, 135
high-yield varieties (HYVs), 314
Holdgate, M. W., 392, 393
homogenization vs. differentiation, 336
human development indices, 239–40
  gender sensitive, 239–40
human immunodeficiency virus (HIV)/
  AIDS, 195, 217, 219
  East Asia, 223
  Eastern Europe, 223
  epidemiology, 222–3
  Pacific, 223
  Sub-Saharan Africa, 223
human–nature relationships, 358–60

changing, 360–3
human rights, cultural factors, 128
humanism, critique of, 301, 306
humankind, environmental impacts, 357,
  358
Hussein, Saddam, 106, 108, 176
hybridity, 449
hybridization, 304
hydropower, 321–2
hypermigration, 253
hypermobility, 29–30

ideas and symbolic meanings, 284
identity
  construction of, 302
  cultural definition of, 132
identity politics, 119
ideology, 302
ideoscapes, 337
imagined communities, 342
immigrants, 196
  and tension, 196
  transcontinental, 338
immutable mobiles, 41, 294
imperialism, 128
India, 134, 314, 321, 378, 420, 433
  Bhopal (Union Carbide plant disaster),
    396
  epistemic violence, 304
  HYV sorgum, 314–15
  Narmada Bachao Andolan (Save the
    Narmada Movement, NBA), 321–2,
    324
  Narmada Valley, 432
indigenous cultures, and knowledge and
  patenting, 433
indigenous/traditional systems, 311, 316
Indonesia, 378
industrial ecosystem, 386
Industrial Revolution, 369, 397, 399
industrialization, 313
  of agri-food system, 59–64
  Asian model, 25
  conditions in South, 313
  pace of, 23
industrialized economies and female
  migrant labor, 243–5
infant mortality, 162, *164*
infectious disease, 217
  and change, 223–4
  global trends, 218–19

information and communication technologies (ICTs), 340
debates about, 341–2
implications of, 340
and reconfiguration of spatial logic, 340
and space–time compression, 340
information processing, 244–5
informational economy, 331
innovation and world cities, 332
insecurity, economic, 24
integrated international production, 22, 26
integration vs. fragmentation, 336
inter-ethnic conflict, 447
intergovernmental organizations (IGOs), 329
Intergovernmental Panel on Climate Change (IGPCC), 408, 420
International Atomic Energy Agency (IAEA), 148
International Bank for Reconstruction and Development see World Bank
International Civil Aviation Organization (ICAO), 148
International Confederation of Free Trade Unions (ICFTU), 80, 81, 84
International Court of Justice (World Court), 147
International Development Association (IDA), 148
International Federation of Chemical, Energy and General Workers' Union (ICEF), 81
International Finance Corporation (IFC), 148
International Fund for Agricultural Development (IFAD), 149
International Labor Organization (ILO), 55
International Maritime Organization (IMO), 149
International Metalworkers' Federation, 82, 84
International Monetary Fund (IMF), 5, 11, 22, 46, 71, 116, 212, 252, 388
and Third World indebtedness, 22, 25
international money, world of, 29, 33, 34–9
international regimes, 427, 438

International Secretariat of Trade Union Centers/International Federation of Trade Unions (IFTU), 79–80
International Trade Organization (ITO), 146
International Trade Secretariat (ITS), 79, 80
International War Crimes Tribunal, 108
internet, 340
investment see foreign direct investment (FDI); foreign investment, role in South
Iraq, 106, 108, 176
Ireland, 140
Islam/Islamism
as anti-globalism movement, 116
Qu'ran Belt, 115
reformism, 115
revival, global, 115–16
Israel/Palestine, 113, 122, 447
issue-attention cycle, 391
Italy, 122
ITU (International Telecommunications Union), 149
IUCN (International Union for Conservation of Nature and Natural Resources), 414

Jameson, F., 39
Japan, 63
high employment growth, 240
life expectancy, 194
Ministry for International Trade and Industry Innovative Technology on the Earth (RITE), 386
part-time worker definition, 238
World War II, economic effects, 71
juxtaposition vs. syncretization, 336

Kennedy, P., 365
Kenya, Green Belt Movement, 321
Keynes, J. M., 162
Keynesianism, 159, 162, 166, 425
knowledge
claims competing, 298
contested and incompatible, 303
deconstruction of, 301
judgment between knowledge claims, 297, 306
situatedness, 300

theory, 297–9
Kosovo, 107, 234
Kuwait, 106, 396
Kyoto Protocol (agreement/accord), 106,
    187, 188, 273, 360, 361, 362, 404,
    436

labor
  in business and financial services,
    242–3
  and development, 314
  effect of Cold War, 78
  female, 240–6
  fragmentation of, 336
  geography of, transnational, 81–4
  history, 78–9
  international/transnational, 78
  organizing, means of, 79–80
  spatial division, 68, 72
  see also new international division of
    labor (NIDL); old international
    division of labor (OIDL)
labor force
  female, skills unrewarded, 237
  global, new divisions in, 245–6
  increased female participation, 245
labor force statistics, 238–9
labor solidarity, international, 81, 82
  accomodatory, 82
  effects of telecommunications
    technologies, 82–4
  transformatory, 82
language
  linguistic cleansing, 277
  linguistic tension, 133–4
  and nationalism, 132
  see also linguistic minorities
language planning, 133
late modernity, 401
Latin America, 11, 22, 23, 239
  transition to democracy, 104, 110,
    122, 123
law and order campaign (1968), 267
Law of the Sea, 363, 438
laws of thermodynamics, biophysical
    resource depletion model, 384
legibility/illegibility, 29–30, 32, 33, 37
legitimacy crisis of the state, 124
legitimation, crisis of, 66–7
Levi's, 74
liberal democracy, 119–20

achievements, 120
and centralized authority, 126
importance of, 122
and the state, 123–4
life expectancy, 194
limits to growth, 409
linguistic cleansing, 277
linguistic minorities, 133–4
linguistic tension, 277
locational advantages, 69, 74
logistic curve, 205
lost decade, 11
lost/stolen years concept, 194
Lovelock, J., 406
  see also Gaia Hypothesis

Mackinder, H., 4–5
  heartland thesis, 4–5
mad cow disease, 359
malaria, 217, 219, 221
  and climate change, 229
Malthus, R., 198, 358, 382
  theory of population growth, 198,
    201–2
Malthusianism see neo-Malthusianism;
    population growth
Man and Nature, 412
Man's Role in Changing the Face of the
    Earth, 365, 412
manufactured risk, 401
  uncertainty, 186
manufacturing
  business services, and women, 240–3
  output, location change, 240
manufacturing, labor costs, role of, 74
Marcos, Subcomandante, 450
  see also Chiapas; Zapatista(s)
market environmentalism, 422–4
market forces, 64
market triumphalism, 278
markets, 24
Marsh, George Perkins, 412
Marxism, 120
Marxist analysis, in geography, 299
Marxist challenge, and development,
    94–7
Marxist–Leninist states, 96
Mauritius, 76
Meadows, D. H. et al., 409
measles, 219
mediascapes, 337

Melucci, A., 16
merger movement, 69
mergers and acquisitions (M&As), 23
methane, 370, 399, 436
metropolitanism, global, 334, 335–6,
    337
  and fantasy as social practice, 337
Mexico, and North American Free Trade
    Agreement, 277
middle classes, 165, 166
migrants in health services, 243–4
military Keynesianism, 266, 267
military, technoscientific, 187
Mintz, S., 288
missing women problem, 237
  see also Sen, A.
mobility, 21
  of international money, 34–7
  see also capital, mobility
modernity, 449
  barometers of, 29–30, 33
  metanarratives, 445
modernization
  as civilizing, 90
  economic, 25
modernization theory, 90, 92–3
  and developing countries, 93
  ethnocentric universalism, 94
  neoliberalism of 1980s, 93–4
modes of regulation
  concept of, 144
  contingent and provisional, 144
  dynamic, 144
  global, 144–57
  new, 331
  post-World War II, 144–51
monetarism, 22, 25
monetary transactions, increasing speed,
    37, 38
monocrops, HYV, 314–15
Montreal Protocol, 404, 421, 437
monuments and nationalism, 137–9
moral responsibility, 309
mortality rates, global, 216–17
Movimento Sem Terra (MST), 323
multiculturalism, 263
multinational corporations (MNCs), 248,
    314
  see also transnational corporations
multi-user domains (MUDs), 344, 347,
    349

murals, 135
mutually assured destruction (MAD),
    175

nation-building, 132–42
  role of public monuments, 132
National Economic Development
    Council, 162
national exceptionalism, 179
National Missiles Defense System, 188
national security/insecurity, 175, 179,
    187
nationalism(s), 130–42, 445
  and gender, 139–41
  and language, 132–43
  and monuments, 137–9
  and the past, 134–7
  reemergence, 130–42
nationalization, 22
nations as imagined communities, 131,
    137
native cultures, impact of colonialism,
    303
natural capital, 423, 424
natural resource dependency, 377
natural resources
  crisis, views/theories of, 381–8
  extraction, 387
  human engineered, 377
  non-renewable, 378
  population pressure, 361
  as private property, 361
  problems and education, 361
  production and consumption trends,
    378–81, 379, 380
  renewable, 381
  water, 378
  wise use, 413
nature
  changing relationships with, 358–60
  changing the changing relationships,
    360–3
  commodification of, 389
nature conservation see World
    Conservation Strategy, The (WCS)
nature myths, 372–3
neo-Fordism, global, 331
neo-Malthusianism, 198, 199–200, 207,
    215, 436
  Erhlichs, P. and A., 199
neoliberalism, 25, 183

agenda, 11, 12
  manifestation of globalism, 13
networks, 36–7
  electronic, 34
  global and local, 37–8
  and industrial agri-food business, 65–6
  of production, transnational, 70
  of social relations, 40–1
  telecommunications, 37–9, *38*
Nevada Test Site, 179
new economy, 388, 389
New Environmental Age, 410
new international division of labor
    (NIDL), 47, 69, 70, 72, 76, 330,
    333
  causes, 73
  characteristics of, 73
  geography of, 73–4
new nature, 377
new world disorder, 105–9, 278, 445
new world order, 103–4, 108, 109,
    151
newly industrializing countries (NICs),
    23, 72, 379
newly industrializing economies (NIEs),
    47
Nicaragua, 91
Nigeria, 7, 431–2
Nike, 74
non-communist manifesto, 312
non-governmental organizations
    (NGOs), 16, 408, 414, 417, 419
  Earth Parliament, 417
  empowerment, 417
  and global civil society, 16
  Global Forum, 417
  influence on *Agenda 21*, 417
  transnational, 331
Non-Proliferation Treaty, 187
North American Free Trade Agreement
    (NAFTA), 14, 23, 26, 27, 55, 450
North Atlantic Treaty Organization
    (NATO), 273
North–South dependency, 25
North–South relations, 89
  and environment, 98
  see also Rio Earth Summit/UNCED
Northern Ireland, 447
  disputed history, 136
Not In My Back Yard (NIMBY), 401
nuclear blackmail, 184

nuclear proliferation, 446
nuclear weapons, 175
  environmental legacy, 180
nursing, and skilled migrants, 243–4

off-farm sectors, 62–4
old international division of labor
    (OIDL), 72
OPEC, 387
open-space, 256
Orientalism, 303
Orientalism as discourse, 303
OS KOSOVO, 84
others, and we, 300
ownership-specific vs. location-specific
    advantages, 75
Our Common Future, 312, 414, 415
  definition of sustainable development,
    414
  see also Brundtland Report
ozone, stratospheric, 399, 403–4
  depletion, 366, 372, 404, *407*
  Treaty, 363

Pakistan, 213, 214
paraspace, 352
part-time work, definitions, 238
particularism, universalization of, 335
  vs. universalism, 306, 336
PAT formula, 373–4
peace dividend, 106
peace-keeping, 147, 157
People's Global Action, 323–4
perestroika, 103
periphery countries, 313
Permanent Court of Arbitration, 147
personal services and female migrant
    labor, 244
pesticides, 315
Philippines *see* Sama Sama squatter
    movement
place
  definition of, 344
  revaluation of, 338–9
  sense of, 338
placelessness, 338, 341, 343
pneumonia, 219, 220
  contributing factors, 220
Polanyi, K., 1
polarization and income/wealth, 25
political ecology, 425–6, 432–4

political movements, 300
political negotiation and sustainable
    development *see* Rio Earth Summit/
    UNCED
political sphere, 118
politically correct behavior, 125
politics
    illiberal, 278
    new, 125
    regional and ethnic, 125–6
politics of difference, 124–5
pollutants, second- and third-generation,
    398
pollution, 391, 392
    Chernobyl nuclear disaster
        (radioactivity release), 395, 396, *397*
    vs. contamination, 393
    cultural definition, 392
    definition, 391–2
    environment as media of, 394–7
    environmental capacities, 393, 394
    and human activity, 393
    human rights aspects, impacts of,
        393–5
    media coverage, 391
    nature and sources, 393–4
    scale of, 400–2
    transboundary problem, 403
    Union Carbide plant, Bhopal, India,
        396
    and urbanization, 397
pollution of poverty, 415
population
    mobility and conflict, 195–6
    mobility and disease, 229, 231, 233–4
    policies, local effects of global, 212–15
    and resources debates, 200–5, 207,
        212
    vulnerability, 211
population growth, 193
    deforestation, 211–12
    see also Malthusianism; neo-
        Malthusianism
post-Cold War era, 103–16, *178*, 181
Post-Development School/anti-
    development theory, 449
post-Fordism, 25
postcolonialism, 303–4
post-industrialism, 388
postmodernism, 130
poststructuralism, 302–3

postwar settlement, 162
    social limitations, 166
potato, commodity chain, 62, *63*
poverty
    and development, 10
    and environment links, 426
    pollution of, 415
    and population growth relationship,
        210, 211–12
    and women, 10
power
    effects of, 30
    in one-party states, 96
    redefined by Foucault, 302
power relations, 302, 307
precautionary principle, 395, 360
Preparing for the Twenty-First Century,
    365
principled positions, 297, 309
prisons, US expansion of, 268
    building boom, 270
    demography, 270
    global implications, 271–3
    local effects, 271
    map of, 269–71
    politics of, 269
    private prison entrepreneurs, 270–1
    and privatization, 272
    violence, 269
privatization, 22, 88
production
    crisis of, 66
    integrated, 22
    new technologies, 330
progress and evolution, theory of, 311
propinquity, 341
public electronic network (PENs),
    347
public expenditure and the welfare
    state, 162, *164*
public services
    social distribution of expenditure,
        163–5
public–private sector relations, 93
Putnam, R., 15

Quebec, 133

race
    economic inequality, 264–5
    and globalization, 261–74

production of, 265
and scale, 264
as structural, 263–4
theorizing, 261–5
racism, 166, 170
rational utilization, 413
rationality, instrumental vs. value, 395
Red International of Labor Unions/
    Profintern, 80
Redclift, M. and Goodman, D., 60
refugees, 195, 233
regime of accumulation, 144
regulation, 144
    crisis of, 66–7
    as regime of accumulation, 144
regulatory apparatus, second
    international food regime, 62
relativism, 306–7, 309
religion revival, 445
remapping, 448
resource crisis, global, 204–5
resource scarcity and conflict, 382
resource use and social justice, 387
resources,
    non-renewable, 378
    rational utilization, 413
    renewable, 381
    reverse transfer, 314
    and technological innovation, 206
    wise use, 413
    see also natural resources
restructuring
    of agriculture, 60, 62, 64–6
    metropolitan, and world city
        formation, 333
    see also economic restructuring
reterritorialization, 338
revanchism, 257
reverse transfer of resources, 314
revolution
    in the South, 96
    working-class, 96
rich–poor income gap, 252
rights
    group vs. individual, 124
    of linguistic minorities, 134
    political vs. social, 122–3
    social, 120
    see also democracy; human rights,
        cultural factors
Rio Declaration on Environment and

Development
Principle 3, 417
Principle 10, 411
principles of, 416–17
US interpretation, 417–18
Rio Earth Summit/UNCED, 363, 401,
    415, 416, 429, 433, 438
achievements, 421
analysis of results, 421–2
Convention on Biological Diversity,
    363, 419–20
Convention on Desertification, 422
failures, 421–2
forest debates, 418–19
Forest Principles, 419
Framework Convention on Climate
    Change, 420–1, 422
global atmospheric change, 417
NGOs, 417, 419
North–South disagreements/debates,
    416–18
Third World fears, 417
World Bank involvement, 423
see also Agenda 21
risk, 400–1, 402
and blame, 373
risk society, 400, 410
    manufactured uncertainty, 186
    and technoscientific modernization,
        186
    world, 186, 187
rogue states, 106–8, 112, 181
US as, 188
Russia, 179, 181, 186
    corruption top-down, bottom-up,
        182
    crony capitalism, 182
    disintegration of, 183
    market capitalism, transition to, 182
    multiple crises, 183
    nuclear dangers and proliferation,
        184–5
shock therapy, 24, 450
    as Third World state, 24
    see also Soviet Union
Russia/US Cooperative Threat Reduction
    Program, 185
Rwanda, 382, 446

Sahabat Alam Malaysia (Friends of the
    Earth in Malaysia), 321

Said, E., 303
St Paul's Cathedral, 135
Sama Sama squatter movement, 320
Sarawak, 320–1
Saro-Wiwa, K., 432
scale(s)
    geographical, 4–7
    see also geography
scale-collapsing, 259
Scalia, A., 118
schistosomaiasis, 219
science, 404–5
    positivist, 297, 299
    and risk management, 411
    and uncertainty, 411, 424
    Western, critiques of, 296
Scottish nationalism, 134–5
sea-level rise, 366
Seattle, 324
    battle in, 447
Second World (former), 450
Secretariat of the UN, and peace-keeping,
    147
Security Council, UN, 147
self, 302
self-determination, national, right of, 126
Sen, A., 195
sense of place, 343, 347
separatist groups, 139
September 11, 2001, 111, 114, 180
    Afghanistan, 112
    Al Quaida, 111
    implications, 111–15
    Osama bin Laden, 111, 112, 113, 114
    war on terrorism, 112–13
Serbia, 107, 447
services
    health services, 243–4
    internationalization of, 243
    see also public services
Seveso, 396
sex ratio, 237
shadow projects, 423
shadow state, 171–2
slash and burn agriculture, 357
slow world, 329
smog, 397, 399
sociability, world cities, 332
social capital, 88, 93
social change, continuity/discontinuity,
    450

social consciousness, 95
social democracy, 120–1
social economy, 171–2
social movements, 197, 310
    civil society, 319
    contemporary, 317
    contested terrain/terrains of resistance,
        319
    demands and goals, 318
    and development, 316
    diversity of, 319
    and environment, 320–2
    new, 124–5, 316–22
    state responses, 320
    and technology, effects of, 324–5
    women in theorization of, 98
social networks, 40, 41, 343
social regulation and the agri-food sector,
    58
social relation networks, 40, 41
social relations, 197
social reproduction, 250–2
    children's playgrounds and school
        yards, 254–5
    contingent, 250
    costs, 251
    public spending, 251
    states and, 251
    US disinvestment, 254
social rights, defense of, 122–3
social sciences and Marxist thought, 94,
    95
social subjectivity, 95
social wage, 161, 251
societal collapse, environment-related,
    368, 369
socioeconomic change and states, 159,
    161
soft law, 424
Son of Star Wars, 106
sorghum, effect of HYV monocrop, 314
South and development discourse, 312
    see also developing countries/world;
        North–South dependency; North–
        South relations; Third World
South Africa, 122
Southeast Asia, 23
    economic crisis, 23, 24–5
    capital flight, 24
Soviet Union, 180
    and the Cold War, 121

former territories, 181
space
   annihilation of by time, 33, 34
   global economic, 30
space of flows, 8, 11, 13
   electronic, 33, 34, 37
   in electronic ghettos, 39
   mobile, 30–1
space and time *see* time–space
      compression; time–space expansion
speculation, 35–6
Spivak, G. C., 302–3
squatter movement (Sama Sama), 320
Sri Lanka, Sarvodaya Shramadana
      Movement, 321
Star Wars, 188
state insurance schemes (healthcare and
      unemployment), 162–3
states (nation-), 6–7, 26, 428, 437, 438,
      448
   administrative power of, 159
   and environmental problems, 428
   fascist, 159
   formation of, twentieth century,
      158–60
   modern, 158–9
   one-party, 96
   postcolonial, language choice, 134
   potential for abuse, 159–60
   problems in, 163–7
   representing whose interests, 123
   responses to new social movements,
      320
   restructuring and reorganization of,
      158, 166, 167–72
   shadow, 171–2
   social-geographical inequality within
      and between, 127
   socialism, 170
   sovereignty, 448
   status, contemporary, 104–5
   strategies for environmental
      protection, 362
   welfare, corporate, 160–3
   workfare, 158–73
   see also nations as imagined
      communities
stewardship, 410
Strategic Arms Reduction Treaty
      (START I), 185, 187
   START II, 183–4

START III, 184
street festivals, 136
structural adjustment programs (SAPs),
      11, 12, 24, 88, 93, 218, 252
   effects on public spending, 252
   and resources, 388
structural violence, 194–5
Subaltern Studies group, 303
subcontracting, 69
substitutionism, 62
supplementarity, strategy of, 305
sustainability, 410
sustainable development, 56, 358, 361–2,
      412, 429
   definitions, 414
   ecocentrism, 412
   efficiency and resource use, 413
   equity issues, 424
   government strategies for
      environmental protection, 362
   mainstream, 414–16
   and market environmentalism, 422–4
   meanings and interpretations, 414
   and natural capital, 423, 424
   political and economic debates,
      416–18
   radicalism vs. reformism, 422
   at Rio (1992), 416–18
   shadow projects, 423
   Stockholm (1972), 415
   strong vs. weak, 424
   technocentrism, 412
   trade-offs, 424–5
Sweden, the Natural Step program, 386
syncretization vs. juxtaposition, 336
synergistic effects, 395

techno-nationalism, geopolitics of, 331
technology, effects on labor, 75, 253
technoscapes, 337
technoscientific modernization, 186
telecommunications
   networks, 37–9
   speed and processing power of, 37
telecommunications satellites, 29
terraforming, 428
terrains of resistance, 319
territorial state and democracy, 123–4,
      319
terrorism
   international, 108–9, 111

terrorism (*cont'd*)
  September 11, 2001, 111–15
  war on terrorism, 112–13
Thailand, resistance to Nam Choan Dam, 319
Thatcher, Margaret, 140
theories, analyses of, 297–8
theorizing, dualistic, 76
theory
  as grand narrative, 300
  nature and purpose, 297–9
  problem of status of Truth/truth, 299
  problems for, 306–7
  as text, 297
third sector, 171
third way, 448
Third World, 16, 57, 252, 379
  IMF response to indebtedness, 150
threshold levels, 393
Tiananmen Square, symbolism of, 139
time–space compression, 7, 21, 29, 30–4, 37, 258, 265, 335, 336
  dealing rooms, 34–6
  electronic ghettos, 39, 41
  leading to time–space depression, 33
time–space convergence, 33
time–space expansion, 258, 265
tourism and national past, 135
toxicity testing, 394–5
trace pollutant releases, 366, 370
trade
  as core–periphery system, 45
  definitions, 43
  East Asia, 50
  environmental issues, 55–6
  evaluation, 53–6
  fair/fairness of, 54
  free trade/free traders, 54
  geography of international, 43, 48–50, 49
  globalizations of, 46
  history, 44–5
  international negotiations, 43
  Japan, 49–50
  labor standards and, 54–5
  protectionism, 54
  types/forms, 44, 46–8
  United States, 50
  WTO and regulation, 43
tradition, invention of, 135
trans-state organization, 143–57

from above, 144–55
from below, 155–6
trans-state organizations, 143, 145, 148–9
transnational corporations, 21, 68, 69, 81, 328, 330
  attraction of world cities, 333
  backward integration, 70
  corporate strategies of, 331
  evolution, 70–2
  forward integration, 68
  vertical linkage, 68–9
transnational producer–service class, 334, 336, 338
transnationalism, 124
Trusteeship Council, 150
truth
  politics of, 299
  regime of, 303
tuberculosis, 217, 219, 220
  drug-resistant, 220
  epidemiology, 222

UN Conference on Environment and Development (UNCED) (1992) *see* Rio Earth Summit
UN Conference on the Human Environment (Stockholm 1972), 415
UN Conference on Human Rights (1993), 128
UN Development Program (UNDP), 150
  human development indices, 239–40, *241*
UN Environment Program (UNEP), 408, 414, 419, 437
uncertainty, 37, 424
underdeveloped world, 311
UNESCO (UN Educational, Scientific and Cultural Organization), *148*, 150
UNFPA (UN Population Fund), 150
UNICEF (United Nations Children's Fund), 150
UNIDO (United Nations Industrial Development Organization), *149*
Union Carbide plant disaster, Bhopal, India, 396
unions
  as transmission belts, 85
  tripartite system, 85–7
United Kingdom *see* Britain/UK

United Nations, 145
  principal organs of, 146–7, 150
  related bodies, *148–9*, 150
  as trans-state organization, 151
United Nations Charter, 146
United States of America
  Americanization of global culture, 335
  crime problem, 267
  dual tracking, 245–6
  employment losses, 240
  hegemony, 5
  manufacturing employment, female,
    240
  merger movement, 69
  military industrial complex, 181
  military spending, 181
  nation-building mythologies, 136
  part-time work categories, 238
  positive identity for the West, 90
  presidential election (2000), 117, 118,
    273, 360
  recession (1970s), 268
  role in Latin America and Caribbean,
    238
  as superpower, 104–6
  and UNCED process, 420
United Steelworkers of America (USWA),
  82, 83
universalism, 445
  ethnocentric, 94
  vs. particularism, 336
universalization of particularism, 335
UPU (Universal Postal Union), *149*
urbanization and globalization, 329–32

Vietnam memorial, 138–9
Vietnam War, 92, 107
virtual place-making, 345
virtual worlds, 349
Vision of Britain, A (Prince of Wales),
  135

war memorials, 137–9
  military leaders, 137
  Vietnam memorial, 138–9
  World Wars I and II, 138–9
we and others, 300
Weale, A., 409
weapons
  black market, 110
  of mass destruction, 176

and security, 175
  technoscientific role, 176
web landscape, spatialization of
  mapping, 350–3, *351*
welfare provision, 195
  changing role of, 160
  domestic unpaid, 165
welfare state, 121, 266
  challenged, 122
  crisis in, 160
  economic and political effects, 161–2
  increased costs, 164
  limited and unusual, 160
  mitigating social problems, 162
  and neoliberalism, 169–70
  problems and limitations, 163–7
  success of, 162
  working-class support for, 165
West, the
  construction of colonial Other, 303
  construction of Oriental Other, 303
  develop and rule, 90–4
West–non-West encounter, 92
Western orthodoxy, 88
whipsawing, 81
WHO (World Health Organization),
  148, 217, 234
WIPO (World Intellectual Property
  Organization), *149*
wise use, 413
WMO (World Meteorological
  Organization), *149*
Wolin, S., 123
Wollstonecraft, M., 301
women
  and broad-banding, 246
  in business and financial services,
    242–3
  as care givers, 238
  in export-industry workforce, 242
  in family enterprises, 238
  in the informal sector, 238–9
  in information processing, 244–5
  invisibility of, 237–9
  in labor market, periphery of, 238
  in management, 242–3
  mobility issues, 243
  in non-standard contracts, 238
  as other, 301
  as part-time workers, 238
  problems in status measurement,

women (*cont'd*)
  237–9
  roles in international flows of people
    and information, 243–5
  status and economic contribution,
    237–8
  status in economic situations, 246
  and the theorization of social
    movements, 98
  unpaid welfare provision by, 165–6
  see also feminism
Woolf, V., 33
workfare, 167–9
workfare state, 269, 276
working class and welfare state, 165
World Bank, 5, 11, 22, 116, 122, 127,
    145–6, *148*, 150, 315, 316, 432
world cities, 279, 328, 329
  authority, 331
  and capital accumulation, 329
  control centers, 328–9
  cosmopolitanism of, 339
  as cultural spaces, 339
  functional components, 332–4
  global hierarchy of, 333–4
  innovation, 332
  interface between global and local, 329
  and metrocentric global cultures,
    334–8
  as nodal points, 330–1, 332
  as places, 338–9
  sociobility, 332
  and the transnational producer–service
    class, 334, 336, 338
World Conservation Strategy, The
    (WCS), 414, 415, 418, 419
World Federation of Trade Unions
    (WFTU), 80, 84
world-system theory and world cities,
    333
World Trade Organization (WTO), 12,
    15, 22, 26, 43, 52, 55, 312, 315,
    316, 323, 360, 437, 447
World Wildlife Fund (WWF), 414

Yugoslavia, 125–6

Zapatista(s), 323, 325
  Ejercito Zapatista Liberacion Nacional
    (EZLN), 277, 318
  Marcos, Subcomandante, 450
  rebellion, 450
  and worldwide web, 344
  see also Chiapas
zero-sum game, 50